Lacustrine Basin Exploration

Case Studies and Modern Analogs

Edited by
Barry J. Katz
Texaco E&P Technology Division

AAPG Memoir 50

Published by
The American Association of Petroleum Geologists
Tulsa, Oklahoma 74101, U.S.A.

ISBN: 0-89181-328-4
ISSN: 0065-731X

Dust jacket photos by Andrew S. Cohen, University of Arizona, Tucson. The major photo shows stromatolite and lacustrine carbonate sand from El Mojorral Springs, Cuatro Cienegas, Coahuila, Mexico. The inset photo is of calcite cemented colluvial rockfall along the Kigoma Escarpment, North Tanzania, Lake Tanganyika. The fish, *Lamprologus sexifaciatus*, is approximately 100 mm long.

Association Editor: Susan Longacre
Science Director: Gary D. Howell
Publications Manager: Cathleen P. Williams
Special Projects Editor: Anne H. Thomas
Project Production: Custom Editorial Productions, Inc.,
                   Cincinnati, Ohio

About the Editor

Barry Jay Katz is currently a Research Associate in Texaco's Exploration and Production Technology Department in Houston, Texas.

He received his B.S. in geology from Brooklyn College, and his Ph.D. in marine geology and geophysics from the Rosenstiel School of Marine and Atmospheric Sciences, University of Miami, in 1979. He was the recipient of the F. G. Walton Smith Prize for his dissertation, "An Application of Amino Acid Racemization—The Determination of Paleoheat Flow." He joined Texaco's Bellaire Research Laboratories in 1979 and has held various positions within Texaco's research organization.

He is currently an Associate Editor of the *AAPG Bulletin*, Chairman of the AAPG Research Activities Subcommittee, and a member of the AAPG Marine Geology Committee. He is also a member of the JOIDES's (Joint Oceanographic Institutions for Deep Earth Sampling) Pollution Prevention and Safety Panel.

The major themes of his research have been related to the processes controlling the deposition of sedimentary organic matter and the characterization of organic facies and how this information may be applied to petroleum exploration. Over the past few years much of this work has focused on lacustrine systems. This interest resulted in his co-convening of the AAPG Research Conference on lacustrine basin exploration and the editing of this proceedings volume.

Table of Contents

Introduction

Barry J. Katz

A subtle but significant shift has occurred in exploration strategies over the last two decades. Central to this shift is the realization of the importance of lacustrine basins in the overall exploration framework; witness the activity in Angola, Australia, Brazil, China, Indonesia, Sudan, and regions within the United States. As exploration progresses within these settings, it becomes increasingly evident that geologic concepts, models, and tactics applied to marine settings are not fully satisfactory. The need for new exploration strategies specifically tailored to lacustrine settings led to Bruce Rosendahl and myself convening an AAPG Research Conference, *Lacustrine Basin Exploration—Case Studies and Modern Analogues*, in Snowbird, Utah, September 7-9, 1988.

The timing of this conference was ideal. Researchers not only had made substantial advances in the understanding of lacustrine processes but also realized that lacustrine stratigraphic sequences often provide detailed sedimentary records of initial rifting and continental breakup as well as better understanding of global climate history. Substantial information had also become available through IGCP Project 219, *Comparative Lacustrine Sedimentology in Space and Time*. In addition, this conference took place after the Geological Society of London's *Symposium on Lacustrine Petroleum Source Rocks* (Fleet et al., 1988).

The Snowbird conference was attended by 70 individuals representing industry, academia, and government. That their backgrounds, levels of experience, and scientific specialties were highly varied led to interesting and at times heated discussions over the 28 oral and poster presentations and in many cases an expansion of ideas and modification of views.

AAPG research conferences typically do not publish formal proceedings; consequently, such meetings foster presentation of material that may be considered sensitive. At the conclusion of this conference, however, the authors collectively decided that a proceedings volume should be assembled. Twenty of the 28 presentations herein are presented.

Following the format of the conference, this volume is divided into five sections. The first section deals with distribution of lakes through time and space. One of the major factors controlling distribution of lakes and the character of their sediments is climate. In the first chapter Eric Barron discusses the potential utility of numeric climate modeling with its application to lacustrine source rock prediction. Unlike other work on paleoclimate modeling and marine source rock prediction, which has emphasized the potential for mapping regions of paleo-upwelling and high levels of organic productivity, Barron's work emphasizes the utility of climate models to high-grade regions for potential lake development and to assess the possibility of enhanced levels of organic preservation. This chapter is followed by Francoise Gasse's paper on the interrelationships of tectonism and climate on lacustrine stratigraphic sequence development in the Afar region of east Africa. One reviewer noted that this paper contains material that until now has had only limited accessibility through obscure publications of the government of Djibouti. In the last paper of this section Michael Smith reviews the distribution of lacustrine oil shale throughout the geologic record. This work presents an annotated atlas of Precambrian through Cenozoic lake sequences.

The second section of the book also contains three chapters. The first chapter by Barry Katz is an overview of the processes that control distribution of lacustrine petroleum source rocks through time and space. Important aspects of basin origin, climate, organic productivity, and preservation are discussed. A source rock model is proposed based on sediment data from four modern east African lakes. The second chapter, by M. R. Mello and J. R. Maxwell, describes geochemical variability of source rocks and their products from lacustrine basins along the Brazilian continental margin. This work emphasizes organic geochemical variabilities one may expect within lacustrine settings as a result of variations in water chemistry. It also shows the potential utility of organic geochemical data obtainable from a produced crude oil to establish the nature of an unsampled source rock system. The third chapter, by M. R. Talbot and K. Kelts, focuses on the use of carbon and oxygen isotope data to infer temporal changes in lake water characteristics. They suggest that by examining stable isotopic compositions of primary lacustrine carbonates, one can distinguish between open and closed hydrologic systems, and that a similar examination of early diagenetic carbonate phases permits an assessment of the relative sulfate abundance in the water column.

The third section contains five papers devoted to geologic, geophysical, and geochemical work that has been conducted in the east African rift system. These lake basins often have been studied as modern analogs to ancient lacustrine systems (e.g., Demaison and Moore, 1980). The first chapter, by Thomas Johnson and Patrick Ng'ang'a, details the impacts of structure, climate, and limnology on sedimentation in Lake Malawi. The authors also discuss the implications of their observations on lacustrine rift-basin oil potential. They conclude that the greatest unknown in such basins is the extent and quality of reservoir beds. This is followed by Andrew Cohen's chapter describing a tectono-stratigraphic model for Lake Tanganyika. Cohen concludes that Lake Tanganyika does provide a viable modern analog for exploration in ancient rifts and that certain synrift exploration targets should be examined—platform-margin sand bodies and deep-water turbidites or contourites derived from platform or axial sources. The third chapter in this section, by Christopher Scholz and Bruce Rosendahl, analyzes seismic reflection data to define potential coarse-grained depositional patterns in Lakes Malawi and Tanganyika. They describe several settings in which sand bodies may have developed, including fan deltas adjacent to border faults, channels within subla-custrine fans near point sources, debris flows associated with rift shoulders, and sand waves formed by contour currents and in reworked lowstand deposits. They note that lake-level changes, because of their magnitude and frequency, are more significant than sea-level changes in controlling depositional facies patterns. Scholz and Rosendahl conclude that half-graben shoaling margins make poor exploration targets because of their extensive amounts of erosion during lowstands. Better targets more likely are to be found in sublacustrine fans and lowstand deltas. The next chapter, by A. Y. Huc and others, continues with a depositional model for organic matter in the northern subbasin of Lake Tanganyika. They note that variability in concentration and type of organic matter in lake-basin sediments is largely controlled by rift-basin morphology and the relative importance of pelagic/hemipelagic sedimentation vs. gravity-transport mechanisms. In the last chapter in this section John Halfman and Paul Hearty describe cyclical sedimentation patterns in Lake Turkana. The observed cyclicity reflects the region's climatic record and its impact on deliverability of fluvial detritus into the lake.

Note that, although the first four chapters in this section suggest that east African rift lakes provide important analogs for ancient rift systems with respect to rift architecture, sedimentary geochemistry, and depositional process, they are not representative of all lacustrine systems. These rift lakes tend to be narrower relative to their depth than most other lake types. This ratio influences both behavior of the water column and sedimentary dynamics of the lakes. One should therefore be cautious when translating the information available from these modern analogs into more general lacustrine exploratory models.

The fourth section presents four case studies from North America, in stratigraphic order. The first, by Scott Imbus and others, describes the middle Proterozoic Nonesuch Formation. The petroleum industry has been interested in the Nonesuch because it seeps oil at the White Pine copper mine in Michigan (Eglinton et al., 1964) and may extend throughout sections of the central North American rift (Seglund, 1989). In the second chapter Paul Olsen describes the Triassic Newark Supergroup basins of eastern United States and Canada. These basins also have received renewed exploration interest (Bowman et al., 1987). Olsen notes that the depositional sequence observed (i.e., a basal fluvial interval overlain by a deeper water lacustrine sequence and a shallow-water lacustrine/fluvial environment) is typical of many lacustrine basins. This point also is made later by Lambiase. Olsen further notes that "Newark lakes" can be broadly classified according to magnitude and frequency of lake-level changes. The lake types are the Newark type, in which water inflow was closely balanced with outflow; Richmond type, which suggests more persistently humid conditions where inflow often was greater than evaporation; and the Fundy type, in which water influx rarely matched evaporation. Next is a discussion by Louis Liro and Yvonna Pardus on use of seismic facies analysis to establish depositional framework of the upper Fort Union Formation (Paleocene), Wind River basin, Wyoming. Their work shows the utility of seismic sequence and attribute analysis in an ancient lacustrine environment to delineate changes in lake morphology and depositional setting through time. The final chapter in this section is by James Castle, who describes the Eocene Green River Formation in the Uinta basin, Utah. He concludes that in lacustrine systems reservoir quality is largely controlled by depositional process. In Lake Uinta the more laterally continuous reservoirs were deposited along a paleoshoreline as part of a barrier-island complex and have been extensively reworked. Less continuous reservoirs were developed as part of fluviodeltaic complexes.

The last section includes five chapters of case studies exclusive of North America. The first chapter, by Joseph Lambiase, presents data from a series of rift basins (Mombasa basin, central Sumatra, Morondava basin, Recôncavo basin, southern Sudan, a west African composite, the Newark basin, and the Mid-Continent rift) to support a tectono-stratigraphic model for rift sedimentation. The model concludes that a major control on lacustrine sequence stratigraphy is changing topography during the sedimentary basin's evolution. In fact, Lambiase states that topographic factors are of equal importance to water availability in controlling development of lacustrine sequences.

The remaining four chapters are presented in stratigraphic succession as were those of North America. In a discussion of the Carboniferous of east Greenland, Lars Stemmerik and others describe the conditions of deposition and organic geochemical character of an oil-prone lacustrine sequence. This work suggests the presence of a major secondary hydrocarbon source sequence in a region that has experienced no significant exploration activity. The next chapter, by Dirceu Abrahão and John Warme, describes the Lower Cretaceous Lagao Feia Formation of the Campos basin, Brazil. This unit has acted as both a source and a reservoir. The authors note that pelecypod-rich coquinas rather than sandstones are the only major reservoirs within the four fields producing from Lagao Feia. This is followed by T. R. McHargue's chapter about the stratigraphic sequence of Cabinda, offshore Angola. Included with a description of stratigraphic type sections is a series of isopach maps and stratigraphic cross sections. The final chapter, by Li Desheng and Luo Ming, describes small Mesozoic and Cenozoic petroliferous basins in southeastern China. The authors note that conditions favorable for the accumulation of hydrocarbons exist in more than 140 minor lacustrine basins with areal extents of only 800–1000 km^2. Stacked reservoirs with multiple pay potential are common within these basins, including both fractured and karstified basement potential and lacustrine sands. Wells with flow rates of approximately 2900 and 7000 bbl oil/day have been completed in the Damintun and Baishe basins, respectively.

Perhaps the most important contribution of this conference was the realization that lacustrine settings seem to be far more complex in an exploration sense than their marine counterparts. There is no single lacustrine exploration model. Acceptable commercial success in lacustrine settings will require fully integrated exploration strategies that consider all major exploration components. This will be the challenge for the explorationists of the future.

This volume reflects not only the efforts of the authors but also those of the following reviewers, who provided constructive comments and suggestions:

R. Anderson	L. M. Liro
W. B. Ayers, Jr.	M. D. Matthews
P. Baker	J. McKenzie
J. W. Castle	M. R. Mello
A. S. Cohen	P. E. Olsen
W. C. Dawson	M. A. Perlmutter
J. A. Curiale	C. R. Robison
G. Demaison	V. D. Robison
L. W. Elrod	D. J. Schunk
J. M. N. Evans	M. A. Smith
L. E. Frostick	F. A. Street-Perrott
P. J. W. Gore	C. P. Summerhayes
J. S. Janks	J. J. Tiercelin
J. J. Lambiase	D. Worsley

The editor also would like to thank Texaco Inc. for making time available to organize the research conference and for providing travel funds for Dr. Francoise Gasse and Mr. Luo Ming. Assistance also was provided by J. Toalson, G. Mayfield, G. Griffith, and Mary Hill.

REFERENCES

Bowman, H. E., R. C. Sheppard, and D. G. Ziegler, 1987, The Early Mesozoic rift system—Hydrocarbon frontier on the east coast: Oil & Gas Journal, v. 85, no. 41, p. 57-58, 63.

Demaison, G. J., and G. T. Moore, 1980, Anoxic environments and oil source bed genesis: AAPG Bulletin, v. 64, p. 1179-1209.

Eglinton, G., P. M. Scott, T. Belsky, A. L. Burlingame, and M. Calvin, 1964, Hydrocarbons of biological origin from a one-billion-year-old sediment: Science, v. 145, p. 263-264.

Fleet, A. J., K. Kelts, and M. R. Talbot, eds., 1988, Lacustrine petroleum source rocks: Geological Society Special Publication 40, 391 p.

Seglund, J. A., 1989, Midcontinent rift continues to show promise as petroleum prospect: Oil & Gas Journal, v. 87, no. 20, p. 55-58.

Climate and Lacustrine Petroleum Source Prediction

Eric J. Barron
Earth System Science Center
Pennsylvania State University
University Park, Pennsylvania, U.S.A.

Lacustrine environments are a major contributor of petroleum source rocks. Lacustrine source rock prediction is, however, influenced by numerous, complex variables governing lake sedimentation. Current predictive capability can be improved by attempting to map essential climatic variables to limit in space and time the area of lacustrine source rock exploration. Climatic characteristics that govern lake occurrence and the potential for stratification have been investigated with a General Circulation Model of the atmosphere for the present and for the mid-Cretaceous. In this analysis, the distribution of areas with a positive water balance first is used as an indicator of the distribution of areas conducive to lake formation. Second, the distribution of areas that experience large annual climatic variations is used as an indicator of the distribution of lakes that are less likely to be stratified and, hence, less likely to be sites of high organic-carbon preservation. Four factors used to define large climatic variations include (1) seasonal temperature cycle in excess of 40°C; (2) seasonal temperature extreme of less than 4C°; (3) average seasonal differences in precipitation minus evaporation balance in excess of 5 mm/day; and (4) distribution of mid-latitude winter storms. Evidence is presented to support the capability of climate models that add insight into lacustrine source rock prediction by simulating geographic regions conducive to lake development and to stratification and organic-carbon preservation.

INTRODUCTION

Many factors, both climatic and tectonic, govern the distribution of lakes and influence lake productivity and preservation of organic matter. Lacustrine sedimentation may reflect a complex interplay of diverse morphologic and environmental factors. Consequently, efforts to investigate deposition of organic-rich sediment necessitates defining those factors (basin tectonics, drainage area, topography, rate of subsidence, nutrient level, light availability, turbidity, oxygen availability, depth, and sedimentation rate) that influence modern lakes or that appear to explain the known record of lacustrine source rocks (Katz, this volume). Although lacustrine source rock prediction is limited by the importance of so many variables, one can improve predictive capability by mapping a selected governing variable to limit the scope of exploration, either spatially or temporally. Therefore, understanding climate becomes important in assessing many aspects of the geologic record, including lacustrine sequences (Summerhayes, 1988).

Large-scale global climate models potentially can improve predictive capability in two ways. First, the occurrence of a lake implies a positive water balance, even if only for a relatively brief geologic time.

Consequently, a positive regional precipitation-evaporation (P-E) balance can be a key (although not exclusive) predictive variable for lake occurrence. Second, climatic variability often governs the potential for lake stratification and oxygen availability and, hence, the preservation of organic matter.

General Circulation Models (GCMs) of the atmosphere are state-of-the-art tools that help define climatic water balance and climatic variability associated with diverse geographies and diverse global climates of the past. In demonstrating the potential of this approach, Kutzbach and Street-Perrot (1985) illustrated a remarkable correspondence of North African lake levels to GCM-simulated changes in moisture balance associated with evolution of the Earth's orbit during the last 18000 yr.

This study, intended for application over much longer time periods, is considerably less specific. The objective is to map "regimes" of lake occurrences and characteristics for different geographies, first by describing regions of positive moisture balance, and then by describing the subset of those regions with (a) high seasonal temperature variation (potential seasonal lake overturn through development and destruction of a seasonal thermocline); (b) high seasonal variation in P-E balance (potential for development and destruction of a seasonal pycnocline); and (c) highly seasonal changes in surface energy (e.g., winter storms that influence lake mixing and oxygenation). Model-generated lake regimes first are predicted for the present as a guide to model capability. Then, regimes are predicted for the mid-Cretaceous as a guide to potential variations in lake character and distribution for different geographies and for comparison with the rock record.

a soil moisture capacity of 15 cm (excess after 15 cm is simulated as runoff), is overly simplistic.

This experiment then should be viewed as a sensitivity test, but even this preliminary assessment demonstrates the CCM's potential to provide useful information about lacustrine source rock deposition. This study therefore will provide a first-order view of climatic controls that influence lake distribution and character.

MODEL CHARACTERISTICS

The two simulations described in this study are based on a version of the National Center for Atmospheric Research's (NCAR) Community Climate Model (CCM). The CCM, which has evolved from the spectral climate model of Bourke et al. (1977) and McAvaney et al. (1978), consists of nine levels and an associated grid of 40 latitudes (~4.4 resolution) and 48 longitudinal grid points (7.5 resolution).

The model includes atmospheric dynamics based on fluid-motion equations and includes radiative processes, convective processes, and evaporation and condensation as described by Ramanathan et al. (1983) and Pitcher et al. (1983). The radiation-cloudiness formulation introduced by Ramanathan et al. (1983) and the surface hydrology formulation introduced by Washington and Williamson (1977) are two major modifications from earlier versions.

This atmospheric GCM is coupled with an ocean formulation consisting of an energy-balance mixed layer that includes heat storage. Specifying a 50-m-thick mixed layer allows realistic simulation of the full annual cycle. Sea ice forms and grows when surface temperatures fall below –1.2°C. The present-day control simulation is that of Washington and Meehl (1984), who compared favorably the results of the present-day annual-cycle simulation with present-day observations.

The model's capability to simulate atmospheric moisture balance is the most essential factor to consider for this discussion. In their comparison of CCM simulations for January and July using specified sea-surface temperatures, Pitcher et al. (1983) successfully simulated many primary features, including the intense equatorial rainfall belt, the subtropical desert zone, and mid-latitude precipitation maxima. Figure 1 compares observed annual average continental precipitation (after Walter, 1973) with annual averages simulated by the version of the CCM used here. Although precipitation associated with storm tracks and with monsoonal circulations were well simulated, two problems are, however, evident. First, precipitation rates, particularly in the tropics, tend to be larger than observed in simulations with an energy-balance ocean. Second, precipitation characterized by high spatial and temporal variability (e.g., convective precipitation) is poorly simulated. In addition, the scheme for surface hydrology, dependent only on the P-E balance with

MODEL PREDICTIONS FOR THE PRESENT

Large-scale characteristics of precipitation and evaporation patterns generally are well simulated by the CCM (Figure 1, see also details in Pitcher et al., 1983; Washington and Meehl, 1984; Barron et al., 1989). Regions of seasonal high precipitation, illustrated in Figure 2, are defined by (1) the basic structure of the atmospheric general circulation (i.e., tropical low pressure and high precipitation and subtropical high pressure and low precipitation); and (2) the primary role of geography and land-sea thermal contrasts in disrupting zonal circulation. Therefore, development of monsoonal regimes (e.g., New Guinea, India-Indonesia, Somalia, and eastern Africa) and distribution of mid-latitude precipitation maxima associated with the position of the jet stream and winter storm track can be well simulated. However, model capability is decidedly poorer in regions where precipitation depends on atmospheric processes that operate at a resolution finer than the model's (e.g., summer convective precipitation in southeastern United States). The model is remarkably successful in simulating the large-scale structure of precipitation patterns, given its coarser resolution, but magnitudes are less well simulated. Evaporation rates (Figure 3) reflect temperature (latitude), nature of seasonal changes in the general circulation, and on continents, availability of moisture. Regions of high continental rainfall tend to be regions of high continental evaporation.

The difference between annual average precipitation and evaporation then should define a first-order prediction for the occurrence of lakes. Because regions whose balance approaches zero are not, however, predictable with confidence. Figure 4 illustrates only those regions whose moisture balance exceeds 0.5 mm/day. Under these conditions the model delineates desert regions, which are eliminated from the scope of lacustrine exploration. However, other regions characterized by a near-zero balance (e.g., the Great Lakes, which has a model-predicted winter excess but a summer deficit) or by convective precipitation (southeastern United States) also are eliminated. Removing such regions from the "predicted" distribution of lakes, as defined solely by annual average atmospheric moisture balance, obviously is incorrect. Although imperfect, particularly in regions of near-zero moisture balance, model-

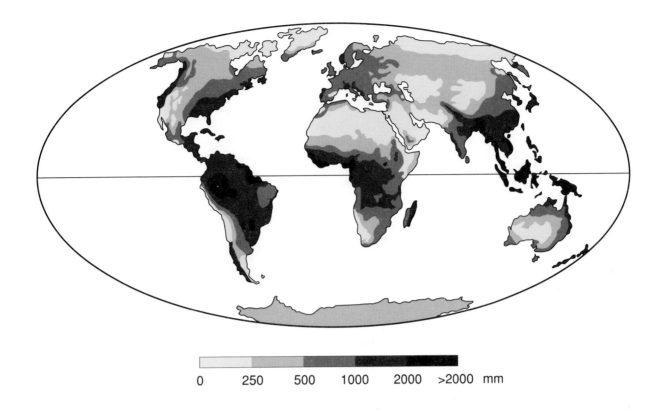

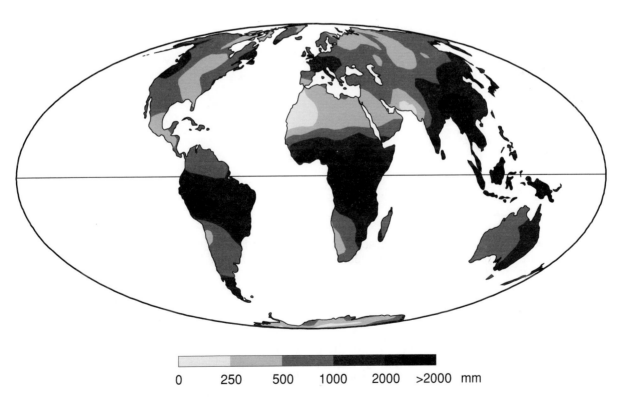

Figure 1. Comparison of (a) observed annual average precipitation (after Walter, 1973) and (b) CCM-predicted annual average precipitation (below).

Lacustrine Basin Exploration: Case Studies and Modern Analogs

3

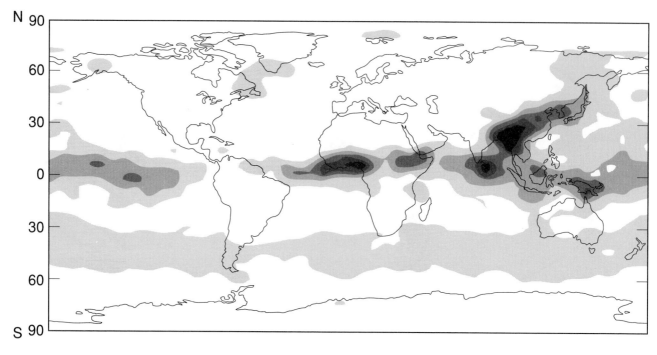

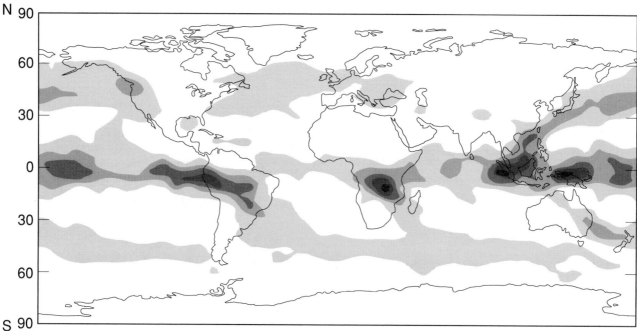

Figure 2. Present-day CCM-simulated precipitation (mm/day) for (a) average June, July, and August and (b) average December, January, and February (below). Shading indicates precipitation in excess of 3 mm/day (contour interval 3 mm/day).

predicted patterns of positive moisture balance generally are reasonable.

Predicted occurrence of lakes can be further analyzed by considering seasonal variability. Figure 5 illustrates seasonal characteristics of the P-E balance as the difference in the values from Figures 2 and 3. In particular, the overall pattern of meridional circulation, the importance of monsoons, and winter storm distribution become evident by defining regions of strong (>1.0 mm/day) moisture deficit and excess and regions of high seasonal variability. Strong seasonal variations in P-E (Figure 6) become evident where the differences between Figures 5a and 5b exceed 5 mm/day.

A second major source of variation, amplitude of the annual cycle of temperature based on the present-day simulation, is illustrated in Figure 7. In particular, continental interiors at high to middle

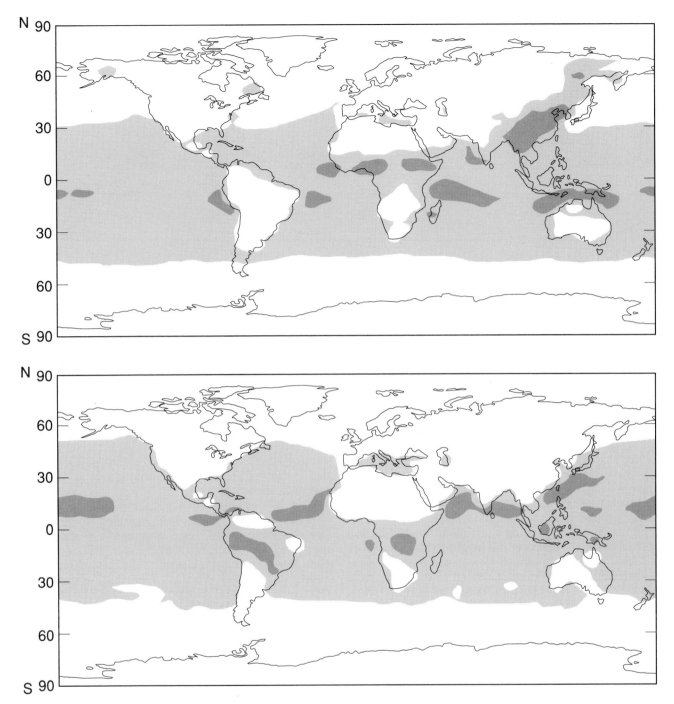

Figure 3. Present-day CCM-simulated evaporation (mm/day) for (a) average June, July, and August and (b) average December, January, and February. Shading indicates evaporation in excess of 3 mm/day (contour interval 3 mm/day).

latitudes experience considerable temperature variation due to the annual solar insolation cycle and the lack of thermal inertia of the continents. Continental size and latitude are the primary controls on amplitude of the annual temperature cycle.

Large seasonal variation in heating or moisture supply promotes seasonal overturn and likely poor organic carbon preservation. For example, Lake Baikal (southern Soviet Union) lies in a region of model-predicted moisture excess but would be eliminated from the scope of lacustrine source rock exploration on the basis of an annual temperature cycle greater than 40C°. Incidentally, if the Great Lakes had been simulated as a region of positive moisture balance, they also would have been characterized by large seasonality and therefore removed from exploration consideration. The distribution of winter air temperatures below 4°C,

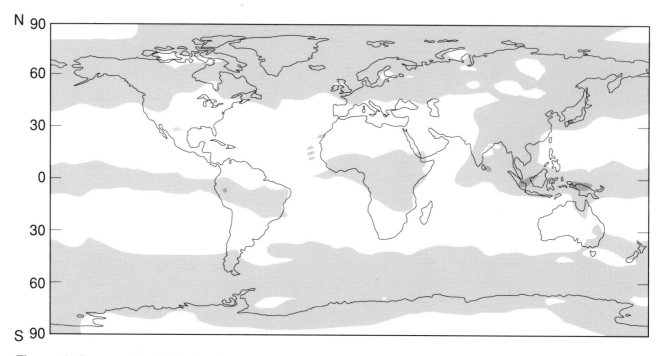

Figure 4. Present-day CCM-simulated precipitation minus evaporation (mm/day) for a given annual average. Light shading indicates positive balance in excess of 0.5 mm/day; dark shading indicates excess of 5.0 mm/day. Hatchuring indicates extreme deficit (<−5.0 mm/day).

because of the relationship between density and temperature in fresh water, becomes another temperature indicator of seasonal overturn.

The model-predicted distribution of winter storms (Figure 8), as defined by the time-filtered standard deviation of the geopotential height field—a measure of how often high- and low-pressure systems pass (Barron, 1989)—is another reason to discount some middle- to high-latitude lakes as source rock environments. The position of the storm track's main axis is closely tied to land-sea thermal contrasts and can be related to the annual cycle of temperature (Figure 7).

The large seasonal differences in P-E balance (Figure 6), regions of large temperature variation (Figure 7), and the position of the winter 4°C isotherm provide a basis for subdividing the model-predicted distribution of positive moisture balance illustrated in Figure 4. Lake distribution then can be defined as three "regimes" of positive annual moisture balance but with (1) large annual temperature variation, (2) large annual variation in P-E balance, or (3) no large annual variation (Figure 9). This third case represents model-predicted regions conducive to lake development with the potential for organic-rich sedimentary deposition, as defined solely by large-scale climatic criteria. Although such a prediction clearly is inaccurate because so many factors influence lake sedimentation, it reasonably serves to limit the spatial scope of investigation by eliminating

desert regions and by characterizing regions of large seasonal variation.

The lacustrine "target" regime is primarily tropical or subtropical but includes some high-latitude continental regions subject to moderate maritime influence. However, the 4°C isotherm eliminates most high-latitude sites based on probability of seasonal overturn. Distribution of winter storms or seasonal sea-ice cover could be used to further characterize high-latitude lacustrine environments as suitable or unsuitable. Evidently the regions most conducive to source rock deposition in lacustrine environments occur within the low-latitude climatic regime, which includes Lake Tanganyika, often cited as a type example for high organic-carbon deposition (Degens and Stoffers, 1976).

As partial confirmation, six of ten modern lakes containing more than 1% organic carbon content (Bradley, 1966; Swain, 1970; Kelts, 1988; Talbot, 1988; and Binjie et al., 1988) lie within the target regime in Africa. A shallow Florida lake (Bradley, 1966), a lake in Nicaragua (Swain, 1970), and one African lake within the monsoonal belt (Talbot, 1988) all lie outside model-predicted regions of conducive environments. Substantially higher predictive capability could be gained by integrating the climatic information from Figure 9 with a knowledge of topography, drainage divides, interior drainage, and basin distribution.

Barron

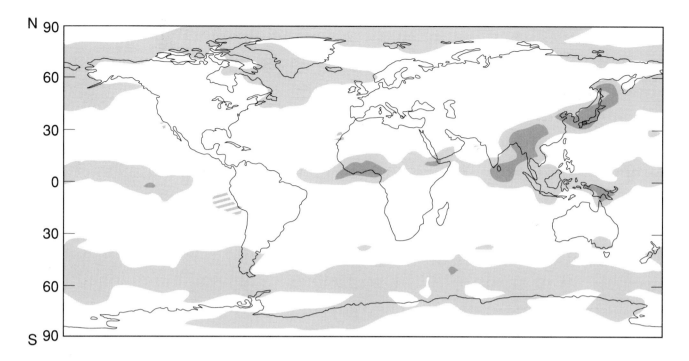

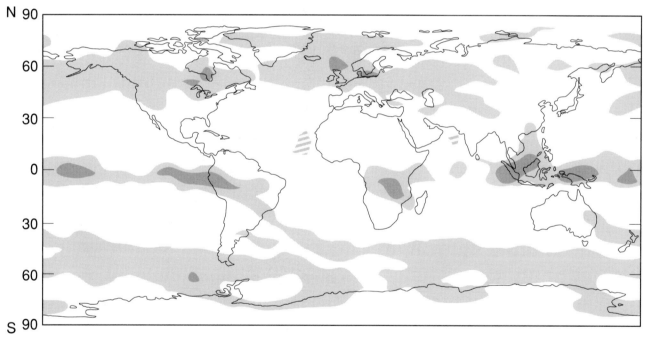

Figure 5. Present-day CCM-simulated precipitation minus evaporation (mm/day) for (a) average June, July, and August and (b) average December, January, and February. Light shading indicates positive balance in excess of 1.0 mm/day; dark shading indicates positive balance in excess of 5.0 mm/day. Hatchuring indicates extreme deficit ($<$-5.0 mm/day).

MODEL PREDICTIONS FOR THE MID-CRETACEOUS

The CCM next was applied to mid-Cretaceous geography described by Barron and Washington (1984). Although the general structure of Cretaceous precipitation maxima and minima (Figure 10) with respect to latitude is similar to present-day control (Figure 2), the simulations differ in several important respects. First, the mid-Cretaceous hydrologic cycle intensity was substantially greater, reflecting conditions for a warmer planet and large oceanic areas within the subtropical evaporation region (Barron et al., 1989). Second, the zonal nature of the

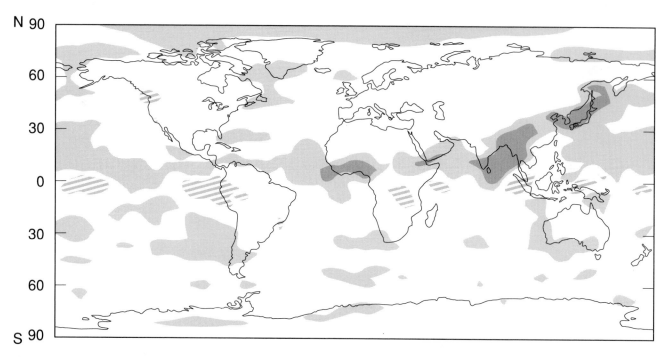

Figure 6. Present-day CCM-simulated differences in precipitation-minus-evaporation balance calculated by subtracting the December, January, and February P-E balance (Figure 5b) from the June, July, and August P-E balance (Figure 5a). Light shading indicates positive balance in excess of 1.0 mm/day; dark shading indicates positive balance in excess of 5.0 mm/day. Hatchuring indicates extreme deficit (<-5.0 mm/day).

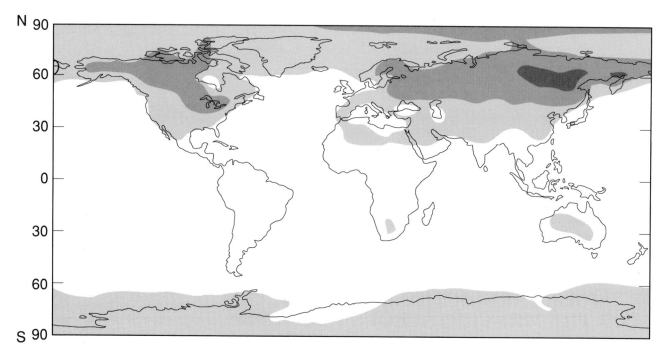

Figure 7. Present-day CCM-simulated temperature difference between June, July, and August and December, January, and February. Shading indicates annual temperature contrast in excess of 20C° (contour interval 20C°).

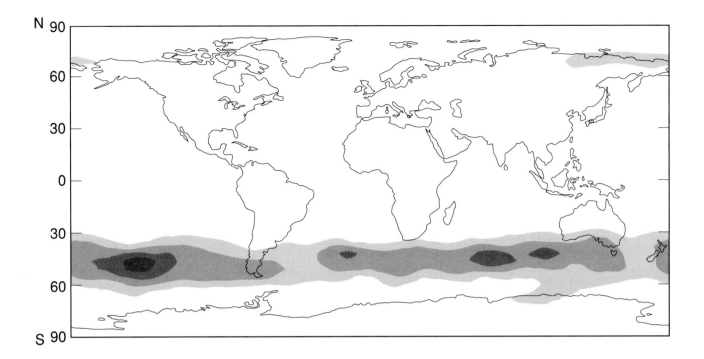

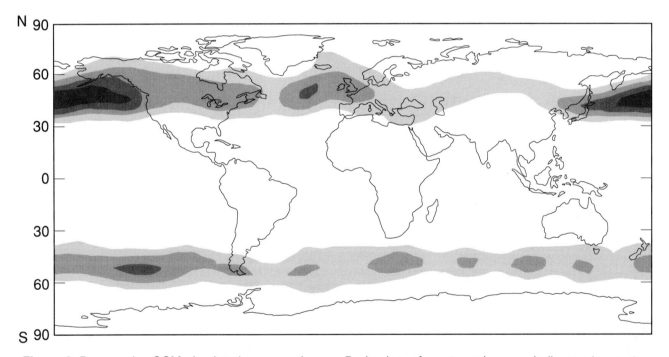

Figure 8. Present-day CCM-simulated storm tracks as represented by the standard deviation of the geopotential height field for (a) average June, July, and August and (b) average December, January, and February. Darkening of contoured areas indicates increasing standard deviation and more frequent, higher magnitude storms.

Tethys Ocean exerted strong control on distribution of high precipitation, which occurs on both the northern and southern continental boundaries of Tethys. Third, northern hemisphere mid-latitude winter precipitation maxima were poorly developed during the mid-Cretaceous, reflecting, in part, major differences in the characteristics of winter-storm distribution (Barron, 1989).

Simulated evaporation rates (Figure 11) reflect temperature (latitude), nature of seasonal changes in general circulation, and on continents, the availability of moisture.

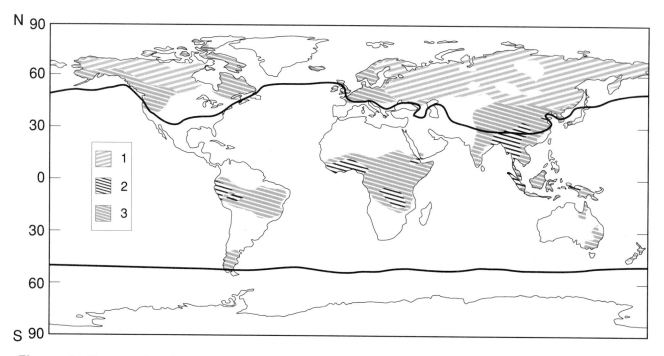

Figure 9. Present-day CCM-defined regimes of lacustrine conditions. Shaded regions are characterized by positive annual moisture balance in excess of 0.5 mm/day (after Figure 4). Within these regions, pattern 1 represents large annual temperature variations (>40C°, after Figure 7); pattern 2 represents large annual P-E variations (>5 mm/day seasonal average, after Figure 6); and pattern 3 represents regions with no substantial variation. Heavy lines indicate interseasonal average positions of the 4°C isotherm.

By the same procedure used to derive Figure 4, differences between annual average precipitation and evaporation were computed to define a first-order prediction of mid-Cretaceous lake occurrence, shown in Figure 12, which similarly illustrates only regions with a moisture balance greater than 0.5 mm/day.

Comparing the mid-Cretaceous and present-day control simulations (Figures 12 and 4) reveals several interesting differences. Most Cretaceous continental areas are characterized by positive moisture balance; smaller areas of evolving southern Africa, Argentina, and Tibet-western China are predicted as either primary deserts or areas of near-zero moisture balance. These results are entirely consistent with the Cretaceous record of continental humid and arid zones, distribution of which is defined by coals and evaporites, respectively. Those described by Hallam (1984) for Aptian to Cenomanian time are remarkably consistent with Figure 12.

By the procedure described earlier, mid-Cretaceous regions predicted to be conducive for lake formation are further subdivided by considering seasonal variability. In Figure 13—seasonal characteristics of P-E balance as a difference between values in Figures 10 and 11—the overall pattern of meridional circulation and the importance of monsoons are evident, although position of the Tethys Ocean clearly dominates distribution of regions of strong moisture excess and deficit. Figure 14, calculated as the difference in seasonal P-E (Figures 13a and 13b),

identifies regions of strongest seasonal difference in moisture balance. Tethys Ocean and its margins again are the most apparent areas of seasonal contrast.

The second major source of variation, amplitude of the annual cycle of temperature, is illustrated in Figure 15 for the mid-Cretaceous simulation. Most importantly, Cretaceous geography also is characterized by high-latitude regions of substantial seasonal temperature variation (most significantly Greenland, Siberia, and Antarctica). A similar argument applies—the combination of seasonal variation in solar insolation and lack of thermal inertia of the continents results in a large annual cycle of temperature. This appears to be a robust aspect of model simulation (Crowley et al., 1986; Schneider et al., 1985; Sloan and Barron, 1990).

Model-predicted distribution of mid-Cretaceous winter storms (Figure 16), as defined by time-filtered standard deviation of the geopotential height field, provides additional information on annual variability. Northern hemisphere winter storms are shifted poleward compared with the present day and, because the North Atlantic is not well developed, the pattern is dominated by the North Pacific (Barron, 1989). Southern hemisphere winter storms are predicted to be weaker.

Large seasonal differences in P-E balance (Figure 14), regions of large seasonal temperature difference (Figure 15), and position of the winter 4°C isotherm

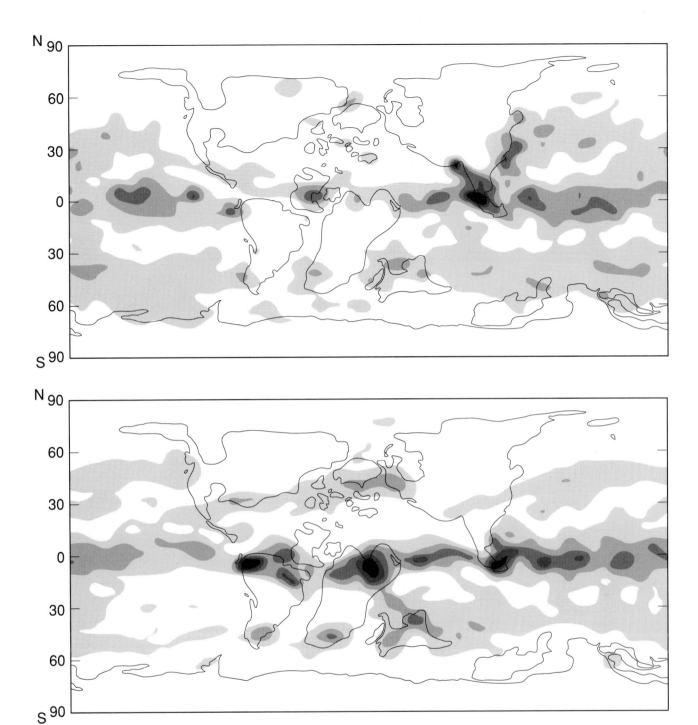

Figure 10. Mid-Cretaceous model-simulated precipitation (mm/day) for (a) average June, July, and August and (b) average December, January, and February.

Shading indicates precipitation in excess of 3 mm/day (contour interval 3 mm/day).

provide the basis for subdividing the model-predicted distribution of positive moisture balance illustrated in Figure 12. Lake regimes in Figure 17 then can be defined as for the present day. Regions of positive annual moisture balance that lack large annual variation correspond to model-predicted regions conducive to lake development with potential for organic-rich sedimentary deposition, as defined solely by large-scale climate criteria.

The target regime again is primarily tropical or subtropical. Greater areas within the middle latitudes are characterized by relatively small seasonal climatic variations. This may reflect greater potential at middle and higher latitudes because of characteristics of mid-Cretaceous geography and greater winter warmth compared with the present. However, position of the winter 4°C isotherm may eliminate many mid-latitude regions from exploration

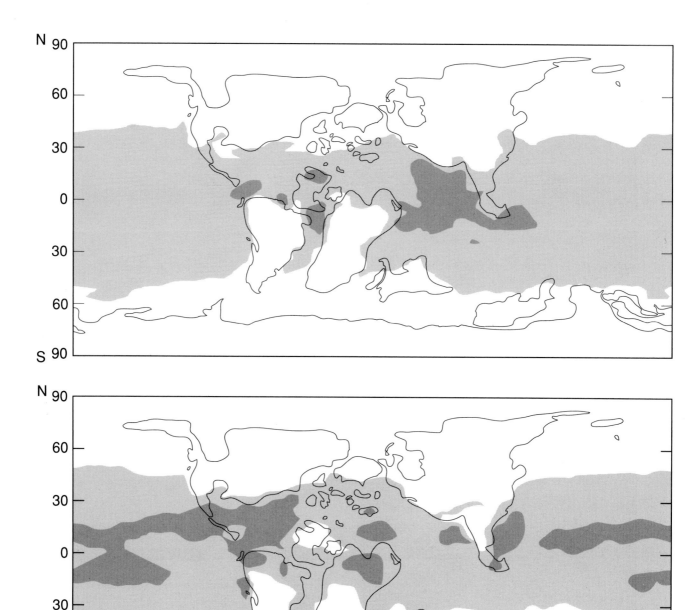

Figure 11. Mid-Cretaceous model-simulated evaporation (mm/day) for (a) average June, July, and August and (b) average December, January, and February.

Shading indicates evaporation in excess of 3 mm/day (contour interval 3 mm/day).

unless they lie close to the continental margin. In terms of climate, the mid-Cretaceous is somewhat more conducive to development of lacustrine source rocks than is the present. Figure 17 includes an initial comparison with a data set of the distribution of Cretaceous lacustrine source rocks. Although not perfect or comprehensive, most Cretaceous lacustrine source rocks fall within predicted "conducive" environments. However, if the 4°C isotherm is a good

criterion for lake turnover and oxygenation, then more than one-half of Cretaceous lacustrine source rocks are eliminated, but this may reflect the fact that Cretaceous climates were warmer than simulated by the model (see Barron and Washington, 1985). Conversely, if Cretaceous undifferentiated sites are eliminated, most remaining sites lie near coastlines and may experience strong maritime influences under conditions of warm ocean temper-

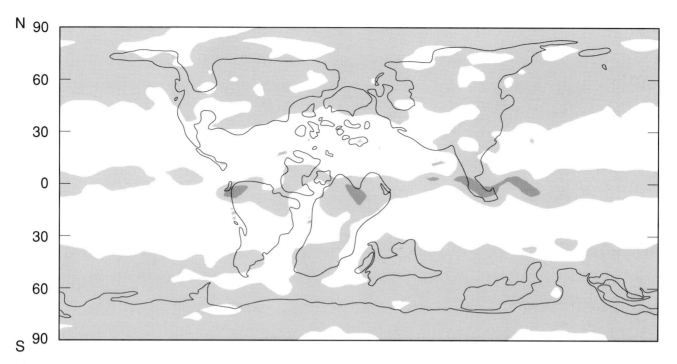

Figure 12. Mid-Cretaceous model-simulated precipitation minus evaporation (mm/day) for a given annual average. Light shading indicates positive balance in excess of 0.5 mm/day; dark shading indicates excess of 5.0 mm/day. Hatchuring indicates extreme deficit (<−5.0 mm/day).

atures. Further research will be needed to examine sequences case by case and to compare model results for higher atmospheric CO_2 concentrations (warmer Cretaceous simulation).

DISCUSSION AND CONCLUSIONS

Lakes are dynamic environments, and lake sedimentation represents the interaction of many complex physical, biological, and chemical variables. The complexity of this interaction limits predictability of lake distribution and lake characteristics. Consequently, the initial approach described here was an attempt to map key governing variables— in this case, climate—that then can be used to limit the scope of exploration either spatially or temporally. This investigation of climatic control on lake distribution and characteristics was highly simplified, based on a twofold assumption that positive moisture balance is a limiting factor in lake distribution and that large annual climatic contrasts are incompatible with stable stratification and the preservation of organic matter.

Applications of the Community Climate Model to the present day and to the mid-Cretaceous illustrate the first-order climate relationships that may aid in investigating lacustrine source rock distribution through geologic time. The simulations suggest substantial variations based on differences in geography and global climate. In addition, several general relationships are apparent.

First, the tropics are the primary region consistently predicted to be a conducive environment; however, this relationship may not apply throughout Earth history. For example, Kutzbach and Gallimore's (1989) simulations of Pangean megacontinents suggest extensive continental aridity even in tropical continental regions. Continental size apparently influences tropical moisture balance.

Second, the cores of middle- to high-latitude, and particularly large, continental regions are likely to experience substantial annual climatic variability and seasonal overturn and, therefore, are less likely to promote high organic-carbon deposition.

Third, environments conducive to lake formation and lake stratification may occur in the middle latitudes depending on the relationship to monsoonal circulations, position with respect to winter storm tracks and intensity of winter storms, and the extent to which climate is modulated by oceanic regions. These regions less easily follow simple rules that can be qualitatively applied to different paleogeographic reconstructions.

The importance of the model results is demonstrated by the CCM's capability to simulate modern climate and by the apparent match with mid-Cretaceous arid and humid climatic indicators. Evidently, the models provide a reasonable determination of arid and humid environments and, therefore, a reasonable estimation of potential lake distribution. In addition, they reasonably simulate

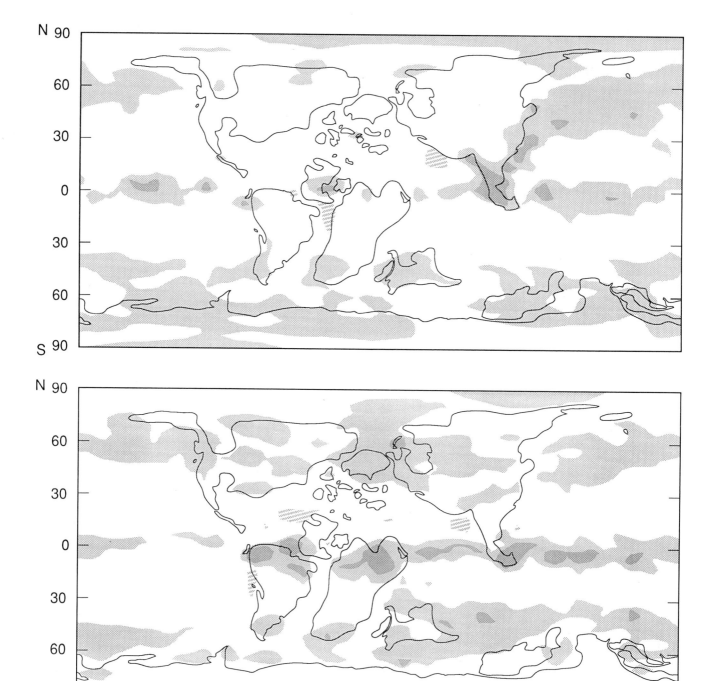

Figure 13. Mid-Cretaceous model-simulated precipitation minus evaporation (mm/day) for (a) average June, July, and August and (b) average December, January, and February. Light shading indicates positive balance in excess of 1.0 mm/day; dark shading indicates positive balance in excess of 5.0 mm/day. Hatchuring indicates extreme deficit (<-5.0 mm/day).

climate variation through the annual cycle. The most significant test of these predictions will be a comprehensive reconstruction of mid-Cretaceous lacustrine source rocks. The question here is the extent to which climatic variability explains source rock distributions. These ideas are largely untested for Earth history.

Coincidence of model predictions and actual lacustrine source rock distribution cannot be expected for three reasons. First, many aspects of the hydrologic cycle, particularly as they relate to model resolution, are not well simulated. Second, experiments are not comprehensive and therefore may not have addressed all the driving factors that

Figure 14. Mid-Cretaceous model-simulated differences in precipitation-minus-evaporation balance calculated from subtracting the December, January, and February P-E balance (Figure 13b) from the June, July, and August P-E balance (Figure 13a). Light shading indicates positive balance in excess of 1.0 mm/day; dark shading indicates positive balance in excess of 5.0 mm/day. Hatchuring indicates extreme deficit (<-5.0 mm/day).

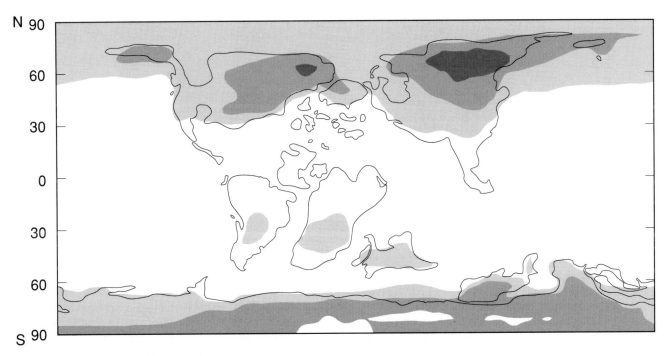

Figure 15. Mid-Cretaceous model-simulated temperature difference between June, July, and August, and December, January, and February. Shading indicates annual temperature contrast in excess of 20°C (contour interval 20°C).

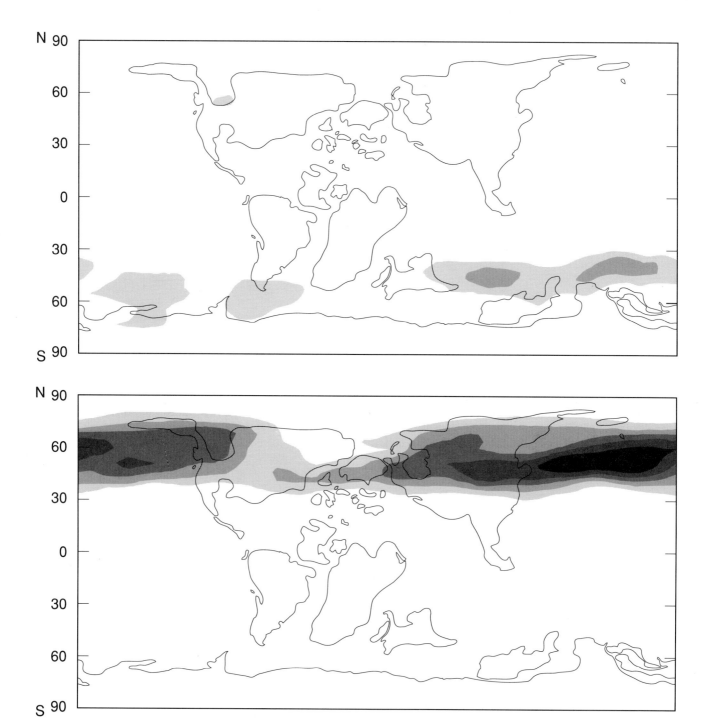

Figure 16. Mid-Cretaceous model-simulated storm tracks as represented by the standard deviation of the geopotential height field for (a) average June, July, and August and (b) average December, January, and February. Stippling indicates greater standard deviation and more frequent, higher magnitude storms.

may have influenced Cretaceous climate (e.g., CO_2 levels). Third, lake location, size, shape, productivity, and organic-matter preservation are governed by climate, tectonism, and various biological and chemical factors. For example, elevation may be a factor in lake overturn even in tropical regions (Håkanson and Jansson, 1983; Katz, this volume).

Rather than expect perfect correspondence, if climatic analysis can help successfully limit the scope of exploration, without inadvertent loss of potential source rocks from consideration, then GCM prediction provides a foundation for lacustrine source rock prediction. The next step should be to combine climatic data and knowledge of basin formation and

Barron

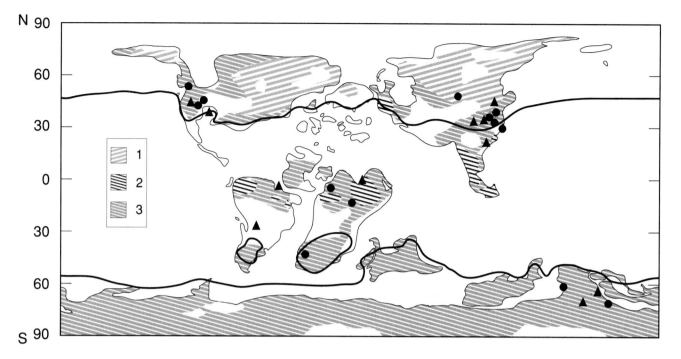

● Late Cretaceous

▲ (Undifferentiated) Cretaceous

Figure 17. Mid-Cretaceous model-defined regimes of lacustrine conditions. All shaded regions are characterized by positive annual moisture balance in excess of 0.5 mm/day (after Figure 12). Of these regions, large annual temperature variations (>40°C after Figure 15) are indicated by pattern 1; large annual P-E variations (>5 mm/day seasonal average, after Figure 14) are indicated by pattern 2; and regions without substantial annual variation are indicated by pattern 3. Winter seasonal average position of the 4°C isotherm is indicated by heavy line. Locations of Cretaceous lacustrine source rocks after Smith (1990) are plotted. (Circle indicates Late Cretaceous; triangle indicates undifferentiated Cretaceous.)

distribution to identify specific regions of lake deposition under conducive environmental conditions.

ACKNOWLEDGMENTS

The author gratefully acknowledges the invitation by B. J. Katz and B. Rosendahl to present this manuscript at the AAPG Workshop. Research was supported in part by National Science Foundation (NSF) Grants ATM-87-15499 and EAR-8720551. Computer time was provided by the NSF-supported National Center for Atmospheric Research (NCAR) in Boulder, Colorado. The author also gratefully acknowledges the use of the present-day control simulation, as provided by Warren Washington, the aid of James Sloan, Peter Fawcett, and Peter Schultz in completing this contribution, and the helpful reviews by V. Robison, C. Summerhayes, and A. Cohen.

REFERENCES CITED

Barron, E. J., 1989, Severe storms during Earth history: GSA Bulletin, v. 101, p. 601-612.

Barron, E. J., W. W. Hay, and S. Thompson, 1989, The hydrologic cycle—A major variable during Earth history: Global and Planetary Change, v. 75, p. 157-174.

Barron, E. J., and W. M. Washington, 1984, The role of geographic variables in explaining paleoclimates—Results from Cretaceous climate model sensitivity studies: Journal of Geophysical Research, v. 89, no. D1, p. 1267-1279.

Barron, E. J., and W. M. Washington, 1985, Warm Cretaceous climates—High atmospheric CO_2 as a plausible mechanism, in E. Sundquist, and W. Broecker, eds., Chapman Conference on Natural Variations in Carbon Dioxide and the Carbon Cycle: American Geophysical Union, p. 546-553.

Binjie, L., X. Yang, H. Lin, and G. Zheng, 1988, Characteristics of Mesozoic and Cenozoic non-marine source rocks in Northwest China, in A. J. Fleet, K. Kelts, and M. R. Talbot, eds., Lacustrine petroleum source rocks: Geological Society Special Publication 40, p. 291-298.

Bourke, W., B. McAvaney, K. Puri, and R. Thurling, 1977, Global modeling of atmospheric flow by spectral models, in J. Chang, ed., General circulation models of the atmosphere: New York, Academic Press, Methods in Computational Physics, v. 17, p. 267-324.

Bradley, W. H., 1966, Presidents farewell address—Tropical lakes, copropel and oil shale: GSA Bulletin, v. 77, p. 1333–1338.

Chen, P-J., 1987, Cretaceous paleogeography in China: Palaeogeography, Palaeoclimatology, Palaeoecology, v. 59, nos. 1–3, p. 49–56.

Crowley, T. J., D. A. Short, J. G. Mengel, and G. R. North, 1986, Role of seasonality in the evolution of climate during the last 100 million years: Science, v. 231, p. 579–584.

Degens, E. T., and P. Stoffers, 1976, Stratified waters as a key to the past: Nature, v. 263, p. 22–27.

Håkanson, L., and M. Jansson, 1983, Principles of lake sedimentology: New York, Springer-Verlag, 316 p.

Hallam, A., 1984, Continental humid and arid zones during the Jurassic and Cretaceous: Palaeogeography, Palaeoclimatology, Palaeoecology, v. 47, p. 195–223.

Katz, B. J., Controls on distribution of lacustrine source rocks through time and space, this volume.

Kelts, K., 1988, Environments of deposition of Lacustrine petroleum source rocks—An introduction, in A. J. Fleet, K. Kelts, and M. R. Talbot, eds., Lacustrine petroleum source rocks: Geological Society Special Publication 40, p. 3–26.

Kutzbach, J. E., and R. G. Gallimore, 1989, Pangean climates—Megamonsoons of the megacontinent: Journal of Geophysical Research, v. 94, p. 3341–3358.

Kutzbach, J. E., and F. A. Street-Perrott, 1985, Milankovitch forcing of fluctuations in the level of tropical lakes from 18 to 0 kyr BP: Nature, v. 317, p. 130–134.

McAvaney, B. J., W. Bourke, and K. Puri, 1978, A global spectral model for simulation of the general circulation: Journal of Atmospheric Science, v. 35, p. 1557–1582.

Pitcher, E. J., R. C. Malone, V. Ramanathan, M. L. Blackmon, K. Puri, and W. Bourke, 1983, January and July simulations with a spectral general circulation model: Journal of Atmospheric Science, v. 40, p. 580–604.

Ramanathan, V., E. J. Pitcher, R. C. Malone, and M. L. Blackmon, 1983, The response of a spectral general circulation model to refinements in radiative processes: Journal of Atmospheric Science, v. 40, p. 605–630.

Schneider, S. H., S. L. Thompson, and E. J. Barron, 1985, Mid Cretaceous, continental surface temperatures—Are high CO_2 concentrations needed to simulate above freezing winter conditions, in E. Sundquist, and W. Broecker, eds., Chapman Conference on Natural Variations in Carbon Dioxide and the Carbon Cycle: American Geophysical Union, p. 554–560.

Sloan, L. C., and E. J. Barron, 1990, "Equable" climates during Earth history?: Geology v. 18, p. 489–492.

Smith, M. A., Lacustrine oil shales in the geologic record, this volume.

Summerhayes, C. P., 1988, Predicting paleoclimates (abs.), in A. J. Fleet, K. Kelts, and M. R. Talbot, eds., Lacustrine petroleum source rocks: Geological Society Special Publication 40, p. 77–78.

Swain, F. M., 1970, Non-marine organic geochemistry: Cambridge University Press, 445 p.

Talbot, M. R., 1988, The origins of Lacustrine oil source rocks—Evidence from the lakes of tropical Africa, in A. J. Fleet, K. Kelts, and M. R. Talbot, eds., Lacustrine petroleum source rocks: Geological Society Special Publication 40, p. 29–43.

Walter, H., 1973, Vegetation of the Earth in relation to climate and the eco-physiological conditions: New York, Springer-Verlag, 237 p. [Translated from the German 2d ed.]

Washington, W. M., and G. A. Meehl, 1984, Seasonal cycle experiment on the climate sensitivity due to a doubling of CO_2 with an atmospheric general circulation model coupled to a simple mixed-layer ocean model: Journal of Geophysical Research, v. 89, no. D6, p. 9475–9503.

Washington, W. M., and D. L. Williamson, 1977, A description of the NCAR global circulation models, in J. Chang, ed., General circulation models of the atmosphere: New York, Academic Press, Methods in Computational Physics, v. 17, p. 111–172.

Tectonic and Climatic Controls on Lake Distribution and Environments in Afar from Miocene to Present

F. Gasse
Laboratoire d'Hydrologie et de Geochimie Isotopique
Université Paris-Sud
Orsay Cedex, France

Located at the triple junction of the African, Arabian, and Somalian plates, the Afar depression has been subjected to intense and complex volcano-tectonic activity from the Miocene into the Holocene. The region also has experienced significant climatic changes. The chronology of ancient lacustrine sedimentation is based on K-Ar dating of associated volcanics, diatom biostratigraphy, and ^{14}C ages.

Spreading of the Afar depression accounts for the migration of active sedimentary basins from the margins of the Afar triangle to the modern active segments (Lakes Asal and Afrera). Tectonic conditions have been favorable for the settlement of deep, closed lakes in central Afar since the late Pliocene (2.5 Ma). There, ancient lakes may not be related to present topography. Sediment distribution shows a southwest-to-northeast progression in age of opening of the modern basins and a change in direction of major active faults from about east-west to northwest at ~1.4 Ma. Deformation of Holocene sediments and shorelines allows the direction and rate of recent vertical displacement to be estimated.

Rifting also has increased elevation differences and hydraulic gradients between the depression and surrounding highlands, thereby increasing the effectiveness of both surface waters and groundwater circulation. This explains the general increase in water depth since the Miocene.

The effects of climatic fluctuations have been superimposed onto the structural evolution of the depression. Because climate governs chemical weathering, it is partly responsible for the sedimentary facies, specifically whether detrital, biogenic silica, or chemical carbonate fractions dominate. At any given time, the ratio of precipitation to evaporation in the Afar and on neighboring plateaus is responsible for the presence or absence of lakes in the deepest basins. Major lacustrine stages are correlated both with evolution of other tropical lakes and with global climatic events.

INTRODUCTION

This paper illustrates changes in spatial and temporal distribution and in environmental conditions of Neogene (Miocene and Pliocene) and Quaternary lakes in the Afar depression by reconstructing their physiography. The depression ranges in elevation from approximately 700 m above sea level to 155 m below sea level.

The Afar occupies an exceptional geological setting—the triple junction of the African, Arabian, and Somalian plates (Figure 1; Tazieff et al., 1972). It has been subjected to intense volcano-tectonic activity from the Miocene to the present. As identified through the volcanic Afar Series, rifting has induced downfaulting of tectonic basins where lakes then developed during episodes of favorable hydrologic conditions. Sedimentary strata become progressively older away from the present active axes as a result of subcontinuous spreading (CNR-CNRS, 1971, 1975).

The Afar also lies in an unusual hydrologic setting. The depression currently is arid because it is situated in the rain shadow of the Ethiopian and southeastern plateaus (Figure 1). However, Afar basins do receive water from these areas through surface runoff and groundwater flow. The deepest basins record climatic fluctuations in the highlands as well as in the Afar itself; thus, they amplify the regional climatic signal. What few small lakes occur today (Figure 2) are relicts of larger, deeper lakes that experienced enormous changes in water level and water chemistry during the late Quaternary (Gasse, 1975).

This discussion focuses on the central Afar (Republic of Djibouti), where optimal structural

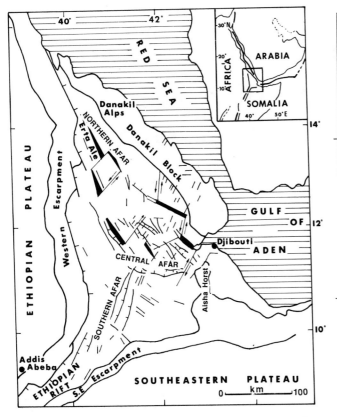

Figure 1. Major tectonic features of the Afar depression (after Behre, 1986). Heavy black lines represent present active axes.

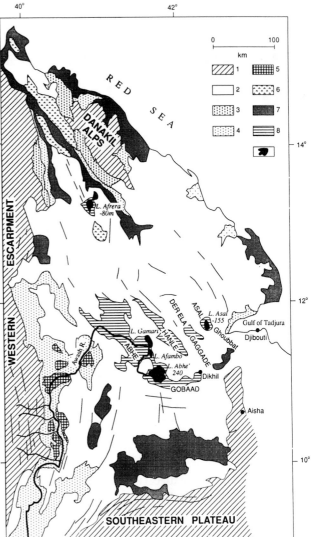

Figure 2. Generalized geology of Afar depression (after CNR-CNRS, 1971, 1975; Gasse, 1975). 1, basement (metamorphic rocks, Mesozoic marine deposits, Ethiopian Plateau Trap Series); 2, Afar Volcanic Series (mainly basalts); 3, Miocene-Pliocene sandstones, sands, red and polychrome clays, and associated volcanites; 4, Pliocene to Holocene alluvial fans; 5, lower Pliocene fluviolacustrine deposits (>4.0-2.5 Ma); 6, middle to upper Pleistocene marine evaporites, principally halite and gypsum (200–80 ka); 7, Pleistocene and Holocene basin deposits, sands, silts, and pedal clays; 8, upper Pliocene to upper Holocene lacustrine deposits; 9, present-day lakes.

situations for lacustrine sedimentation have persisted from 3–2.5 Ma to the present, and in particular during the late Pliocene and early Pleistocene, periods of major topographical changes. Miocene and early Pliocene lakes also are briefly discussed in their general structural setting.

GENERAL GEOLOGICAL AND ENVIRONMENTAL SETTING

Volcano-Tectonic Setting

The Afar depression is part of the Afro-Arabian Rift System (Pilger and Rösler, 1975, 1976; Barberi et al., 1975; Carrelli, 1980; Behre, 1986). The continental Ethiopian rift, which separates the African and Somalian plates, terminates in southern Afar (Figure 1). In central Afar active segments of the Gulf of Tadjura and the Asal-Ghoubbat rift (Figure 2) trend northwest and represent a continuation of the Gulf of Aden Ridge system that separates the Arabian and Somalian plates. Northern Afar is characterized by several active volcanic chains, such as Erta Ale (Figure 1), that parallel the accretion axis

of the Red Sea Rift, which separates the African and Arabian plates.

Both the time of early opening of the Afar—Eocene (Girdler and Styles, 1978) to Miocene-Pliocene (Christiansen et al., 1975)—and causes of its distension are still under debate (Behre, 1986). However, the major stages of volcanic activity after 25 Ma in the Afar and along its margins are well

Gasse

known (Behre, 1986) and provide a general background for paleolake chronology. Table 1 and Figures 3 and 4 summarize the chronology of lacustrine sedimentation and its correlation with structural setting. Chronology of ancient lakes is based mainly on K-Ar dates on volcanics associated with lacustrine deposits. The K-Ar method is not, however, the most appropriate for paleolimnological studies because (1) the standard error in K-Ar dates (100 k.y.) may exceed a lake's lifetime, as shown in the example of ^{14}C-dated late Quaternary lakes; (2) dating lava flows underlying and/or overlying paleolake deposits provides only a maximum time range for lacustrine episodes; and (3) lava flows immediately beneath lacustrine strata often are altered and cannot be dated. In favorable cases, diatom biostratigraphy complements radiometric chronology. Upper Quaternary stratigraphy (<40 ka) is supported by several hundred ^{14}C dates with a time resolution of 100–1000 yr.

Present-Day Physiography, Climate, and Hydrology

Climate and hydrology of the Afar (United Nations, 1965, 1971; Müller, 1982) depend on regional topography. A marked elevation gradient and a discrete south-north gradient of increasing aridity are observed.

The steep western escarpment (Figure 2) receives attenuated rainfall from the Atlantic and Indian ocean monsoons (600–1500 mm/yr in the south, 400–1200 mm/yr in the north). Surface runoff from its abrupt slopes is responsible for enormous detrital input along its base. Meteoric water enters this dislocated zone and circulates toward the deep basins of the central Afar. The southeastern plateau and its northern extension, the Aisha horst (Figure 2), receive rainfall from the Indian Ocean monsoon. The Aisha horst also acts as a catchment area for wadis flowing toward the central Afar. The Afar depression itself is water deficient during all seasons because of high evaporation rates (>2000 mm/yr in the deepest zones).

The southern Afar slopes gently to the north from elevations of about 700 to 400 m; precipitation decreases from about 600 to 400 mm/yr. Draining the area is the Awash River, which flows from the Ethiopian Plateau into Lake Abhé in central Afar.

Central Afar is characterized by alternating plateaus and depressions separated by north-northwest- to east-west-trending fault scarps several hundred meters high. From west to east, the major lake basins become narrower and descend in elevation as follows (Figure 2): Abhé-Gobaad (240 m), Hanlé (100 m), Gaggadé-Der Ela (80 m), and Asal (155 m below sea level). Central Afar receives less than 300 mm/yr of precipitation. In the Abhé basin the Awash River supplies a chain of freshwater lakes (e.g., Gamari and Afambo) before flowing into closed Lake Abhé where TDS (total dissolved solids) content reaches 160‰ (Gasse, 1975; Fontes et al., 1985). Lake Abhé is a shallow hyperalkaline lake (Na^+ and $CO_3^=$ dominant) presently shrinking. The lowest basin is occupied by hypersaline Lake Asal (TDS 300‰, Na^+ and Cl^- dominant), whose present steady state and chemistry are attributable to infiltration by marine waters circulating through the Asal-Ghoubbat rift (Gasse and Fontes, 1989).

The hyperarid northern Afar (<50 mm/yr rainfall) is characterized by Pleistocene marine evaporites (Figure 2; Lalou et al., 1970) and contains the small hypersaline Lake Afrera (TDS, 160‰; Martini, 1969), which is maintained by inflow of groundwater circulating from the western escarpment through the evaporites.

MIOCENE AND EARLY PLIOCENE LAKES (10–2.5 Ma)

No lacustrine deposits are known prior to 10 Ma. Three stages of lacustrine sedimentation have been recognized between 10 and 2.5 Ma.

Stage I (14–10 Ma)

Major separation along the Afar and Ethiopian rifts started 14 Ma ago with formation of the Mabla Rhyolite Series and lateral equivalents (Behre, 1986). Lacustrine deposits occur in the upper part of these silicic volcanics, which form the margins of the Ethiopian rift and crop out north and south of the Gulf of Tadjura in central Afar (Figure 3b).

Sedimentation

Along the foot of the southeastern escarpment, the fluviolacustrine Ch'orora Formation, about 60 m thick, filled a basin about 130 km long (Sickenberg and Schönfeld, 1975). Its position between volcanic flows allowed K-Ar dating at ~10 Ma (Table 1). The formation consists of alternating volcanic ash, conglomerate and gravel, and clay and diatomite (Tiercelin, 1981). Diatom floras (Table 2) contain upper Miocene biostratigraphic indicators (*Mesodictyon* spp.; Fourtanier, 1987) and reflect shallow, well mixed and turbid environments with silica-rich, eutrophic fresh waters. However, taphocoenoses of several levels reflect strongly alkaline conditions (abundance of such diatoms as *Thalassiosira faurii* and *Navicula elkab*) and indicate short-term stagnation and evaporation episodes in generally high-energy environments. Multiple changes in environments may reflect climatic instability and/or local changes in the hydrologic network that were induced by uplift and volcanism.

In central Afar, only a few outcropping lenses of fine-grained, laminated strata (millimetric layers of smectite and of tridymite-cristobalite assumed to represent diagenetic transformation of diatomite) were observed between Mabla basaltic flows dated at about 10 Ma (Table 1). That these deposits are

Table 1. Correlation of Volcanic Units, Opening of Tectonic Basins, and Lacustrine Sedimentation in Afar Depression

Lacustrine Stage	Age	Volcanic Units	Rifting and Spreading Events: Opening of Sedimentary Basins	Lacustrine Profile Locality	References
Early-Middle Miocene					
	<25 Ma	Trap series of Ethiopian Plateau			Behre, 1986
	25–15 Ma	Alkaline granites Adolei Basalt	Northern Afar		Barberi et al., 1975
I	14–10 Ma	Mabla rhyolites and associated basalts	Afar southeastern margin / Mabla massif: Small basins independent of present topography	Ch'orora (10.5–9.05) (10.7–10.5 Ma) / Alay Daba (~9.9–9.5 Ma)	Küntz et al., 1975 / Tiercelin, 1981 / Gasse et al., 1985
Late Miocene–Early Pliocene					
II	8.6–4 Ma	Dahla Basalts	Central Afar: Small basins independent of present topography	Garenlé, W. Afay, Unda Hemed, W. Galammoudla (5.9–4 Ma) W. Kalou (<5.3–>2.4 Ma)	Gasse et al., 1985, 1987 / Gasse, Varet et al., 1986 / Boucarut et al., 1980
Early Pliocene					
III	4.4–2.5 Ma	Stratoid Basalts (lower flows)	Middle Awash Valley: Southern / Northern	Bodo (>3.8–4.0 Ma) / Hadar (4.2–2.9 Ma)	Williams et al., 1986 / Tiercelin, 1986
			Central Afar: Small basins independent of present topography	Southern scarp of the Dikhil (>2.1 Ma) and Gobaad basins (>2.7 Ma)	Gasse, Richard et al., 1980 / Boucarut et al., 1980
		Initial basalts (eastern flows)	Gulf of Tadjura (eastern part)		Gasse et al., 1983
Late Pliocene–Early Pleistocene					
IV	2.5–1.4 Ma	Stratoid Basalts (uppermost flows)	Central Afar: Dikhil-Gobaad	Dikhil (<2.1–1.6 Ma)	Gasse, Richard et al., 1980 / Chavaillon et al., 1987
			Large basins independent of present topography	Dika (>2.3 Ma) / Kori (>2.3 Ma) / Bodo le Dabba (2.1–1.4 Ma) / Dimbir (>1.4 Ma) / Esa Gegalou (>1.2 Ma)	Gasse, Richard et al., 1980 / Gasse, Richard et al., 1980 / Gasse, Richard et al., 1980 / Geothermica Italiana, 1983

Gasse

Table 1. (continued)

Lacustrine Stage	Age	Volcanic Units	Rifting and Spreading Events: Opening of Sedimentary Basins	Lacustrine Profile Locality	References
			Early Pleistocene		
V	1.4–0.8 Ma	Initial basalts (western flows)	Gulf of Tadjura: Small basins independent of present topography	Tadjura (>1.0 Ma: Va–b) (1.0 Ma–>0.125 Ma: Vc)	Gasse and Fournier, 1983 Gasse et al, 1985
			Central Afar: Lower Awash Valley Hanlé Gaggadé-Der Ela	Abhé (<1.4 Ma) Hanlé (<1.4 Ma: Va–c) Der Ela (<1.4 Ma: Va–b) Gaggadé (<1.4 Ma: Vb–c)	
	1.0–0.8 Ma	Axial volcanic ranges	Central Afar: Small basins independent of present topography	Unda Hemed (<1.0–>0.8 Ma: Vc) W. Kalou (<2.7–>1.0/0.8 Ma: Vc)	Gasse et al, 1987 C.E.G.D., 1974
			Middle Pleistocene		
	0.8 Ma–130 ka	Axial volcanic ranges	Ghoubbat-al-Kharab Northern Afar (Dallol-Afrera)	Marine	
			Late Quaternary		
VI	130–10 ka (Late Pleistocene)	Axial volcanic ranges Asal Basalts	Northern Afar Central Afar: Asal	Dallol-Afrera: Marine Abhé, Asal (100–70 ka?: VIa) Abhé, Asal (40 ka?: VIb) Abhé, Hanlé, Gaggadé, Asal (30–20 ka: VIc)	Lalou et al., 1970 Fontes and Pouchan, 1975 Fontes et al, 1973 Fontes et al, 1985 Gasse, 1975, 1977 Gasse and Delibrias, 1976
	10–0 ka (Holocene)	Axial volcanic ranges Asal Basalts	Northern Afar Central Afar: Asal Asal	Afrera (10–4 ka: VId–e) Abhé, Hanlé, Gaggadé, Asal (10–4.5 ka: VId–e) Abhé, Hanlé, Asal (<4 ka: VIf)	Gasse and Fontes, 1989 Gasse and Street, 1978 Gasse et al, 1974 Gasse, Rognon, and Street, 1980b

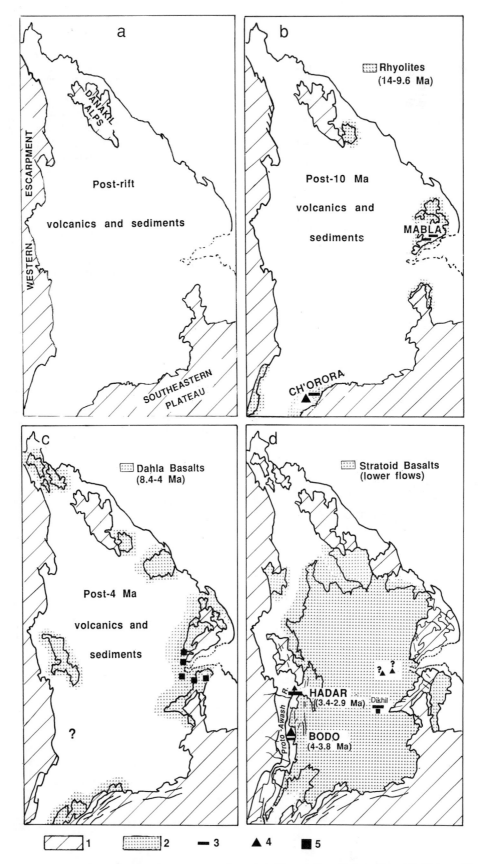

Figure 3. Changes in distribution of volcano-tectonic activity and lacustrine sedimentation in the Afar from 25 to 2.5 Ma. (a) 25 Ma (prerift). (b) Stage I, ~10 Ma. (c) Stage II, ~8.4-4 Ma. (d) Stage III, ~4-2.5 Ma. 1, basement; 2, active volcanic zones; dominant lacustrine lithofacies—3, clay and silt; 4, diatomite; 5, carbonate.

Gasse

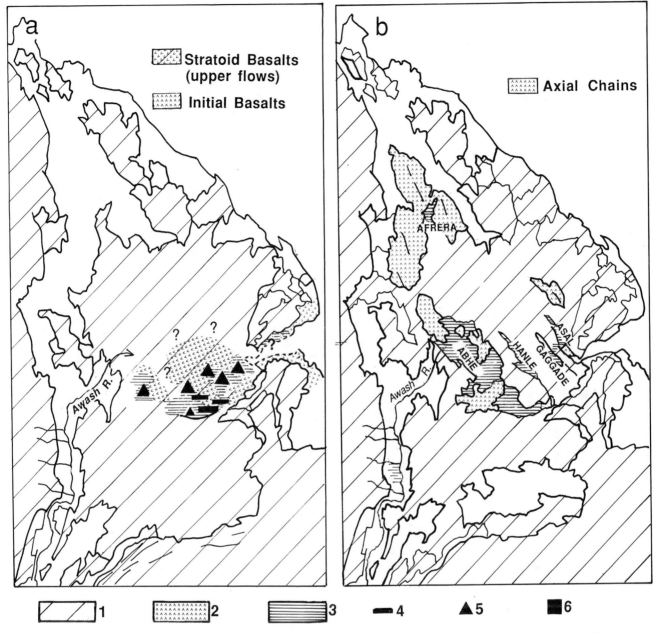

Figure 4. Volcano-tectonic activity and lacustrine sedimentation in the Afar from (a) Stage IV, ~2.5 to ~1.4 Ma and (b) Stage VI, late Quaternary. 1, basement; 2, active volcanic zone; (3-5) lacustrine sediments— 3, visible extent of the sediments; dominant lithofacies— 4, clay and silt; 5, diatomite; 6, carbonate.

devoid of coarse-grained detrital material suggests deposition in small basins with low topographic and hydrologic gradients.

Afar Physiography ~10 Ma

The major sedimentation zone appears to have been located along the southeastern escarpment, as a result of rifting of the southwestern Afar margin. The Ch'orora depression may be regarded as the narrow Miocene protorift supplied by surface runoff from the southeastern protoescarpment. No lacustrine sequence has been recorded for the period from about 10 Ma to <5.9 Ma probably because of climatic factors, although neither erosion nor lack of observation can be excluded.

The Miocene rhyolites were subjected to normal faulting and were considerably eroded before being covered by the Dahla Basalts.

Stage II (<~6Ma)

The Dahla Basalts (~8.6-4 Ma; Table 1) and laterally correlatable basalts (Behre, 1986) can be traced from the Ethiopian Rift margins to the northern extremity of the Afar (Figure 3c). In central

Table 2. Diatoms as Biostratigraphic and Paleoecologic Indicators in Afar Lakes

Lacustrine Stage	Age	Locality	Lithofacies	Dominant (D) and/or Characteristic (C) Taxa	Biological Significance*
I	10 Ma	Ch'orora[1]	Clay, diatomite	**Miocene** D *Aulacoseira granulata* + var. *Aulacoseira agassizii* *Fragilaria* spp. Pennates C *Mesodictyon* spp. (last occ.) *Thalassiosira faurii* *Navicula elkab*	Shallow, well mixed. Turbid. Fresh water. Slightly alkaline. Very high Si and P levels. Evaporated water. Hypersaline, Na-carbonate water.
II	5.94 Ma	Central Afar Dahla Series	No diatoms	**Early Pliocene**	
III IIIa	4–3.8 Ma	Bodo[1]	Diatomite	**Early Pliocene** D *Aulacoseira agassizii* *Aulacoseira granulata* + var. *Aulacoseira nyassensis* *Stephanodiscus minutulus* C *Cyclostephanos* spp. (first occ.) *Stephanodiscus carconensis* *Stephanodiscus subtransilvanicus*	Large lakes. Generally deep mixing. Fresh water. Slightly alkaline. Moderate Si loading, high P level.
IIIb	3.65–2.9 Ma	Hadar	Silicified clay	D *Aulacoseira granulata* C *Thalassiosira faurii* (rare)	Well mixed Turbid. Fresh water to mesosaline. Alkaline, Na-carbonate water.
IV	>2.3 Ma >2.3 Ma >2.1–>1.6 Ma >1.4 Ma 2.1–1.6 Ma >1.4 Ma >1.2 Ma	Dika[2,3] Kori[2,3] Bodo le Dabba[4] Dimbir[2,3] Dikhil[4] Dimbir[2,3] Esa Gegalou[4]	**Late Pliocene–Early Pleistocene** Diatomite Clay and diatomite Diatomite Clay and diatomite Calcareous diatomite Calcareous diatomite Diatomite	D *Aulacoseira granulata* C *Stephanodiscus minutulus* D *Stephanodiscus niagarae* + var. C *Stephanodiscus carconensis* + var. D *Epithemia argus* *Campylodiscus clypeus* *Mastogloia elliptica* C *Nitzschia granulata, N. elegantula* *Amphora* spp. *Cyclotella subatomus*	Generally well mixed. Fresh water. Deep, translucent lakes. Fresh water. Cool conditions? Shallow. Evaporated water. Oligosaline to eusaline. Chloride or sulfate water.
V Va	<1.2 Ma	Der Ela	**Early Pleistocene** Clayey diatomite	D *Stephanodiscus niagarae* *Stephanodiscus carconensis* + var.	Deep, translucent lakes. Fresh water. Cool conditions?
Vb	<1.2 Ma >1.0 Ma 1.0–0.9 Ma	Der Ela Tadjura Unda Hemed	Silty diatomite Calcareous diatomite Marls	D *Aulacoseira granulata* + var. *Stephanodiscus* aff. *minutulus* C *Stephanodiscus niagarae* (last occ.) *Stephanodiscus carconensis* (last occ.)	Generally well mixed. Fresh water. Moderate Si loading, high P level.

Table 2. (continued)

Lacustrine Stage	Age	Locality	Lithofacies	Dominant (D) and/or Characteristic (C) Taxa	Biological Significance*
Vc	<1.0 Ma	Tadjura	Carbonate	D *Aulacoseira granulata*	Well mixed, Turbid. Fresh water.
	0.8 Ma	Unda Hemed	Carbonate		
VI				**Late Pleistocene**	
VIa	100–70 ka?	Abhé, Asal[4,5]	Laminated diatomite	D *Aulacoseira granulata*	Meromictic conditions. Fresh water. Si and P loadings moderate to high.
	40 ka?	Abhé, Asal[4,5]	Laminated diatomite	*Synedra acus* var. *Nitzschia aequalis* *Stephanodiscus hantzschii var.*	
VIb	31 ka	Abhé[5]	Sandy clay	D *Thalassiosira faurii*	Mesosaline, hyperalkaline. Na-carbonate water.
VIc	30–20 ka	Abhé[5]	Clayey diatomite	D *Aulacoseira granulata*	Well mixed. Turbid. Fresh water. Circumneutral pH. Relatively low nutrient content.
		Hanlé[4]	Clayey diatomite	C *Cyclotella ocellata*	
		Gaggadé[4]	Silty diatomite		
		Asal[4]	Silt	No diatoms	
				Holocene	
VId	10–8 ka	Abhé[5]	Low Mg-calcite	D *Stephanodiscus minutulus*	Deep, translucent lake. Fresh to oligosaline water. Alkaline. Low Si, high P supply rates.
VIe	7.5–4.5 ka	Abhé[5]	Low Mg-calcite	C *Nitzschia lacuum*	
VId–e	10–4.5 ka	Hanlé[4]	Low Mg-calcite	*Navicula scutelloides*	
VId	10–8.6 ka	Asal[6]	High Mg-calcite Aragonite	*Campylodiscus clypeus* *Epithermia argus* *Mastogloia elliptica*	Shallow. Oligosaline to eusaline. Chloride or sulfate water.
	8.6–6 ka	Asal[6]	High Mg-calcite	D *Aulacoseira granulata* *Synedra acus* var. *Nitzschia aequalis*	Deep. Well mixed. Fresh water. Si loading moderate to high.
	2.5–1.7 ka	Abhé[4]	Low Mg-calcite Smectite	D *Stephanodiscus minutulus* *Thalassiosira faurii*	Fresh to mesosaline conditions. Hyperalkaline, Na-carbonate water.
	<4.2 ka	Asal[6]	Gypsum, halite	No diatoms	Sulfate, chloride water.

* Salinity classes: fresh water, <0.5°/oo; oligosaline, 0.5–<5°/oo; mesosaline and polysaline, 5–<30°/oo; eusaline, 30–40°/oo; hypersaline, >40°/oo. Biological significance mainly based on Gasse (1986) and Kilham et al. (1987).

[1]Fourtanier (1987)
[2]Robbe (1979)
[3]Gasse, Richard et al. (1980)
[4]Gasse (1975)
[5]Gasse (1977)
[6]Gasse and Fontes (1989)

Afar the upper part of the series (<5.9 Ma) contains intercalated lacustrine deposits relatively rich in carbonate (Boucarut et al., 1980, 1985; Gasse et al., 1986, 1987). The profiles show an alternation of ash layers frequently interrupted by coarse-grained detrital levels and by surfaces marked with desiccation cracks. No microfossils were found. Molluscan fauna (recrystallized shells of Unionidae and *Melanoides*) indicate fresh to oligosaline waters (salinity <5‰). Discontinuity of the outcrops allows neither correlation between sections nor extension of the paleolakes. These lower Pliocene deposits suggest shallow, unstable environments supplied by surface runoff.

The Dalha Series has been intensely eroded, faulted, and tilted before being covered, commonly unconformably, by the Stratoid Basalts.

Stage III (~4.4–2.5 Ma)

Most of the Afar has a dominantly basaltic base, namely the Afar Stratoid Series (4.4–1.2 Ma). Marked changes in the distribution of lakes occurred during this 3-m.y. interval of intense rifting, especially at ~2.5 Ma.

Southern Afar

Thick fluviolacustrine deposits were laid down in the trough occupied today by the Middle Awash River valley, which has been downfaulted between the western escarpment and Stratoid Basalts (Figure 3d; Table 1).

In the Bodo sector (Figure 3d; Table 1) horizontally bedded strata, at least 45 m thick, are capped by a primary air-fall tuff, the Cindery Tuff, which was dated by $^{40}Ar/^{39}Ar$ and by zircon fission-track analyses at 3.8–4.0 Ma (Hall et al., 1984). These strata reflect alternating lake transgression (1- to 6-m-thick diatomite beds) and regression (pedal brown clays devoid of microfossils); fluviatile deposition occurred westward after 4.0–3.8 Ma (Clark et al., 1984; Williams et al., 1986). The base of the Bodo deposits contains no characteristic Miocene diatoms, although *Cyclostephanos*, which appears at the Miocene-Pliocene boundary, is abundant (Fourtanier and Gasse, 1988; Table 2). The entire sequence would therefore be of early Pliocene age. The diatom assemblages are typical of plankton in freshwater eutrophic lakes. Lacustrine conditions were generally turbid with deep mixing and a high silica level.

The Hadar Formation (Figure 3d; Table 1) consists of 150–300 m of coarse- to fine-grained sand, silt, and clay (Taieb, 1974; Tiercelin, 1986). The Hadar region experienced a complex succession of fluvial, marshy, and lacustrine environments ranging from fresh water to mesosaline-alkaline conditions. Swampy environments dominated from 4.2 to 3.65 Ma, while a lake occupied the entire region from 3.65 to 2.9 Ma. Lacustrine sedimentation ended at 2.9 Ma and gave way to fan-gravel accumulation from 2.9 to 2.4 Ma until complete infilling of the basin. Plio-Pleistocene sediments have been affected by numerous north-northwest- to south-southeast-trending faults and tilted 2°–4°N.

Central Afar

Fluviolacustrine deposits (several tens of meters of conglomerate, silt, clay, and lacustrine carbonate) associated with hyaloclastites overlie, or are interbedded with, lower Stratoid Basalts (e.g., along the southeastern margin of the Gobaad and Dikhil basins; Figure 3d, Table 1). Although the age of diatomites underlying the last Stratoid flows (dated at about 2.3 Ma) is uncertain, they have been tentatively assigned to Stage IV on the basis of marked similarities with sediments dated between 2.4–2.2 and 1.4 Ma.

Afar Physiography during Early Pliocene

Lacustrine deposition centered along the western escarpment reflects rifting in southwestern Afar during the first stages of Stratoid magmatic activity. Fluviolacustrine deposits occupying the present-day Middle Awash River valley may have accumulated in the terminal basin of an ancestral Awash River, which became dammed by fault scarps that affected the Stratoid Basalts. The abundance of detrital material is easily explained by surface waters flowing from the Awash River and western escarpment. The deepest area of the sedimentary basin apparently was situated at Bodo at ~4–3.8 Ma and at Hadar at ~3.65–2.9 Ma. This migration of the depocenter may be linked to progressive opening of the Middle Awash basin and/or to its northward tilting during sedimentation. The disappearance of large lakes in the southern Afar between early and late Pliocene coincides with a drought in east Africa at ~2.5–2.35 Ma, as deduced from palynological studies (Bonnefille, 1983). This disappearance was not, however, climatically induced because lakes developed farther east after 2.5 Ma. Instead, it is attributed to uplift of the Afar margins and downfaulting of the central Afar basins.

LATE PLIOCENE TO EARLY PLEISTOCENE LAKES (~2.5–0.8? Ma)

Large lakes of late Pliocene to early Pleistocene age are observed in central Afar only.

Stage IV (~2.5–1.4 Ma)

Contemporaneous lacustrine deposits are either exposed on the floors of the Dikhil and Abhé-Gobaad basins or are interbedded with the latest basaltic flows of the present-day plateaus (Figures 4a, 5, 6; Table 1).

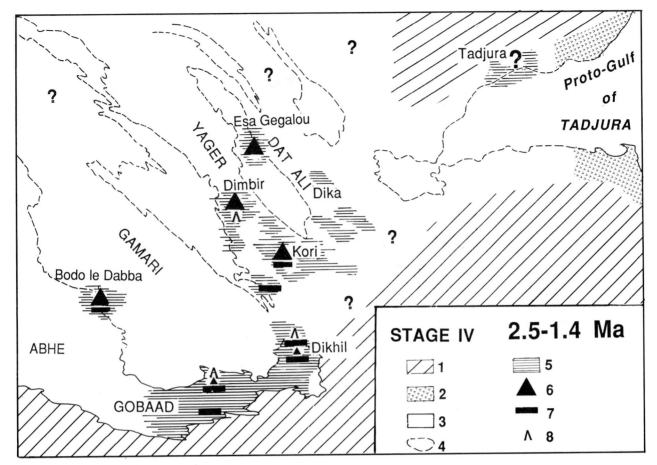

Figure 5. Volcano-tectonic activity and lacustrine sedimentation in central Afar from ~2.5 to ~1.4 Ma (Stage IV). 1, basement (Afar Volcanic Series older than 2.5 Ma); 2, initial basalts (eastern flows); 3, zones affected by volcanic flows <2.5 Ma; 4, tectonic structures nonexistent prior to 1.4 Ma; 5, visible extent of lacustrine sediments; dominant lacustrine lithofacies—6, diatomite; 7, clay; 8, evaporites.

Dikhil and Abhé-Gobaad Basins

Tens of meters of clay overlain by fine- to coarse-grained sandstone lie conformably on the Stratoid Basalts (2.1–2.0 Ma; Gasse, Richard et al., 1980), which form the southern scarp and floor of the Dikhil basin. The sandstones are composed mainly of quartz and Jurassic limestone fragments from the Afar basement, and their deposition probably followed uplift of the Aisha horst. This accumulation, which subsequently was faulted and tilted with the basalts, is attributed to a river-fed lake occupying a preformed single Dikhil-Gobaad basin and may have been deposited either before or during the tectonic events that created the basin's southern scarp.

The lacustrine Dikhil Formation (Gasse, 1975) lies unconformably on either the sandstones or Stratoid Basalts. This 40-m-thick accumulation of clay contains an occasional diatomitic episode (Figure 6a) and grades upward into gypsiferous clay and gypsum. The lacustrine strata have been truncated by conglomerates or gravel fans (Dikhil sector; Figure 6a), or by sands bearing mammal bones and stone artifacts (Gobaad). ESR dating of tooth enamel from

Elephas recki ileterensis gives a date of 1.6–1.3 Ma, which agrees with ages deduced from paleontology and artifacts (Chavaillon et al., 1987). Thus, the Dikhil Formation would have been deposited between 2.1–2.0 Ma (age of the floor basalt) and 1.6–1.3 Ma (Table 1).

The lower part of the diatomite bed contains floras that reflect freshwater, turbid environments (Table 2), which suggests that the Dikhil and Abhé-Gobaad basins first were occupied by a shallow river-fed lake. The diatomites, which contain extinct species (e.g., *Amphora tertiara*), are then characterized by a littoral assemblage of chloride- or sulfate-type saline water (Table 2). These anions may have been derived from weathering of Mesozoic marine sedimentary rocks in the Aisha horst rather than from the Afar volcanics. The disappearance of diatoms in the upper gypsiferous clays may have been prompted by excess salinity due to evaporative concentration.

The Dikhil basin experienced no other lacustrine episodes. Its strata have been slightly tilted (2°WSW) and broken into northwest-southeast–trending blocks that rise stepwise toward the northeastern

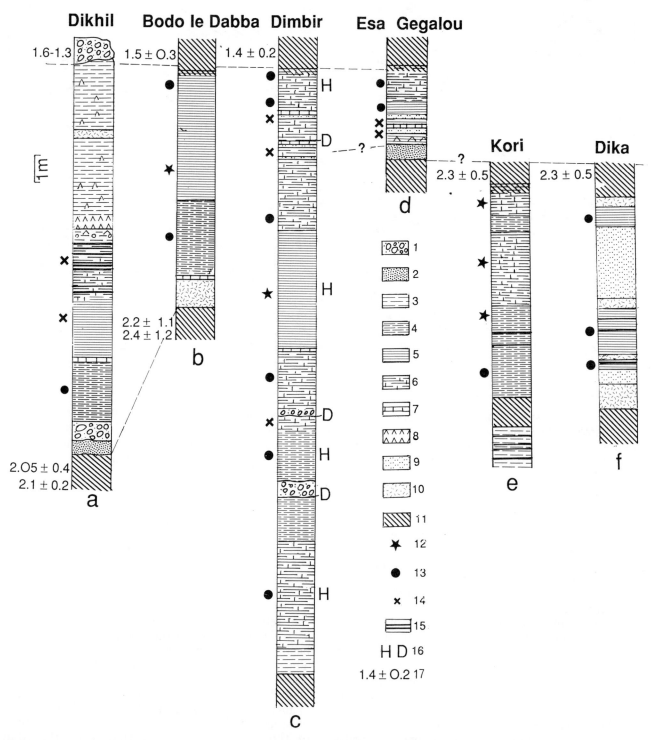

Figure 6. Stratigraphic profiles of lacustrine sediments associated with the last flows of the Afar Stratoid Basalts (Stage IV). Location on Figure 5. 1, conglomerate; 2, sand; 3, clay; 4, clayey diatomite; 5, diatomite; 6, calcareous diatomite; 7, carbonate bed; 8, gypsum; 9, hyaloclastite; 10, pumice, ignimbrite; 11, basalt (Stratoid Series); diatom assemblages—12, *Stephanodiscus carconensis* and var., *S. niagarae* and var. abundant; 13, *Aulacoseira granulata* and var. abundant; 14, littoral saline flora; 15, silicified layers; 16, H—high water level, D—desiccation (fluvial deposits, mud cracks); 17, K-Ar ages (Ma) except for the top of section a, which is by ESR dating.

margin and then pass beneath the last Stratoid basalt flows (Figure 5). Lacustrine sedimentation occurred later in the Abhé-Gobaad basin.

Gamari, Yager, and Dat Ali Plateaus

Clays, diatomites, and carbonates capped by the last Stratoid basalt flows crop out at numerous sites along the tops of fault scarps and in wadi cuts on the plateaus that separate the central Afar basins (Figures 5, 6b–f). They lie up to 500 m above the floors of the modern Abhé, Hanlé, and Gaggadé-Der Ela basins. K-Ar ages of overlying flows range from 2.3 to 1.4 Ma (Table 1; Figure 6).

Although floras in the clayey and calcareous diatomites are similar to those of the Dikhil deposits, pure diatomites are characterized by the dominance of large *Stephanodiscus* (*S. niagarae* and *S. carconensis*), which had disappeared from Africa by the middle Pleistocene (Gasse, 1980; Fourtanier and Gasse, 1988). This planktonic assemblage registers deep, transparent, very dilute water with a low silica level and possibly cool temperature. At Dimbir (Figures 5, 6c) changes in lithology and in diatom floras along the section show four episodes of high water level alternating with shallow saline water or droughts. Detailed correlations between sections are therefore risky because similar ecological conditions have occurred several times during Stage IV.

Afar Physiography from 2.5 to 1.4 Ma

Lacustrine sediments in central Afar reflect generally humid conditions, although lake-level oscillations may indicate a more complex climatic evolution. Subcontinuity of outcrops from the Dikhil and Gobaad-Abhé basins to the Dat Ali Plateau (Figure 5) suggests the existence of a single depression, with a marked river influence in the Dikhil area (clay dominant) and a maximum depth in the vicinity of the present-day plateaus, where pure pelagic diatomite accumulated. However, limits of the paleolake(s) and general direction of the corresponding basin(s) are unknown.

Based on sediment distribution, the Dikhil and Abhé-Gobaad basins formed about 2.0 Ma and since have experienced only minor topographical changes. In contrast, magmatic and tectonic activity occurred to the north and northwest, where sedimentary strata and overlying basalt flows have been faulted and tilted westward and southwestward before and during the subsequent lacustrine episode.

Stage V (~1.4–0.8 Ma)

Tectonism responsible for the downwarping of the Hanlé and Gaggadé-Der Ela basins (Figure 7) also affected the last volcanic flows of the Yager Plateau (1.4 Ma at Dimbir; Figure 6c; Table 1) and the Baba Alou rhyolitic massif (1.2 Ma; Figure 7; Geothermica Italiana, 1983).

In the Hanlé basin the oldest visible lacustrine deposits consist of clay and diatomite (Stages Va and Vb), overlain disconformably by paludal shelly

carbonate (Stage Vc). These deposits are truncated by conglomerate fans attributed to the middle Pleistocene (Gasse et al., 1987).

In Gaggadé-Der Ela basin rifting has progressed from north-northwest to south-southeast. At least 150 m of Pleistocene strata (Figure 8) crop out at the foot of the Baba Alou massif along the western margin of this asymmetrical basin and lie unconformably on tilted Stratoid Basalts. The oldest deposits (70 m of fluviolacustrine silt), observed to the north, contain lacustrine clays that thicken and become diatomitic toward the southeast (Stage Va; Figure 8a–e). That Stage Va still is characterized by the diatoms *Stephanodiscus niagarae* and *S. carconensis* (Table 2) indicates that no drastic change in climate occurred during the first phase of Gaggadé-Der Ela rifting. In the center of the basin, a second lacustrine episode (Stage Vb, Figure 8e–f) of silty, clayey diatomite occurred in which lithofacies and diatoms indicate several water-level fluctuations and where archaic *Stephanodiscus* was replaced by modern forms (*S. minutulus*, *S.* aff. *rotula*; Table 2).

Another unit, about 70 m thick, developed mainly southward. Detrital material first predominates, but intercalated chemical carbonate, shelly beds (with freshwater *Melanoides* and Unionidae) and gypsum mark temporary reestablishment of aquatic environments with fluctuating salinity (Stage Vc; Figure 8g–h). The entire sequence through the basin finally was truncated by a 1- to 10-m-thick cobble and gravel mantle (unit G, Figure 8).

Sedimentation in the Gaggadé-Der Ela basin has been largely controlled by climate, which was generally wetter than today but progressively deteriorated. Continuous magmatism, rifting, and tilting, probably related to accretion of the Baba Alou massif, also influenced sedimentation and the present distribution of outcrops. The total offset between the diatomite of the Yager Plateau and its tilted equivalent in the Der Ela basin reaches 500 m, reflecting a mean rate of downfaulting of 83 cm/k.y. between 1.4 and 0.8 Ma. Hyaloclastite surges and ash beds are intercalated in the lower units. Spatial distribution of successive sedimentary units indicates migration of the depocenter from north-northwest to south-southeast. Synsedimentary normal faults, trending N30°–60°W, with visible vertical offset up to 20 m, cross the sequence. Block tilting most commonly follows a direction opposite to fault strike, and the sedimentary sequence was generally tilted 5°–20°WSW before truncation by the gravel fans. The fan surface has been displaced by normal faults trending N30°–60°W (initial direction) and N30°E. The fluviolacustrine sequence disappears abruptly northward along a N30°E fault that marks the southern limit of the present Der Ela basin. Toward the Gaggadé, the strata dip progressively beneath upper Pleistocene deposits.

Basins Marginal to Oceanic Structures

Along the southwestern margin of the Asal-Ghoubbat rift, volcanic occurrences of trachyte and

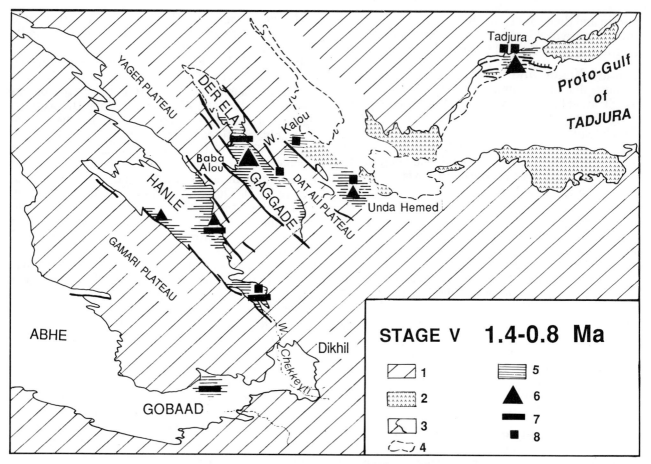

Figure 7. Distribution of volcano-tectonic activity and lacustrine sedimentation in central Afar from ∼1.4 Ma to ∼0.8 Ma (Stage V). 1, basement (Afar Stratoid Basalts and older Volcanic Series); 2, Initial Basalts; 3, limit of lacustrine basins; 4, tectonic structures nonexistent before ∼0.8 Ma; 5, visible extent of lacustrine strata; dominant lithofacies—6, diatomite (Stages Va-b); 7, clay (Stages Va–b); 8, carbonate (Stage Vc).

rhyolite (1.4–0.8 Ma) and basalt (1.0–0.8 Ma) mark the first manifestations of Asal-Ghoubbat rifting (Gasse et al., 1987).

Sedimentary strata crop out at elevations up to 430 m in the cliffs along Wadi Kalou (Figure 7), which follows a N30°E-trending fault between the Gaggadé and Asal basins. These rocks, which lie between Stratoid Basalts (2.4 Ma) and a basaltic flow dated at 1.0–0.8 Ma (Table 1), contain abundant faunal remains that indicate an early Pleistocene age and reflect freshwater to brackish environments (Boucarut et al., 1980). The deposits are analogous to upper fluviolacustrine units of the Gaggadé-Der Ela basin (Stage Vc).

The basaltic flows at Unda Hemed (Figure 7), dated at ∼1–0.9 Ma to ∼0.8 Ma. (Table 1), are interrupted at their base by a 20-m-thick volcano-sedimentary sequence that includes diatomites assigned to Stages Va and Vb. Toward the top, calcareous beds (Stage Vc) lie between hyaloclastites and conglomerates.

Along the northern margin of the Gulf of Tadjura, Plio-Pleistocene time was characterized by accumulation of several hundred meters of conglomerate,

intensely faulted and tilted slightly northward (Gasse and Fournier, 1983). These enormous alluvial fans reflect uplift of the Mabla Mountains under generally arid conditions. However, at least two lacustrine episodes interrupted this coarse detrital sedimentation, indicating (1) existence of tectonic basins downfaulted through the conglomerate; and (2) sufficient rainfall over the Mabla Mountains to ensure filling and permanent water supply to basins underlain by highly permeable strata.

The first lacustrine episode is represented by stratified, clayey, slightly calcareous diatomites and is attributed to the early Pleistocene (Stages Va-b) based on diatom flora (Figure 9; Table 2). These diatomites crop out near Tadjura (Figure 9c) and were observed at a depth of 40 m in a 130-m core taken 5 km to the east (Figure 9a). To the north, sediments of Stages Va and Vb pass beneath a basaltic flow of the Initial Series (1.0 Ma; Table 1; Figure 9d–e). Younger lacustrine deposits of Stage Vc—shelly marls with a poorly preserved diatom flora devoid of archaic forms—overlie the Initial Basalts (Figure 9e–f) or crop out between conglomerates (Figure 9b).

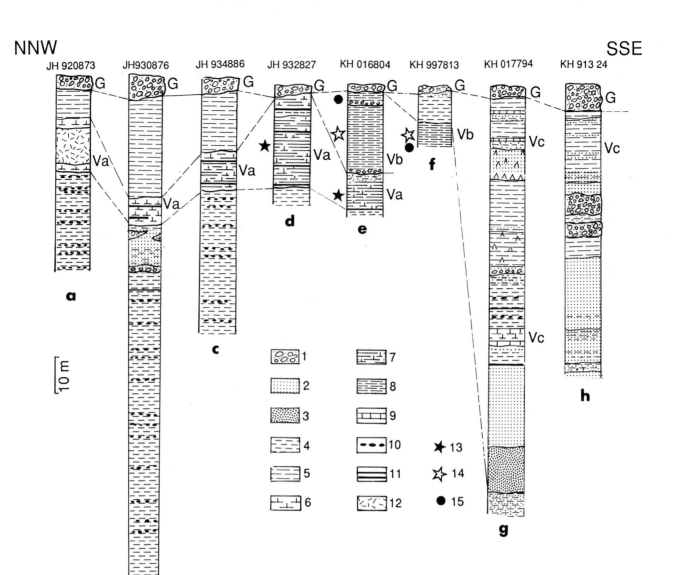

Figure 8. Stratigraphic profiles of lower Pleistocene fluviolacustrine sedimentation in the Gaggadé-Der Ela basin. Profiles are located by UTM coordinates. 1, conglomerate; 2, fluvial sand; 3, eolian sand; 4, silt; 5, clay; 6, marl; 7, calcareous diatomite; 8, clayey diatomite; 9, calcareous bed; 10, calcareous concretions; 11, silicified layers; 12, hyaloclastite; diatom assemblages—13, *Stephanodiscus carconensis* and var., *S. niagarae* and var. abundant; 14, *Stephanodiscus aff. rotula, S. minutulus* abundant; 15, *Aulacoseira granulata* and var. dominant; G, pebble or gravel fans.

To the south, the entire sequence steps down, through a series of faults, to the ocean. Coral reefs, Th-U dated at 140–100 ka, formed on the conglomerate, the lacustrine sediments, and the Initial Basalts (Gasse and Fournier, 1983). The southern part of the Tadjura lacustrine basin(s) has been below sea level since at least the late Pleistocene.

Afar Physiography from ~1.4–1.2 to 0.8 Ma
The time interval 1.4–0.8 Ma was a period of important climatic fluctuations. A generally humid phase was first registered by freshwater diatomites (Stages Va–b), species of which still live in large temperate lakes (*Stephanodiscus* spp.). The age of Stage Va–b appears to be <1.4–1.2 Ma in the Gaggadé-Der Ela, >1.0 Ma at Tadjura, and close to 1.0–0.9 Ma at Unda Hemed. Establishment of the long, middle Pleistocene arid period then led to disappearance of the archaic *Stephanodiscus*, as observed elsewhere in tropical Africa (Fourtanier and Gasse, 1988); and replacement of the diatomite facies by increased detrital (fluvial and even eolian) influx. Humid recurrences are, however, indicated by lacustrine or paludal carbonate (Stage Vc) dated at ~0.8 Ma at Unda Hemed and <1.0 at Tadjura.

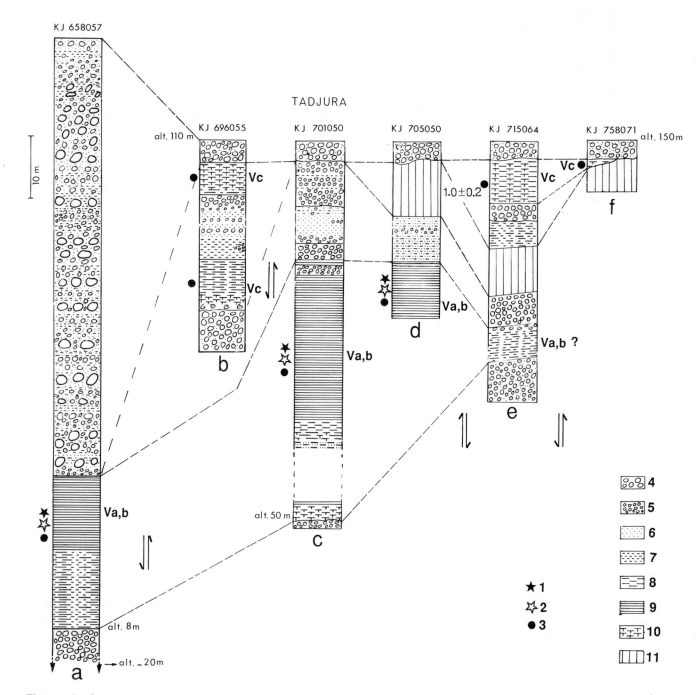

Figure 9. Stratigraphic profiles of lower Pleistocene fluviolacustrine sediments in the Tadjura area. Profiles are located by UTM coordinates. Diatom assemblages—1, *Stephanodiscus carconensis* and var., *S. niagarae* and var. present; 2, *Stephanodiscus aff. rotula*, *S. minutulus* abundant; 3, *Aulacoseira granulata* and var. dominant; 4, pebble; 5, gravel; 6, sand; 7, silt; 8, clay; 9, diatomite; 10, marl; 11, Initial Basalt.

Distribution of these sediments and the tectonic effects upon them show that the Hanlé and Gaggadé-Der Ela basins were preformed at that time but still subsiding, faulting, and tilting. Paleolakes with no relation to modern topography also extended into areas of intensive magmatism, concentrated along preoceanic structures of the Asal-Ghoubbat rift, and into the western Gulf of Tadjura, which had not opened at that time. The limits and trends of these paleolakes cannot, however, be traced.

No lacustrine formation can be attributed with certainty to the arid middle Pleistocene; erosion and

34   Gasse

coarse-grained fluvial deposition prevailed. While the Gobaad and the Dikhil basins were relatively stable, subsidence continued in the Hanlé and Gaggadé-Der Ela basins as a result of the activity of faults trending N30°–60°W, N30°E (Der Ela) and N70°E (Hanlé, Gaggadé). The Tadjura lacustrine basin had partly downfaulted below sea level and no longer existed. Two major tectonic events occurred during this long arid period—opening of the western Gulf of Tadjura and downfaulting of the Asal basin.

LATE QUATERNARY LAKES

Stage VI (130?–0 ka)

Late Quaternary lakes (Figure 4b) developed in topography similar to that of today. The Afar lakes have been the focus of intensive paleoclimatic studies that are briefly summarized here (see Table 1 for references). The respective roles of climatic changes and of tectonism on lake evolution are easier to detect than for older periods because of the relatively brief and well dated time interval. Differences observed in the behavior of individual basins arise from their particular structural settings. In some favorable cases, reconstructing the tectonic events allows one to estimate rates of vertical movement.

Stages VIa–c (130?–20 ka)

Late Pleistocene lakes developed in all the basins of central Afar—Abhé-Gobaad, Hanlé, Gaggadé, and Asal—which are characterized by diatomitic sedimentation in deep, freshwater environments. Three successive highstands (Stages VIa–c) have been documented in Lake Abhé and Lake Asal (Table 2). The youngest is ^{14}C dated at ~30 to 20–19 ka; intermediate events ended at ~31 ka; and the oldest may have occurred about 70–100 ka (Gasse, 1975, 1977). During the last highstand, water depth, as deduced from maximum elevation of lacustrine remnants, reached at least 160 m in Abhé, and probably several hundred meters in Asal.

An arid episode (~20–10 ka) followed during which time all the lakes were at least as low as today. This episode, which corresponds to the last glacial stage, was a time of intense erosion.

In the Asal basin, late Pleistocene was a time of active rifting. Synsedimentary faults affect the Stage VIa diatomite, which contains pyroclastic levels. A major volcano-tectonic event occurred after Stage VIb in the southern part of the basin. Along the margin of the Asal rift, upper Pleistocene diatomites were downfaulted toward the active axis and covered by marginal basaltic flows of the Asal Series, thereby obscuring the southward extension of Lake Asal. However, occurrence of deformed reefs, Th-U dated at ~125 ka (Gasse and Fournier, 1983), along the northern shore of the Ghoubbat indicates that Lake Asal was at that time separated from the marine Gulf of Tadjura.

Stages VId–f (10–0 ka)

Example of a Holocene River-Fed Lake

Lake Abhé (Figure 10) is an excellent climatic indicator because it directly registers rainfall fluctuations on the Ethiopian Plateau and western escarpment through flow in the Awash River. The lake was high from ~9.5 to ~4.5 ka (Stages VId-e), except during a brief episode centered on 8 ka (Figure 11). A minor positive oscillation (Stage VIf) occurred about 2–1.5 ka.

Lake Abhé highstands were characterized by chemical deposition of low-Mg calcite associated with freshwater diatoms. Clay remained abundant, however, close to the mouth of the Awash River. During lowstands, deposits rich in river-transported detritus prevailed with organisms with affinities for concentrated Na-carbonate water left by evaporation of the Awash River (Gasse, 1975, 1977).

Neighboring Lakes Hanlé and Asal-Gaggadé

During the early and middle Holocene wet episode, a southwest-to-northeast progression developed in water depths and elevations of Lakes Abhé (160 m deep, 400 m elevation), Hanlé (200 m deep, 300 m elevation), and the single lake Asal-Gaggadé (315 m deep, 160 m elevation) (Figures 10–12). The latest two lakes are hydrologically connected with the Awash River system by groundwater flow through (tectonically induced) highly permeable lake bottoms. Lake depth increases with decreased elevation.

Despite the hydrologic connection, the lakes exhibit marked differences in their histories. For example, in Lake Asal, the terminus of the system, the effects of a short-term drought at ~8 ka, apparent in river-fed Lake Abhé, were damped. However, a drastic lake-level drop of about 300 m in Asal at about 6 ka (Figure 11) may be explained not only by climate, which began to deteriorate at that time, but also by a dramatic tectonic event that triggered sudden seepage from Lake Asal to the sea through recently opened fractures (Gasse and Fontes, 1989).

Proximity to oceanic rifting activity also explains differences in sedimentation between the lakes. In Lake Asal, marine influence from the opening Gulf of Tadjura during low-level stages explains why salinity never dropped to freshwater levels. High-Mg calcite was precipitated during highstands, while aragonite, gypsum, and halite characterize low-level episodes (Gasse and Fontes, 1989).

Holocene Volcano-Tectonic Activity

The degree of volcano-tectonic activity decreases westward through the basins from Lake Asal to Lake Abhé.

The effects are most pronounced in the Asal rift area. Holocene deposits there have been partly covered by basaltic flows of the Afar Series and crossed by numerous faults and open fissures. The

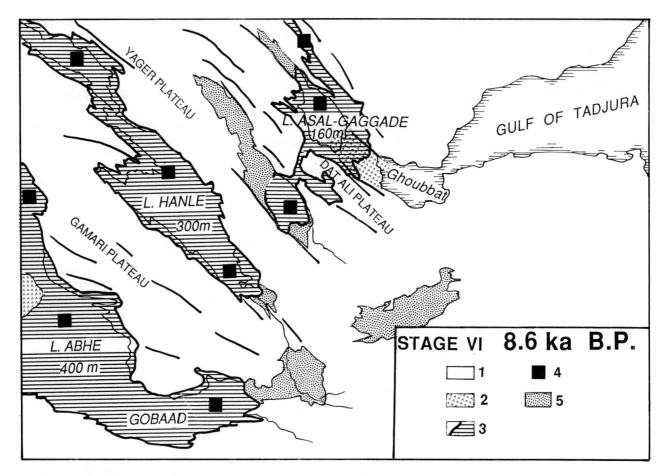

Figure 10. Distribution of volcano-tectonic activity and lacustrine sedimentation in central Afar during the early Holocene (Stage VId). 1, basement (Afar Volcanic Series); 2, axial chains (Asal Basalts and Dama Ale flows); 3, paleolakes; 4, carbonate dominant; 5, sand, silt, and clay.

most recent rifting event occurred in 1978 (Abdallah et al., 1979), and since 1979, 40 cm of spreading has occurred, at a maximum rate of 60 mm/yr from 1983 to 1987. Vertical movement on normal faults reached 8 mm/yr (Ruegg and Kasser, 1987). The paleoshoreline at the time of its maximum extent (8.6–6.2 ka) is clearly seen at about 160 m elevation around the Asal basin, except in the rift zone where its lower present elevation indicates 70 m of subsidence along the axial zone at a mean rate of about 1 mm/yr during the last 6000 yr (Gasse and Fournier, 1983).

In the Gaggadé, Der Ela, and Hanlé basins, although no recent volcanic events have been observed, recent or subrecent fumaroles occur along the western basin margins. Surface drainage patterns reflect structural influence. Three major tectonic directions are active today:

1. N30°W–N60°W—major basin-limiting faults, particularly active along the western and eastern bases of the Baba Alou massif.
2. N10°E–N30°E—documented by faults and lineaments in the Der Ela and northern Gaggadé and Hanlé basins.

3. N70°E–S80°E—responsible for downfaulting (6–10 m) Holocene sediments in the center of the Gaggadé plain.

No volcano-tectonic effects have been observed in the Abhé-Gobaad basin, but to the west, basalt flows from the axial chains (Dama Ale system; Figure 10) cover lower and upper Holocene lacustrine strata, which have been crossed by N70°E-trending open fissures.

CONCLUSIONS

Spatial and temporal distribution of Neogene and Quaternary lakes and their environments have been controlled by climatic fluctuations superimposed on complex structural evolution of the Afar depression.

Tectonic Control on Lake Distribution

General lake distribution was governed by structural evolution of the Afar depression. Spread-

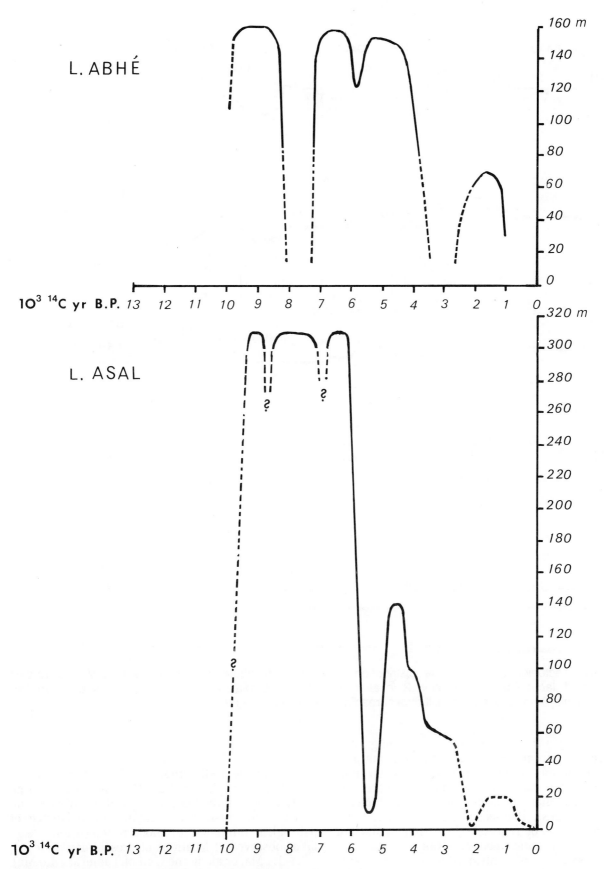

Figure 11. Lake level fluctuations in Lake Abhé (after Gasse, 1977) and Lake Asal (after Gasse and Fontes, 1989) from 10 ka to present.

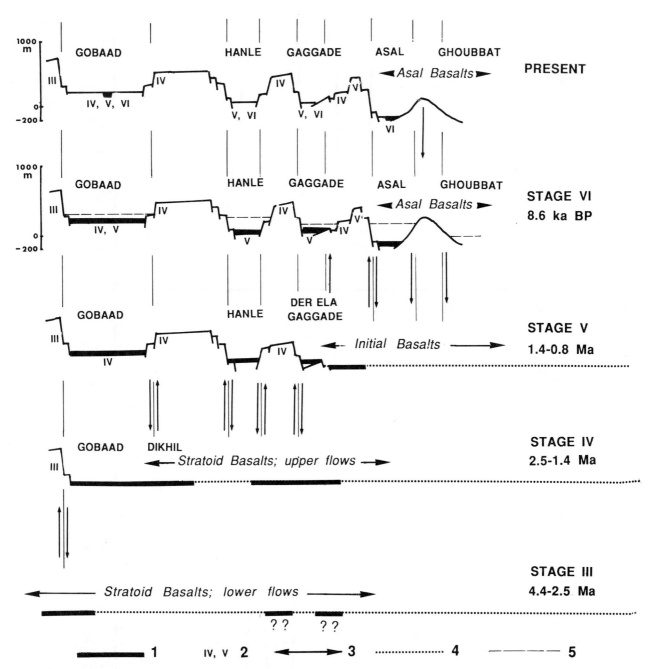

Figure 12. Tectonic control on lake distribution in central Afar. 1, lacustrine sedimentation; 2, location of older sediments; 3, volcanic activity; 4, further spreading activity; 5, water level (Stage VI only). Vertical arrows represent relative vertical movements between two lacustrine stages.

ing beginning in the Miocene explains the migration of active sedimentary basins from the Afar margins to the present-day active segments. At a given time, water, and thus lacustrine sedimentary materials, tends to accumulate in the deepest basins, which are also the youngest. The sediments accumulated in a given basin do not represent long time spans but rather reflect tectonic stages of rifting.

In the example of central Afar, modern basins exhibit a south-southwest to north-northeast progression in their age of opening, from ~2.5–2.0 Ma (Gobaad) to the present (Asal). During this interval, the directions of active faults have varied from S80°E to N30°–60°W at ~1.4 Ma. This agrees with a change in direction of extension from N40°E to N20°E from 1.7 to 1.25 Ma., as deduced from analysis of tectonic features. Variations in direction of extension also may explain the existence of paleolakes unrelated to modern topography (e.g., lakes observed at locations of present-day plateaus, ~2.4–1.4 Ma, or along the Gulf of Tadjura, ~1.0 Ma). Sediment deformations document the activity of faults and lineaments along N30°E and N70°–80°E during the late Quaternary in the Hanlé and Gaggadé basins and allow estimation of a mean vertical displacement rate of 1 cm/yr in the Asal rift.

Distribution of sediments in individual basins also was tectonically induced. Those filling the Gaggadé-Der Ela basin show that the opening migrated north to south between ~1.4 and 0.8 Ma.

Tectonic Control on Regional Hydrology and Lake Environments

Rifting increased the elevational, and thus the hydraulic gradient, between the Afar depression and the Ethiopian and southeastern plateaus. This increased the effectiveness of both the surface drainage network and groundwater circulation, which itself was facilitated by numerous faults and fissures crossing the escarpments and the Afar floor.

Miocene and early Pliocene lakes were rather shallow, and their sedimentary infill was notably characterized by detritus supplied by rivers flowing directly from the escarpments. In contrast, after ~2.5 Ma, lakes of the central Afar, located away from the escarpments, were deep and characterized by biogenic or chemical deposition. They were supplied either by groundwater or by rivers that had deposited their coarse-grained loads farther upstream.

Climatic Control on Temporal Lake Distribution and Water-Level Fluctuations

At a given time, climate is responsible for the presence or absence of lakes in the deepest basins in the Afar depression. The climatic influence is clearly shown for the late Quaternary. Lake-level fluctuations that occurred over a time span of a thousand years in the central Afar basins readily correlate with the evolution of other tropical African lakes (Street-Perrott and Roberts, 1983). This evolution is linked to worldwide climatic events, such as changes in solar radiation (Kutzbach and Street-Perrott, 1985) and ice-volume fluctuations. Low-level lake stages or lowstands correspond to glacial episodes.

Global effects probably are responsible also for the numerous lake-level fluctuations inferred from Pliocene and lower Pleistocene sedimentary rocks. The absence of lacustrine strata of late Miocene age tentatively is attributed to a generally dry climate. The oldest lacustrine episode recorded in Afar, ~10 Ma, coincides with the worldwide major unconformity between the middle and upper Miocene. This chronostratigraphic boundary is marked by a "short-term" (1-m.y.) rise in sea level (Haq et al., 1987) and thus to a warm climatic event.

Climate and Sedimentary Facies

Climate partly explains the type of sedimentation, either dominantly detrital, biological, or chemical.

Rainfall regime and vegetation cover obviously control the river's coarser grained solids load and thus the input of detrital material to the lakes. From analysis of the late Quaternary lakes described here, the following hypothesis is proposed to explain the occurrence of diatomitic or carbonate sedimentation during lake highstands.

A major sedimentological change is observed between upper Pleistocene stages (diatomitic) and Holocene stages where carbonates prevailed. According to Gasse and Fontes (1989), the predominance of biogenic siliceous material indicates that chemical weathering was well advanced and that the dissolved load was limited to alkaline ions and silica. During the Holocene, carbonate sedimentation indicates a supply of cations, particularly of alkaline earths. Gasse and Fontes (1989) assumed that this supply was derived through mechanical erosion that prevailed during the terminal Pleistocene arid episode (~20–10 ka), which had exposed bedrock to intense chemical weathering during the subsequent humid phase. This hypothesis also can be applied to older lacustrine episodes. Diatomitic sedimentation characterizes Stage I (10 Ma) and Stages III to Va-b (4.4–1.0 Ma). Carbonate deposition dominated during Stages II (<5.9–4.4 Ma) and Vc (1.0–0.8 Ma). These major sedimentological changes may therefore have been indirectly controlled by climatic changes.

ACKNOWLEDGMENTS

This work was supported by the Centre National de la Recherche Scientifique and by the Ministère des Affaires Etrangére (France). I am also indebted to the Institut Supérieur d'Etudes et de Recherches Scientifiques et Techniques (Djibouti) and to the Ministère des Armées (France) for providing field facilities in Djibouti.

Anis Abdallah, M. Fournier, F. Recroix-Lejeune, and P. Vellutini are gratefully acknowledged for their constructive discussions. F. A. Street-Perrott also made helpful comments in reviewing the manuscript. I am very grateful to B. Katz for inviting me to participate in the AAPG conference in Salt Lake City.

REFERENCES CITED

Abdallah, A., V. Courtillot, V. M. Kasser, A. Y. Le Dain, J. C. Lépine, B. Robineau, J. C. Ruegg, P. Tapponnier, and A. Tarantola, 1979, Relevance of Afar seismicity and volcanism to the mechanics of accreting plate boundaries: Nature, v. 282, p. 17-23.

Barberi, F., G. Ferrara, R. Santacroce, and J. Varet, 1975, Structural evolution of the Afar triple junction, in A. Pilger, and A. Rösler, eds., Afar depression of Ethiopia—Proceedings of international symposium on the Afar region and related rift problems: Stuttgart, E. Schweizerbartsch Verlagsbuchhandl, v. 1, p. 38-54.

Behre S. M., 1986, Geologic and geochronologic constraints on the evolution of the Red Sea-Gulf of Aden and Afar depression: Journal of African Earth Sciences, v. 5, no. 2, p. 101-117.

Bonnefille, R., 1983, Evidence for a cooler and drier climate in the Ethiopian uplands towards 2.5 Myr ago: Nature, v. 303, p. 487–491.

Boucarut, M., R. Chessex, M. Clin, R. Dars, F. Debon, M. Delaloye, J. C. Fontes, J. P. Hauquin, R. Languth, C. Moussié, J. Muler, P. Pouchan, P. Roger, M. Seyler, and C. Thibault, 1980, A stratigraphic scale of the volcanic and sedimentary formations of the Republic of Djibouti, in A. Carrelli (president), Geodynamic evolution of Afro-Arabian rift system: Academia Nazionale dei Lincei, Proceedings no. 47, p. 515–526.

Boucarut, M., M. Clin, P. Pouchan, and C. Thibault, 1985, Impact des événements tectono-volcaniques plio-pléistocènes sur la sédimentation en République de Djibouti (Afar Central): Geologische Rundschau, v. 74, no. 1, p. 123–137.

Carrelli, A. (president), 1980, Geodynamic evolution of Afro-Arabian Rift System: Academia Nazionale dei Lincei, Proceedings no. 47, 705 p.

C.E.G.D., 1974, Carte géologique du T.F.A.I.—Feuille d'Asal: Université de Bordeaux, III, 1:100,000.

Chavaillon, J., J. L. Boisaubert, M. Faure, C. Guérin, J. L. Ma, B. Nickel, P. Poupeau, P. Rey, and S. A. Warsama, 1987, Le site de dépecage à Elephas recki de Barogali (République de Djibouti)—Nouveaux résultats et datation: Comptes Rendus de l'Académie des Sciences de Paris, v. 305, p. 1259–1266.

Clark, J. D., B. Asfaw, G. Assefa, J. W. K. Harris, H. Kurashina, R. C. Walter, T. D. White, and M. A. J. Williams, 1984, Palaeoanthropological discoveries in the Middle Awash Valley, Ethiopia: Nature, v. 307, p. 423–428.

Christiansen, R. B., H. U. Schaefer, and M. Shönfeld, 1975, Geology of Southern and Central Afar, Ethiopia, in A. Pilger, and A. Rösler, eds., Afar depression of Ethiopia—Proceedings of international symposium on the Afar region and related rift problems: Stuttgart, E. Schweizerbartsch Verlagsbuchhandl, v. 1, p. 259–277.

CNR-CNRS, 1971, Carte géologique de la dépression des Danakil (Afar Septentrional, Éthiopie): La Celle-Saint-Cloud, Geotechnip, BEICIP.

CNR-CNRS, 1975, Geological map of central and meridional Afar: La Celle-Saint-Cloud, Geotechnip, BEICIP, 1:500,000.

Fontes, J. C., F. Gasse, E. Camara, B. Millet, J. F. Saliége, and M. Steinberg, 1985, Late Holocene changes in Lake Abhé hydrology: Zeitschrift für Gletscherkunde und Glazialgeologie, v. 21, p. 89–96.

Fontes, J. C., and P. Pouchan, 1975, Les cheminées du lac Abhé (TFAI)—Stations hydroclimatiques de l'Holocène: Comptes Rendus de l'Académie des Sciences de Paris, v. 280, p. 383–386.

Fourtanier, E., 1987, Diatomées néogènes d'Afrique—Approche biostratigraphique en milieux marins (Sud-Ouest africain) et continental: Thèse, Université Paris, VI, 365 p.

Fourtanier, E., and F. Gasse, 1988, Premiers jalons d'une biostratigraphie et évolution des diatomées lacustres d'Afrique depuis 11 millions d'années: Comptes Rendus de l'Académie des Sciences de Paris, v. 306, p. 1401–1408.

Gasse, F., 1975, L'évolution des lacs de l'Afar Central (Ethiopie et T.F.A.I.) du Plio-Pléistocène à l'Actuel—Reconstitution des paléomilieux lacustres à partir de l'étude des diatomées: Thèse, Université Paris, VI, 406 p.

Gasse, F., 1977, Evolution of Lake Abhé (Ethiopia and TFAI) from 70,000 B.P.: Nature, v. 265, p. 42–45.

Gasse, F., 1980, Les diatomées lacustres plio-pléistocènes du Gadeb (Ethiopie)—Systématique, paléoécologie, biostratigraphie: Revue Algologique, Mémoire Hors-Série 3, 249 p.

Gasse, F., 1986, East African diatoms—Taxonomy, ecological distribution: Bibliotheca Diatomologica, v. 11, 201 p.

Gasse, F., J. Dagain, M. Fourtanier, G. Mazet, and O. Richard, 1987, Carte géologique de la République de Djibouti—Dikhil: Bondy, Éditions de L'ORSTOM, 85 p., 1:100,000.

Gasse, F., and G. Delibrias, 1976, Les lacs de l'Afar Central (Ethiopie et TFAI) au Pléistocène supérieur, in S. Horie, ed, Paleolimnology of Lake Biwa and the Japanese Pleistocene: Kyoto, v. 4, p. 529–575.

Gasse, F., and J. C. Fontes, 1989, Palaeoenvironments and palaeohydrology of a tropical closed lake (L. Asal, Djibouti) since 10,000 yr B.P.: Palaeogeography, Palaeoclimatology, Palaeoecology, v. 69, p. 67–102.

Gasse, F., J. C. Fontes, and P. Rognon, 1974, Variations hydrologiques et extension des lacs holocènes du désert Danakil: Palaeogeography, Palaeoclimatology, Palaeoecology, v. 15, p.109–148.

Gasse, F., and M. Fournier, 1983, Sédiments plio-quaternaires et tectoniques en bordure du Golfe de Tadjoura (République de Djibouti), in Rifts et fossés anciens—Tectonique, volcanisme, sédimentation: Bulletin des Centres de Recherche, Exploration-Production Elf-Aquitaine, v. 7, no. 1, p. 285–300.

Gasse, F., M. Fournier, J. C. Lépine, O. Richard, and J. C. Ruegg, 1983, Carte géologique de la République de Djibouti—Djibouti: Bondy, Éditions de L'ORSTOM, 70 p., 1:100,000.

Gasse, F., M. Fournier, O. Richard, and J. C. Ruegg, 1985, Carte géologique de la République de Djibouti—Tadjoura: Bondy, Éditions de L'ORSTOM, 115 p., 1:100,000.

Gasse, F., O. Richard, D. Robbe, P. Rognon, and M. A. J. Williams, 1980, Evolution tectonique et climatique de l'Afar Central d'après les sédiments plio-pléistocènes: Bulletin de la Société Géologique de France, Série 7, v. 22, no. 6, p. 987–1001.

Gasse, F., P. Rognon, and F. A. Street, 1980, Quaternary history of the Afar and Ethiopian Rift lakes, in M. A. J. Williams, and H. Faure, eds., The Sahara and the Nile—Quaternary environments and prehistoric occupation in northern Africa: Rotterdam, A. A. Balkema, p. 361–400.

Gasse, F., and F. A. Street-Perrott, 1978, Late Quaternary lake-level fluctuations and environments of the northern rift valley and Afar region (Ethiopia and Djibouti): Palaeogeography, Palaeoclimatology, Palaeoecology, v. 24, p. 279–325.

Gasse, F., J. Varet, G. Mazet, F. Recroix, and J. C. Ruegg, 1986, Carte géologique de la République de Djibouti—Eali Sabih: Bondy, Éditions de L'ORSTOM, 104 p., 1:100,000.

Geothermica Italiana, 1983, Preferability of geothermal resources in the Hanlé-Gaggadé zone of Djibouti: Institut Supérieur d'Etudes et de Recherches Scientifiques et Techniques internal report, 55 p.

Girdler, R. W., and P. Styles, 1978, Seafloor spreading in the western Gulf of Aden: Nature, v. 271, p. 615–617.

Hall, C. M., R. C. Walter, J. A. Westgate, and D. York, 1984, Geochronology, stratigraphy and geochemistry of Cindery Tuff in Pliocene hominid-bearing sediments of the Middle Awash, Ethiopia: Nature, v. 308, p. 26–31.

Haq, B. U., J. Hardenbol, and P. R. Vail, 1987, Chronology of fluctuating sea levels since the Triassic: Science, v. 235, p. 1156–1167.

Kilham, P., S. S. Kilham, and R. E. Hecky, 1986, Hypothesized resource relationships among African planktonic diatoms: Limnology and Oceanography, v. 31, no. 6, p. 1169–1181.

Küntz, K., H. Kreuzer, and P. Müller, 1975, Potassium-argon ages determinations of the trap basalt of the south eastern part of the Afar Rift, in A. Pilger, and A. Rösler, eds., Afar depression of Ethiopia—Proceedings of international symposium on the Afar region and related rift problems: Stuttgart, E. Schweizerbartsch Verlagsbuchhandl, v. 1, p. 370–374.

Kutzbach, J. E., and F. A. Street-Perrott, 1985, Milankovitch forcing of fluctuations in the level of tropical lakes from 18 to 0 kyr B.P.: Nature, v. 317, p. 130–134.

Lalou, C., H. V. Nguyen, H. Faure, and L. Moreira, 1970, Datation par la méthode U/Th des hauts niveaux de coraux de la dépression de l'Afar (Ethiopie): Revue de Géographie Physique et de Géologie Dynamique, v. 12, no. 1, p. 3–8.

Martini, M., 1969, La geochimica del lago Giulieni (Ethiopia): Rendiconti della Societa Italiana di Mineralogia e Petrologia, v. 25, p. 1–16.

Müller, W., 1982, Rapport Coopération Hydrologique Allemande—Inventaire et mise en valeur des ressources en eau de la République de Djibouti: Hanover, Bundesanstalt für Geowissenschaften und Rohstoffe, Coopération Technique Projet 78/2233/1, 5 v.

Pilger, A., and A. Rösler, eds., 1975, Afar depression of Ethiopia—Proceedings of international symposium on the Afar region and related rift problems: Stuttgart, E. Schweizerbart Verlagsbuchhandl, v. 1, 416 p.

Pilger, A., and A. Rösler, eds., 1976, Afar between continental and oceanic rifting—Proceedings of international symposium on the Afar region and related rift problems: Stuttgart, E. Schweizerbart Verlagsbuchhandl, v. 2, 216 p.

Robbe, D., 1979, Les diatomites plio-pléistocènes de l'Afar Central (paléoécologie, biostratigraphie, sédimentologie): Thèse, Université Paris, VI, 70 p.

Ruegg, J. C., and M. Kasser, 1987, Deformation across the Asal-Ghoubbat rift, Djibouti, uplift and crustal extension 1979-1986: Geophysical Research Letters, v. 14, no. 7, p. 745–748.

Sickenberg, O., and M. Schönfeld, 1975, The Ch'Orora Formation—Lower Pliocene limnical sediments in southern Afar (Ethiopia), *in* A. Pilger, and A. Rösler, eds., Afar depression of Ethiopia—Proceedings of international symposium on the Afar region and related rift problems: Stuttgart, E. Schweizerbartsch Verlagsbuchhandl, v. 1, p. 277-284.

Street-Perrott, F. A., and N. Roberts, 1983, Fluctuations in closed-basin lakes as an indicator of past atmospheric circulation patterns, *in* F. A. Street-Perrott, M. Beran, and R. A. S. Ratcliffe, eds., Variations in the global water budget: Boston, D. Reidel Publishing Co., p. 331-345.

Taieb, M., 1974, Evolution quaternaire du bassin de l'Awash (Rift Ethiopien et Afar): Thèse, Université Paris, VI, 390 p.

Tazieff, H., J. Varet, F. Barberi, and G. Gighia, 1972, Tectonic significance of the Afar (or Danakil) depression: Nature, v. 235, p. 144-147.

Tiercelin, J. J., 1981, Rifts continentaux, tectonique, climats, sédiments—Exemples, la sédimentation dans le Nord du Rift Gregory (Kenya) et dans le Rift de l'Afar (Ethiopie) depuis le Miocène: Thèse, Université Aix-Marseille, II, 260 p.

Tiercelin, J. J., 1986, The Pliocene Hadar Formation, Afar depression of Ethiopia, *in* L. E. Frostick, R. W. Renaut, I. Reid, and J. J. Tiercelin, eds., Sedimentation in the Africa rifts: Geological Society Special Publication 25, p. 221-240.

United Nations, 1965, Programme des Nations-Unies pour le développement—Report on survey of the Awash River Valley: Roma, 5 v.

United Nations, 1971, Rift Valley geothermal report (UNO, Ethiopia 171. Report on the geology, geochemistry and hydrology of hot springs of East African rift system in Ethiopia).

Williams, M. A. J., G. Assefa, and D. A. Adamson, 1986, Depositional context of Plio-Pleistocene hominid-bearing formations in the Middle Awash valley, southern Afar Rift, Ethiopia, *in* L. E. Frostick, R. W. Renaut, I. Reid, I., and J. J. Tiercelin, eds., Sedimentation in the Africa rifts: Geological Society Special Publication 25, p. 241-251.

Lacustrine Oil Shale in the Geologic Record

Michael A. Smith
Texaco E&P Technology Division
Houston, Texas, U.S.A.

Oil shales have been deposited across the globe in a variety of lacustrine and marine settings for more than a billion years. These finely laminated rocks originated as some of the most kerogen-rich clastic sedimentary sequences in the world and are therefore of great interest and importance as petroleum source rocks. Paleolatitude, paleoclimate, and tectonic setting were all important factors in the distribution of lakes and their organic-rich sedimentary facies, but major lakes most commonly were associated with rift, intracratonic, or intermontane basins.

Precambrian and early Paleozoic lacustrine petroleum source rocks contain mostly type I algal kerogen, but because the development of vascular plants during the Silurian provided additional types of vegetation, younger organic shales also can produce gas and waxy crude oils. Large lacustrine oil shale accumulations of late Paleozoic age are found in South America, eastern Canada, Greenland, Great Britain, western China, and east Africa.

Lacustrine oil shales continued to accumulate along continental margins in rift basins associated with the breakup of Pangaea during the Mesozoic. Large long-lived lakes formed in southeastern Africa and western Australia, as well as in many Newark Supergroup basins in eastern North America and the conjugate African margin. During the Early Cretaceous, lacustrine deposits were associated with rifting between South America and Africa. Thick late Mesozoic and Cenozoic lacustrine sequences with extensive oil shales are prevalent throughout China, southeastern Asia, and western United States.

The distribution of lacustrine oil shales through geologic time reflects deposition in both shallow and deep lakes in all types of continental sedimentary basins. Therefore, one finds no shortage of productive analogs for lacustrine systems around the world. Through improved understanding of geologic controls on deposition and preservation of source and reservoir facies, petroleum exploration in lacustrine basins also will become more effective.

INTRODUCTION

The petroleum source rock potential of lacustrine strata in the geologic record is the focus of considerable research (Powell, 1986; Fleet et al., 1988; Talbot and Kelts, 1989; Smith, 1989). For example, Kelts (1988) reviewed the depositional environments of these source rocks in ancient lakes, models that have been proposed for accumulation of abundant oil-prone organic matter, and factors controlling the carbon cycle in lakes. Permanent lake stratification favors accumulation and preservation of organic matter in anoxic waters (Demaison and Moore, 1980; Talbot, 1988) of the hypolimnion—a lake's deepest, and generally coldest, dense, oxygen-poor layer of water. Finally, a complete inventory and a geological database of major Phanerozoic lake deposits of the world are currently being compiled by Gierlowski-Kordesch (in preparation) as the end product of International Geological Correlation Program (IGCP) Project 219, Comparative Lacustrine Sedimentology in Space and Time.

In the broadest sense oil shales are sedimentary rocks that contain kerogen and yield hydrocarbon material when heated (Taylor, 1987). As such, they can be considered equivalent to petroleum source rocks. They typically yield substantial amounts of hydrocarbons upon destructive distillation, and they have been deposited worldwide in a variety of lacustrine and marine settings for more than a billion years. Coals that form in marginal lacustrine and paludal environments also yield hydrocarbons during pyrolysis but are distinguished from oil shale by lower (<33%) ash content. Oil shale yields no appreciable oil by solvent extraction, but an excellent correlation is found between the S_2 peak area from Rock-Eval pyrolysis and Fischer assay oil yield. Because oil shales are too thermally immature to have generated significant hydrocarbons, they are of commercial interest only if they provide substantially more shale-oil energy than is required to process the rock. Tissot and Welte (1984) defined economic oil shales as having a kerogen content of at least 5% by weight, corresponding to an oil yield of 6 gal/ton (25 L/MT), and rocks that contain more than 15 gal/ton (62.6 L/MT) are considered to be rich oil shales (Taylor, 1987).

Some large modern lakes (Figure 1) and several ancient ones formed in fluviatile or coastal settings, but almost all long-lived lacustrine basins are of tectonic origin. These geologic origins—rift valleys and regions of cratonic uplift or downwarp—and lake distribution through geologic time are shown here on paleogeographic reconstructions of Ziegler et al. (1985). Paleolatitude, paleoclimate, and tectonic setting are all factors in distribution of lakes and their organic-rich sedimentary facies, but occurrences of lacustrine oil shale suggest that major lakes are most commonly associated with rift, intracratonic, or intermontane basins of variable size.

Precambrian lacustrine sedimentary sequences may be difficult to recognize and need to have escaped intense metamorphism and erosion for a long period of the Earth's history. It is not surprising then that only about one percent of lacustrine deposits reported in the literature formed during Precambrian time (Clemmey, 1978). Bitumen and oil seeps from several Proterozoic lacustrine shales in Australia and North America demonstrate that these rocks still have a sufficient pyrolytic yield to be classified as oil shale. Precambrian and early Paleozoic lacustrine petroleum source rocks contain mostly type I algal kerogen. However, development of vascular plants during the Silurian provided additional types of vegetation so that younger organic shales also can produce gas and waxy crude oils.

Oil shales became most abundant in Gondwana during the Permian. The most complete record of Paleozoic lacustrine deposition is found in Australia, but other large late Paleozoic nonmarine oil shale accumulations are found in South America, eastern Canada, Greenland, Great Britain, large interior basins of western China, and in the Karroo graben fill of east Africa. Many are associated with coal seams, and well-defined sedimentary cycles are displayed in their lacustrine record.

Lacustrine organic shales continued to accumulate along continental margins in rift basins related to the breakup of Pangaea during the Mesozoic. Large, long-lived lakes formed at this time in southeastern Africa and western Australia, as well as in Newark Supergroup basins of eastern North America and the conjugate African margin. During the Early Cretaceous, synrift lacustrine deposits were widely associated with the rifting between South America and Africa.

Cenozoic lacustrine oil shales occur in diverse tectonic settings, including rifts, continental sag basins, and compressional belts with significant uplift. Thick late Mesozoic and Cenozoic lacustrine sequences with extensive oil shales are especially prevalent throughout China, southeastern Asia, and western United States. However, even though lacustrine strata are common and well preserved in many Cenozoic basins, the lakes frequently were shallow and holomictic (subject to complete mixing of waters during overturn) and did not necessarily produce oil shale accumulations.

This paper demonstrates that nonmarine oil shales with remaining hydrocarbon-generation potential have been deposited in both shallow and deep lakes in all types of continental sedimentary basins throughout geologic time. Many petroleum-producing lacustrine basins around the world can provide models of deposition and preservation of high-quality source and reservoir facies as well as analogs for successful exploration.

DISTRIBUTION OF LARGE LAKES THROUGH TIME

Precambrian Lacustrine Deposits— The First Oil Shales?

Well-preserved lacustrine sedimentary sequences of Precambrian age have been reported from a number of localities, but, as Clemmey (1978) noted, they comprise only about one percent of all lacustrine deposits that have been described. Recognizing ancient lacustrine rocks in Precambrian strata is notably difficult (Picard and High, 1972), and unequivocal identification may be impossible in many cases, especially for those deposits that have survived for 2 b.y. or more. In Figure 2 the distribution of the oldest known lacustrine oil shales includes three Proterozoic units for which geochemical data are available.

Indigenous live oil in reservoirs more than 1.4 b.y. old has been reported in a shallow Australian Bureau of Mineral Resources drill hole in the McArthur basin, Northern Territory. The importance of these petroleum source and reservoir rocks, which were deposited in lacustrine and marine environments, and results of biomarker studies were discussed by Jackson et al. (1986). Additional geochemical data that support the classification of these source rock horizons as oil shales containing types I and II kerogen were presented in a series of articles by Jackson et al. (1988), Crick et al. (1988), and Summons et al. (1988). Womer (1986) described tarry bitumen in the 1.5-b.y.-old Roper Group, where liquid hydrocarbons were discovered, and in the 1.65-b.y.-old lacustrine Batten Subgroup of the McArthur River area. Details of the sedimentary facies in these units, which also host stratiform lead-zinc deposits, were provided by Muir (1983) and by Donnelly and Jackson (1988), who suggested a quiet distal lacustrine setting for Middle Proterozoic varved, organic-rich black shales.

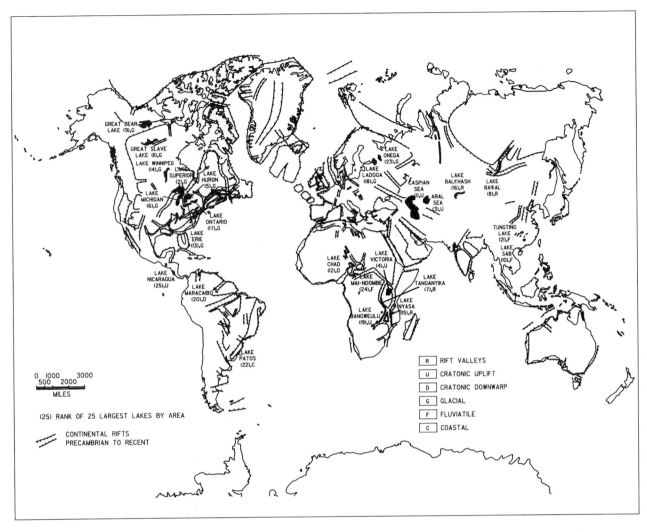

Figure 1. Worldwide distribution of large modern lakes in relation to the continental rift system. Of the 25 largest lakes by area, four are located in rift-valley basins (R), five are associated with cratonic uplift (U), and two with downwarp in sag basins (D). In addition, ten lakes have a glacial origin (G), three occur in a fluviatile setting (F), and one is coastal (C).

Two Precambrian organic-rich lacustrine deposits in North America are undergoing testing as oil exploration targets. A source rock assessment of the Middle Proterozoic Nonesuch Formation, part of an unmetamorphosed fluviolacustrine system deposited along the Mid-Continent Rift 1.05 b.y. ago during Late Keweenawan time (Dickas, 1986), is given by Imbus et al. (this volume). Total organic carbon (TOC) values range from near zero to nearly 3.0%, but only two of 25 samples they analyzed have a Rock-Eval $S_1 + S_2$ total generation potential exceeding 2 mg of hydrocarbon per g of rock (mg HC/g rock), which Tissot and Welte (1984) consider to be moderate to good source rocks. Apparently indigenous oil seeps from the Nonesuch Formation at the White Pine copper mine in northern Michigan controlled mineralization (Eglinton et al., 1964; Hoering, 1976; Kelly and Nishioka, 1985).

The Chuar Group, a younger sequence exposed on the northern side of the Colorado River in the eastern Grand Canyon, consists predominantly of organic-rich black mudstone that appears to have accumulated in a subsiding lacustrine basin. Palacas and Reynolds (1989) noted that the Walcott Member of the Kwagunt Formation in this sequence has considerably higher organic content than Nonesuch shales, with TOC values up to 8.0% and $S_1 + S_2$ potentials as high as 16 mg HC/g rock.

Paleozoic Oil Shales

Gondwanian Lacustrine Deposits

Lacustrine environments in Gondwana became more prevalent at the end of the Paleozoic, and most large lakes of this hemisphere were still in existence

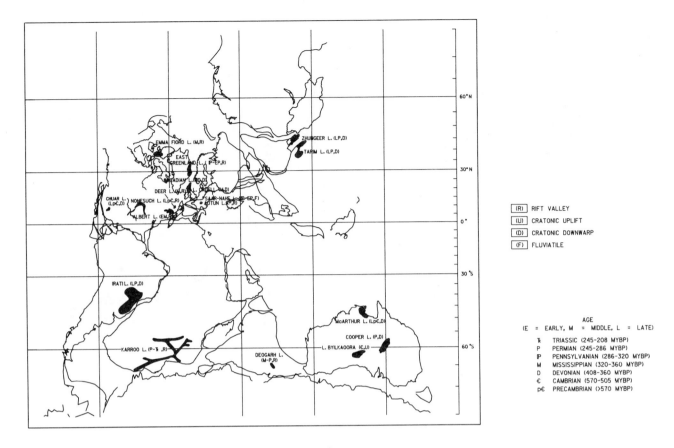

Figure 2. Early Permian (270 Ma) reconstruction of major Precambrian and Paleozoic lacustrine oil shale deposits, classified by origin.

during the Permian. Many that formed at high paleolatitudes contain glacial materials. One exception to the more abundant high-latitude, late Paleozoic lacustrine deposits occurs in the northeastern Officer basin of South Australia (Figure 2). Here, the Cambrian Observatory Hill Beds, penetrated by the Byilkaoora 1 stratigraphic test well, have been interpreted as the world's oldest Phanerozoic alkali playa-lake deposits and are analogous to Devonian lacustrine sequences in the Orcadian basin of Scotland and to the Eocene Green River Formation in Wyoming (White and Youngs, 1980). Australia, which had the most complete geologic record of lake systems of any continent during the Paleozoic, was drifting southward during the Cambrian, and Lake Byilkaoora had a near-equatorial setting at that time, similar to Lake Magadi in Kenya today. Depositional cycles in the Observatory Hill Beds reflect arid-climate playa lakes periodically inundated by fresh water. The presence of Magadi-type cherts and evaporites indicate a highly alkaline, high-bicarbonate, evaporative environment that was characterized by increased humidity during the middle playa-lake depositional sequence. Cyanobacteria were abundant in the shallow lakes, with methanogenic and halophilic bacteria also present in the benthic biota (McKirdy and Morton, 1985).

Organic-rich bottom muds, which were covered and preserved by evaporative chert crusts, have produced oil seeps rich in isoprenoid alkanes.

Deogarh Lake deposits (Carboniferous) in the state of Bihar, India, represent the oldest example of a lacustrine sequence within a rift-basin setting in the southern hemisphere. Several dozen coalfields occur in three small basins that tectonically and sedimentologically resemble some Newark Supergroup basins of the eastern United States. The Talchir Formation, 200–225 m thick in this region, contains glacial and fluviolacustrine rocks that were deposited in narrow grabens at paleolatitudes of 45°S to 75°S. The overlying Karharbari Formation (Permian) consists of up to 235 m of fluviolacustrine rocks deposited at a paleolatitude of about 65°S (Figure 2). Both units contain abundant trace fossils and were deposited as shallow- and deep-water facies in fresh-water lakes (Mukherjee and Guha, 1984).

The Karroo graben system of southeastern Africa existed as a failed rift from the Early Permian to the close of the Triassic and is about the same size as the northern part of the western Australian rift system (Veevers and Cotterill, 1976). Up to 4000 m of fluviolacustrine and other graben fill were deposited in these rift valleys, which from their position near paleolatitude 60°S during the Permian

Smith

drifted northward with the eastern Gondwana continent to about 40°S by the end of the Triassic. Evidence for glaciation in Karroo beds along Lake Nyasa, including varved glaciolacustrine siltstone, was presented by Wopfner and Kreuser (1986). In the Republic of South Africa, a small industry has produced shale oil from the coal-bearing Karroo Series (Duncan and Swanson, 1965). The possibility of additional oil accumulations sourced from Karroo coal measures within the extension of these grabens offshore from Mozambique was suggested by DeBuyl and Flores (1986).

The Cooper/Eromanga basin of South Australia and Queensland contains fluviolacustrine rocks ranging in age from Pennsylvanian through Cretaceous, with oil and gas production and shows throughout most of the section (Lavering et al., 1986). The Merrimelia Formation at the base of the Cooper basin stratigraphic section consists of glaciolacustrine and fluvioglacial rocks deposited near paleolatitude 60°S (Figure 2). Glaciolacustrine mudrock may have included a rich algal flora and possibly represents the source rock for eolianite sandstone reservoirs in the unit. In the thermally mature Lower Permian Patchawarra Formation (Kantsler et al., 1984), both deep and shallow lakes were the depositional sites for a productive sequence of thick coals and sediments in addition to fluvial and backswamp facies. From a study of the organic petrography of these and other formations in the Permian Gidgealpa Group, Smyth (1983) reached the problematic conclusion that dispersed inertinite could be a source of liquid hydrocarbons in the Cooper basin.

Lacustrine deposits of the Irati Shale (Upper Permian) in the Parana basin of southern Brazil, Uruguay, and eastern Paraguay comprise the largest known shale-oil resource in the world besides the Green River Formation (Duncan and Swanson, 1965). They contain bacterially degraded algal matter that is still somewhat immature because of its shallow burial history. In their description of Irati geochemistry and depositional environment, Da Silva and Cornford (1985) concluded that it formed in nonconnected fresh- or brackish-water lakes under extremely reducing conditions at mid-latitudes (about 40°S). The basin was open to the sea to the south, and Late Permian climate varied from warm and dry to cold and rainy.

Laurasian Lacustrine Deposits

All the Laurasian lacustrine oil shales shown in Figure 2 are late Paleozoic (Middle Devonian and younger) in age, but their paleoenvironments ranged from equatorial to subpolar in various tectonic settings. Rich lacustrine oil shales with abundant algal matter are common in the northern hemisphere. Lacustrine cycles and coaly sequences in many of these deposits may reflect climatic changes during this time.

In the Orcadian basin of northeastern Scotland the Old Red Sandstone (Middle Devonian) may reach a thickness of up to 10000 m (Anderton et al., 1979). These fluviolacustrine sediments were deposited in a fault-bounded intermontane lake basin covering more than 30000 km² at about paleolatitude 15°N. Laminites of alternating thin (<1 mm) carbonate and siltstone suggest a meromictic stratified lake, like Todilto Lake (Middle Jurassic) in New Mexico, with algal blooms in warm, dry seasons and fluvial clastic deposition during wet seasons (Donovan, 1975). The section also contains lake-margin stromatolites (Fannin, 1969) and Magadi-type cherts (Parnell, 1985, 1988a).

Hydrocarbon distribution in bituminous flagstone and organic shale, deposited in the occasionally hypersaline Orcadian Lake, was discussed by Hall and Douglas (1983). Duncan and Hamilton (1988) presented evidence that these source rocks have contributed a Middle Devonian component to the Beatrice crude oil toward the original basin center at Moray Firth. Although geochemical data indicate that most of the rocks are still in an early stage of hydrocarbon generation, hydrocarbon shows occur throughout the formation and basin (Marshall et al., 1985).

The Dinantian Oil Shale Group, deposited in Mississippian Lake Cadell in the eastern Midland basin, Scotland, is a lacustrine sequence whose cyclic sedimentation reflects periods of lake expansion and contraction. The rocks reportedly have distillation yields as high as 52.8 gal/ton (220.3 L/MT), S_2 yields of 180 mg/g, and up to 30% TOC (Parnell, 1988b). Lake Cadell occurred near paleolatitude 5°N in a wet tropical climate, and contained abundant algal oozes. Loftus and Greensmith (1988) concluded that these oil shales were deposited in a shallow (<100 m), freshwater meromictic lake (incomplete mixing of a bottom, noncirculating water mass and upper, circulating layer) that was intermittently open to the sea on the east.

Algal carbonates in the small Permo-Pennsylvanian Saar-Nahe basin in Germany (Schafer and Stapf, 1978) were compared with those of modern Lake Constance in the Rhine River system. More than 9000 m of fluviolacustrine sediment and associated lacustrine coal measures accumulated under humid conditions in Lake Saar-Nahe, which was near the equator during the Pennsylvanian. By the end of the Paleozoic, the basin had moved north to a subtropical location at paleolatitude 20°N, where it became part of a northeast-trending series of basins and mountains in western Europe. Farther southwest, the Permian coal-bearing section near Autun, one of the largest oil shale deposits in France, was deposited around paleolatitude 12°N in paludal and lacustrine environments in a fault-bounded intermontane basin. As in Green River oil shale, algal organic matter is abundant, and the paleoclimate varied from semiarid to humid.

The oldest lacustrine rocks in China were deposited in the Tarim and Zhungeer basins during the Late Permian near paleolatitude 45°N (Figure 2). Tarim, the country's largest interior basin, contains a largely marine platform sequence that in places is more than 6000 m thick. By the end of the Paleozoic, about 400 m of lacustrine mudstone, shale, and oil shale—possible petroleum source rocks—had been deposited in the northern and western foredeeps and possibly in the eastern depression (Lee, 1985b). In the Zhungeer basin more than 1000 m of Upper Permian lacustrine black shales and oil shales that accumulated in a wet climate have generated some of the oil discovered to date (Lee, 1985a; Ulmishek, 1986). Alluvial fans and playa lakes also began to develop in the Permian, and as much as 2500 m of coarse clastics were deposited in intramontane foredeeps near the Karamay oil field. Chang (1981) noted that two-thirds of the oil in this giant field occurs along the basin margins in Permo-Triassic alluvial fan reservoirs.

The Spillway Member of the Rocky Brook Formation (Mississippian) in western Newfoundland represents lacustrine deposition (with some oil shale) in an equatorial rift valley (Hyde and Ware, 1985; Kalkreuth and Macauley, 1989). Cyclic sedimentation was characterized by carbonate deposition during highstands and mudflat deposits as a result of lowstands (Hyde, 1984). Deer Lake lithofacies suggest a shallow perennial lake with moderately saline, alkaline water high in sodium and magnesium.

Other Mississippian and Pennsylvanian lacustrine sequences in Canada, shown on Figure 2 as Albert Lake, formed near the equator in what is now New Brunswick and Nova Scotia. More than 1300 m of lake sediments accumulated in the 4000-km^2 Albert rift valley (Howie and Cumming, 1963; Carter and Pickerill, 1985). Those in the Moncton subbasin may include the richest oil shales in Canada (Duncan and Swanson, 1965; Kalkreuth and Macauley, 1984; Macauley et al., 1984). Two indicators of organic richness—total organic carbon content and Rock-Eval S_1 and S_2 total generation potential—are cross-plotted in Figure 3. High TOC values and good to excellent hydrocarbon-generation potential are indicated for some Albert Formation lithotypes; the organic-lean samples are dolomite marl or laminated mudstone (Smith and Gibling, 1987). Altebaeumer (1985) found that these oil shales are thermally mature in places and have generated hydrocarbons from lacustrine algal kerogen. Figure 4 is a modified Van Krevelen diagram that shows types of kerogen and their maturation pathway as defined by the Rock-Eval hydrogen index (S_2/TOC) and oxygen index (S_3/TOC). Oil shale samples from the Albert Mines Zone contain a mixture of oil-prone types I and II kerogen, whereas samples with a lower hydrogen index are the less organic-rich dolomite marl and laminated mudstone. Lacustrine deposition of organic-rich marlstone continued for several hundred thousand years during the Early Mississippian and ended with an evaporite sequence (Greiner, 1974). The Lower Mississippian lacustrine Horton Bluff Formation and thick Pennsylvanian lacustrine coals and clastics in the Pictou coalfield, northern Nova Scotia, are considered part of Albert Lake. Pictou coals represent stream-transported vegetation, and as many as 45 seams as much as 14 m thick are interbedded in the 2500-m-thick lacustrine black-shale sequence (Lawton, 1985).

Farther north, the Mississippian Emma Fiord lacustrine oil shale of the Sverdrop basin was discussed by Davies and Nassichuk (1988). Alginite-rich marlstone contains type I and II kerogen deposited in a meromitic lake system located from 10°N to 15°N paleolatitude. Surlyk et al. (1986) described Permian oil-prone lacustrine source rocks in central East Greenland. A Pennsylvanian-Early Permian rift sequence includes black shale deposited in backswamps and lakes around paleolatitude 30°N as well as flood-plain and lacustrine carbonate. At some localities, this organic shale, which contains mostly algal material, averages 12.3% TOC and is thermally mature for oil generation.

Mesozoic Oil Shales

Gondwanian Lacustrine Deposits

Lacustrine sequences in Australia extend from the Permian well into the Mesozoic in several basins. The downwarped intracratonic Cooper/Eromanga basin in east-central Australia has oil and gas production and shows from numerous intervals within fluviolacustrine strata that accumulated over a period of 230 m.y. (Lavering et al., 1986). Most of the hydrocarbons produced to date have come from older Cooper basin rocks (see Figure 2), but an additional 60 million bbl (9.54 million m^3) of recoverable oil in the Eromanga basin have been discovered in the largely fluviolacustrine, Jurassic-Cretaceous section that unconformably overlies the Cooper deposits (Kantsler and et al., 1984). McKirdy and Morton's (1985) geochemical analyses of Eromangan lacustrine crude oils suggest a temperate depositional paleoenvironment with large, oligotrophic, fresh-water lakes. This basin remained at paleolatitude 45°S to 60°S throughout the Mesozoic (Figure 5).

In the Bowen/Surat basin, another Australian intracratonic basin, both Permian and Mesozoic lacustrine strata were deposited at fairly high paleolatitudes. Although its primary oil source rocks occur in the marine Permian section, the Middle Triassic Moolayember Formation on the Roma Shelf contains a nearshore to lacustrine sequence with limited generation potential. Following a period of fluvial deposition, organic shale of the Evergreen Formation (Lower Jurassic) accumulated in a lacustrine environment (Cosgrove and Mogg, 1985) at a paleolatitude of about 50°S.

Younger lacustrine sediments are associated with several rift basins that formed along the southern margin of Australia during the late Mesozoic. Geochemical analyses of lacustrine crude oils from the Otway basin (McKirdy et al., 1988) indicate that

Smith

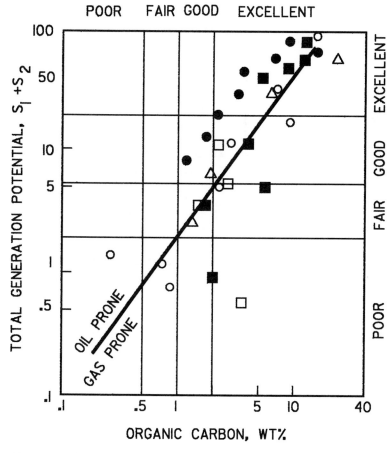

Figure 3. Organic richness and hydrocarbon source potential of lacustrine oil shales as indicated by Rock-Eval pyrolysis generative capacity and organic-carbon content.

○ Mississippian Albert Formation (Mossman et al., 1987)

□ Northern early Mesozoic Newark rift basins (Pratt et al., 1986)

△ Southern Newark basins (Smith and Robison, 1988; Smith, unpublished)

● Eocene Green River oil shale (Katz, 1988)

■ Modern East African rift lakes (Katz, 1988)

a deep, stratified, fresh-water to saline rift lake existed in a temperate climate at this time, even though its paleolatitude was somewhat south of 60°S. These studies of bitumen from oil seeps and noncommercial reservoirs in the western Otway basin suggest an Early Cretaceous setting similar to that in productive Brazilian and west African basins, where lacustrine oils of algal and bacterial origin have been generated. Otway basin oils, in which botryococcane, an algal biomarker, is highly concentrated, are similar to several families of Sumatran crude oils (McKirdy et al., 1986). To the east in the Bass basin, the Eastern View Coal Measures are possible source rocks for commercial oil discoveries (Williamson et al. 1985). This unit, which formed during Late Cretaceous and Paleocene time, consists of sandstone, oil shale, and coal that were deposited in alluvial fan, flood-plain, and lacustrine settings.

Several lacustrine systems of various tectonic origins occurred on the African continent throughout the Mesozoic. Other sequences with significant petroleum potential developed during the late Mesozoic and continued to be active into the early Cenozoic. In the central African rift system of Chad,

the Chari basin contains "effective" or "mature" lacustrine source rocks, which have generated the oil in at least eight discoveries reported by Conoco (Baker and Derksen, 1984). Lacustrine sediments in another part of the rift, the Sudd basin of Sudan (Schull, 1984), may have produced up to 10 billion bbl (1.59 million m³) of recoverable oil. Farther north, the Nubian Sandstone (Cretaceous-lower Tertiary) was deposited in a tropical, humid climate. Because the name "Nubian" has been used for rocks of many different ages throughout northern Africa and the Arabian Peninsula, its use as a stratigraphic term was discouraged by Klitzsch and Squyres (1990). Prasad et al. (1986) described a newly discovered flora from Sudan that includes aquatic plants from a lacustrine facies.

Lower Cretaceous lacustrine deposits are widely associated with the rifting episode between South America and Africa (Figure 5) and include some of the most productive sequences in the southern hemisphere. These continents rotated apart from south to north during the breakup of Gondwana, initially separating away from a near-equatorial spreading pole in northeastern Brazil (Szatmari et al., 1985). Rift and postrift sedimentation along the

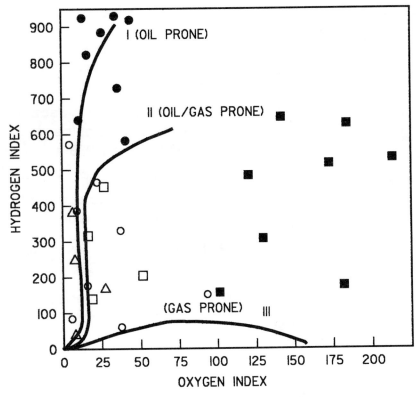

Figure 4. Modified van Krevelen diagram showing kerogen type and relative thermal maturity as indicated by Rock-Eval hydrogen and oxygen indices.

Legend:
- ○ Mississippian Albert Formation (Mossman et al., 1987)
- □ Northern early Mesozoic Newark rift basins (Pratt et al., 1986)
- △ Southern Newark basins (Smith and Robison, 1988; Smith, unpublished)
- ● Eocene Green River oil shale (Katz, 1988)
- ■ Modern East African rift lakes (Katz, 1988)

Brazilian continental margin was discussed by Ponte et al. (1977) and by Ojeda (1982). Oil shale occurs in the lacustrine Guaratiba Formation in the Santos basin, a rift-phase deposit equivalent to the productive Lagoa Feia Formation of the Campos basin. More than 1500 m of Lagoa Feia limestones were deposited in alkaline saline lakes that alternated between playa and pluvial stages (Bertani and Carozzi, 1985a, 1985b).

In the Espirito Santo basin, Upper Jurassic prerift fluviolacustrine rocks of the Sergi Formation underlie rift-phase lacustrine deposits (Ojeda, 1982). Estrella et al. (1984) identified two younger marine shales as the basin's major petroleum source rocks, but geochemical data confirm the lacustrine origin of other oil shales within the rift section. On the other side of the Atlantic Ocean, equivalent lacustrine rocks occur in the conjugate Cabinda basin (Brice et al., 1982). In west Africa an Upper Jurassic prerift fluviolacustrine sequence more than 1000 m thick is overlain by as much as 2000 m of Lower Cretaceous synrift lacustrine rocks. The deep lake system filled with turbidites and dolomitic oil shale containing up to 20% organic matter (Brice et al., 1980). This shale grades upward into shallow lacustrine carbonate and sandstone with alluvial-plain deposits. Algal limestone banks formed along the lake margins in late synrift time.

The Sergipe-Alagoas basin in Brazil contains Upper Jurassic prerift fluviolacustrine rocks in the Serraria and lower Barra de Itiuba Formations. Elongate lacustrine subbasins formed in the Early Cretaceous during the active rifting phase, which continued through deposition of the Paripueira evaporites (Ojeda, 1982). Lacustrine prodeltaic organic shale from this stage is the principal petroleum source rock and has generated oil in the Camorim, Caioba, and other fields (Ponte et al., 1977). In some fields, prerift reservoir sandstones have been faulted upward against the oil source. In the conjugate west African Gabon basin, southern continuations of grabens can be traced from the Sergipe-Alagoas basin (Szatmari et al., 1985). Rifting in the Gabon basin was contemporaneous with opening of the Brazilian Bahia Sul basin, and Ghignone and de Andrade (1970) suggested that in their early history, these rifts may have been contiguous with the Reconcavo basin. Gabon lakes are interpreted as having been brackish to saline during deposition of Lower Cretaceous oil shales (Brink, 1974).

Oil-producing Upper Jurassic and Lower Cretaceous lacustrine strata are well documented in the Reconcavo basin of Brazil. In the Bahia Supergroup, which has a maximum thickness of 6500 m, the Itaparica, Candeias, and Sao Sabastiao Formations are predominantly lacustrine (Ghignone and de

Smith

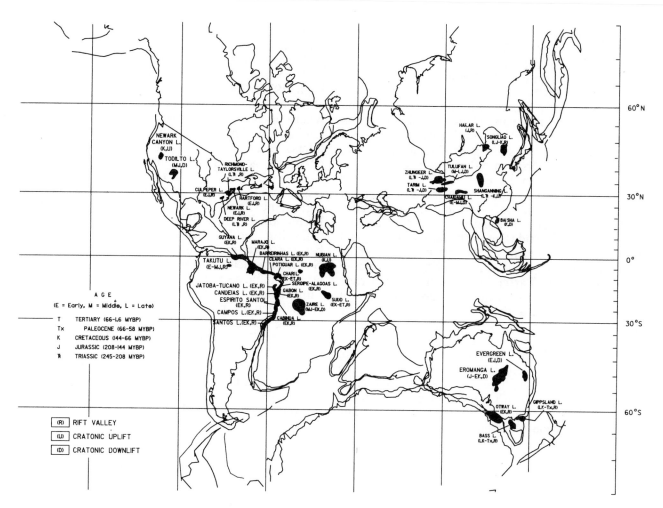

Figure 5. Early Cretaceous (120 Ma) reconstruction of major Mesozoic lacustrine oil shale deposits, classified by origin.

Andrade, 1970). The primary source rocks in this basin are considered to be dark-gray to green, alga-rich, calcareous organic shales in the Candeias Formation. The geologic setting of Candeias Lake may have been similar to that of Lockatong Lake (Newark Lake, Figure 5) of the early Mesozoic Newark basin in the eastern United States. These rift-valley lakes formed in an asymmetrical basin at moderately high elevations where adiabatic cooling of maritime air occurred in a humid climate. Similar, but less productive, Lower Cretaceous lacustrine sequences are found north of Reconcavo in the Tucano and Jatoba rift basins.

During the Early Cretaceous, lacustrine sedimentation was widespread in rift basins that now occupy the Brazilian equatorial continental margin (Figure 5). A well-developed prerift fluviolacustrine system in the Barreirinhas basin is overlain by rift-phase deposits, including lacustrine oil shales in elongate subbasins. Similar lacustrine strata are found to the east in the Acu Formation of the Ceará basin (Ojeda, 1982). In the Potiguar basin, the rift-phase lacustrine

sequence includes volcanics, whereas the overlying transitional Pendencia and Acu Formations consist of fluvial-deltaic-lacustrine rocks deposited in half-grabens (Ojeda, 1982; Falkenhein et al., 1984).

Farther north in the Marajo basin, Lower Cretaceous fluvial-deltaic-lacustrine rocks of the Cassipore Formation are found around Marajo Island (Ojeda, 1982; Mesner and Wooldridge, 1964). These stratigraphic units are overlain by the Upper Cretaceous Limoeiro Formation, which also reportedly contains organic-rich lacustrine deposits (W. Manspeizer, pers. comm.).

Laurasian Lacustrine Deposits

Large lakes continued to exist in western China where Upper Permian lacustrine oil shales and reservoirs are shown on Figure 2. The rest of the Chinese mainland and southeast Asia developed at about the end of the Paleozoic through the accretion of microcontinents that had moved northward across the Tethys Ocean. Lacustrine deposition in this region proliferated during the late Mesozoic and throughout most of the Tertiary.

In the Zhungeer basin, lacustrine deposits continued to accumulate in the Urumchi foredeep near paleolatitude 60°N with deposition of Triassic clastics and some fresh-water limestone. As this basin moved south to paleolatitude 30°N during the late Mesozoic (Figure 5), the south Urumchi foredeep continued to be a locus for lacustrine organic shale and siltstone deposition into the Late Jurassic.

South of Zhungeer Lake up to 1000 m of Upper Triassic lacustrine sediment were deposited in foredeeps on the southwestern and northern sides of the Tarim basin. The paleoenvironment at this time was more humid than in the late Mesozoic, and more than 3000 m of Jurassic lacustrine organic shale and other rocks accumulated in the Kashgar foredeep in southwestern Tarim basin near paleolatitude 45°N. As the basin continued to move southward during the Late Jurassic, well-preserved lacustrine strata developed on its northern edge in the Kucha foredeep.

Between the Tarim and Zhungeer basins, late Mesozoic lacustrine oil shales also are found in the Tulufan basin. Although this basin is small compared with others in western China, its lacustrine rocks appear to have generated large quantities of oil. The Cretaceous section representing lakes in the Shanganning basin farther to the east (Figure 5) contains, in places, more than 2000 m of lacustrine oil shales, deltaic sequences, and other continental deposits.

One of the most well known examples of a prolific lacustrine source-reservoir rock sequence is found in the Songliao basin of northeastern China. This basin formed by rifting during the Middle Jurassic Yanshanian orogeny (Chen, 1985), and Upper Jurassic to Cretaceous basin fill includes organic-rich, oil-generating shales and deltaic sandstone reservoirs that prograded into Songliao Lake (Ma, 1985). The entire basin periodically flooded during wet climatic cycles, and the area underlain by deep-water oil shale covers most of the basin. Daqing field, the world's largest nonmarine oil accumulation—with average production of one million b/d (159000 m^3/d)—is largely sourced from land-plant remains in black oil shale of the Qingshankou Formation (Lower Cretaceous).

Hundreds of small rift basins on the South China Platform contained large, shallow lakes throughout the Cretaceous and Tertiary, and many hydrocarbon occurrences have been reported in them. Red, brown, and dark-gray lacustrine clastics and organic shale deposited in a tropical to subtropical climate in the Bai Sha basin on Hainan Island, for example, are typical of Upper Cretaceous strata in this intermontane region.

As in China and southeast Asia, extensive lacustrine systems in western North America became more widespread during the Mesozoic and Cenozoic, and oil shales in the western United States are among the world's largest and most well documented examples. The Middle Jurassic Todilto Limestone in the Four Corners region was deposited during a geologically brief, 20-k.y. interval in a shallow lake that covered 89600 km^2 (Tanner, 1970; Picard and High, 1972). This unit is thermally mature in part of the San Juan basin in northwestern New Mexico, where it is characterized by dark, kerogen-rich laminae in limestone interbedded with anhydrite (Meissner et al., 1984). The Todilto also has been cited as a good-quality source rock in parts of the Rio Grande Rift. Later in the Jurassic, in the same region and to the north, similar but less organic-rich lacustrine deposits occur within the Morrison Formation (Tanner, 1970).

During the Cretaceous, lacustrine sequences became even more abundant throughout western North America. At the base of the Great Basin stratigraphic section in Nevada, part of the Newark Canyon lacustrine section contains several meters of lipid-rich oil shale (Fouch et al., 1979).

Along the East Coast of North America and in conjugate rift systems of northwest Africa (Figure 5), lacustrine deposits include organic shale and mudstone with varying degrees of petroleum source potential. Recent U.S. Geological Survey and other studies of the Newark Supergroup rift basins were summarized in Froelich and Robinson (1988), Manspeizer (1988), and Olsen et al. (1989). The great lateral continuity of lacustrine deposits in these basins suggests large perennial lakes, hundreds of kilometers long and more than 100 m deep, with well-developed, transgressive-regressive 23-k.y. cycles (Olsen, 1985, 1986). Important oil shows have been reported from several basins in the Newark rift system, and significant hydrocarbon potential has been attributed to them (Ziegler, 1983; Ziegler and Cornet, 1985).

In the southern Newark rift system, a series of early Mesozoic rift basins occurs near the eastern edge of the Piedmont province where it borders the Atlantic coastal plain. Other Triassic and Jurassic basins occur as half-grabens on the west where the Piedmont borders the Blue Ridge or Valley and Ridge province. In one of the eastern half-grabens, the Deep River basin, North Carolina, continental rocks of the Chatham Group (Upper Triassic) formed in lacustrine, paludal, fluvial, and alluvial fan environments (Gore, 1986). The Pekin and Sanford red-bed sequences are separated by coal, sandstone, and lacustrine organic shale of the Cumnock Formation. In this area coal swamps, located about paleolatitude 7°N, were flooded as the lake expanded over an area of about 770 km^2 during wet cycles of the arid to humid Triassic climate (McCarn and Mansfield, 1985). The sample with the highest TOC value on Figure 3 is from the Cummock Formation. It is more gas prone than the Albert or Green River oil shale because of the association of Cummock shale with lake fringe swamps (Robbins et al., 1988). On the modified Van Krevelen diagram (Figure 4), however, samples from the Deep River basin contain predominantly oil-prone type I kerogen at moderate to advanced levels of thermal alteration. Elsewhere in the eastern Newark Supergroup trend, the Richmond-Taylorsville basins contain rocks that may be as old as Middle Triassic deposited in a humid climate. This section includes coal measures and fluvial, paludal and lacustrine sediments; hydrocarbon shows occur in organic shale of the lower Vinita beds. Cornet (1985) believed that the great extent of these lacustrine-deltaic deposits suggests that a much larger rift system existed at that time.

Upper Triassic and Lower Jurassic lacustrine units were deposited in the central and western Culpeper basin of Virginia during a 25-m.y. period and are interbedded with fluvial red beds (Gore, 1988). The Bull Run Formation (Upper Triassic) contains thermally overmature flood-plain lacustrine deposits, but the overlying Buckland and Waterfall Formations (Lower Jurassic) consist of black and gray lacustrine oil shales and associated sandstone facies (Smith and Robison, 1988). Most Culpeper oil shale displays good to excellent organic richness (Figure 3) and appears to be slightly oil prone when its Rock-Eval hydrocarbon generative capacity and TOC are compared. Kerogen from the Waterfall Formation has a higher hydrogen content on Figure 4 than that of other southern Newark basin samples, whereas the Buckland Formation has a mixture of kerogen types at a higher degree of thermal maturation as indicated by its position on the kerogen evolution pathway.

The northern Newark rift system is represented on Figure 5 by lacustrine deposits of the Newark and Hartford basins. Organic shale in the Lockatong and Passaic Formations (Upper Triassic) of the Newark basin has been thermally altered beyond the oil window, but Lower Jurassic units, as in the Culpeper basin, demonstrate a lower maturity level and are still capable of oil generation (Katz et al., 1988). Samples on Figure 3 from the Towaco Formation (Lower Jurassic) are oil prone with fair to good organic richness, whereas the Lockatong Formation has negligible or poor hydrocarbon yields, even though TOC measurements of nearly 10% have been reported. In the Hartford basin, organic shale of the East Berlin and Portland formations (Lower Jurassic) is more gas prone and exhibits fair to good pyrolysis yields with good to excellent organic richness indicated by TOC content (Figure 3). Lower Jurassic shale from the northern Newark basins displays a mixture of kerogen types on Figure 4, with the Hartford samples reaching a higher level of maturation.

Cenozoic Oil Shales

Lacustrine Deposits of the Southern Hemisphere

Several late Mesozoic lacustrine oil shale sequences in the central African rift system continued developing into the early Cenozoic, but the younger rocks tend to be of poorer hydrocarbon-source quality and are not shown on Figure 6. Active rifting in east Africa replaced extension along the central African rift system at the beginning of the Miocene, and lacustrine deposition in many half-grabens, which alternate polarity along strike of the rift (Rosendahl, 1987), has been fairly continuous throughout the late Cenozoic. Lake distribution and morphology vary in the eastern and western rift arms of east Africa (Yuretich, 1982). Extensive volcanism in the eastern (Gregory) rift has produced small, shallow, closed basins with saline, alkaline lakes that occupy subbasins over parts of the main graben. In contrast, western-arm rift lakes are much larger and deeper, receive high runoff, and cover most of the rift floor. Most are long lived and have surface outlets with stratified waters low in dissolved solids.

In the east Africa rift valleys are the most well known examples of tectonically controlled lakes (Rosendahl, 1987), although climate and volcanism also significantly influences lake character. At least 2400 m of Miocene to mid-Pleistocene Kaiso-Kisegi lacustrine beds accumulated in Lake Albert (McConnell, 1972), perhaps the oldest east African lake. Oil seeps have been reported there for many years, and some unsuccessful petroleum exploration was done in the 1950s. A similar graben-fill sequence is Lake Edward, included with the Lake Albert oil shale deposit on Figure 6. Lake Kivu, a third small lake in the western rift arm, lies south of Lake Albert and is separated from it by a chain of volcanoes that crosses the Edward-Kivu Trough (Degens et al., 1973). Still farther south, surface seeps of heavy crude have been detected on the western side of Lake Tanganyika. Katz (1988) compared geochemical source rock data for these four modern, largely clastic rift lakes with data from the lower Tertiary, carbonate-dominated Lake Uinta system. Indicators of organic enrichment, plotted on Figure 3, are quite variable and suggest less oil potential for the African lake samples where the kerogen is diluted by allochtonous, hydrogen-poor organic matter. The high oxygen indices for these samples on the modified Van Krevelen diagram (Figure 4) are caused by humic-acid concentrations and will decrease at higher burial temperatures (Degens, 1965). At the southern end of the rift system, Lake Nyasa (Lake Malawi) has undergone a depositional history similar to that of other rift-valley lakes in the western arm (Crossley, 1984).

Lacustrine Deposits of the Northern Hemisphere

Examples of lower Tertiary basins with thick continental, largely lacustrine deposits are most common in China and southeastern Asia. A high rate of subsidence in the Zhungeer basin of northwestern China continued throughout the Tertiary with the most extensive lacustrine organic shale facies accumulating toward the basin center. To the southeast a thick lower Tertiary section occupies the

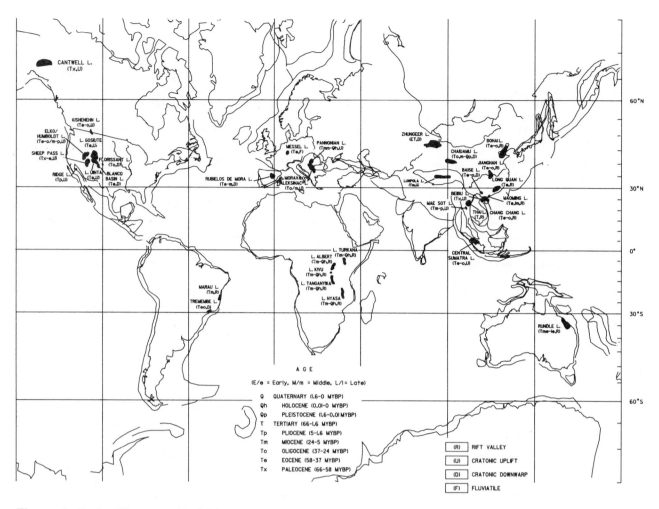

Figure 6. Early Oligocene (35 Ma) reconstruction of major Cenozoic lacustrine oil shale deposits, classified by origin.

Chaidamu basin, in which Oligocene lacustrine deposits of the Chen-chen-shan Suite include both thermally mature, organic-rich mudstone and reservoir sandstones in several fields (Lee, 1984). The Cenozoic section is as much as 12000 m thick because of extensive subsidence related to Himalayan orogenic compression. Productive Miocene and Pliocene sandstone in other Chaidamu oil fields contain hydrocarbons that were generated by deep burial and thrusting and subsequently sealed by Pliocene evaporites.

The Lunpola basin in Tibet is a third western China-type basin that was the site of numerous small, shallow Cenozoic lakes that developed at much lower elevations than today. Mio-Pliocene climate also was wetter and supported widespread forests. Part of the 4000-m-thick Paleocene to Pliocene fluviolacustrine section is thermally mature because of a high geothermal gradient in the region during much of the Tertiary (Burke and Lucas, 1985). The lacustrine Niubao Formation yields low-maturity crude oils in

places, and oil shale and reservoir rocks have been found in the Dingqing Formation (Xu, 1984).

In northern China pull-apart rifting resulting from transform motion in the Bohai basin persisted from 80 to 30 Ma. Eocene and Oligocene alluvial fan and lacustrine deposits account for 5000 m of the graben fill. Lacustrine deposition was prevalent during the middle Eocene, and evaporites and economically important oil shales developed in the somewhat drier late Eocene (Zhao and Liu, 1984). Widespread lacustrine organic shales accumulated in the early Oligocene, and the overlying deltaic sediments form excellent reservoirs. The giant Gudao oil field occurs in a small, faulted Bohai subbasin with source and reservoir rocks of Oligocene age (Chen and Wang, 1980). Hu (1985) documented similar production from the small Damintun depression of the Bohai basin.

Numerous Jurassic to lower Tertiary grabens in southeastern China are filled with shallow-water lacustrine and fluvial sediments. The lower Tertiary Jianghan basin contains at least 5500 m of interbedded mudstone and salt that have been deformed into possible diapiric traps (Wang et al., 1985). Lacustrine organic shale facies in the small Biyang depression were deposited in deep, anoxic lake waters. This oil-producing subbasin contains sediment as thick as 8000 m within an area of less than 1000 km² (Hu, 1985).

The Long Quan region is typical of the South China Platform (Figure 6) in that it contains several of the hundreds of small continental depressions or rifts that developed in the Late Cretaceous. Salt lakes within them usually became depocenters for thick organic shales. Tropical to subtropical plants were abundant and during arid times, such as much of the Eocene, extensive evaporites were deposited with the shale sequences. In the Long Quan basin an Eocene section more than 2300 m thick includes significant oil shales.

The coastal Maoming rift basin of south China also contains up to 150 m of Eocene lacustrine clastics, organic shale, and lignite deposited in an area dominated by tropical rain forests and coal swamps. Biomarker studies of these rocks (Brassell et al., 1988) verify that peats were overlain by lacustrine deposits rich in dinoflagellates and by a horizon with evidence of botryococcoidal algal blooms. In a large late Miocene lake that occupied this basin, more than 500 m of lacustrine strata accumulated, including oil shale, oil-bearing mudstone, and evaporites. On the west side of Leichou Peninsula, the Beibu basin contains more than 4000 m of nonmarine Upper Cretaceous and lower Tertiary rocks, as well as a younger marine section. Although this region has a high geothermal gradient, which is probably related to volcanic activity, Paleocene lacustrine organic shales are not thermally overmature but, like the Maoming deposits, have good hydrocarbon source potential.

Many characteristics in stratigraphy and petroleum geology of Eocene-Oligocene lacustrine sequences are similar in the Chang Chang basin on Hainan Island and in the small Baise basin depression (Hu, 1985). Part of a stratigraphic section in the Chang Chang basin contains a 300-m-thick interval of oil shale, clastics, dolomite lenses, and occasional thin evaporites. The Baise basin rift contains lacustrine Eocene and Oligocene oil reservoirs within a stratigraphic section up to 5000 m thick. Tar sands on its flanks were mapped more than 50 yr ago.

The Mae Sot basin in northern Thailand, typical of the area's intermontane Cenozoic basins, has been active since the late Eocene. Here Gibling et al. (1985) described high-grade lacustrine organic shales with excellent source rock potential. They were deposited in perennially stratified lakes and share many geochemical similarities with the Green River Formation in the western United States and the Rundle deposits of Australia. Other rock types in this lacustrine assemblage include marl from subaerially exposed mudflats and shoreline, deltaic, and turbidite sandstone complexes.

Williams et al. (1985) described the productive Pematang Group (Eocene-Miocene?) in the Central Sumatra basin of Indonesia. Brown Shale Formation oil shales, its major source rocks, were deposited in deep, fresh-water tropical lake systems that experienced no annual turnover during the rifting of this basin. Furthermore, about 5 to 10 m.y. before hydrocarbons were emplaced, organic acids released during thermal maturation may have caused dissolution of rock fragments, thereby enhancing reservoir porosity by 50%.

Several small Tertiary basins in northeastern Spain were filled with organic-rich sediment deposited in deep-water meromictic lakes. A typical Miocene sequence in the Rubielos de Mora basin (Anadon et al., 1988) includes Green River-type oil shale containing kerogen of both algal and macrophytic origin.

Many lacustrine oil shales in western North America accumulated during the Paleocene and Eocene. In northeastern Nevada (Figure 6) the Sheep Pass Formation consists of more than 1000 m of freshwater limestone, oil shale, and other lacustrine facies in which oil shows have been reported. The lower Sheep Pass is equivalent to the lacustrine Newark Canyon Formation (Figure 5) in the eastern Great basin (Fouch et al., 1979). At the Eagle Springs oil field in Railroad Valley, Nevada, the Sheep Pass Formation is a potential petroleum source rock with thermally mature kerogen, and in Utah it is considered to be a member of the Green River Formation.

The most thoroughly studied lacustrine system in the geologic record is the series of lake beds associated with the Green River Formation, including the underlying Wasatch Formation and overlying Bridger and Uinta Formations. These strata cover an area of more than 65000 km² (Figure 6) with the lacustrine sequence averaging about 600 m thick. Oil shale and evaporite beds typically alternate according to 23-k.y. and 100-k.y. Milankovitch cycles (Olsen, 1986). The Green River Formation was first described in 1859, and pioneering research on its paleolacustrine environment, extent, sedimentology, and oil shale was done later by W. H. Bradley (1931, 1948). Lake Uinta (Johnson, 1985) occupied the basin south of the Uinta Mountains in Utah and Colorado during much of the Paleocene and Eocene, at which time Lake Gosiute (Sklenar and Andersen, 1985) was present in the largest Green River basin of west-central Wyoming. In northeastern Utah more than 3000 m of siliciclastic and carbonate sediment accumulated during "Green River time" (Ryder et al., 1976), and Lake Uinta reached its maximum extent in the middle Eocene when the Mahogany Zone oil shale was deposited. Total generation potentials (Figure 3) for the thermally immature, carbonate-dominated Lake Uinta oil shale indicate good to excellent hydrocarbon yields from oil-prone kerogen. Green River kerogen has been used to define the Van Krevelen type I reference curve, and the high hydrogen indices of these samples are apparent on Figure 4.

The depositional environments and extent of anoxic conditions for the Green River Formation, as well as origin of its oil shale and dark, organic-rich dolomite, have long been debated. In the playa-lake model (Surdam and Wolfbauer, 1975; Eugster and Hardie, 1975) a complex system of organic shale, dolomite, and trona accumulated in shallow lakes surrounded by large playa flats. An alternative model of the perennial stratification (Desborough, 1978; Boyer, 1982) requires deep, biologically productive lakes in this intermontane region. Both paleoenvironments appear to have existed at various times, changing in response to climatic and, to a lesser extent, tectonic processes (Sullivan, 1985). Increased fresh-water inflow enlarged the earlier salt lakes and allowed deposition of oil shale laminites in water depths of 60 m or more in the deepest parts of the basins.

Other Eocene lacustrine deposits are found in the southern Rocky Mountains, Colorado Plateau, and Basin and Range provinces. For example, the Blanco Basin Formation in the San Juan sag of southwestern Colorado, west of the San Luis basin, contains black, carbonaceous lacustrine oil shale (Gries, 1985).

Although lacustrine oil shales are more common in the older Paleogene section of western North America, one well-known Oligocene sequence is the Florissant lake beds of central Colorado, which includes abundantly fossiliferous oil shale, water-laid tuff, and gypsum. These lacustrine sediments accumulated in a setting characterized by warm climate, thick forests, moderate relief, and active volcanoes to the west (MacGinitie, 1953).

Large lacustrine systems in western North America became even more common during the Neogene, when numerous small lake basins developed in landscapes similar to today's. In his description of 242 Precambrian to Pleistocene lake deposits in the western United States, Feth (1964) observed that 70% formed during the late Cenozoic, which suggests that many older lacustrine units may be poorly preserved or difficult to recognize. Almost one-half of the younger lakes originated in the Pleistocene, whereas 21% are Miocene in age. However, most Pleistocene lakes were short-lived and have limited petroleum source potential; none of the lacustrine oil shales discussed here are limited solely to the Pleistocene.

The Basin and Range province contains numerous Mio-Pliocene lake deposits, such as the Humboldt Formation, northern Nevada (Figure 6), with its 2000-m-thick fluviolacustrine section. These beds overlie lacustrine oil shale of the Elko Formation (Eocene-Oligocene), which is 633 m thick at its type locality (Palmer, 1984). Humboldt Lake covered more than 100000 km², and the lower member of the formation includes oil shale and fresh-water molluscan limestone (Sharp, 1939).

Although their areal extent is only 500 km², late Miocene to Pleistocene lake beds in the Ridge basin of southern California represent one of the world's thickest (nearly 5000 m) lacustrine sections. The youngest unit of the Ridge Basin Group, the Ridge Route Formation (Pliocene), includes finely laminated oil shale, fresh-water stromatolites, and other fossiliferous lacustrine sediments (Link et al., 1978). The section shows that lake water depths became shallower during the Pleistocene.

SUMMARY AND CONCLUSIONS

Oil shales include some of the most organic-rich potential petroleum source rocks in the world. Nonmarine oil shales have been deposited in both shallow and deep lakes in all types of continental sedimentary basins for more than one billion years. The largest shale-oil resources occur in Cenozoic basins of western North America, but conditions generally were more favorable for accumulation and preservation of lacustrine organic matter during the Paleozoic. Although paleolatitude, paleoclimate, and tectonic setting are all factors in the distribution of lakes and their organic-rich facies, major lakes are most commonly associated with rift, intracratonic, or intermontane basins.

Important lacustrine oil shales accumulated along continental borders in rift basins associated with the breakup of Pangea during the Mesozoic. Large, long-lived lakes were present in southeastern Africa and Australia, as well as in Newark Supergroup basins of eastern North America and the conjugate African margin. Early Cretaceous lacustrine deposition was widely associated with rifting between South America and Africa. Finally, thick late Mesozoic and Cenozoic lacustrine sequences with extensive oil shales developed throughout China, southeastern Asia, and western United States.

Lacustrine oil shales represent rich petroleum source rocks with remaining hydrocarbon-generation potential. Organic-rich shale deposits like the Green River Formation contain abundant autochthonous planktonic organic matter and are characterized by extremely high hydrocarbon yields. Worldwide many productive lacustrine basins of various geologic ages can provide analogs for successful petroleum exploration. As our understanding of geologic controls on deposition and preservation of source and reservoir facies improves, wildcat drilling strategies in many areas can become more effective.

REFERENCES

Altebaeumer, A. M., 1985, Organic geochemical investigation of selected oil shales of the Albert Formation, New Brunswick: Bulletin of Canadian Petroleum Geology, v. 33, p. 427–445.

Anadon, P., L. Cabrera, and R. Julia, 1988, Anoxic-oxic cyclical sedimentation in the Miocene Rubielos de Mora Basin, Spain, in A. J. Fleet, K. Kelts, and M. R. Talbot, eds., Lacustrine petroleum source rocks: Geological Society Special Publication 40, p. 353–367.

Anderton, R., P. H. Bridges, M. R. Leeder, and B. W. Sellwood, 1979, A dynamic stratigraphy of the British Isles—A study in crustal evolution: Boston, Allen and Unwin, 301 p.

Baker, D. G., and S. J. Derksen, 1984, Hydrocarbon potential of intracratonic rift basins (abs.): AAPG Bulletin, v. 68, p. 1199–1200.

Bertani, R. T., and A. V. Carozzi, 1985a, Lagoa Feia Formation (Lower Cretaceous), Campos Basin, Offshore Brazil—Rift valley stage lacustrine carbonate reservoirs—I: Journal of Petroleum Geology, v. 8, no. 1, p. 37–58.

Bertani, R. T., and A. V. Carozzi, 1985b, Lagoa Feia Formation (Lower Cretaceous), Campos Basin, Offshore Brazil—Rift valley stage lacustrine carbonate reservoirs—II: Journal of Petroleum Geology, v. 8, no. 2, p. 199–220.

Boyer, B. W., 1982, Green River laminites—Does the playa-lake model really invalidate the stratified-lake model?: Geology, v. 10, p. 321–324.

Bradley, W. H., 1931, Origin and microfossils of the oil shale of the Green River formation of Colorado and Utah: USGS Professional Paper 168, 58 p.

Bradley, W. H., 1948, Limnology and the Eocene lakes of the Rocky Mountain region: GSA Bulletin, v. 59, p. 635–648.

Brassell, S. C., G. Sheng, J. Fu, and G. Eglinton, 1988, Biological markers in lacustrine Chinese oil shales, in A. J. Fleet, K. Kelts, and M. R. Talbot, eds., Lacustrine petroleum source rocks: Geological Society Special Publication 40, p. 299–308.

Brice, S. E., M. D. Cochran, G. Pardo, and A. D. Edwards, 1982, Tectonics and sedimentation of the South Atlantic rift sequence: Cabinda, Angola, in J. S. Watkins, and C. L. Drake, eds., Studies in continental margin geology: AAPG Memoir 34, p. 5–18.

Brice, S. E., K. R. Kelts, and M. A. Arthur, 1980, Lower Cretaceous lacustrine source beds from early rifting phases of south Atlantic (abs.): AAPG Bulletin, v. 64, p. 680–681.

Brink, A. H., 1974, Petroleum geology of Gabon basin: AAPG Bulletin, v. 58, p. 216–235.

Burke, K., and L. Lucas, 1985, Thrusting on the Tibetan Plateau within the last 5 Ma (abs.): Eos, v. 66, p. 375.

Carter, D. C., and R. K. Pickerill, 1985, Algal swamp, marginal and shallow evaporitic lacustrine lithofacies from the late Devonian-early Carboniferous Albert Formation, southeastern New Brunswick, Canada: Maritime Sediments and Atlantic Geology, v. 21, p. 69–86.

Chang, C., 1981, Alluvial-fan coarse clastic reservoirs in Karamay, in J. F. Mason, ed., Petroleum geology in China: Tulsa, PennWell, p. 154–170.

Chen, C., 1985, Paleolacustrine processes and sequences of the Songliao basin of northeastern China (abs.): GSA Abstracts with Programs, v. 17, p. 11.

Chen, S., and P. Wang, 1980, Geology of Gudao oil field and surrounding areas, in M. T. Halbouty, ed., Giant oil and gas fields of the decade 1968-1978: AAPG Memoir 30, p. 471–486.

Clemmey, H., 1978, A Proterozoic lacustrine interlude from the Zambian Copper belt, in A. Matter and M. E. Tucker, eds., Modern and ancient lake sediments: International Association of Sedimentologists Special Publication 2, p. 259–278.

Cornet, B., 1985, Structural styles and tectonic implications of Richmond-Taylorsville rift system, eastern Virginia (abs.): AAPG Bulletin, v. 69, p. 1434–1435.

Cosgrove, J. L., and W. G. Mogg, 1985, Recent exploration and hydrocarbon potential of the Roma Shelf, Queensland: Australian Petroleum Exploration Association Journal, v. 25, pt. 1, p. 216–234.

Crick, I. H., C. J. Boreham, A. C. Cook, and T. G. Powell, 1988, Petroleum geology and geochemistry of Middle Proterozoic McArthur Basin, Northern Australia, part 2—Assessment of source rock potential: AAPG Bulletin, v. 72, p. 1495–1514.

Crossley, R., 1984, Controls on sedimentation in the Malawi rift valley, central Africa: Sedimentary Geology, v. 40, p. 33–50.

Da Silva, Z. C. C., and C. Cornford, 1985, The kerogen type, depositional environment and maturity, of the Irati Shale, Upper Permian of Parana Basin, southern Brazil: Organic Geochemistry, v. 8, p. 399–411.

Davies, G. R., and W. W. Nassichuk, 1988, An Early Carboniferous (Visean) lacustrine oil in Canadian arctic archipelago: AAPG Bulletin, v. 72, p. 8–20.

DeBuyl, M., and G. Flores, 1986, The southern Mozambique Basin—The most promising hydrocarbon province offshore

East Africa, in M. T. Halbouty, ed., Future petroleum provinces of the world: AAPG Memoir 40, p. 399–425.

Degens, E. T., 1965, Geochemistry of sediments—a brief survey: Englewood Cliffs, NJ, Prentice-Hall, 342 p.

Degens, E. T., R. P. Von Herzen, H. K. Wong, W. G. Deuser, and H. W. Jannasch, 1973, Lake Kivu—Structure, chemistry, and biology of an East African rift lake: Geologische Rundschau, v. 62, p. 245–277.

Demaison, G. J., and G. T. Moore, 1980, Anoxic environments and oil source bed genesis: AAPG Bulletin, v. 64, p. 1179–1209.

Desborough, G. A., 1978, A biogenic-chemical stratified lake model for the origin of oil shale of the Green River Formation—An alternative to the playa-lake model: GSA Bulletin, v. 89, p. 961–971.

Dickas, A. B., 1986, Comparative Precambrian stratigraphy and structure along the Mid-Continent Rift: AAPG Bulletin, v. 70, p. 225–238.

Donnelly, T. H., and M. J. Jackson, 1988, Sedimentology and geochemistry of a mid-Proterozoic lacustrine unit from northern Australia: Sedimentary Geology, v. 58, p. 145–169.

Donovan, R. N., 1975, Devonian lacustrine limestones at the margin of the Orcadian Basin, Scotland: Journal of the Geological Society of London, v. 131, p. 489–510.

Duncan, A. D., and R. F. M. Hamilton, 1988, Palaeolimnology and organic geochemistry of the Middle Devonian in the Orcadian Basin, in A. J. Fleet, K. Kelts, and M. R. Talbot, eds., Lacustrine petroleum source rocks: Geological Society Special Publication 40, p. 173–201.

Duncan, D. C., and V. E. Swanson, 1965, Organic-rich shale of the United States and world land areas: USGS Circular 523, 30 p.

Eglinton, G., P. M. Scott, T. Belsky, A. L. Burlingame, and M. Calvin, 1964, Hydrocarbons of biological origin from a one-billion-year-old sediment: Science, v. 145, p. 263–264.

Estrella, G., M. Rocha Mello, P. C. Gaglianone, R. L. M. Azevedo, K. Tsubone, E. Rossetti, J. Concha, and I. M. R. A. Bruning, 1984, The Espirito Santo Basin (Brazil) source rock characterization and petroleum habitat, in G. Demaison, and R. J. Murris, eds., Petroleum geochemistry and basin evaluation: AAPG Memoir 35, p. 253–271.

Eugster, H. P., and L. A. Hardie, 1975, Sedimentation in an ancient playa-lake complex—The Wilkins Peak Member of the Green River Formation of Wyoming: GSA Bulletin, v. 86, p. 319–334.

Falkenhein, F. U. H., R. M. Barros, I. G. Da Costa, and C. Cainelli, 1984, Potiguar Basin—Geologic model and habitat of oil of a Brazilian equatorial basin (abs.): AAPG Bulletin, v. 68, p. 475.

Fannin, N. G. T., 1969, Stromatolites from the Middle Old Red Sandstones of Western Orkney: Geological Magazine, v. 106, p. 77–88.

Feth, J. H., 1964, Review and annotated bibliography of ancient lake deposits (Precambrian to Pleistocene) in the western states: USGS Bulletin 1080, 119 p.

Fleet, A. J., K. Kelts, and M. R. Talbot, eds., 1988, Lacustrine petroleum source rocks: Geological Society Special Publication 40, 391 p.

Fouch, T. D., J. H. Hanley, and R. M. Forester, 1979, Preliminary correlation of Cretaceous and Paleogene lacustrine and related nonmarine sedimentary and volcanic rocks in parts of the eastern Great Basin of Nevada and Utah, in G. W. Newman, and H. D. Goode, eds., Basin and Range Symposium and Great Basin Field Conference 1979: Rocky Mountain Association of Petroleum Geologists, and Utah Geological Association, p. 305–312.

Froelich, A. J., and G. R. Robinson, Jr., eds., 1988, Studies of the early Mesozoic basins of the eastern United States: USGS Bulletin 1776, 423 p.

Ghignone, J. I., and G. de Andrade, 1970, General geology and major oil fields of Reconcavo Basin, Brazil, in M. T. Halbouty, ed., Geology of giant petroleum fields: AAPG Memoir 14, p. 337–358.

Gibling, M. R., C. Tantisukrit, W. Uttamo, T. Thanasuthipitak, and M. Haraluck, 1985, Oil shale sedimentology and geochemistry in Cenozoic Mae Sot Basin, Thailand: AAPG Bulletin, v. 69, p. 767–780.

Gierlowski-Kordesch, E., ed., (in preparation), Global geological record of lake basins: Cambridge University Press.

Gore, P. J. W., 1986, Depositional framework of a Triassic rift basin: the Durham and Sanford sub-basins of the Deep River basin, North Carolina, *in* D. A. Textoris, ed., SEPM field guidebooks, southeastern United States: SEPM 3d Annual Midyear Meeting, Raleigh, N. C., p. 55-115.

Gore, P. J. W., 1988, Lacustrine sequences in an early Mesozoic rift basin: Culpeper Basin, Virginia, USA, *in* A. J. Fleet, K. Kelts, and M. R. Talbot, eds., Lacustrine petroleum source rocks: Geological Society Special Publication 40, p. 247-278.

Greiner, H. R., 1974, The Albert Formation of New Brunswick—A Paleozoic lacustrine model: Geologische Rundschau, v. 63, p. 1102-1113.

Gries, R. R., 1985, San Juan sag—Cretaceous rocks in a volcanic-covered basin, south central Colorado: Mountain Geologist, v. 22, p. 167-179.

Hall, P. B., and A. G. Douglas, 1983, The distribution of cyclic alkanes in two lacustrine deposits, *in* M. Bjoroy, and others, eds., Advances in organic geochemistry 1981: New York, John Wiley, p. 576-587.

Hoering, T. C., 1976, Molecular fossils from the Precambrian Nonesuch Shale: Carnegie Institution of Washington Year Book 75, p. 806-813.

Howie, R. D., and L. M. Cumming, 1963, Basement features of the Canadian Appalachians: Geological Survey of Canada Bulletin 89.

Hu, Chaoyuan, 1985, Geologic characteristics and oil exploration of small depressions in eastern China: Geology, v. 13, p. 303-306.

Hyde, R. S., 1984, Geologic history of the Carboniferous Deer Lake Basin, west-central Newfoundland, Canada, *in* H. H. J. Geodsetzer, ed., Atlantic Coast basins: 9th International Congress on Carboniferous Stratigraphy and Geology, v. 3, p. 85-104.

Hyde, R. S., and M. J. Ware, 1985, Fluvial-lacustrine cycles in the Lower Carboniferous Rocky Brook Formation, western Newfoundland, Canada (abs.): GSA Abstracts with Programs, v. 17, p. 26.

Imbus, S. W., M. H. Engel, and R. D. Elmore, Sedimentology and organic geochemistry of Middle Proterozoic Nonesuch Formation—Hydrocarbon source rock assessment of a lacustrine rift deposit, this volume.

Jackson, M. J., T. G. Powell, R. E. Summons, and I. P. Sweet, 1986, Hydrocarbon shows and petroleum source rocks in sediments as old as 1.7×10^9 years: Nature, v. 322, p. 727-729.

Jackson, M. J., I. P. Sweet, and T. G. Powell, 1988, Studies on petroleum geology and geochemistry of the Middle Proterozoic McArthur basin, northern Australia, part 1—Petroleum potential: Australian Petroleum Exploration Association Journal, v. 28, pt. 1, p. 283-302.

Johnson, R. C., 1985, Early Cenozoic history of the Uinta and Piceance Creek Basins, Utah and Colorado, with special reference to the development of Eocene Lake Uinta, *in* R. M. Flores, and S. S. Kaplan, eds., Cenozoic paleogeography of the west-central United States: SEPM Rocky Mountain Section, 3d Rocky Mountain Paleogeography Symposium, p. 247-276.

Kalkreuth, W., and G. Macauley, 1989, Organic petrology and Rock-Eval studies on oil shales from the Lower Carboniferous rocky Brook Formation, western Newfoundland: Bulletin of Canadian Petroleum Geology, v. 37, p. 31-42.

Kantsler, A. J., T. J. C. Prudence, A. C. Cook, and M. Zwigulis, 1984, Hydrocarbon habitat of the Cooper/Eromanga Basin, Australia, *in* G. Demaison, and R. J. Murris, eds., Petroleum geochemistry and basin evaluation: AAPG Memoir 35, p. 373-390.

Katz, B. J., 1988, Clastic and carbonate lacustrine systems—An organic geochemical comparison (Green River Formation and East African lake sediments), *in* A. J. Fleet, K. Kelts, and M. R. Talbot, eds., Lacustrine petroleum source rocks: Geological Society Special Publication 40, p. 81-90.

Katz, B. J., C. R. Robison, T. Jorjorian, and F. D. Foley, 1988, The level of organic maturity within the Newark basin and its associated implications, *in* W. Manspeizer, ed., Triassic-Jurassic rifting, continental breakup and the origin of the Atlantic Ocean and passive margins: New York, Elsevier, v. 2, p. 693-696.

Kelly, W. C., and G. K. Nishioka, 1985, Precambrian oil inclusions in late veins and the role of hydrocarbons in copper mineralization at White Pine, Michigan: Geology, v. 13, p. 334-337.

Kelts, K., 1988, Environments of deposition of lacustrine petroleum source rocks—An introduction, *in* A. J. Fleet, K. Kelts, and M. R. Talbot, eds., Lacustrine petroleum source rocks: Geological Society Special Publication 40, p. 3-26.

Klitzsch, E. H., and C. H. Squyres, 1990, Paleozoic and Mesozoic geological history of northeastern Africa based upon new interpretation of Nubian strata: AAPG Bulletin, v. 74, p. 1203-1211.

Lavering, I. H., V. L. Passmore, and I. M. Paton, 1986, Discovery and exploitation of new oilfields in the Cooper-Eromanga Basins: Australian Petroleum Exploration Association Journal, v. 26, pt. 1, p. 250-259.

Lawton, D. C., 1985, Seismic facies analysis of delta-plain coals from Camrose, Alberta, and lacustrine coals from Pictou coalfield, Nova Scotia: AAPG Bulletin, v. 69, p. 2120-2129.

Lee, K. Y., 1984, Geology of the Chaidamu Basin, Qinghai Province, northwest China: USGS Open-File Report 84-413, 39 p.

Lee, K. Y., 1985a, Geology of the petroleum and coal deposits in the Junggar (Zhungaer) Basin, Xinjiang Uygur Zizhiqu, Northwest China: USGS Open-File Report 85-230, 53 p.

Lee, K. Y., 1985b, Geology of the Tarim Basin with special emphasis on petroleum deposits, Xinjiang Uygur Zizhiqu, Northwest China: USGS Open-File Report 85-616, 55 p.

Link, M. H., R. H. Osborne, and S. M. Awramik, 1978, Lacustrine stromatolites and associated sediments of the Pliocene Ridge Route Formation, Ridge basin, California: Journal of Sedimentary Petrology, v. 48, p. 143-158.

Loftus, G. W. F., and J. T. Greensmith, 1988, The lacustrine Burdiehouse Limestone Formation—A key to the deposition of the Dinantian Oil Shales of Scotland, *in* A. J. Fleet, K. Kelts, and M. R. Talbot, eds., Lacustrine petroleum source rocks: Geological Society Special Publication 40, p. 219-234.

Ma, Li, 1985, Subtle oil pools in Xingshuang delta, Songliao basin: AAPG Bulletin, v. 69, p. 1123-1132.

Macauley, G., F. D. Ball, and T. G. Powell, 1984, A review of the Carboniferous Albert Formation oil shales, New Brunswick: Bulletin of Canadian Petroleum Geology, v. 32, p. 27-37.

MacGinitie, H. D., 1953, Fossil plants of the Florissant beds, Colorado: Carnegie Institute Publication 599, 198 p.

Manspeizer, W., ed., 1988, Triassic-Jurassic rifting, continental breakup and the origin of the Atlantic Ocean and passive margins, 2 v.: New York, Elsevier, 998 p.

Marshall, J. E. A., J. F. Brown, and S. Hindmarsh, 1985, Hydrocarbon source rock potential of the Devonian rocks of the Orcadian Basin: Scottish Journal of Geology, v. 21, p. 301-320.

McCarn, S. T., and C. F. Mansfield, 1985, Petrographically deduced Triassic climate for the Deep River Basin, eastern Piedmont of North Carolina (abs.): GSA Abstracts with Programs, v. 17, p. 657.

McConnell, R. B., 1972, Geological development of the rift system of eastern Africa: GSA Bulletin, v. 83, p. 2549-2572.

McKirdy, D. M., R. E. Cox, and J. G. G. Morton, 1988, Biological marker, isotopic and geological studies of lacustrine crude oils in the western Otway Basin, South Australia (abs.), *in* A. J. Fleet, K. Kelts, and M. R. Talbot, eds., Lacustrine petroleum source rocks: Geological Society Special Publication 40, p. 327.

McKirdy, D. M., R. E. Cox, J. K. Volkman, and V. J. Howell, 1986, Botryococcane in a new class of Australian non-marine crude oils: Nature, v. 320, p. 57-59.

McKirdy, D. M., and J. G. G. Morton, 1985, Lacustrine crude oils in south Australia—Biotic and palaeoenvironmental inferences from petroleum geochemistry (abs.): Geological Society Lacustrine Petroleum Source Rocks Conference, Programme and Abstracts.

Meissner, F. F., J. Woodward, and J. L. Clayton, 1984, Stratigraphic relationships and distribution of source rocks in the greater Rocky Mountain region, *in* J. Woodward, F. F. Meissner, and J. L. Clayton, eds., Hydrocarbon source rocks of the greater Rocky Mountain region: Rocky Mountain Association of Geologists, p. 1-34.

Mesner, J. C., and L. C. P. Wooldridge, 1964, Maranhao Paleozoic basin and Cretaceous coastal basins, north Brazil: AAPG Bulletin, v. 48, p. 1475-1512.

Mossman, D. J., J. F. Macey, and P. D. Lemmon, 1987, Diagenesis in the lacustrine facies of the Albert Formation, New Brunswick, Canada—A geochemical evaluation: Bulletin of Canadian Petroleum Geology, v. 35, p. 239-250.

Smith

Muir, M. D., 1983, Depositional environments of host rocks to northern Australian lead-zinc deposits, with special reference to McArthur River, in D. F. Sangster, ed., Sediment-hosted stratiform lead-zinc deposits: Mineralogical Association of Canada Short Course Handbook, v. 8, p. 141–174.

Mukherjee, B. C., and P. K. S. Guha, 1984, Biogenic sedimentary structures from the fluviatile Gondwana basin, Deogarh coalfield, Santhal Parganas District, Bihar, India, in E. S. Belt, and R. W. Macqueen, eds., Sedimentology and geochemistry: 9th International Congress on Carboniferous Stratigraphy and Geology, v. 3, p. 517–522.

Ojeda, H. A. O., 1982, Structural framework, stratigraphy, and evolution of Brazilian marginal basins: AAPG Bulletin, v. 66, p. 732–749.

Olsen, P. E., 1985, Significance of great lateral extent of thin units in Newark Supergroup (lower Mesozoic, eastern North America) (abs.): AAPG Bulletin, v. 69, p. 1444.

Olsen, P. E., 1986, A 40-million-year lake record of Early Mesozoic orbital climatic forcing: Science, v. 234, p. 842–848.

Olsen, P. E., R. W. Schlische, and P. J. W. Gore, 1989, Tectonic, depositional, and paleoecological history of Early Mesozoic rift basins, eastern North America: 28th International Geological Congress, Field Trip Guidebook T351, 174 p.

Palacas, J. G., and M. W. Reynolds, 1989, Preliminary petroleum source rock assessment of Upper Proterozoic Chuar Group, Grand Canyon, Arizona (abs.): AAPG Bulletin, v. 73, p. 397.

Palmer, S. E., 1984, Hydrocarbon source potential of organic facies of the lacustrine Elko Formation (Eocene/Oligocene), Northeast Nevada, in J. Woodward, F. F. Meissner, and J. L. Clayton, eds., Hydrocarbon source rocks of the greater Rocky Mountain region: Rocky Mountain Association of Geologists, p. 491–511.

Parnell, J., 1985, Hydrocarbon source rocks, reservoir rocks and migration in the Orcadian Basin: Scottish Journal of Geology, v. 21, p. 321–335.

Parnell, J., 1988a, Significance of lacustrine cherts for the environment of source rock deposition in the Orcadian basin, Scotland, in A. J. Fleet, K. Kelts, and M. R. Talbot, eds., Lacustrine petroleum source rocks: Geological Society Special Publication 40, p. 205–217.

Parnell, J., 1988b, Lacustrine petroleum source rocks in the Dinantian Oil Shale Group, Scotland—A review, in A. J. Fleet, K. Kelts, and M. R. Talbot, eds., Lacustrine petroleum source rocks: Geological Society Special Publication 40, p. 235–246.

Picard, M. D., and L. R. High, Jr., 1972, Criteria for recognizing lacustrine rocks, in J. K. Rigby, and W. K. Hamblin, eds., Recognition of ancient sedimentary environments: SEPM Special Publication 16, p. 108–145.

Ponte, F. C., J. D. R. Fonseca, and R. G. Morales, 1977, Petroleum geology of eastern Brazilian continental margin: AAPG Bulletin, v. 61, p. 1470–1482.

Powell, T. G., 1986, Petroleum geochemistry and depositional setting of lacustrine source rocks: Marine and Petroleum Geology, v. 3, p. 200–219.

Prasad, G., A. Lejal-Nical, and N. Vaudois-Mieja, 1986, A Tertiary age for Upper Nubian Sandstone Formation, central Sudan: AAPG Bulletin, v. 70, p. 138–142.

Pratt, L. M., A. K. Vuletich, and C. A. Shaw, 1986, Preliminary results of organic geochemical and stable isotope analyses of Newark Supergroup rocks in the Hartford and Newark basins, eastern U.S.: USGS Open-File Report 86–284, 29 p.

Robbins, E. I., G. P. Wilkes, and D. A. Textoris, 1988, Coal deposits of the Newark rift system, in W. Manspeizer, ed., Triassic-Jurassic rifting, continental breakup and the origin of the Atlantic Ocean and passive margins: New York, Elsevier, v. 2, p. 649–682.

Rosendahl, B. R., 1987, Architecture of continental rifts with special reference to East Africa: Annual Review of Earth and Planetary Sciences, v. 15, p. 445–503.

Rosendahl, B. R., D. J. Reynolds, P. M. Lorber, C. F. Burgess, J. McGill, D. Scott, J. J. Lambiase, and S. J. Derksen, 1986, Structural expressions of rifting— Lessons from Lake Tanganyika, Africa, in L. E. Frostick, R. W. Renaut, I. Reid, and J. J. Tiercelin, eds., Sedimentation in African rifts: Geological Society Special Publication 25, p. 29–43.

Ryder, R. T., T. D. Fouch, and J. H. Elison, 1976, Early Tertiary sedimentation in the western Uinta basin, Utah: GSA Bulletin, v. 87, p. 496–512.

Schafer, A., and K. R. G. Stapf, 1978, Permian Saar-Nahe Basin and Recent Lake Constance (Germany)—Two environments

of lacustrine algal carbonates, in A. Matter, and M. E. Tucker, eds., Modern and ancient lake sediments: International Association of Sedimentologists Special Publication 2, p. 83–107.

Schull, T. J., 1984, Oil exploration in nonmarine rift basins of interior Sudan (abs.): AAPG Bulletin, v. 68, p. 526.

Sharp, R. P., 1939, The Miocene Humboldt Formation in northeastern Nevada: Journal of Geology, v. 47, p. 133–160.

Sklenar, S. E., and D. W. Andersen, 1985, Origin and early evolution of an Eocene lake system within the Washakie basin of southwestern Wyoming, in R. M. Flores, and S. S. Kaplan, eds., Cenozoic paleogeography of the west-central United States: SEPM Rocky Mountain Section, 3d Rocky Mountain Paleogeography Symposium, p. 231–245.

Smith, M. A., 1989, Lacustrine oil shales—Their controlling processes and distribution through geologic time (abs.): 28th International Geological Congress, Abstracts, v. 3, p. 140.

Smith, M. A., and C. R. Robison, 1988, Early Mesozoic lacustrine petroleum source rocks in the Culpeper basin, Virginia, in W. Manspeizer, ed., Triassic-Jurassic rifting, continental breakup and the origin of the Atlantic Ocean and passive margins: New York, Elsevier, v. 2, p. 697–709.

Smith, W. D., and M. R. Gibling, 1987, Oil shale composition related to depositional setting—a case study from the Albert Formation, New Brunswick, Canada: Bulletin of Canadian Petroleum Geology, v. 35, p. 469–487.

Smyth, M., 1983, Nature of source material for hydrocarbons in Cooper basin, Australia: AAPG Bulletin, v. 67, p. 1422–1426.

Sullivan, R., 1985, Origin of lacustrine rocks of Wilkins Peak Member, Wyoming: AAPG Bulletin, v. 69, p. 913–922.

Summons, R. E., T. G. Powell, and C. J. Boreham, 1988, Petroleum geology and geochemistry of the Middle Proterozoic McArthur basin, northern Australia, part 3—Composition of extractable hydrocarbons: Geochimica et Cosmochimica Acta, v. 52, p. 1747–1763.

Surdam, R. C., and C. A. Wolfbauer, 1975, Green River Formation, Wyoming—A playa-lake complex: GSA Bulletin, v. 86, p. 335–345.

Surlyk, F., J. M. Hurst, S. Piasecki, F. Rolle, P. A. Scholle, L. Stemmerik, and E. Thomsen, 1986, The Permian of the western margin of the Greenland Sea—A future exploration target, in M. T. Halbouty, ed., Future petroleum provinces of the world: AAPG Memoir 40, p. 629–659.

Szatmari, P., E. Milani, M. Lana, J. Conceicao, and A. Lobo, 1985, How South Atlantic rifting affects Brazilian oil reserves distribution: Oil and Gas Journal, v. 83, no. 2, p. 107–113.

Talbot, M. R., 1988, The origins of lacustrine oil source rocks— Evidence from the lakes of tropical Africa, in A. J. Fleet, K. Kelts, and M. R. Talbot, eds., Lacustrine petroleum source rocks: Geological Society Special Publication 40, p. 29–43.

Talbot, M. R., and K. Kelts, eds., 1989, The Phanerozoic record of lacustrine basins and their environmental signals: Palaeogeography, Palaeoclimatology, Palaeoecology, v. 70, p. 1–304.

Tanner, W. F., 1970, Triassic-Jurassic lakes in New Mexico: Mountain Geologist, v. 7, p. 281–289.

Taylor, O. J., compiler, 1987, Oil shale, water resources, and valuable minerals of the Piceance basin, Colorado—The challenge and choices of development: USGS Professional Paper 1310, 143 p.

Tissot, B. P., and D. H. Welte, 1984, Petroleum formation and occurrence, 2d ed.: New York, Springer-Verlag, 699 p.

Ulmishek, G., 1986, Petroleum basins of western China (abs.): AAPG Bulletin, v. 70, p. 657.

Veevers, J. J., and D. Cotterill, 1976, Western margin of Australia— A Mesozoic analog of the East African rift system: Geology, v. 4, p. 713–717.

Wang, X.-P., Q. Fei, and J.-H. Zhang, 1985, Cenozoic diapiric traps in eastern China: AAPG Bulletin, v. 69, p. 2098–2109.

White, A. H., and B. C. Youngs, 1980, Cambrian alkali playa-lacustrine sequence in the northeastern Officer Basin, South Australia: Journal of Sedimentary Petrology, v. 50, p. 1279–1286.

Williams, H. H., P. A. Kelley, J. S. Janks, and R. M. Christensen, 1985, The Paleogene rift basin source rocks of central Sumatra, in The past, the present, the future: Indonesian Petroleum Association 14th Annual Convention, Proceedings, v. 2, p. 57–90.

Williamson, P. E., C. J. Pigram, J. B. Colwell, A. S. Scherl, K. L. Lockwood, and J. C. Branson, 1985, Pre-Eocene stratigraphy, structure, and petroleum potential of the Bass Basin: Australian Petroleum Exploration Association Journal, v. 25, pt. 1, p. 362–381.

Womer, M. B., 1986, Hydrocarbon occurrence and diagenetic history within Proterozoic sediments, McArthur River area, Northern Territory, Australia: Australian Petroleum Exploration Association Journal, v. 26, pt. 1, p. 363–374.

Wopfner, H., and T. Kreuser, 1986, Evidence for late Palaeozoic glaciation in southern Tanzania: Palaeogeography, Palaeoclimatology, Palaeoecology, v. 56, p. 259–275.

Xu, Z., 1984, Tertiary System and its petroleum potential in the Lunpola basin, Xizang (Tibet): USGS Open-File Report 84-420, 5 p.

Yuretich, R. F., 1982, Possible influences upon lake development in the East African rift valleys: Journal of Geology, v. 90, p. 329–337.

Zhao, Z., and M. Liu, 1984, Facies model of the sublake-fan and its application to oil and gas exploration: Journal of the East China Petroleum Institute, v. 8, p. 323–334.

Ziegler, A. M., D. B. Rowley, A. L. Lottes, D. L. Sahagian, M. L. Hulver, and T. C. Gierlowski, 1985, Paleogeographic interpretation, with an example from the mid-Cretaceous: Annual Review of Earth and Planetary Sciences, v. 13, p. 385–425.

Ziegler, D. G., 1983, Hydrocarbon potential of the Newark rift system—Eastern North America: Northeastern Geology, v. 5, p. 200–208.

Ziegler, D. G., and B. Cornet, 1985, Newark rift system—A potentially prolific hydrocarbon province (abs.): AAPG Bulletin, v. 69, p. 1452.

Controls on Distribution of Lacustrine Source Rocks through Time and Space

Barry J. Katz
Texaco E&P Technology Division
Houston, Texas, U.S.A.

Exploration in many nonmarine sequences requires an understanding of what conditions permit the development of lacustrine hydrocarbon source rocks. Although quantitative predictions are not yet possible, qualitative assessments of the probability of source presence may be made.

To establish the presence of a lacustrine water body requires an understanding of the distribution of topographic depressions and paleoclimatic conditions. The lacustrine water body needs to be areally significant and long-lived to permit the development of substantial volumes of organic-rich rocks. This typically means that "commercial" volumes of lacustrine source rocks can only develop in lakes of tectonic origin.

Paleoclimate and paleogeography not only play major roles in controlling distribution of lake bodies but also influence water chemistry. Saline lakes develop when evaporation exceeds precipitation and during geologic episodes of maximum continentality. Fresh-water lakes develop when precipitation exceeds evaporation and along continental margins, even during times of high continentality. Water chemistry controls the nature and level of organic productivity and influences preservation by altering water density and oxygen solubility and determining availability of other chemical oxidizing agents (e.g., sulfates and nitrates).

The most favorable conditions for development of hydrocarbon sources occur in lakes of moderate water depth (50–400 m) at low latitudes and altitudes. Such lakes typically exhibit elevated levels of organic productivity and preservation.

INTRODUCTION

Exploration strategies have largely invoked organic-rich marine shales as the primary source for commercial hydrocarbons. In recent years, however, as source rock attributes have become better defined and more geochemical data have become available, the significance of lacustrine source rocks has become recognized (Figure 1). These sources may account for much of a region's reserve base (Figure 2). Individual accumulations may be quite substantial. Chen (1980) reported that a Jurassic-Cretaceous lacustrine shale sequence sourced 14 billion bbl (2.27 billion m³) of recoverable oil in the Sungliao basin, China.

Lacustrine source rocks exhibit the same geochemical characteristics with respect to source criteria as do marine sediments. They contain above-average organic carbon (TOC >1.0 wt. %; Bissada, 1982) and if thermally immature yield, upon pyrolysis, above-average quantities of hydrocarbons ($S_1 + S_2 > 2.5$ mg HC/g rock; Bissada, 1982). The kerogens contain elevated levels of organically bound hydrogen (hydrogen index >400 mg HC/g TOC and atomic H/C ratios >1.2). The principal difference between lacustrine and marine petroleum source rocks is the character of the generated crude oil. Lacustrine source rocks typically yield a high-wax crude (i.e., large proportion of C_{22+}), whereas marine kerogens, unless strongly influenced by terrestrial input, yield nonwaxy products (Figure 3).

Lacustrine sedimentary rocks, however, comprise only a relatively minor part of the preserved stratigraphic record (Picard and High, 1981), and of that, only a small percentage is of hydrocarbon-source quality. This suggests that such rocks form and are preserved under special conditions. This paper focuses on these conditions by analyzing three controlling factors—(1) distribution of lakes through time and space, (2) lacustrine primary productivity, and (3) organic preservation in lacustrine settings. In light of these general considerations, the paper then examines the source potential of four east African lakes.

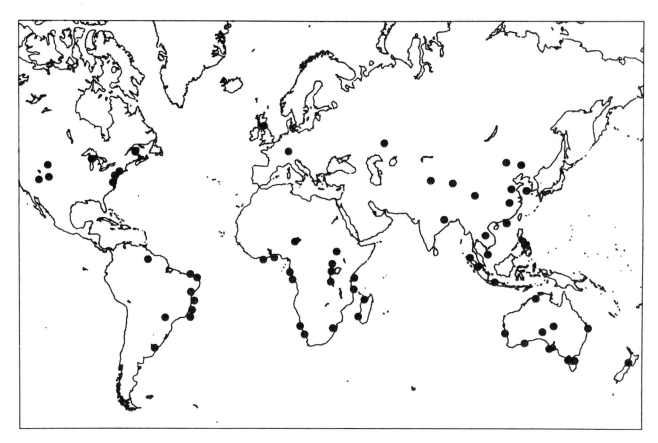

Figure 1. Distribution of known lacustrine source rock sequences.

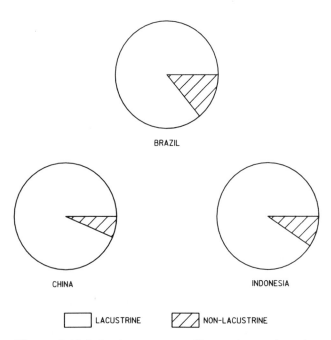

BRAZIL

CHINA

INDONESIA

☐ LACUSTRINE   ▨ NON-LACUSTRINE

Figure 2. Relative importance of lacustrine and marine source rocks in Brazil (85% lacustrine), China (95% lacustrine), and Indonesia (90% lacustrine) (after Mello and Maxwell, this volume; Halbouty, 1980; Katz and Kahle, 1988).

CONTROLS ON LACUSTRINE WATER BODIES

A lacustrine water body is determined by both the presence of a topographic depression and the availability of water to fill it. Of the eleven principal genetic lake types described by Hutchinson (1957)—anthropogenic, eolian, fluvial, glacial, landslide (or debris flow), meteoritic, organic, shoreline, solution, tectonic, volcanic—only tectonic lakes are considered both areally extensive and temporally persistent enough for substantial volumes of sediment to be deposited.

Tectonic lakes form in both extensional and compressional settings. They may develop in rift valleys, gentle sags, foreland basins, pull-apart basins, intermontane basins, "Chinese-type" basins, and those basins formed by capture through epeirogenic movement. In general, lakes formed in rift basins, pull-apart basins, and "Chinese-type" basins have low maximum width-to-maximum depth ratios, commonly less than 50. This contrasts with high ratios, typically greater than 100 and in some cases exceeding 2000, for sag, foreland, and inter-montane basins as well as captured water bodies.

Katz

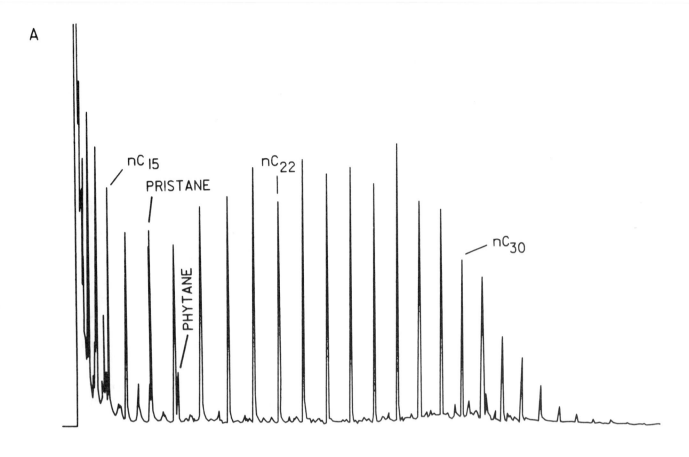

A

nC 15
PRISTANE
PHYTANE
nC 22
nC 30

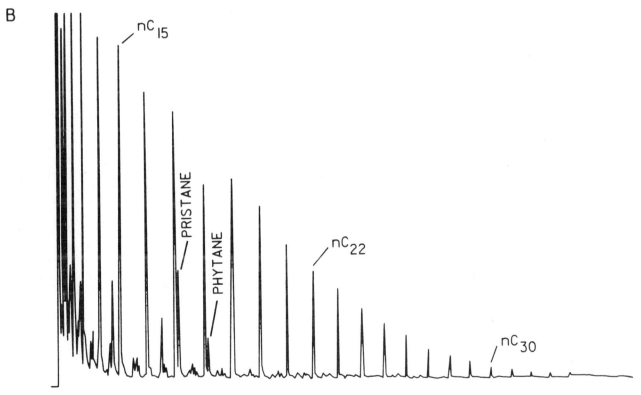

B

nC 15
PRISTANE
PHYTANE
nC 22
nC 30

Figure 3. Representative C₁₅₊-saturated-fraction chromatograms obtained on (A) lacustrine-source oil from South China Sea, and (B) marine-source oil from Gulf of Mexico.

Lacustrine Basin Exploration: Case Studies and Modern Analogs

These ratios influence organic preservation potential by affecting the wind's ability to mix the water column. Lower ratios (deeper lakes or narrower lakes) correlate with greater water-column stability.

Climate is the primary control on availability and character of water. Lake levels are a direct expression of climate, reflecting changes in relative amounts of precipitation and evaporation. Examination of the global hydrologic cycle reveals, as expected, a net transport of water from the oceans to the continents. At the present time, the greatest excess precipitation on the continents occurs within the tropical belt between 15°N and 15°S. Such conditions favor lake development.

Paleogeographic reconstructions have shown that, at times during the Phanerozoic, most of the landmass was concentrated away from the equatorial belt (Ziegler et al., 1979). Such periods, as in the Silurian, where the landmass was largely concentrated near the South Pole, would not favor development of an extensive lacustrine network.

An examination of tropical lakes indicates that even the tropics experienced periods of greater aridity. Tropical aridity appears to be associated with glacial episodes. Several possible explanations have been proposed for this association, including reduction of net water transport from the ocean to the continents because of decreased evaporation (Manabe and Hahn, 1977) or a shift in the intertropical convergence zone (Newell, 1973).

In addition to latitudinal position, the degree of continentality plays a role in establishing general rainfall patterns. During periods of high continentality (e.g., Permian), climatic conditions appear more arid, even in tropical regions, with precipitation largely restricted to continental margins. Landward transport of moisture would be controlled by prevailing wind directions and topography; consequently, lower topography correlates with greater potential for landward transport of water.

Water quality also is controlled by climate. Saline lakes, either permanent or ephemeral, are more prone to develop in semiarid regions. Lakes formed through interior drainage tend to exhibit very large fluctuations in volume in response to both seasonal (wet vs. dry) and annual variations in runoff. In many cases such lakes owe their existence solely to groundwater recharge.

Although arid conditions typically are not associated with tropical regions, certain geographic constraints, such as elevation and wind direction, may result in saline lake development within rain shadows. Near the equator in east Africa, a series of such lakes has developed within the Ethiopian rift system, where evaporation exceeds precipitation below 1700 m. Fresh-water lake bodies developed above 1700 m. The relative amounts of precipitation and evaporation here are controlled by adiabatic heating and cooling of the air and by the relationship between temperature and the air's capacity to retain water.

LACUSTRINE PRODUCTIVITY

Productivity in lacustrine systems is controlled largely by the availability of nutrients from either external or internal sources (recycling) and by light, salinity, and temperature. Lakes may be classified according to their level of productivity, that is, their trophic level (Table 1).

Although it is generally assumed that the principal control on lacustrine productivity is nutrient availability, Brylinsky and Mann (1973) have indicated that solar input variables may be more important in establishing productivity. In general their findings indicate that productivity decreased with increasing latitude in direct response to shortening of the growing season, the length of day, and the sun's position.

However, not all low-latitude lakes are necessarily productive. Lake turbidity may restrict productivity. Turbidity may be caused by the influx of suspended load from streams or by the biomass itself.

In latitudinally restricted lake populations, most productivity variation may, however, be related directly or indirectly to nutrient levels. High-elevation lakes generally exhibit lower productivity levels than low-elevation lakes because fewer nutrients drain from their higher and, hence, smaller drainage areas.

Nutrient recycling rather than external input is more important in maintaining high productivity levels. Nutrient recycling is a function of lake mixing type (i.e., frequency of water-column overturn; Table 2) and the actual rate of remineralization. The frequency of overturn is a response to both elevation and latitude. External nutrient sources are a minor factor in most lacustrine systems (Bloesch et al., 1977); exceptions may include immature lakes prior to establishment of an internal nutrient pool (Dean, 1981) and those in which no regular mixing occurs.

In lakes where no mixing occurs, a large nutrient pool accumulates in the hypolimnion and becomes effectively unavailable to primary producers. Consequently, primary productivity is low, and the lakes commonly become oligotrophic. Exceptions occur where the drainage basin is largely phosphatic or where hot springs supply large nutrient loads.

When overturn does occur in normally stratified lakes, however, the effects are a marked increase in standing phytoplankton crop and successive trophic levels (Green et al., 1976). Recycling nutrients back into the photic zone does not require mixing or overturn of the entire water mass. A permanently

Table 1. Lake productivity classification

Trophic Level	Primary Productivity Level ($gC/m^2/yr$)
Oligotrophic	<30
Mesotrophic	30-60
Eutrophic	60-200
Hypertrophic	>200

Katz

Table 2. Classification of lake water-column overturn

Mixing Type	Frequency of Overturn
Amictic	Never circulates, remains frozen
Monomictic	Once per year
Dimictic	Twice per year, generally fall and spring
Polymictic	Frequent overturn or circulation
Oligomictic	Irregular and rare circulations

stratified lake still may exhibit partial seasonal erosion of the thermocline, which returns nutrient-enriched water from the hypolimnion into the epilimnion and photic zone.

This partial recycling often is observed in deep tropical lakes such as Tanganyika and Kivu, where approximately 10% of the hypolimnion is recycled annually (Coulter, 1970), as a result of seasonal upwelling due to temperature changes and wind patterns. During periods of stable stratification these lakes commonly are oligotrophic, exhibiting high transparency and a blue color. During mixing periods (June through September for Lake Tanganyika) transparency decreases, and biomass increases to eutrophic or hypertrophic levels (Coulter, 1970; Hecky and Kling, 1981).

Lake basin geometry also is a major factor in controlling the effectiveness of lake basin overturn. Rawson's (1955) data indicated a decrease in overall productivity with increasing lake depth. Hecky and Kling (1981) also noted that lakes with low bathymetric gradients, such as Lake Victoria, more effectively recirculate nutrient loads than can high-gradient systems such as Lake Tanganyika. Recognizing this, Talling (1985) stated that denser phytoplankton blooms are associated with shallow basins or shallow epilimnions that would be more readily mixed.

The ability to recirculate water from the hypolimnion alone is insufficient to insure nutrient recycling; the biomass must be at least partly remineralized. Bacteria are effective in the remineralization process. Because of the influence of temperature, bacterial mineralization is more rapid in the tropics (Serruya and Pollingher, 1983). Remineralization occurs throughout the water column (Burns and Ross, 1971) and in the sediment (Håkanson and Jansson, 1983). Bacterial processes are active under both aerobic and anaerobic conditions and appear most intense at the sediment-water interface (Menon et al., 1971).

Salinity and water chemistry also influence primary productivity but apparently exert greater control on the nature of primary and secondary producers than on absolute levels of productivity (Pearsall, 1921). In general as salinity increases, organic diversity decreases at all trophic levels (Warren, 1986). In many saline lakes productivity may be limited to phytoplankton and photosynthetic bacteria. At elevated salinities (hypersaline conditions) unialgal blooms are common, whereas macrophytes commonly are absent or insignificant. This contrasts with some fresh-water systems wherein macrophytes may dominate, and algae are minor.

In fresh-water lakes the nature of the primary producers appears to be related to ionic speciation. The water's ionic composition is controlled by rock character of the drainage basin and by the type and quantity of suspended load. Alkali-rich waters tend to develop over igneous terranes, whereas in drainage basins underlain by sedimentary rocks, carbonate-type waters develop. In alkali-dominated systems, green algae dominate, and macrophytes are nearly absent. In carbonate-dominated systems, silica and diatoms typically are abundant, with macrophytes common. In lakes with high concentrations of dissolved organic matter, blue-green algae (cyanobacteria) dominate.

Saline and hypersaline lakes are dominated by blue-green algae, certain bacteria, and a single genus of green algae, *Dunaliella*. These organisms can tolerate salt concentrations as high as 350 ppt (Borowitzka, 1981). In saline lakes with salt concentrations below 100 ppt, diatoms and filamentous green algae also may contribute to the biomass.

ORGANIC PRESERVATION

The degree and extent of organic preservation are controlled by chemical character of the water and sedimentary columns, nature of the biosphere, and physical dynamics of the lake system itself.

Destruction of organic matter in both the water and sedimentary columns is a stepwise process. Organic carbon and hydrogen enrichment will develop where each diagenetic reaction is minimized. As a result of early diagenesis, organic hydrogen initially is reduced, followed by organic nitrogen and then carbon (Koyama et al., 1973). Consequently, as in other sedimentary environments, sediments may develop that are carbon rich but hydrogen poor and therefore principally gas-prone as a result of diagenesis.

Oxidation may occur as a result of both biologic (respiration) and abiologic processes. The extent of oxidation in either the water and sedimentary columns is controlled by free-oxygen content, exposure time (settling time, burial rate, and degree of bioturbation), biologic activity, and grain size.

The amount of oxygen in the water column is a function of temperature, salinity, and rate of renewal. Oxygen solubility in water decreases with increasing temperature (Figure 4) and salinity (Figure 5). Oxygen renewal is largely the result of lake mixing rather than molecular diffusion, except near the air-water interface (Hutchinson, 1957).

As noted previously, the frequency of lake overturn is controlled by the interrelationship between latitude and elevation. Overturn is an attempt by the lake to establish a dynamically stable water column. Although density contrasts are minor in tropical

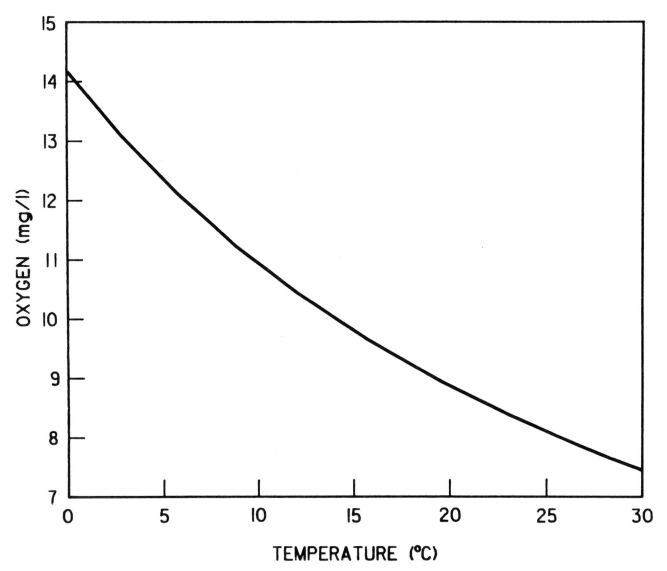

Figure 4. Oxygen solubility as a function of temperature (after Mortimer, 1956).

regions, the most stable stratification develops at low latitudes because seasonal temperature variations and, hence, surface-density variations are minimal and less than those required to displace denser bottom waters.

Wind, however, is a more active agent in both mixing and overturn. With respect to its ability to alter water-column stability and energy balance, one must consider three aspects—velocity, direction, and fetch (area available for wave generation by wind). The importance of wind in the mixing process is greatest in the equatorial region because of the absence of seasonally induced thermal mixing.

In large lakes, particularly those with high maximum width/depth ratios (e.g., Lake Victoria), wind stresses may effectively induce overturn and/ or increase the rate of mixing. In such cases the water column is well oxygenated, and organic preservation efficiencies are low. In addition, in regions with strong, frequent winds, such as those associated with Lake Turkana, the water mass may be mixed daily, although fetch is limited.

In contrast, in small lakes where fetch is limited and where the lake is topographically sheltered, mixing rates are reduced. If dissolved mineral content is high, stratification actually may develop (e.g., Green Lake, New York). High dissolved mineral loads may aid in establishing a strong density contrast between the upper and lower water masses. High-density contrasts require larger wind energy input for overturn.

In addition to potential oxygen saturation level and renewal rate, consumptive demand will affect water-column oxygen content. High organic-carbon content and high levels of biologic activity reduce the lake system's oxygen content. Furthermore, if reduced

Katz

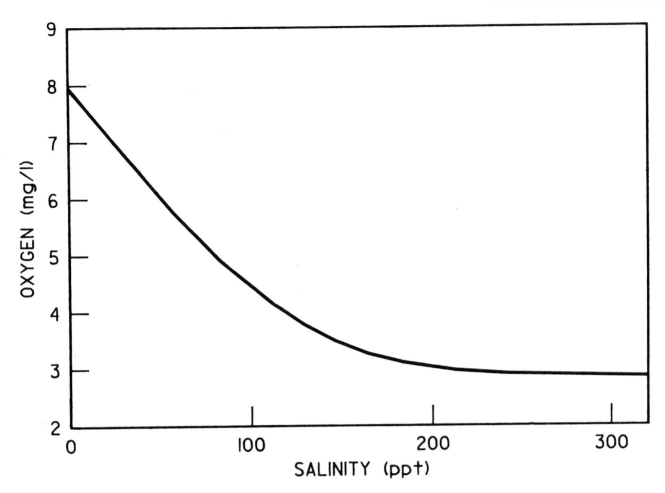

Figure 5. Oxygen solubility as a function of salinity (after Kinsman et al., 1973).

chemical species (e.g., S^{2-}, CH_4) are introduced into the water column, through either internal or external sources, their oxidation also will reduce free-oxygen content. If demand is greater than supply, anoxic conditions become established at least temporarily within the water column. Biological oxygen demand may be so great that even if complete overturn and water-column mixing occurs, oxygen depletion within the hypolimnion can be reestablished within several weeks (Green et al., 1976; Verbeke, 1957).

If oxygen is present throughout the water column, it will be transferred to the sedimentary column by incorporation of initially oxygen-bearing water into the sediment, with resupply occurring by diffusion near the sediment-water interface. Resupply is enhanced by the presence of organisms, which may resuspend sediment or act as burrowers, allowing oxygen-bearing waters to irrigate the sediment to a depth of several centimeters. This process is most effective wherever the rate of bioturbation is greater than the sedimentation rate. Oxygen transfer is enhanced in coarse-grained sediments.

The extent of oxidation is not only controlled by oxygen content but also by exposure time, the time

organic matter is associated with oxygen-bearing fluids and consuming organisms. In an oxic water column, exposure time includes the entire settling time through the water column and the initial burial period. This initial diagenetic period will vary, depending upon sedimentation rate, and has been shown to be a major factor controlling the abundance of organic hydrogen (oil-proneness). At high sedimentation rates, organic matter is rapidly removed from the sediment-water interface, where benthic organisms preferentially remove organic hydrogen, increasing preservation efficiency. In partially anoxic water columns, exposure time is limited to the time for settling through the oxic part of the column. Therefore, under such conditions sedimentation rate is not as critical in establishing preservation efficiency.

Settling or sinking time of organic detritus is a function of water depth, density contrast between the organic matter and lake waters (which is controlled by temperature and dissolved load), as well as the physical nature of the organic matter. Much autochthonous organic detritus is nearly neutrally buoyant because of phytoplankton's need to remain in the photic zone. In lacustrine settings algal bodies commonly achieve neutral buoyancy by incorporating high lipid concentrations, although other mechanisms exist. Settling rate also is reduced by the large surface area-to-volume ratios typical of most algal debris (Waples, 1983). Consequently, much of the organic detritus may be recycled within the oxic zone. Settling time may be enhanced in hypersaline-evaporitic systems or where a sharp pycnocline exists (Katz et al., 1987).

Sinking rate may increase substantially when organic detritus is incorporated into fecal pellets. The increase in pellet size, as well as compaction (limited dewatering) and addition of mineral matter, may counter the buoyancy effects of the low-density organic matter and effect sinking rates of tens to hundreds of meters per day (Porter and Robbins, 1981). Such rates would substantially reduce settling time, particularly in deeper lacustrine bodies. Furthermore, pellets are typically protected by a mucilaginous or chitinous membrane which aids in the preservation of their contents. The survival of this membrane and the potential for the pellet to remain intact decrease with increasing temperature as a result of microbial activity.

As oxygen content decreases, the mechanisms of organic decomposition change (Figure 6). As oxygen saturations approach 5% (dysaerobic conditions), nitrates become the principal oxidizing agent (Berner, 1980). As noted previously, denitrification is an important process in nutrient recycling as it is bacterially driven and occurs in both the water and sedimentary columns.

As the supply of nitrate is reduced, denitrification nears completion, and reduction of MnO_2 and ferric iron begins. These reactions, in part microbial (Nealson, 1982), typically occur early in the sediment's burial and diagenetic history, commonly within the upper tens of centimeters (Robbins and Callender, 1975) and lead to formation of a suite of authigenic minerals.

In a marine setting, following reduction of iron and manganese, sulfate would be reduced principally by bacterial activity. However, unlike marine environments wherein sulfate is abundant, sulfate may be absent or present at only trace levels in lacustrine settings; therefore, sulfate reduction is only a minor process in destruction of organic matter. However, Kelts (1988) indicated that in anoxic lacustrine environments with high sulfate concentrations, the rate of organic-matter degradation may be as high as that in a strongly oxic environment, and little organic matter may be preserved. If present, sulfate reduction occurs mostly in the upper several tens of centimeters.

The last major early diagenetic reaction that consumes sedimentary organic matter is methanogenesis (formation of methane). In lacustrine settings methanogenesis and oxidation by free oxygen probably are the primary consumers of organic matter. Methane-producing bacteria require strongly anaerobic conditions. Because of lower sulfate levels, methane production is more common in anoxic lacustrine settings than in marine settings. Significant methane accumulation and generation typically do not occur with dissolved-sulfate levels above 1.0 millimole (Whiticar et al., 1986).

Methane concentrations in lacustrine sediment rapidly increase in the upper few centimeters as a result of both production and migration. Interstitial waters may become methane saturated, leading to bubble formation and discharge into the water column, as observed in Lake Kivu (Jannasch, 1975). Unlike the reactions discussed above, methanogenesis does not depend on downward transport of chemical species into the sediment. Bacterial production of methane continues as long as organic matter is available and as long as the thermal regime is suitable. This suggests that in organic-rich sediments the limiting criterion probably is temperature. Coleman et al. (1979) suggested that this thermal regime occurs at depths of 2000–3000 m or at temperatures between 60°C and 80°C. Conditions favoring minimum biological methanogenesis would, therefore, include high sedimentation rate and/or high geothermal gradient, as expected within a rift setting.

EAST AFRICAN LAKES— CASE STUDY

Lakes within the east African rift system exhibit significant morphometric (area, depth, elevation) and chemical variability (Table 3). As a result, they provide an excellent natural laboratory for testing and examination of the processes and principles discussed above. To a large extent the only common attributes among the lakes are that they owe their existence in one way or another to rifting and that they are located in the tropics. This study will specifically examine four lakes from which organic material was available for analysis—Lakes Albert, Edward, Kivu, and Tanganyika.

All the lakes are situated within 9° of the equator (Figure 7). Water depths vary from 58 m for Lake Albert to 1470 m for Lake Tanganyika, and elevations range from 616 m for Lake Albert to 1463 m for Lake Kivu (Table 3).

Lakes Tanganyika and Kivu are permanently stratified with anoxic hypolimnions. The depth to the thermocline varies as a result of seasonal erosion of the hypolimnion (Serruya and Pollingher, 1983). In contrast, Lake Albert is nonstratified principally because of frequent and intense winds, often exceeding 50 km/hr (Serruya and Pollingher, 1983). Lake Edward experiences seasonal overturn during August and possibly a second overturn in February (Verbeke, 1957). However, because of the availability of organic matter, available oxygen is consumed

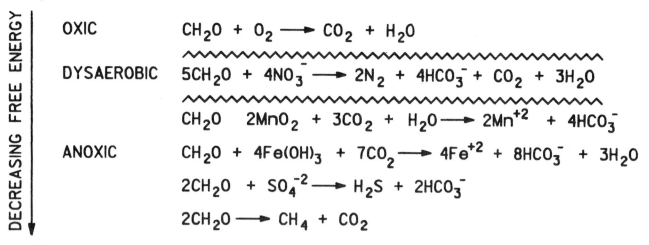

Figure 6. Organic-matter diagenetic reaction sequence.

Table 3. Morphometric and chemical variability of east African lakes

Lake	Elevation (m)	Area (km²)	Mean Depth (m)	Maximum Depth (m)	Dissolved Ion Concentration (ppm)	Dominant Chemical Species
Albert	616	6800	25	25	597	Na-K-CO₃
Edward	91	2325	34	117	789	Na-K-HCO₃
Kivu	1463	2700	240	489	1115	Na-Mg-HCO₃
Tanganyika	773	33000	~700	1470	530	Na-Mg-HCO₃

during nonmixing periods, and the water column is anoxic below 40 m.

The absolute concentrations of salts are low in these four lakes (Table 3). Lake waters are dominated by sodium, potassium, and magnesium due to drainage through a largely volcanic terrane.

Annual productivity in all four lakes may be considered eutrophic to hypertrophic. Productivity in Lakes Tanganyika and Kivu has been estimated to be approximately 370 gC/m²/yr (Degens et al., 1973; Hecky and Kling, 1981). However, considerable seasonal variation occurs as a result of water-column mixing (Hecky and Kling, 1981), and much of the productivity is associated with lake-basin margins. Productivity of Lakes Albert and Edward is estimated to be nearly three times that of the two deeper lakes (Hecky and Degens, 1973). The difference relates to loss of nutrients from the epilimnion in Lakes Tanganyika and Kivu (Hecky and Degens, 1973).

Cores from the deeper parts of Lake Kivu reveal that recent sediments typically are laminated—diatom-rich (light) vs. largely organic (dark) (Degens et al., 1973). Nonlaminated brown sapropelic units also are present within the upper part of the section.

Material recovered from Lake Tanganyika is similar to that from Lake Kivu. Sediments are finely laminated with as many as 50 layers/cm (Degens et al., 1971). The Tanganyika cores contain, however, a greater mixture of clays (kaolinite, illite, and chlorite). Turbidite deposition also appears to have been more common in the deeper parts of Lake Tanganyika, compared to either Lake Kivu or the shallow areas of Lake Tanganyika.

The deeper Lake Edward cores also were laminated, but the light layers varied in thickness, with several between 1 and 2 cm. The high organic-pigment content gave the cores a pronounced green color (Hecky and Degens, 1973). Core material from water depths shallower than 80 m was nonlaminated but homogeneous brown in color and organic rich. The absence of laminations has been attributed to seasonal turbulence, which can destroy fine structures below the permanent thermocline independently of benthic bioturbators.

Sediment recovered from Lake Albert was markedly different from that recovered from the other three lakes—nearly uniformly gray, dominated by clay, and apparently nonlaminated (Hecky and

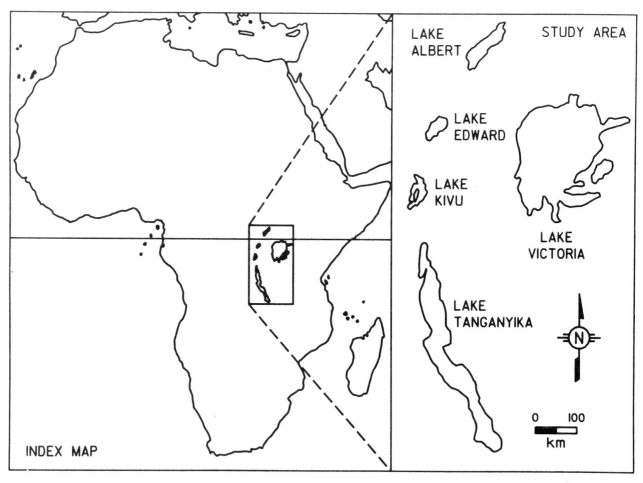

Figure 7. Locations of four rift-basin lakes in east African rift system.

Degens, 1973). Organic content, although elevated, was significantly less than that of the other lakes (Figure 8).

Organic geochemical analyses of the lake sediments indicate that organic input was a mixture of algal (largely diatom) detritus, higher land-plant debris, and natural charcoal (Katz, 1988). All samples examined exhibit above-average levels of organic enrichment relative to all fine-grained sedimentary rocks (TOC >1.0 wt. %) and range from 1.22 to 12.78 wt. % TOC (Figure 9). The lowest levels of organic enrichment were associated with the Lake Albert data, which were typically below 3.0 wt. % TOC. Similarly, most of the samples yielded above-average quantities of hydrocarbons upon pyrolysis ($S_1 + S_2$ > 2.5 mg HC/g rock), with $S_1 + S_2$ values ranging from 0.60 to 75.62 mg HC/g rock. As with the organic-carbon data, the lowest hydrocarbon yields also were associated with Lake Albert. The most elevated and uniform hydrocarbon yields were obtained from the Lake Edward samples.

The total pyrolysis yields and the kerogen character, determined through whole-rock pyrolysis (Figure 10) and elemental analysis of isolated kerogens (Figure 11), reveal significant scatter, but some trends are discernible. The lowest levels of organic-hydrogen enrichment are associated with Lake Albert sample material; the highest levels are associated with Lake Edward.

The low levels of both organic-hydrogen and organic-carbon enrichment in the Lake Albert material clearly results from the oxidized nature of the water column. Above-average enrichment levels reflect principally elevated levels of lake productivity. Higher levels of enrichment in Lake Edward are the combined result of the elevated productivity level, relative water depth (reduced settling time), and generally stratified nature of the water column, which favors hypolimnion anoxia for most of the year. The intermediate source quality of the deeper lakes (Tanganyika and Kivu) apparently reflects lower levels of productivity and possibly, therefore, a greater proportion of organic matter derived from allochthonous sources (i.e., less reactive and hydrogen-depleted).

The higher source-quality sediments developed in an anoxic lake with the highest levels of productivity. Intermediate-quality sources were deposited in those

70   Katz

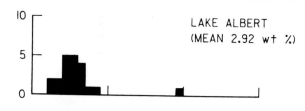

LAKE ALBERT
(MEAN 2.92 wt %)

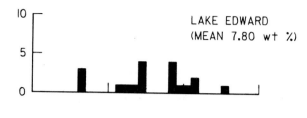

LAKE EDWARD
(MEAN 7.80 wt %)

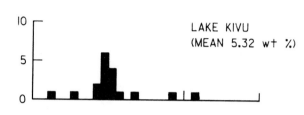

LAKE KIVU
(MEAN 5.32 wt %)

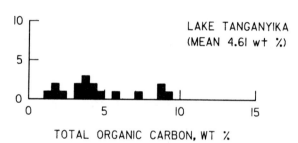

LAKE TANGANYIKA
(MEAN 4.61 wt %)

TOTAL ORGANIC CARBON, WT %

Figure 8. Comparison of organic enrichment in east African lake sediments.

erodes the hypolimnion. In shallower lakes (water depths <400 m), higher levels of productivity are maintained by regular water-column overturn.

DISCUSSION AND CONCLUSIONS

Clearly the prediction of long-lived lacustrine bodies through time and space is complex. However, several first-order principles permit an assessment of their potential presence. The first control is clearly tectonic. Long-lived basins form principally as a result of tectonic processes. Permanent fresh-water lakes tend to form at or near the equator as long as they are not on the lee side of a mountain belt, where more arid conditions prevail even in the tropics. Arid conditions also may develop at low elevations on the windward side as a result of adiabatic reduction of relative humidity. During times of high continentality, such as the Permian, fresh-water lake systems will be largely limited to continental margins. When much of the landmass was concentrated near one of the poles (Silurian), lake sequences will be of minor stratigraphic importance.

Numerical climate modeling should provide a means to address runoff and the probability of lake formation (Barron, this volume). As would be expected, lacustrine bodies associated with more arid regions and continental interiors tend to be more saline. These lakes may, in fact, be maintained solely by groundwater recharge. Because many of these lakes are ephemeral, their prediction is more complex, and their significance with respect to source rock deposition is questionable.

Except during early stages of lake development, productivity generally does not appear to be a common limiting factor in development of lacustrine source sequences. Lakes with the highest levels of productivity are found at low latitudes, which allow maximum solar energy input when turbidity levels are low. This suggests that the highest levels of primary productivity are favored where chemical weathering dominates as opposed to physical weathering (e.g., Europe is more prone to chemical weathering than Asia). A high chemical load also aids in resupplying essential nutrients to the system, although runoff alone apparently cannot supply the nutrient levels necessary for eutrophic or hypertrophic conditions except where the drainage basin is largely phosphatic.

Within latitudinal belts, lakes with large drainage areas generally can support higher productivity levels, which commonly favors development of source-quality lakes at low elevations. Chemical recycling of the internal nutrient load is critical in establishing and maintaining elevated levels of productivity. Nutrient recycling is controlled by both biologic and abiologic processes, including the

lakes whose productivity was reduced but organic preservation was high. The poorest source material was developed when productivity was high, and preservation efficiency was low.

The data from these lakes clearly show that organic preservation is a key to source rock development. Preservation potential is limited in shallow lakes principally because of the availability of free oxygen from water-column overturn. At the low latitudes of these lakes and at their low-to-intermediate elevations, frequent water-column overturn appears to be restricted to water depths less than 50 m.

Productivity, which appears to be a secondary control, reflects the ability of the nutrient pool to be returned to the photic zone. Therefore, in deeper lakes, which have developed permanent stratification, lower levels of phytoplankton productivity may be expected, and much of it may be associated with lake margins and seasonal upwelling, which partly

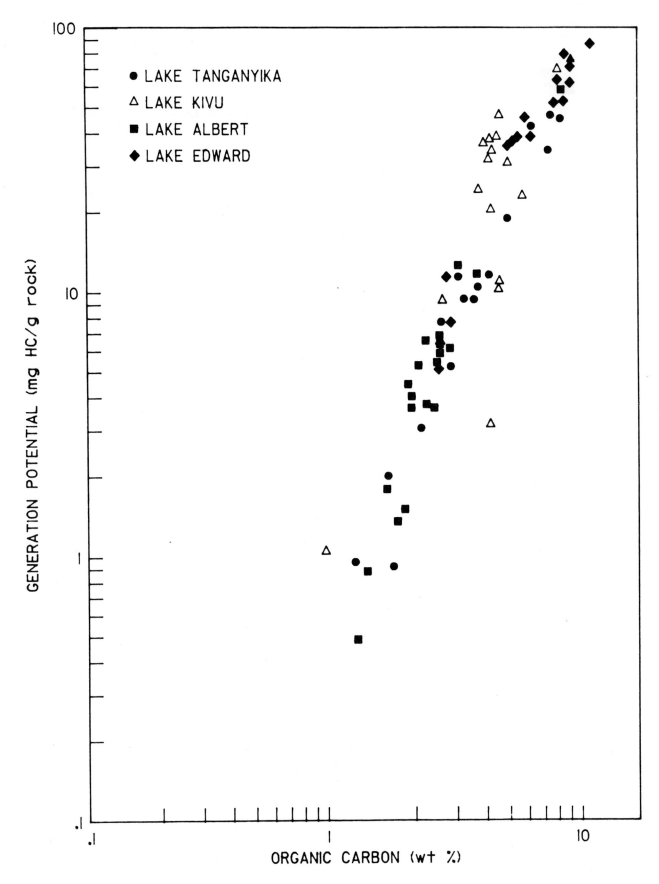

Figure 9. Total hydrocarbon generation potential vs.
organic enrichment in east African lake sediments.

Katz

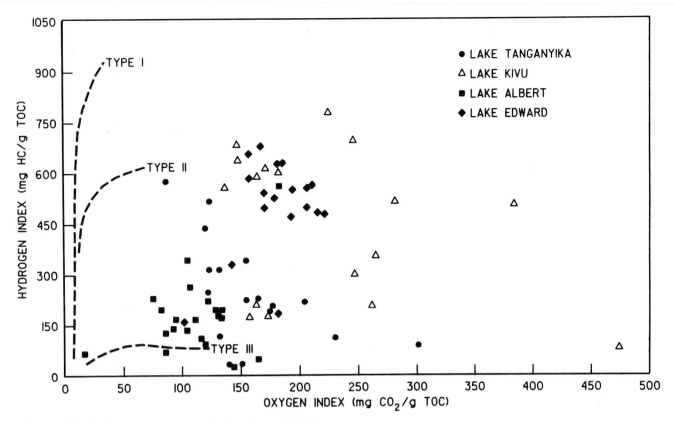

Figure 10. Hydrogen vs. oxygen indices in east African lake sediments.

frequency of overturn; however, the turnover need not be complete to impact productivity. Water chemistry and salinity are more instrumental in establishing the nature of the biomass rather than its size. For example, the green alga *Botryococcus* may dominate in fresh-water systems, while blue-green algae or the green alga *Dunaliella* may dominate in saline systems.

Organic-matter preservation may be the major limiting factor with respect to development of hydrocarbon source rocks. Because of their complexity, interrelationships among the processes controlling preservation of sedimentary organic matter cannot at this time be predicted quantitatively. They may, however, be described qualitatively. Within lacustrine settings organic-matter destruction typically results from oxidation and methanogenesis. Preservation efficiencies are, therefore, highest when conditions favoring these processes are minimized. Oxidation is minimized where both water-column oxygen content and rate of resupply are low, such as in hypersaline lakes, low-elevation tropical lakes, and lakes with small fetches relative to their total depths. Oxidation is further minimized by reducing exposure time in the oxic environment (water and/ or sedimentary columns), either by establishing an anoxic hypolimnion, by pelletizing organic matter to increase settling rate, or by high sedimentation rates. Methanogenesis, which is independent of the downward transport of chemical species into the sedimentary column, will be reduced in settings with elevated geothermal gradients or high sedimentation rates where bacterial activity is minimal. Methanogenesis also will be reduced in lacustrine settings with high sulfate contents. However, sulfate reduction consumes large quantities of organic matter and does not aid organic preservation. Consequently, preservation efficiencies are higher in sulfate-poor systems.

A case study of east African rift-basin lakes illustrates how these processes interact to influence formation of hydrocarbon source rocks. An examination of their sediments reveals that the most favorable conditions for source rock development occur where both productivity and preservation are maximized. Intermediate-quality source rocks can be expected to develop where preservation is high, but productivity is reduced (although elevated). Poor-quality hydrocarbon sources develop where preservation is low even if productivity is high. Present lake morphometries suggest that both preservation and productivity in tropical lakes are maximized at water depths between 50 and 400 m.

ACKNOWLEDGMENTS

The author wishes to thank the Woods Hole Oceanographic Institution for supplying sample

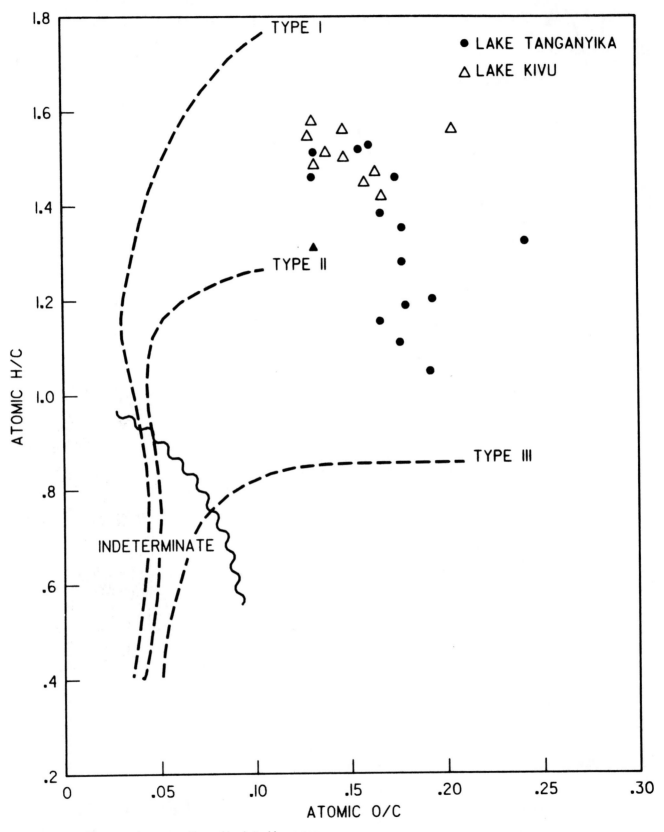

Figure 11. Elemental composition of isolated kerogens
in east African lake sediments.

material from the east African rift lakes. The author thanks Dr. Mary Hill for her assistance in preparing this manuscript and Drs. Marcio Mello and Vaughn Robison for their critical review. Analytical support was provided by the Organic Geochemical Laboratories of Texaco's E & P Technology Division. Permission for publication was granted by Texaco Inc.

REFERENCES CITED

Barron, E. J., Climate and lacustrine petroleum source prediction, this volume.

Berner, R. A., 1980, Early diagenesis—A theoretical approach: Princeton University Press, 241 p.

Bissada, K. K., 1982, Geochemical constraints on petroleum generation and migration—A review: ASCOPE '81, Proceedings, p. 69-87.

Bloesch, J., P. Stadelman, and H. Buhrer, 1977, Primary production, mineralization, and sedimentation in the euphotic zone of two Swiss lakes: Limnology and Oceanography, v. 22, p. 511-526.

Borowitzka, L. J., 1981, The microflora—Adaptions to life in extremely saline lakes: Hydrobiologa, v. 81, p. 33-46.

Brylinsky, M., and K. H. Mann, 1973, An analysis of factors governing productivity in lakes and reservoirs: Limnology and Oceanography, v. 18, p. 1-14.

Burns, N. M., and C. Ross, 1971, Nutrient relationships in the stratified eutrophic lake, in Proceedings, 14th Conference on Great Lakes Research: University of Michigan Great Lakes Research Division Publication, p. 749-760.

Chen, C. 1980, Nonmarine setting of petroleum in the Sungliao basin of northeastern China: Journal of Petroleum Geology, v. 2, p. 233-264.

Coleman, M. L., C. D. Curtis, and H. Irwin, 1979, Burial rate a key to source and reservoir potential: World Oil, v. 188, no. 4, p. 83-92.

Coulter, G. W., 1970, Hydrological changes in relation to biological production in southern Lake Tanganyika: Limnology and Oceanography, v. 15, p. 463-377.

Dean, W. E., 1981, Carbonate minerals and organic matter in sediments of modern north temperate hard-water lakes, in F. G. Ethridge, and R. M. Flores, eds., Recent and ancient nonmarine depositional environments—Models for exploration: SEPM Special Publication 31, p. 213-231.

Degens, E. T., R. P. Von Herzen, and H. W. Wong, 1971, Lake Tanganyika—Water chemistry, sediments, geological structure: Naturwissenschaften, v. 58, p. 29-241.

Degens, E. T., R. P. Von Herzen, H. W. Wong, W. G. Deuser, and H. W. Jannasch, 1973, Lake Kivu—Structure, chemistry and biology of an east African rift lake: Geologische Rundschau, v. 62, p. 245-277.

Green, J., S. A. Corbet, E. Watts, and O. B. Lan, 1976, Ecological studies on Indonesian lakes—Overturn and restratification of Ranu Lamongan: Journal of Zoology, v. 180, p. 315-353.

Håkanson, L., and M. Jansson, 1983, Principles of lake sedimentology: New York, Springer-Verlag, 316 p.

Halbouty, M. T., 1980, Methods used, and experience gained in exploration for new oil and gas in highly explored (matured) areas: AAPG Bulletin, v. 64, p. 1210-1222.

Hecky, R. E., and E. T. Degens, 1973, Late Pleistocene-Holocene chemical stratigraphy and paleolimnology of the rift valley lakes of Central Africa: Woods Hole Oceanographic Institution Technical Report 73-28, 93 p.

Hecky, R. E., and H. J. Kling, 1981, The phytoplankton and protozooplankton of the euphotic zone of Lake Tanganyika—Species composition, biomass, chlorophyll content and spatio-temporal distribution: Limnology and Oceanography, v. 26, p. 548-564.

Hutchinson, G. E., 1957, A treatise on limnology—Geography, physics and chemistry: New York, John Wiley and Sons, v. 1, 1015 p.

Jannasch, H. W., 1975, Methane oxidation in Lake Kivu (Central Africa): Limnology and Oceanography, v. 20, p. 860-864.

Katz, B. J., 1988, Clastic and carbonate lacustrine systems—An organic geochemical comparison (Green River Formation and East African lake sediments, in A. J. Fleet, K. Kelts, and M. R. Talbot, eds., Lacustrine petroleum source rocks: Geological Society Special Publication 40, p. 81-90.

Katz, B. J., K. K. Bissada, and J. W. Wood, 1987, Factors limiting potential of evaporites as hydrocarbon source rocks (abs.): AAPG Bulletin, v. 71, p. 575.

Katz, B. J., and G. M. Kahle, 1988, Basin evaluation—A supply-side approach to resource assessment: Indonesian Petroleum Association 17th Annual Convention, Proceedings, p. 135-168.

Kelts, K., 1988, Environments of deposition of lacustrine petroleum source rocks—An introduction, in A. J. Fleet, K. Kelts, and M. R. Talbot, eds., Lacustrine petroleum source rocks: Geological Society Special Publication 40, p. 3-26.

Kinsman, D. J. J., M. Boardman, and M. Borcsik, 1974, An experimental determination of the solubility of oxygen in marine brines, in A. H. Coogan, ed., Proceedings, Fourth Symposium on Salt: Northern Ohio Geological Society, v. 1, p. 325-327.

Koyama, T., M. Nikaido, T. Tomino, and H. Hayakawa, 1973, Decomposition of organic matter in lake sediments, in Proceedings, Symposium on Hydrogeochemistry and Biogeochemistry: Washington, Clarke Co., v. 2, p. 512-535.

Manabe, S., and D. G. Hahn, 1977, Simulation of the tropical climate of an ice age: Journal of Geophysical Research, v. 82, p. 3889-3911.

Mello, M. R., and J. R. Maxwell, Organic geochemical and biological marker characterization of source rocks and oils derived from lacustrine environments in Brazilian continental margin, this volume.

Menon, A. S., C. V. Marion, and A. N. Miller, 1971, Microbiological studies of oxygen depletion and nutrient regeneration in the Lake Erie central basin, in Proceedings, 14th Conference on Great Lakes Research: University of Michigan Great Lakes Research Division Publication, p. 768-780.

Mortimer, C. H., 1956, The oxygen content of air-saturated freshwaters, and aids in calculating percentage saturation, in Verhandlungen der Internationalen Vereinigung für Theoretische und Angewandte Limnologie: Stuttgart, E. Schweizbartsch Verlagsbuchhandl, v. 6, 20 p.

Nealson, K. H., 1982, Microbiological oxidation and reduction of iron, in H. D. Holland, and M. Schidlowski, eds., Mineral deposits and the evolution of the biosphere: New York, Springer-Verlag, p. 51-64.

Newell, R. E.,1973, Climate and the Galapagos Islands. Nature, v. 245, p. 91-92.

Pearsall, W. H., 1921, A suggestion as to factors influencing the distribution of free-floating vegetation: Journal of Ecology, v. 9, p. 241-253.

Picard, M. D., and L. R. High, Jr., 1981, Physical stratigraphy of ancient lacustrine deposit, in F. G. Ethridge, and R. M. Flores, eds., Recent and ancient nonmarine depositional environments—Models for exploration: SEPM Special Publication 31, p. 213-231.

Porter, K. G., and E. I. Robbins, 1981, Zooplankton fecal pellets link fossil fuel and phosphate deposits: Science, v. 212, p. 931-933.

Rawson, D. S., 1955, Morphometry as a dominant factor in the productivity of large lakes, in Verhandlungen der Internationalen Vereinigung für Theoretische und Angewandte Limnologie: Stuttgart, E. Schweizbartsch Verlagsbuchhandl, v. 12, p. 164-175.

Robbins, J. A., and E. Callender, 1975, Diagenesis of manganese in Lake Michigan sediments: American Journal of Science, v. 275, p. 512-533.

Serruya, C., and U. Pollingher, 1983, Lakes of the Warm Belt: Cambridge University Press, 569 p.

Talling, J. F., 1985, Modern phytoplankton in African lakes (abs.), in Lacustrine petroleum source rocks: Joint Meeting of IGCP Project 219 and Petroleum Group of the Geological Society, Programme and Abstract.

Verbeke, J., 1957, Recherches cologiques sur la faune des grand lacs de l'Est Congo Belge, in Exploration hydrobiologique des Lacs Kivu, Edouard et Albert (1952-54)—Résultats scientifiques: Institut Royal des Science Naturelles, v. 3, 177 p.

Waples, D. W., 1983, Reappraisal of anoxia and organic richness with emphasis on Cretaceous of North America: AAPG Bulletin, v. 67, p. 963–978.

Warren, J. K., 1986, Shallow-water evaporite environments and their source rock potential: Journal of Sedimentary Petrology, v. 56, p. 442–454.

Whiticar, M. J., E. Faber, and M. Schoell, 1986, Biogenic methane formation in marine and freshwater environments—CO_2 reduction vs. acetate fermentation-isotope evidence: Geochimica et Cosmochimica Acta, v. 50, p. 693–709.

Ziegler, A. M., C. R. Scotese, W. S. McKerrow, M. E. Johnson, and R. K. Bambach, 1979, Paleozoic paleogeography: Annual Review of Earth and Planetary Sciences, v. 7, p. 473–502.

Organic Geochemical and Biological Marker Characterization of Source Rocks and Oils Derived from Lacustrine Environments in the Brazilian Continental Margin

M. R. Mello
Petrobras/Cenpes/Divex
Ilha do Fundao
Rio de Janeiro, Brazil

J. R. Maxwell
Organic Geochemistry Unit, School of Chemistry
University Of Bristol
Bristol, England

Geochemical studies, together with paleogeographical and geological evidence, suggest that most of the organic-rich Neocomian to Aptian rift-stage succession in the Brazilian continental margin was deposited in lacustrine environments. It is possible to differentiate two lacustrine systems responsible for about 85% of Brazilian oil discovered to date—a relatively large, deep, fresh-water type of basin, ranging in age from early Neocomian to Aptian, and a shallow lower to upper Neocomian saline system.

Fresh-water systems are characterized by thick beds of dark-gray to black shale (TOC <6%). In the oils and source rocks abundant high molecular-weight n-alkanes, low S and V/Ni values, low $\delta^{13}C$ values, high Pr/Ph ratios, absence of dinosterane and C_{30} desmethyl steranes, and low concentrations of steranes and porphyrins characterize the fresh-water depositional environment.

Saline systems are composed of thick beds of calcareous black shales (TOC up to 9%). The oils and rocks, in addition to showing diagnostic evidence of a nonmarine environment, are characterized by features typical of deposition under saline conditions. These include higher V/Ni ratios, presence of β-carotane, high concentration of C_{30} $\alpha\beta$-hopane, moderately abundant gammacerane, Ts/Tm < 1, and high $\delta^{13}C$ values.

INTRODUCTION

The importance of lacustrine basins and their hydrocarbon potential has increased substantially in the last decade. Recent data from oil fields in China, Brazil, Australia, and Indonesia, and the realization that most hydrocarbons in these countries come from lacustrine source rocks, opened a new era in hydrocarbon exploration (Powell, 1986; Wang Teiguan et al., 1988; Hu Chaoyuan and Quiao Hanseug 1983; Yang Wanli et al., 1985; Philp and Gilbert, 1986; Fu Jiamo et al., 1988; McKirdy et al., 1988; Katz, this volume).

Recently many authors have shown that both geochemical evidence and biological marker distributions can provide diagnostic criteria for distinguishing and characterizing organic-rich sedimentary rocks deposited in a variety of lacustrine environments, such as fresh water, fresh-brackish water, saline, and hypersaline in China (Fu Jiamo et al., 1986; Brassel et al., 1988; Wang Tieguan et al., 1988); fresh water in Australia, Sudan, and Chad (McKirdy et al., 1986; Philp and Gilbert, 1986; Moldowan et al., 1985); and fresh water and saline water in China, United States, and Brazil (Powell, 1986; Mello, Gaglianone et al., 1988; Mello, Telnaes et al., 1988).

This study summarizes a multidisciplinary approach (geochemical, geological, paleontological) to differentiating and characterizing depositional environments of lacustrine source rocks in Brazilian marginal basins (Figure 1) and extends several earlier preliminary studies (Mello, Estrella, and Gaglianone, 1984; Mello, Soldan, and Brüning, 1984; Estrella et al., 1984; Pereira et al., 1984; Cerqueira et al., 1984; Babinski and Santos, 1987; Rodrigues et al., 1988; Mello, Gaglianone et al., 1988; Mello, Telnaes et al., 1988). In this study, 2000 rock samples recovered from lower Neocomian to Aptian sedimentary successions and 40 oil samples were analyzed. Although they cover a wide range of maturity values (0.45% to 0.90% R_o and 20° to 33° API), only sediment samples with R_o values between 0.45% and 0.75% and oils with 20° to 30° API are discussed in terms of molecular data because of maturation effects on the concentrations of biological markers (Rullkotter et al., 1984; Mello, Gaglianone et al., 1988; Mello, 1988).

GENERAL GEOLOGY

The Brazilian marginal basins are directly related to separation of the African and South American plates. They originated as new accretionary plate boundaries, but once formed they ceased to be plate boundaries and now mark the junction between oceanic and continental crust within plate interiors. The nearly 8000-km-long line of basins (Figure 1) can be classified as components of a typical divergent, mature, Atlantic-type continental margin (Ponte and Asmus, 1978; Ojeda, 1982; Estrella et al., 1984). Based on their tectonosedimentary sequence, they can be linked to a single evolutionary geological history (Figure 2), which can be subdivided into three main stages—prerift, rift, and drift (gulf proto-oceanic and oceanic phases; Asmus, 1975).

The Late Jurassic/Early Cretaceous prerift stage was associated with stretching of the continental crust and lithosphere. This phenomenon resulted in block faulting, sedimentary troughs, and localized mafic volcanism associated with thinning of the underlying crust and mantle and with upwelling of the asthenosphere, producing a thermal anomaly (Bott, 1976).

The Neocomian rift stage (Figure 2A), an evolutionary consequence of the processes above, was a direct result of overall subsidence induced by thinning of the lithosphere. Rifting generally was associated with basement-involved block-rotated faulting and intense, widespread mafic volcanism (Bott, 1976; Mohriak and Dewey, 1987). As a result, a thick succession of fine- to coarse-grained siliciclastics and carbonates was deposited in fresh-water to saline lacustrine environments (e.g., Viana et al.,1971; Bertani and Carozzi, 1985a, 1985b). In places this section overlies or is intercalated with the volcanics (mainly basalt). After rifting, tectonic activity appears to have been restricted to subsidence and basinward tilting, with initiation of gravitational sliding and local reactivation of faults (Ponte and Asmus, 1978).

Rifting ceased with the onset of sea-floor spreading. During the subsequent drifting stage progressive cooling and contraction of the underlying lithosphere induced flexural subsidence of the margin with no conspicuous faulting (Bott, 1976).

The drift stage can be subdivided into two distinct phases—gulf proto-oceanic and oceanic. The gulf proto-oceanic phase (Figure 2B) marked the first marine incursions into the coastal basins during the Aptian. The combination of tectonic quiescence, topographical barriers, and arid climate led to low clastic influx and restricted conditions appropriate for deposition of mixed carbonate and siliciclastics together with evaporites in coastal, shallow continental to marine hypersaline environments.

The oceanic phase (Figure 2C) was a consequence of increasing sea-floor spreading and continuous subsidence of the Brazilian continental margin. Differences in paleoenvironmental settings allow subdivision of this phase into three sequences:

1. In the Albian marine carbonate sequence (Figure 2C) platform and slope carbonates accumulated in a neritic to upper bathyal environment in a shallow and narrow epicontinental sea (Koutsoukos and Dias-Brito, 1987). This succession appears to correlate with tectonic quiescence, although some adiastrophic tectonism was associated with listric detached faults soling out onto the Aptian salt.

2. The Cenomanian to Campanian open-marine shelf-slope system (Figure 2D) is characterized by predominantly siliciclastic and calcareous mudstone deposition in progressively deepening basins (e.g., Koutsoukos, 1987). Maximum water depths occurred toward the end of the period when bathyal/abyssal conditions were established in distal areas.

3. The Maastrichtian to Holocene progradational sequence (Figure 2E) is generally characterized by a proximal coarse-grained siliciclastic facies and distal facies with pelitic and turbiditic deposits. Geochemical and micropaleontological evidence show that oxygenated conditions have prevailed in most Brazilian marginal basins since the Campanian, with the deposition of organic-poor mixed clastic and carbonate sediments (Mello, Gaglianone et al., 1988, and references therein).

Mello and Maxwell

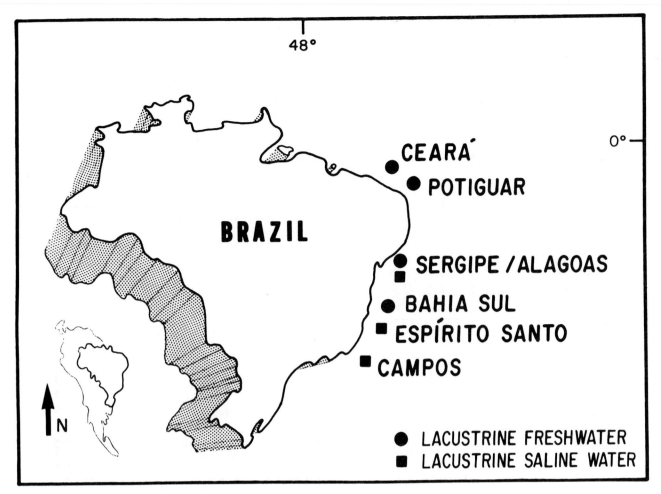

Figure 1. General locations of Brazilian continental-margin lacustrine basins.

Local basaltic flows, progressive basin subsidence, seaward tilting, and large adiastrophic growth faults marked the tectonosedimentary activity of the entire open-marine sequence (Estrella et al., 1984).

ANALYTICAL PROCEDURES

All oils and rock extracts were subjected to bulk, elemental, liquid chromatography, and UV/vis spectrophotometry analysis according to procedures described by Mello (1988). GC-MS analyses of alkanes were carried out with a Finnigan 4000 mass spectrometer coupled to a Carlo Erba 5160 gas chromatograph equipped with an on-column injector and 60-m DB-1701 column. Helium was used as carrier gas with a temperature program of 50°–90°C at 6°C/min and 90°–310° at 4°C/min (cf. Mello, Gaglianone et al., 1988; Mello, Telnaes et al., 1988). Mello et al. (1988) described all procedures related to biological marker ratios and concentrations.

RESULTS AND DISCUSSION

A distinct advantage in examining geochemical and biological marker characteristics of lacustrine source rocks is the availability of samples from many lacustine depositional environments whose geology and paleontology have been well described. Thus the Brazilian marginal basins provide an ideal opportunity for such investigation. Furthermore, where assumptions must be made, other well-documented geochemical and biological marker studies provide a basis for investigation (Powell, 1986; Brassel et al., 1988; Mello, 1988; Mello, Gaglianone et al., 1988; Mello, Telnaes et al., 1988).

Although some of the molecular parameters discussed herein are maturity dependent (e.g., biological marker concentrations), the availability of both immature and mature rock samples that were selected from a relatively narrow maturity range (0.45–0.75% R_o) allows other features to be described as source dependent.

Samples were selected that include not only typical organic-rich lacustrine rocks but also those repre-

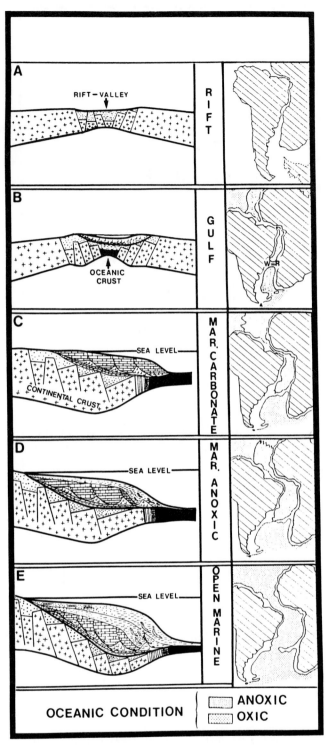

Figure 2. Evolution of Brazilian marginal basins through Cretaceous and Tertiary times, showing distribution of depositional environments (modified from Tissot et al., 1980; Carozzi and Falkenhein, 1985). W = R, Walvis Ridge.

senting nearly the entire range of bulk, elemental, and biological marker properties found within each type of lacustrine environment—fresh water and saline. Because extremes have been included,

properties of samples from either environment may overlap. For convenience the terms "low," "medium," and "high" (defined in Mello, Gaglianone et al., 1988) are used to describe particular measurements, as summarized in Table 1. Hence, not only the actual data are considered but also overall trends in the data. For example, low sulfur contents are considered "typical" of samples from the freshwater environment. However, of the approximately 15 samples examined, three had sulfur contents of 0.3% ("medium" in Table 1) but are still classified in this group on the basis of overall geochemistry.

For source rock correlation, elemental and bulk properties and biological marker distributions were analyzed for 40 oil samples from lower Neocomian to Miocene reservoirs in the major basins shown in Figure 1.

LACUSTRINE FRESHWATER ENVIRONMENT

Organic-rich sediments and oils of the freshwater environment were analyzed from the Ceará, Potiguar, Sergipe/Alagoas, and Bahia Sul basins in the equatorial and central continental margin. The lower Neocomian to Aptian fossil biota include typical freshwater ostracodes, gastropods, and conchostracans (Schaller, 1969; Ghignone and De Andrade, 1970).

Figure 3 shows typical geochemical logs of two important organic-rich horizons that were selected from the lower Neocomian Bahia Sul succession— (1) thick-bedded black shale, and (2) to a less extent, calcareous black shale (5–18% $CaCO_3$). Both typically are rich in organic matter (≤6.5% TOC) with low to medium sulfur content (≤0.4%; Table 1). Based on petrologic data and on the high hydrocarbon source rock potential (S_2 up to 37 mg HC/g of rock), arising largely from type I kerogen (hydrogen index up to 779 mg HC/g organic carbon; Table 1, Figure 3), the organic matter has been identified as mainly lipid-rich material (40–90% amorphous plus liptinitic matter; Table 1).

The excellent hydrocarbon source potential of these shales together with appropriate maturation conditions (R_o values; Table 1, Figure 3) indicate favorable source rock characteristics. Extracts show a trend toward high saturate content (≤66%), dominance of n-alkanes around C_{23}–C_{25}, and pristane greater than phytane.

The tendency toward high saturate content, the odd/even n-alkane predominance, and the bias toward high molecular-weight n-alkanes ($>C_{23}$) indicate major contributions of long-chain lipids from higher terrestrial plants and freshwater algae (Lijmbach, 1975; Didyk et al., 1978; McKirdy et al., 1986). Isotopically light $\delta^{13}C$ values (<−28‰; Table 1, Figure 4) for the whole extracts and their saturate fractions are consistent with a freshwater origin, inasmuch as their principal lipid constituents

Mello and Maxwell

Table 1. Elemental, bulk, and biological marker parameters of rocks and extracts of samples from sediments derived from lacustrine freshwater environment and lacustrine saline environment (values in parentheses) in Brazilian marginal basins

Elemental	Bulk	Alkanes	Seranes	Triterpanes	Porphyrins/- Type Organic Matter
Carbon (%): 1.2–4.1 (1.5–11.2)	TOC (%): 1.0–6.4 (2–5)	n-alkanes maxima: C_{23}–C_{25} (C_{19}–C_{21})	C_{27} sterane (ppm):[6] 0–40 (0–150)	C_{30} $\alpha\beta$ hopane (ppm):[11] 150–470 (200–1520)	Nickel (ppm): Trace (Trace–2831)
Hydrogen (%): 0.37–1.02 (0.36–1.0)	S_2:[1] 4.3–37 (12–38)	Saturates (%): 38–66 (37–49)	C_{27}/C_{29}:[7] 0.7–2.0 (1.4–2.5)	Gammacerane index:[12] 15–50 (13–72)	Vanadyl (ppm): Trace (Trace–95)
Nitrogen (%): 0.10–0.20 (0.07–0.20)	HI (mg/g):[2] 133–779 (300–980)	Pr/Ph: 1.1–2.6 (1.3–2.2)	Diasterane Index:[8] 0–50 (10–45)	Bisnorhopane index:[13] Undetected (3–10)	Amorphous (%): 40–90 (80–90)
Sulfur (%): 0.1–0.4 (0.3–0.9)	R_0 (%): 0.50–0.70 (0.45–0.66)	i-C_{25} + i-C_{30} (ppm):[4] 50–200 (70–1811)	4-Me sterane index:[9] 0–35 (20–165)	Hopane/Sterane:[14] 5–15 (4–14)	Herbaceous (%): 5–50 (5–15)
$CaCO_3$(%): 5–18 (4–49)	$\delta^{13}C$ (°/oo):[3] -28 to -31.0 (-25.5 to -26.9)	β-carotane:[5] Undetected (10–500 ppm)	C_{21} + C_{22} steranes (ppm):[10] 0–5 (8–30)	C_{34}/C_{35} hopanes:[15] >1.0 (1.3–2.1)	Woody/Coaly (%): 5–20 (5–10)

Measurement procedures:

1. Hydrocarbon source potential: mg HC/g rock (Pyrolysis and Rock-Eval)
2. Hydrocarbon Index (Pyrolysis and Rock-Eval)
3. PDB (°/oo)
4. Sum of 2, 6, 10, 14, 18- and/or 2, 6, 10, 15, 19-pentamethyleicosane (i-C_{25}) and squalane (i-C_{30}) peak areas in RIC trace and normalized to added sterane standard.
5. Peak area (β) in RIC trace and normalized to added sterane standard.
6. Sum of peak areas for 20R and 20S 5α, 14α, 17α(H)-cholestane (8 + 10) in m/z 217 chromatogram and normalized to added sterane standard (m/z 221 chromatogram).
7. Peak area of 20R 5α, 14α, 17α(H)-cholestane (10) over peak area of 20R 5α, 14α, 17α(H)-cholestane ethyl-cholestane (16) in m/z 217 chromatogram.
8. Sum of peak areas of C_{27} 20R and 20S 13β, 17α diasteranes (6 + 7) in m/z 217 chromatogram over sum of peak areas of C_{27} 20R and 20S 5α, 14α, 17α(H)-cholestane (8 + 10) × 100.
9. Sum of peak areas of all C_{30} 4-methyl steranes in m/z 231 chromatogram (recognized using mass spectra and m/z 414 chromatogram) over sum of peak areas of C_{27} 20R and 20S 5α, 14α, 17α (H)-cholestane (8 + 10) × 100.
10. Sum of peak areas (1 + 2 + 3 + 5) in m/z 217 chromatogram and normalized to added sterane standard.
11. Peak area of 35 measured in RIC and normalized to added sterane standard.
12. Peak area of gammacerane (40) in m/z 191 chromatogram over peak area of 17α(H), 21β(H)-hopane (35) × 100.
13. Peak area of C_{28}, 30-bisnorhopane (32) in m/z 191 chromatogram over peak area of 17α(H), 21β(H)-hopane (35) × 100.
14. Peak area of C_{30} l7α, 21β(H)-hopane (35) in m/z 191 chromatogram over sum of peak areas of C_{27} 20R and 20S 5α, 14α,17α(H)-cholestane (8 + 10) in m/z 217 chromatogram.
15. Peak areas of C_{34} 22R and 22S 17α, 21β(H)-hopanes (44) in m/z 191 chromatogram over peak areas of C_{35} counterparts (45).

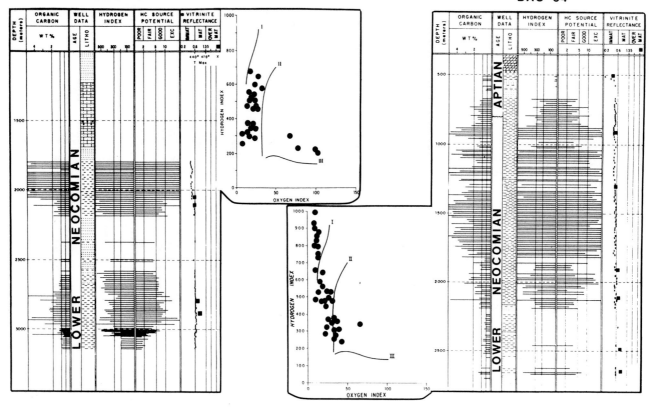

Figure 3. Geochemical logs of two typical wells from Bahia Sul basin, showing stratigraphic position of lacustrine freshwater organic-rich sediments and their hydrogen index (S_2/TOC), presented on van Krevelen-type diagram.

generally are depleted in ^{13}C relative to those of marine or lacustrine saline biota (Galimov, 1973; Tissot and Welte, 1984). The high pristane/phytane (Pr/Ph) ratios probably reflect the relationship between contributing organisms and chemistry of the environment, e.g., low salinity, rather than simply the anoxic/oxic conditions of sedimentation (Didyk et al., 1978). In a freshwater environment photosynthetic organisms and methanogenic bacteria containing lipids that are considered major sources of pristane would be expected to be abundant. With a marked increase in salinity (higher Eh), however, the archaebacterial population—mainly halophiles, which are considered to contain lipids that are a major source of phytane (ten Haven et al., 1988)—might be expected to become more abundant. Thus, the more saline the environment, the greater is the potential for an increase in concentration of phytane precursors. This may help explain the predominance of pristane in freshwater environments compared to a dominance of phytane in hypersaline environments (Table 1; Figures 5, 6; see Mello, 1988 for details).

Moldowan et al. (1985) found carbon isotope ratios and Pr/Ph ratios ineffective in distinguishing nonmarine and marine environments on a global basis. However, carbon isotope values in this study

allowed distinction between lacustrine freshwater and saline water samples.

To illustrate source rock correlation, consider first Figure 5, a typical lithologic log of a lower Neocomian source rock horizon and its corresponding gas chromatogram and m/z 191 and m/z 217 mass chromatograms of the alkane fraction. In Figure 6 gas chromatograms and m/z 217 (sterane) and m/z 191 (terpane) chromatograms of the alkane fraction of oils from two different reservoirs are compared with chromatograms for an upper Neocomian rock sample from the Sergipe/Alagoas basin. Although differences in the n-alkane and sterane distributions are apparent, similarities among the bulk data and other biological marker distributions (terpanes) and concentrations for these samples indicate a reasonable oil-source rock correlation.

Several aspects can be seen in the data that help characterize the lacustrine freshwater environment and its derived oils. Most notable are low concentration (<40 ppm of extract; Table 1) of C_{27} $\alpha\alpha\alpha$ steranes (peaks 8 and 10), medium concentration of C_{30} $\alpha\beta$ hopane (≤470 ppm, peak 35), low to medium relative abundance of gammacerane (peak 40), high pristane/phytane ratio, low sulfur content, low relative abundances of tricyclic terpanes (peaks 19-

Mello and Maxwell

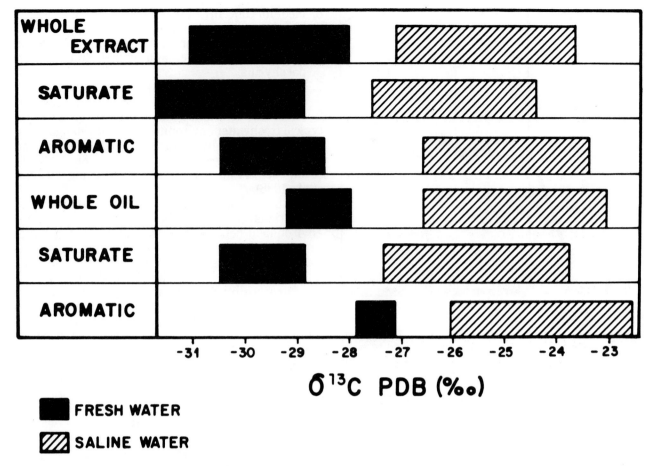

	FRESH WATER
	SALINE WATER

Figure 4. Variation of carbon isotopic data of whole oil and extract and their saturate and aromatic fractions for lacustrine freshwater and saline water samples.

26), and light $\delta^{13}C$ value (–29‰) for the whole extract. First, the paucity of steranes, which Moldowan et al. (1985) and McKirdy et al. (1986) also reported in freshwater samples from Brazil, China, Sudan, and Australia, may be due to organisms using lipids other than sterols as rigidifiers and cell-wall protectors. Possibly the terrestrial and freshwater plants live under higher oxygen conditions than their saline counterparts and therefore require greater cell protection.

Next, the tendency toward a dominance of C_{27} steranes contrasts with previous reports (Huang and Meinschein, 1979; Mackenzie et al., 1984; Hoffmann et al., 1984; Moldowan et al., 1985) of a predominance of C_{29} steranes in nonmarine environments. The interpretation is that these environments receive large contributions of higher plant material whose precursor sterols are mainly C_{29}. However, only two of the samples analyzed showed this tendency. Therefore, interpretation of C_{29} sterane predominance as either an indication of higher plant input or a characteristic of the nonmarine environment must be made cautiously.

Another peculiar feature is low to medium abundance of gammacerane (peak 40). The use of a standard and an efficient GC column allowed its identification, with complete separation from the hopanes (cf. peak 40, Figures 5 and 6).

Moldowan et al. (1985) have asserted that gammacerane cannot be used reliably to distinguish marine and nonmarine samples because it occurs in several different environments. Their evidence suggests the possibility of a bacterial origin, given its widespread occurence in time and space. Our results, together with those of Mello, Gaglianone et al. (1988) and Mello, Telnaes et al. (1988), suggest then that the value of gammacerane as an environmental indicator lies in its relative abundance (and concentration) rather than its mere presence. Furthermore, other characteristics (Table 1, Figures 3–7), such as high hopane/sterane ratios (5–15), absence of β-carotane and 28, 30-bisnorhopane, and traces (or absence) of porphyrins do indeed characterize this environment in the Brazilian marginal basins. All these features, together with the absence of C_{30} desmethyl steranes and dinosteranes, which

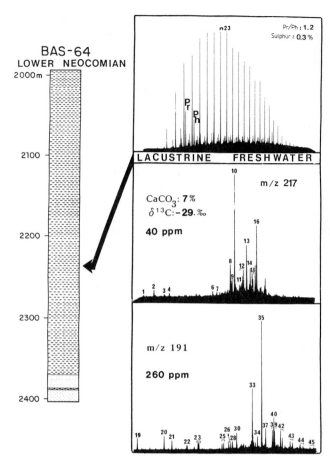

Figure 5. Lithologic log of a typical well from Bahia Sul basin, showing stratigraphic position of organic-rich sample (BAS-64) for which gas chromatograms of total alkanes, bulk and elemental parameters, absolute concentrations of steranes and hopane, and partial m/z 217 and m/z 191 chromatograms are shown. (See Appendices 1 and 2 for peak assignments and quantitation.)

are diagnostic of marine organic matter (Figures 7A, 7C), in the samples derived from the freshwater environment support the nonmarine character of these source rocks (Moldowan et al., 1985; Summons et al., 1987; Goodwin et al., 1988; Mello, 1988).

Integration of the bulk, elemental, and molecular data in Table 1 and Figures 5–7, therefore, suggests for this organic matter a lacustrine freshwater origin whose most marked characteristics include:

1. high Pr/Ph ratio linked with odd n-alkane predominance;
2. tendency toward high saturates content, associated with dominance of high molecular-weight n-alkanes;
3. low to medium sulfur content;
4. $\delta^{13}C$ values of whole extract more negative than $-28\%_{oo}$;
5. very low concentration of steranes and 4-methyl steranes;

6. medium concentration of hopane;
7. high hopane/sterane ratio;
8. absence of C_{30} steranes, dinosterane, β-carotane, 28, 30-bisnorhopane, and nickel and vanadyl porphyrins;
9. low relative abundance of gammacerane.

Similar characteristics have been reported from typical freshwater environments in the Shanganning and Songliao basins, China, and the Otway and Gippsland basins, Australia (Powell, 1986; McKirdy et al., 1986; Philp and Gilbert, 1986; Wang Tieguan et al., 1988).

The block diagram in Figure 8 illustrates the main depositional facies of the lacustrine freshwater environment believed to be typical of rift-stage lakes in the Brazilian marginal basins. Optimal conditions for organic-rich sedimentation here include deep water; warm and wet climate with no seasonal overturn; salinity ranging from fresh water to brackish; low sulfate concentration (fermentation rather than sulfate reduction); abundant dissolved plant nutrients (nitrates and phosphates); negative supply/demand balance of oxygen in the bottom water (anoxic conditions); and moderate to high sedimentation rates (Demaison and Moore, 1980; Fouch and Dean, 1984; Kelts, 1988; Talbot, 1988). Algal blooms, higher plant debris, and nutrient input to the photic and aerobic zones (shallow waters) enhanced anaerobic bacterial activity and anoxic conditions in the bottom waters (deep parts of the lakes), thereby creating ideal biological and chemical conditions for preserving organic matter. The thickest, most organic-rich deposits appear to be associated with the basin's paleodepocenter (Soldan et al., 1987).

LACUSTRINE SALINE-WATER ENVIRONMENT

Organic-rich sediments and oils of the saline-water environment are confined to the Campos, Espirito Santo, and Sergipe/Alagoas basins. The lower Neocomian to Aptian fossil biota here are characterized by nonmarine ostracodes with thick, coarsely reticulated shells, and widespread occurrence of gypsum and anhydrite molds. Heulandite-clinoptilolite-type zeolites and some authigenic minerals such as stevensite/talc/sepiolite indicate a shallow, saline alkaline lacustrine depositional environment (Castro and Azambuja, 1980; Bertani and Carozzi, 1985a, 1985b; De Dekker, 1988).

Specifically, three source rock systems have been identified—lower and upper Neocomian dark-gray, calcareous shales in the Campos and Espirito Santo basins and Aptian black shale in the Sergipe/Alagoas basin. Figure 9 shows typical geochemical logs of two selected upper Neocomian organic-rich horizons in the Campos and Espirito Santo basins. Both are

Mello and Maxwell

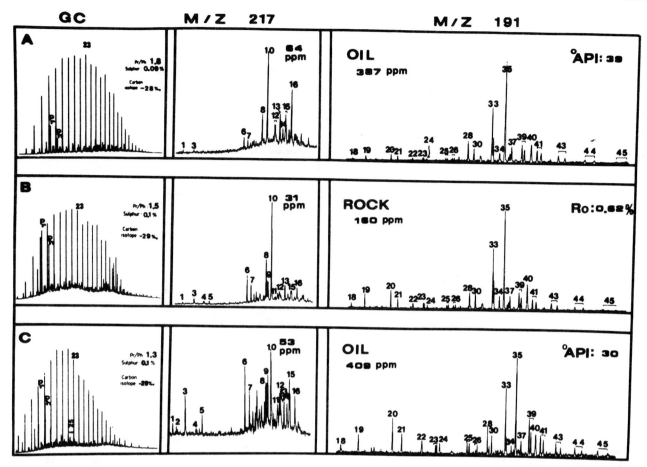

Figure 6. Oil-source rock correlation using gas chromatograms of total alkanes, bulk and elemental parameters, and partial m/z 191 and m/z 217 chromatograms, and absolute concentrations of steranes and hopane for a lacustrine freshwater rock sample from (B) Sergipe-Alagoas basin (CS-1) vs. two typical group I oil samples from (A) Ceará basin (CES-8) and (C) Potiguar (RNS-53) basin. (See Appendices 1 and 2 for peak assignments and quantitation.)

composed of dark-gray to black, thick-bedded, well-laminated, calcareous shale and marl (4–49% CaCO$_3$), rich in organic matter (≤5% TOC) with low to medium sulfur content (≤0.9%) (Table 1). Based on organic petrology and hydrogen indexes (up to 980 mg HC/g organic carbon), the organic matter is almost entirely lipid-rich material (type I kerogen, about 90% amorphous plus liptinitic material; Table 1, Figure 9). The excellent hydrocarbon source potential (S$_2$ up to 38 mg HC/g of rock) together with appropriate maturation conditions (R$_o$ values; Table 1, Figure 9) indicate that they are good source rocks.

Compositional data from rock extracts (Table 1, Figure 10) show a tendency toward high saturates content (≤49%, although slightly reduced relative to the lacustrine freshwater sample; Table 1), n-alkane maxima mostly around C$_{19}$ and C$_{21}$, and high Pr/Ph ratios (≤2.2). The oils (discussed below) show similar results and are consistent with nonmarine characterization (cf. Mello, Telnaes et al., 1988).

The relatively higher sulfur contents of these sediments (Table 1) may reflect a more saline character (high Eh) of the depositional environment. Enhanced salinity might also explain the isotopically heavier δ^{13}C values (>-26.9°/$_{oo}$; Table 1, Figure 4). Plants from saline environments can preferentially use carbonate complexes as carbon sources for photosynthesis. These are richer in ^{13}C than atmospheric carbon dioxide, which is enhanced in ^{12}C (Tissot and Welte, 1984). δ^{13}C data in Figure 4 suggest that carbon isotopes might be useful in discriminating lacustrine saline and freshwater samples.

The high Pr/Ph ratios (1.3–2.2; Table 1) suggest that water salinity in Campos, Espirito Santo, and Sergipe-Alagoas basin lakes was not high enough to achieve the ratios found in hypersaline environments (Pr/Ph < 1; Cerqueira and Santos Neto, 1986; ten Haven et al., 1987; Mello, Gaglianone et al., 1988; Mello, Telnaes et al., 1988). Yet, as proposed in previous studies, such low ratios are useful indicators of hypersalinity in the water column.

To further illustrate the idea of enhanced salinity, consider Figure 10, a lithologic log of an important

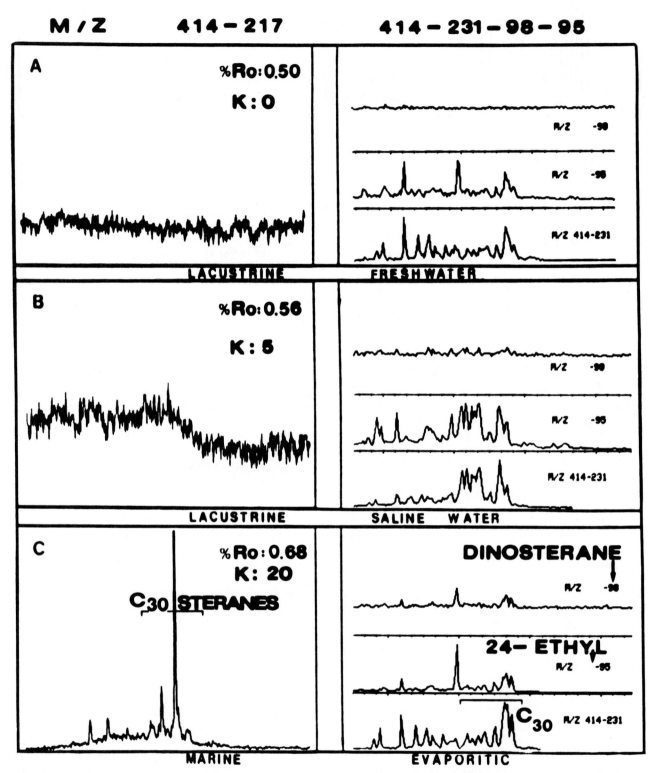

Figure 7. Mass chromatograms from metastable-ion monitoring of transition m/z 414–217 (C_{30} steranes), MS-MS mass chromatogram of transitions m/z 414–231, 414–95, and 414–98 (4-methyl steranes) of alkane fractions, and vitrinite reflectance data from sediment extracts from depositional environments in Brazilian marginal basins—A, lacustrine fresh water (CES-14); B, lacustrine saline water (RJS-71); C, marine evaporitic (CES-42). K values indicate concentration relative to constant amounts of added deuterated sterane standard.

LACUSTRINE FRESHWATER

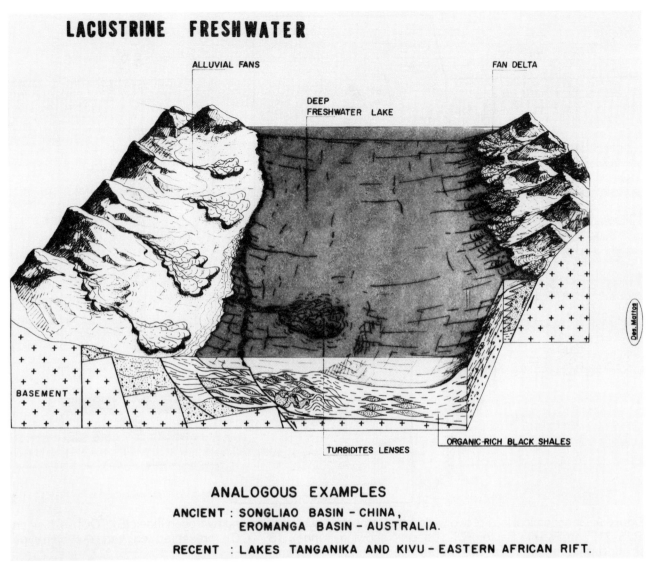

ALLUVIAL FANS

FAN DELTA

DEEP
FRESHWATER LAKE

BASEMENT

TURBIDITES LENSES

ORGANIC-RICH BLACK SHALES

ANALOGOUS EXAMPLES
ANCIENT : SONGLIAO BASIN - CHINA, EROMANGA BASIN - AUSTRALIA.

RECENT : LAKES TANGANIKA AND KIVU - EASTERN AFRICAN RIFT.

Figure 8. Block diagram of sedimentary facies in a deep freshwater lake from the rift stage in Brazilian marginal basins (modified from Azambuja Filho, 1987).

upper Neocomian source rock horizon in the Campos basin. It consists of about 185 m (2890–3175 m depth) of dark-gray, well-laminated, organic-rich (≤5% TOC) calcareous shale with low sulfur content (≤0.3%). The gas chromatogram of total alkanes in a relatively mature (0.56% R_o) sample shows the presence of β-carotane. On the m/z 217 mass chromatogram, steranes show low concentration (83 ppm; peaks 8 and 10), with a dominance of C_{27}. The m/z 191 mass chromatogram shows high concentrations of C_{30} $\alpha\beta$ hopane (320 ppm, peak 35), medium relative abundances of gammacerane (peak 40), and abundant tricyclic terpanes up to C_{35}. These features, together with the absence of C_{30} desmethyl steranes and dinosteranes (sample B, Figure 11), support the idea of enhanced salinity and nonmarine character of these source rocks.

Figures 11 and 12 show in greater detail gas and mass chromatograms of the alkane fractions of three

oils pooled in different reservoirs of the Campos basin compared to the rock sample (RJS–71; Figures 10, 11D, 12D). Although significant differences can be seen in the n-alkane distributions, similarities in bulk data and biological marker distributions and concentrations (e.g., steranes, terpanes, β-carotane, Pr/Ph, and gammacerane relative abundance) among the samples indicate that the sedimentary succession from the upper Neocomian (sample RJS–71) is the source for the oils from Campos basin.

Despite the overall similarities and good oil-source rock correlation, differences in molecular properties exist in different horizons within the succession. Figure 13 compares three organic-rich sediments with similar vitrinite reflectance values. Although they belong to three sedimentary successions (Aptian, upper Neocomian, lower Neocomian) in basins nearly 2000 mi (3200 km) apart, their molecular parameters show overall similarities (e.g., n-alkanes pattern

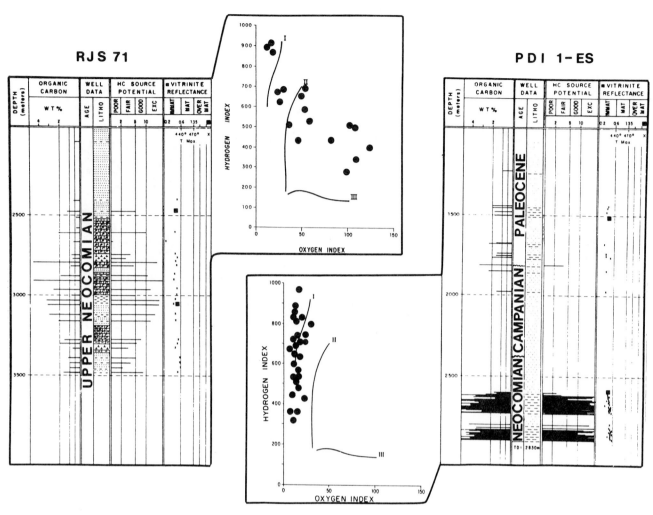

RJS 71

PDI 1-ES

Figure 9. Geochemical logs of two wells from Campos (RJS-71) and Espirito Santo (PDI-1) basins, showing stratigraphic position of two upper Neocomian lacustrine saline-water, organic-rich sedimentary successions and hydrogen index (S₂/TOC) vs. oxygen index (S₂/TOC), presented on van Krevelen-type diagram.

distribution, Pr/Ph ratio, β-carotane relative abundance, Ts/Tm ratio, and terpane pattern distribution). On the other hand, note the differences in gammacerane abundance (peak 40), carbon-isotope ratio, carbonate content, and tricyclic terpane relative abundances (peaks 18–26) among these samples and those in Figures 10 and 12. Inasmuch as all the samples share similar maturity, the observed differences suggest fluctuations (e.g., lake geometry, depth, and salinity) in depositional environment. Indeed, contractions and expansions of lake systems occurred during deposition of the lacustrine saline organic-rich sediments in the Brazilian marginal basins (Castro and Azambuja, 1980; Bertani and Carozzi, 1985a, 1985b; Mello, 1988).

Compared to lacustrine freshwater samples, the saline-water samples show higher concentrations and greater relative abundances of C_{25} isoprenoid and i-C_{30} (squalane; Table 1, Figures 5, 6, 10–12). This may reflect increased salinity in the depositional environment (Waples et al., 1974; ten Haven et al., 1985, 1987; Fu Jiamo et al., 1986; Mello, Gaglianone et al., 1988; Mello, Telnaes et al., 1988).

Another characteristic of the saline nature of these samples is the variable concentrations of β-carotane (10–500 ppm of extract; Table 1, Figures 10–13). Hall and Douglas (1983) suggested that its presence might be related to a lacustrine saline environment. However, Moldowan et al. (1985) regard β-carotane as a terrestrial marker because it had not been reported from marine sources. As supported by evidence from this study, its abundance in samples from lacustrine saline and hypersaline environments, and actually from marine hypersaline environments (Shi Ji-Yang et al., 1982; Jiang and Fowler, 1986; Mello, Gaglianone et al., 1988; Mello, Telnaes et al., 1988), suggests that salinity is a controlling factor of β-carotane concentration. The low concentration of steranes, as in the lacustrine freshwater samples (Table 1, Figures 5, 6, 10, 11), has been considered

Mello and Maxwell

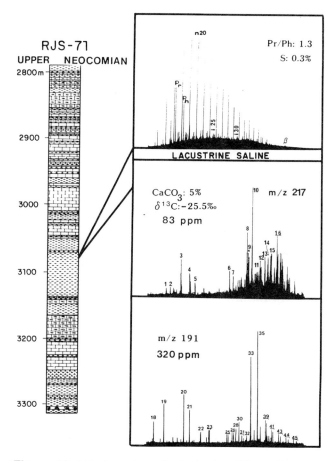

Figure 10. Lithologic log of a typical well from Campos basin, showing stratigraphic position of organic-rich sample (RJS-71) for which gas chromatogram of total alkanes, bulk and elemental parameters, absolute concentrations of steranes and hopane, and partial m/z 217 and m/z 191 chromatograms are shown. (See Appendices 1 and 2 for peak assignments and quantitation.)

characteristic of nonmarine samples from Australia, Sudan, Chad, China, Brazil, and the United States (Moldowan et al., 1985; McKirdy et al., 1986; Mello, Gaglianone et al., 1988; Mello, Telnaes et al., 1988).

Still another characteristic of lacustrine saline samples is low to medium concentration (≤30 ppm; Table 1) of low molecular-weight C_{21} and C_{22} steranes and high relative abundance of 4-methyl-homopregnanes (peak 4 in Figures 5, 6, 10, 11). These compounds are often associated with enhanced salinity in the depositional environment (ten Haven et al., 1985, 1988; Fu Jiamo et al., 1986). The predominance of C_{27} steranes over C_{29} steranes in all samples again demonstrates that C_{29} sterane predominance is not always diagnostic of nonmarine environments. That the concentrations of bacterially derived hopanes are high (Table 1, Figures 10, 12) perhaps reflects the importance of bacterial lipids in saline lakes (hopane/sterane ratios of 4-14; Table 1). In addition, the prominence of tricyclic compo-

nents ranging from C_{20} to C_{35} (tricyclic index of 100-200; Figures 10, 12) also may be the result of saline conditions of such lakes. Because they appear to arise from bacterial precursors, perhaps as specific membrane lipids (Ourisson et al., 1982), their abundance might be expected in saline lakes where bacteria thrived (Mello, 1988). Finally, low Ts/Tm ratios, typically less than 1 (peaks 28 and 30; Figures 10, 12; cf. Seifert et al., 1980; Mello, Gaglianone et al., 1988; Mello, Telnaes et al., 1988), may reflect a specific source input or mineral-matrix effects.

The triterpanes $17\alpha(H), 21\beta(H)$ 28,30-bisnorhopane (Table 1; peak 32 in Figures 10, 12), and 25, 28, 30-trisnorhopane (present only in m/z 177) have been recognized in most samples of this environment. Salinity at the time of deposition (and subsequent increase in anoxicity) possibly influences their occurrence. Indeed, such compounds have been reported to originate from sulfur-reducing bacteria in strongly reduced conditions (Katz and Elrod, 1983).

In summary, analyses of this group of samples show (Table 1, Figures 10-13) that they are diagnostic of a nonmarine environment characterized by the following elemental, isotopic, and biological marker features specific to deposition under lacustrine saline conditions:

1. high Pr/Ph ratio linked with odd n-alkane predominance;
2. tendency toward high saturates content, associated with dominance of medium molecular-weight n-alkanes;
3. medium to low sulfur content;
4. low to high concentration of β-carotene;
5. high to very high concentration of C_{30} $\alpha\beta$ hopane;
6. medium to high concentration of C_{25} and C_{30} isoprenoids;
7. medium to low relative abundance of gammacerane,
8. heavy $\delta^{13}C$ values of whole extract;
9. high relative abundance of low molecular-weight steranes (C_{21}-C_{22});
10. Ts/Tm < 1;
11. high hopane/sterane ratio;
12. low abundance of 28, 30-bisnorhopane and 25, 28, 30-trisnorhopane;
13. abundant tricyclic terpanes up to C_{35};
14. absence of C_{30} steranes and dinosteranes.

Overall, the bulk, elemental, and molecular features for these source rocks are in agreement with paleontological and mineralogical data and extend previous evidence for oils and source rocks from lacustrine saline environments in China and in the Green River Formation (Reed, 1977; Tissot et al., 1978; Powell, 1986; Wang Tieguan et al., 1988). Few analogous cases of ancient, shallow saline lake systems have been reported in the literature. The best comparison with the Brazilian example appears to be the well studied Eocene Green River Formation in the Uinta basin, Utah (Tissot et al., 1978; Demaison

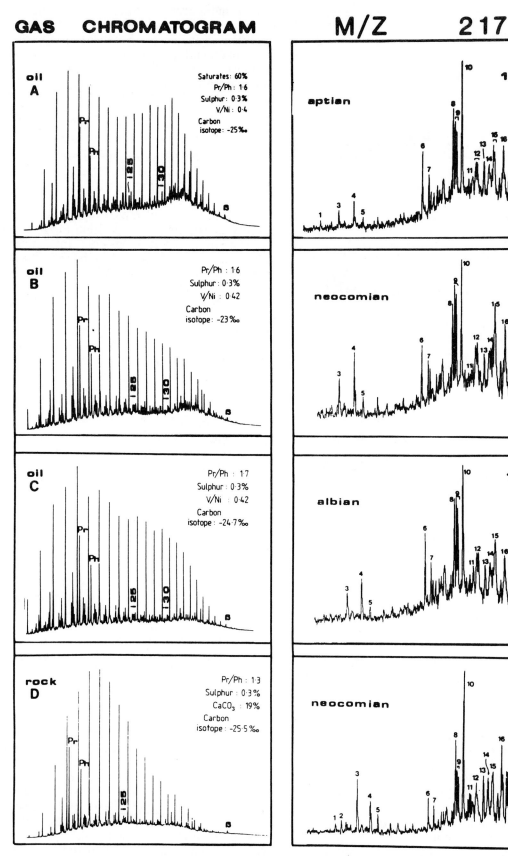

Figure 11. Oil-source rock correlation for (D) Campos basin rock sample (RJS-71) at 3020 m using bulk and elemental data, gas chromatograms, and sterane concentrations vs. three typical oil samples pooled in different reservoirs—A, Aptian; B, Neocomian; C, Albo-Cenomanian.

Mello and Maxwell

M/Z 191

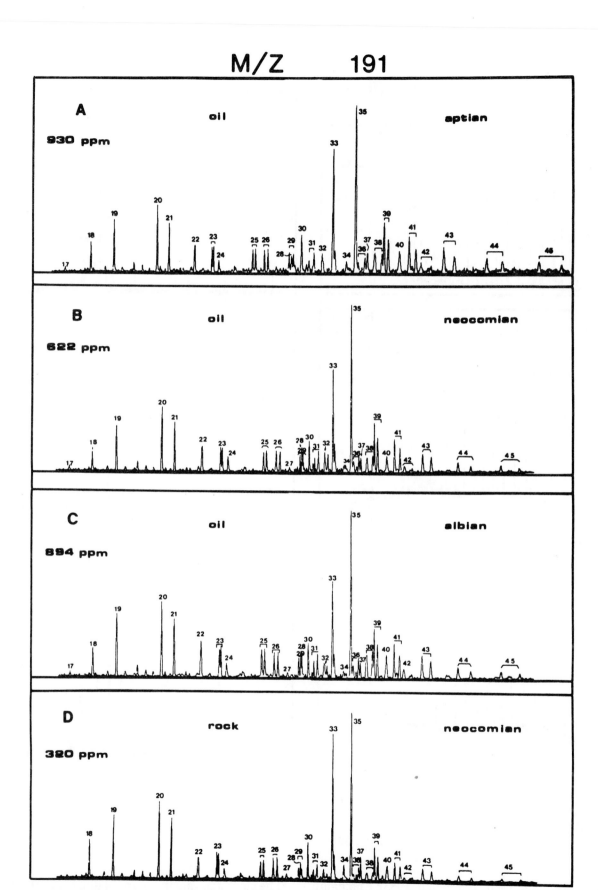

Figure 12. Oil-source rock correlation for (D) Campos basin rock sample (RJS-71) using partial m/z 191 mass chromatograms and absolute concentration of C$_{30}$ $\alpha\beta$ hopane for oils derived from lacustrine saline-water depositional environments—A, Aptian; B, Neocomian; C, Albo-Cenomanian.

Lacustrine Basin Exploration: Case Studies and Modern Analogs

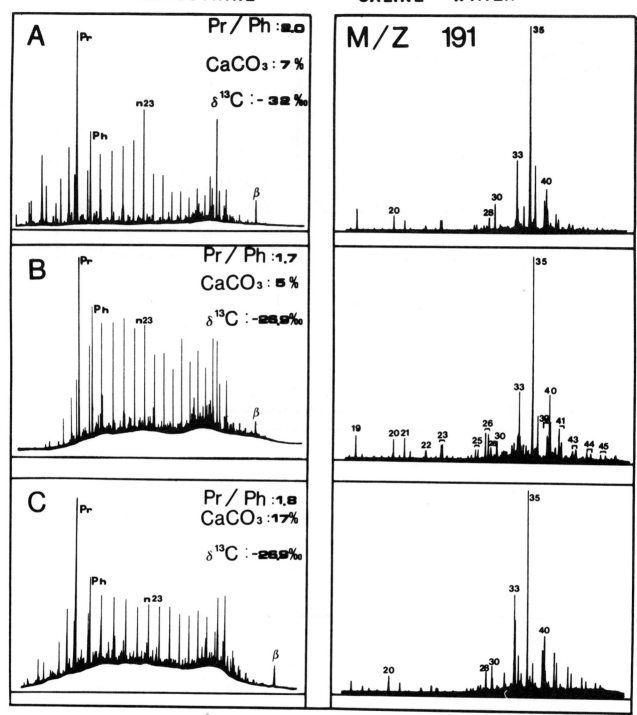

Figure 13. Gas chromatograms of total alkanes, bulk and elemental parameters, and partial m/z 191 mass chromatograms and absolute concentrations of hopane from sediment extracts derived from lacustrine saline-water depositional environments—A, Potiguar basin; B, Espirito Santo basin; C, Campos basin. (See Appendix 1 for peak assignment.)

and Moore, 1980; Dean and Fouch, 1983); Chaidamu and Jianghan basins, China (Chen Changming et al., 1984; Powell, 1986; Fu Jiamo et al., 1986); and Officer basin, Australia (McKirdy et al., 1986).

The block diagram in Figure 14 illustrates the main sedimentological facies of a shallow saline lake system of alkaline affinities, which appears to have dominated the depositional paleoenvironment during

Mello and Maxwell

LACUSTRINE SALINE WATER ENVIRONMENT

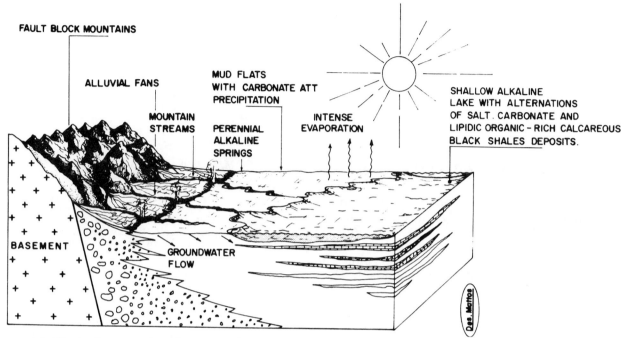

Figure 14. Block diagram of sedimentary facies in a shallow saline to hypersaline lake with alkaline affinities from the rift stage in Brazilian marginal basins (modified from Eugster and Hardie, 1975).

late Neocomian time in the Campos and Espirito Santo basins. In such an environment, which generally occurs in areas of high evaporation (semiarid/moist climates), high lake productivity is directly related to association with alkaline springs. The high pH of the water in alkaline lakes has tremendous effects on carbon cycles—first, in supporting a larger biomass than many neutral pH lakes, and second, in maintaining rather high concentrations of phosphorous and other nutrients, which are highly pH dependent, (Kelts, 1988). Under these conditions well-adapted species show prolific productivity, resulting in high input of algal and bacterial organic matter. Differences in salinity between an upper aerobic, less saline layer and a lower anaerobic, highly saline (denser) and alkaline layer enhance water-column stability, leading to stratification and permanent bottom-water anoxia. These conditions, although they promote anaerobic bacterial activity, are lethal to microfauna and benthic organisms. Low sulfate concentration (alkalinity), associated with extreme bottom-water anoxia, enhances the degree of organic-matter preservation, resulting in deposition of well-laminated, organic-rich calcareous black shale (Kelts, 1988; Talbot, 1988; De Dekker, 1988; Volkman, 1988; Duncan and Hamilton, 1988; Katz, this volume).

CONCLUSIONS

Based on sedimentological, paleontological, and mainly geochemical and biological marker data, it is possible to differentiate and characterize two distinct, lower Neocomian to Aptian, organic-rich lacustrine systems in the rift stage of the Brazilian marginal basins—(1) a relatively large, deep lacustrine freshwater type of basin; and (2) a closed, shallow lake system having saline to hypersaline waters of alkaline affinities.

The use of geochemical and biological marker parameters in oil-source rock correlation indicates that about 85% of the known Brazilian oils originated from lacustrine sequences, and most were sourced by upper Neocomian lacustrine saline organic-rich rocks. That most of the deep, Neocomian to Aptian freshwater paleolakes occur along the equatorial and central Brazilian continental margin suggests that this area was the last part of the Brazilian plate to have disconnected from Africa.

The "end members" of the specific lacustrine environments possess a diagnostic group of characteristics. Where biological marker features were observed to overlap is believed to represent environmental transitions (lacustrine fresh to brackish to saline water). These "overlaps" point out the difficulty in attempting to characterize and distinguish depositional environments in lacustrine basins where fluctuations (contractions and expansions) of the water body frequently take place.

Our results demonstrate the value of biological markers for both lateral and vertical correlations. For example, samples collected from a lacustrine saline depositional environments more than 2000 miles (3200 km) apart show remarkably similar features. The data compilation in Table 1 therefore provides a framework of biological marker and geochemical characteristics that can be used to help characterize lacustrine depositional environments elsewhere in the world.

ACKNOWLEDGMENTS

The authors would like to thank the Geochemistry Section of Petrobras Research Centre for all elemental and bulk analyses, Mr. Jose Roberto Cerqueira and Luiz Freitas for helpful comments, Mr. Gilberto Pereira da Silva for typing the manuscript, and Ms. A. P. Gowar and Ms. L. Dyas for advice during the analytical work. The authors also are grateful to Bristol University for use of its GC-MS facilities and to Petrobras for permission to publish.

REFERENCES

Asmus, H. E., 1975, Controle estrutural da deposicão Mesozóica nas bacias da margem continental Brasileira: Revista Brasileira de Geociencias, v. 5, p. 160-175.

Azambuja F., N. C. De, 1987, Preenchimento sedimentar de bacias do tipo rift, in Rifts Intracontinentais, Seminario: Rio de Janeiro, Petrobrás, DEPEX, p. 17-44.

Babinski, N. A., and R. C. B. Santos, 1987, Origem e classificacao dos hidrocarbonetos da Bacia Sergipe-Alagoas—Caracterizacao geoquimica: Boletim Geociencias da Petrobrás, v. 1, no. 1, p. 87-95.

Bertani, R. T., and A. V. Carozzi, 1985a, Lagoa Feia Formation (Lower Cretaceous), Campos Basin, Offshore Brazil—Rift valley stage lacustrine carbonate reservoirs, I: Journal of Petroleum Geology, v. 8, no. 1, p. 37-58.

Bertani, R. T., and A. V. Carozzi, 1985b, Lagoa Feia Formation (Lower Cretaceous), Campos Basin, Offshore Brazil—Rift valley stage lacustrine carbonate reservoirs, II: Journal of Petroleum Geology, v. 8, no. 2, p. 199-220.

Bott, M. H. P., 1976, Formation of sedimentary basins of graben type by extention of the continental crust: Tectonophysics, v. 36, p. 77-86.

Brassell, S. C., Sheng Guoying, Fu Jiamo, and G. Eglinton, 1988, Biological markers in lacustrine Chinese oil shales, in A. J. Fleet, K. Kelts, and M. R. Talbot, eds., Lacustrine petroleum source rocks: Geological Society Special Publication 40, p. 299-308.

Castro, J., and N. C. Azambuja Filho, 1980, Facies, analise estratigrafica e reservatorios da Formacao Lagoa Feia—Cretaceo inferior da Bacia de Campos: Rio de Janeiro, Petrobrás, unpublished rept., p. 110.

Cerqueira, J. R., and E. V. Santos Neto, 1986, Projeto análise de bacia do Paraná (geoquimica orgânica): Petrobrás/Cenpes internal report, 3 v.

Cerqueira, J. R., A. L. Soldan, M. R. Mello, and C. V. Beltrami, 1984, Identificacao das rochas geradoras de hidrocarbonetos da Bacia do Ceara: XXXIII Congresso Brasileiro de Geologia, Anais Vd. X, p. 4778-4791.

Chen Changming, Huang Jiakuan, Chen Jingshan, and Tian Xingyou, 1984, Depositional models of Tertiary rift basins, eastern China, and their application to petroleum prediction: Sedimentary Geology, v. 40, p. 73-88.

De Dekker, P., 1988, Large Australian lakes during the last 20 million years—Sites for petroleum source rocks or metal, ore deposition, or both, in A. J. Fleet, K. Kelts, and M. R. Talbot, eds., Lacustrine petroleum source rocks: Geological Society Special Publication 40, p. 45-59.

Demaison, G. J., and G. T. Moore, 1980, Anoxic environments and oils source bed genesis: AAPG Bulletin, v. 64, p. 1179-1209.

Didyk, B. M., B. R. T. Simoneit, S. C. Brassel, and G. Eglinton, 1978, Organic geochemical indicators of paleoenviromental conditions of sedimentation: Nature, v. 272, p. 216-222.

Duncan, A. D., and R. F. M. Hamilton, 1988, Palaeolimnology and organic geochemistry of the middle Devonian in the Orcadian Basin, in A. J. Fleet, K. Kelts, and M. R. Talbot, eds., Lacustrine petroleum source rocks: Geological Society Special Publication 40, p. 173-201.

Estrella, G., M. R. Mello, P. C. Gaglianone, R. L. M. Azevedo, K. Tsubone, E. Rossetti, J. Concha, and I. M. R. A. Bruning, 1984, The Espirito Santo Basin (Brazil) source rock characterization and petroleum habitat, in G. Demaison, and R. J. Murris, eds., Petroleum geochemistry and basin evaluation: AAPG Memoir 35, p. 253-271.

Eugster, H. P., and L. A. Hardie, 1975, Sedimentation in an ancient playa-lake complex—The Wilkins Peak Member of the Green River Formation of Wyoming: GSA Bulletin, v. 86, p. 319-334.

Fouch, T. D., and W. E. Dean, 1984, Lacustrine and associated clastic depositional environments, in P. A. Scholle, ed., Sandstone depositional environments: AAPG Memoir 31, p. 87-114.

Fu Jiamo, S. Guoyng, P. Pingan, S. C. Brassel, G. Eglinton, and J. Jigang, 1986, Pecularities of salt lake sediments as potential source rocks in China, in D. Leythaeuser, and J. Rullkotter, eds., Advances in organic geochemistry 1985: Elmsford, N.Y., Pergamon Journals Ltd., p. 119-127.

Galimov, E. M., 1973, Carbon isotopes in oils and gas geology: Washington, NASA TT F-682 [translated from Russian]

Ghignone, J. I., and G. de Andrade, 1970, General geology and major oil fields of Reconcavo Basin, Brazil, in M. T. Halbouty, ed., Geology of giant petroleum fields: AAPG Memoir 14, p. 337-358.

Goodwin, N. S., A. L. Mann, and R. L. Patiente, 1988, Structure and significance of C_{30} 4-methylsteranes in lacustrine shales and oils: Organic Geochemistry, v. 12, p. 405-506.

Hall, P. B., and A. G. Douglas, 1983, The distribution of cyclic alkanes in two lacustrine deposits, in M. Bjoroy, and others, eds., Advances in organic geochemistry 1981: New York, John Wiley, p. 576-587.

Hoffmann, C. F., A. S. Mackenzie, C. A. Lewis, J. R. Maxwell, J. L. Oudin, B. Durand, and M. A. Vandenbroucke, 1984, Biological marker study of coals, shales, and oils from Mahakam Delta, Kalimantan, Indonesia: Chemical Geology, v. 42, p. 1-23.

Hu Chaoyuan, and Qiao Hanseng, 1983, Characteristics of oil and gas distribution in the North China Basin and the adjacent seas: 11th World Petroleum Congress, Proceedings, v. 2, p. 111-119.

Huang, W. Y., and W. G. Meinschein, 1979, Sterols in sediments from Baffin Bay, Texas: Geochimica et Cosmochimica Acta., v. 42, p. 1391-1396.

Jiang, Z. S., and M. G. Fowler, 1986, Carotenoid-derived alkanes in oils from northwestern China, in D. Leythaeuser, and J. Rullkotter, eds., Advances in organic geochemistry 1985: Elmsford, N.Y., Pergamon Journals Ltd., p. 831-839.

Katz, B. J., and L. W. Elrod, 1983, Organic geochemistry of DSDP Site 467, offshore California, middle Miocene to lower Pliocene strata: Geochimica et Cosmochimica Acta, v. 47. p. 389-396.

Katz, B. J., Controls on distribution of lacustrine source rocks through time and space, this volume.

Kelts, K., 1988, Environments of deposition of lacustrine petroleum source rocks—An introduction, in A. J. Fleet, K. Kelts, and M. R. Talbot, eds., Lacustrine petroleum source rocks: Geological Society Special Publication 40, p. 3-26.

Koutsoukos, E. A. M., and D. Dias-Brito, 1987, A área noroeste da Bacia de Campos, Brasil, do Mesocretáceo ao Neocretaceo—Evolucão paleoambiental e paleogeográfica através de estudos de foraminiferos: IX Congresso Brasileiro de Paleontologia, Anais, Fortaleza, Ceará.

Lijmbach, G. W., 1975, On the origin of petroleum: 9th World Petroleum Congress, Proceedings, v. 2, p. 357-369, Applied Science, London.

Mackenzie, A. S., J. R. Maxwell, M. L. Coleman, and C. E. Deegan, 1984, Biological marker and isotope studies of North Sea crude oils and sediments, in Geology, exploration, reserves: 11th World Petroleum Congress, Proceedings, v. 2, p. 45-56.

McKirdy, D. M., R. E. Cox, J. K. Volkman, and V. J. Howell, 1986, Botryococcane in a new class of Australian non-marine crude oils: Nature, v. 320, p. 57-59.

McKirdy, D. M., and J. G. G. Morton, 1988, Biological marker isotopic and geological studies of lacustrine crude oils in the western Otway Basin, South Austalia, in A. J. Fleet, K. Kelts, and M. R. Talbot, eds., Lacustrine petroleum source rocks: Geological Society Special Publication 40, p. 327-329.

Mello, M. R., 1988, Geochemical and molecular studies of the depositional environments of source rocks and their derived oils from the Brazilian marginal basins: Ph.D. Thesis, Bristol University.

Mello, M. R., G. O. Estrella, and P. C. Gaglianone, 1984, Hydrocarbon source potential in Brazilian marginal basins (abs.): AAPG Bull., v. 68, p. 506.

Mello, M. R., P. C. Gaglianone, S. C. Brassel, and J. R. Maxwell, 1988, Geochemical and biologial marker assessment of depositional environment using Brazilian offshore oils: Marine and Petroleum Geology, v. 5, p. 205-223.

Mello, M. R., A. L. Soldan, and I. M. R. Brüning, 1984, Fundamentos da quimica orgânica—Essenciais para a geoquimica do petróleo, in Geoquimica do petróleo: Petobrás, p. 1-14.

Mohriak, W. U., and F. F. Dewey, 1986, Deep seismic reflectors in the Campos basin, offshore, Brazil: Geophysical Journal of the Royal Astronomical Society, v. 89, p. 133-140.

Moldowan, J. M., W. K. Seifert, and E. J. Gallegos, 1985, Relationship between petroleum composition and depositional environment of petroleum source rocks: AAPG Bulletin, v. 69, p. 1255-1268.

Ojeda, H. A. O., 1982, Structural framework, stratigraphy, and evolution of Brazilian marginal basins: AAPG Bulletin, v. 66, p. 732-749.

Ourisson, G., P. Albrecht, and M. Rohmer, 1982, Predictive microbial biogeochemistry from molecular fossils to procaryotic membranes: Trends in Biochemical Sciences, v. 7, p. 236-239.

Pereira, M. J., L. A. F. Trindale, and P. C. Gaglianone, 1984, Origem e evolucao das acumulacoes de hidrocarbonetos na Bacia de Campos: XXXIII Congresso Brasileiro de Geologia, Anais Vd. X, p. 4763-4777.

Philp, R. P., and T. D. Gilbert, 1986, Biomarker distribution in Australian oils predominantly derived from terrigenous source material, in D. Leythaeuser, and J. Rullkotter, eds., Advances in organic geochemistry 1985: Elmsford, N. Y., Pergamon Journals Ltd., p. 73-84.

Ponte, E. C., and H. E. Asmus, 1978, Geological framework of the Brazilian Continental Margin: Geologische Rundschau, v. 68, p. 201.-235.

Powell, T. G., 1986, Petroleum geochemistry and depositional settings of lacustrine source rocks: Marine and Petroleum Geology, v. 3, p. 200-219.

Reed, W. E., 1977, Molecular compositions of weathered petroleum and comparison with its possible source: Geochimica et Cosmochimica Acta, v. 41, p. 237-247.

Rodrigues, R., L. A. F. Trindale, J. N. Cardoso, and F. R. Aquino Neto, 1988, Biomarker stratigraphy of the Lower Cretaceous of Espirito Santo Basin, Brazil, in L. Mattavelli, and L. Novelli, eds., Advances in organic geochemistry 1987: New York, Pergamon Press.

Rullkotter, J., A. S. Mackenzie, D. H. Welte, D. Leythaeuser, and M. Radke, 1984, Quantitative gas chromatography-mass spectrometry analysis of geological samples, in P. A. Schenk, J. W. De Leeuw, and G. W. M. Lijmbach, eds., Advances in organic geochemistry 1983: New York, Pergamon Press, p. 817-827.

Santos Neto, E. V., J. R. Cerqueira, F. J. C. Concha, and V. A. Castelo Branco, 1989, The role of intrusive bodies in the thermal history of Irati Formation, Parana Basin, Brazil—An unusual example of thermal maturation (abs.): 14th International Meeting on Organic Geochemistry, Paris.

Schaller, H., 1969, Revisão bioestratigráfica da bacia de Sergipe-Alagoas: Boletim Técnico da Petrobrás, v. 12, no. 1, p. 21-86.

Seifert, W. K., J. M. Moldowan, and R. W. Jones, 1980, Application of biological marker chemistry to petroleum exploration: 10th World Petroleum Congress, Proceedings, v. 2, p. 425-438.

Shi Ji-Yang, A. S. Mackenzie, R. Alexander, G. Eglinton, A. P. Gowar, G. A. Wolff, and J. R. Maxwell, 1982, A biological marker investigation of petroleums and shales from the Shangli oilfield, Peoples Republic of China: Chemical Geology, v. 35, p. 1-31.

Soldan, A. L., L. C. S. Freitas, P. C. Gaglianone, H. A. F. Chaves, and J. R. Cerqueira, 1987, Quantitative estimation of original hydrocarbon source rock potential—A statistical model of behavior (abs.): 13th International Meeting on Organic Geochemistry, Venice.

Summons, R. E., J. K. Volkman, and C. J. Boreham, 1987, Dinosterane and other steroidal hydrocarbons of dinoflagellate origin in sediments and petroleum: Geochimica et Cosmochimica Acta, v. 51, p. 3075-3082.

Talbot, M. R., 1988, Non-marine oil-source rock accumulation—Evidence from the lakes of tropical Africa, in A. J. Fleet, K. Kelts, and M. R. Talbot, eds., Lacustrine petroleum source rocks: Geological Society Special Publication 40, p. 29-45.

ten Haven, H. L., J. W. De Leeuw, J. Rullkotter, and J. S. Sinninghe Damste, 1987, Restricted utility of the pristane/phytane ratio as a palaeoenvironmental indicator: Nature, v. 330, p. 641-643.

ten Haven, H. L., J. W. De Leeuw, and P. A. Schenk, 1985, Organic geochemical studies of a Messinian evaporitic basin, northern Apennines (Italy), part 1—Hydrocarbon biological markers for a hypersaline environment: Geochimica et Cosmochimica Acta, v. 49, p. 2181-2191.

ten Haven, H. L., J. W. De Leeuw, J. S. Sinninghe Damste, P. A. Schenk, S. E. Palmer, and J. E. Zumberge, 1988, Application of biological markers in the recognition or paleo hypersaline environments, in A. J. Fleet, K. Kelts, and M. R. Talbot, eds., Lacustrine petroleum source rocks: Geological Society Special Publication 40, p. 123-131.

Tissot, B., G. Derro, and A. Hood, 1978, Geochemical study of the Uinta Basin—Formation of petroleum from the Green River formation: Geochimica et Cosmochimica Acta, v. 42, p. 1469-1465.

Tissot, B. P., and D. H. Welte, 1984, Petroleum formation and occurrence, 2d ed.: New York, Springer-Verlag, 699 p.

Waples, D. W., P. Haug, and D. H. Welte, 1974, Occurrence of a regular C_{25} isoprenoid hydrocarbon in tertiary sediments representing a lagoonal-type, saline environment: Geochimica et Cosmochimica Acta, v. 38, p. 381-387.

Yang Wanli, 1985, Daqing oil field, People's Republic of China—A giant field with oil of nonmarine origin: AAPG Bulletin, v. 69, p. 1101-1111.

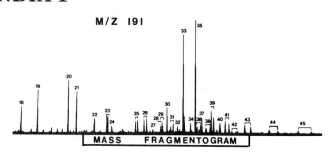

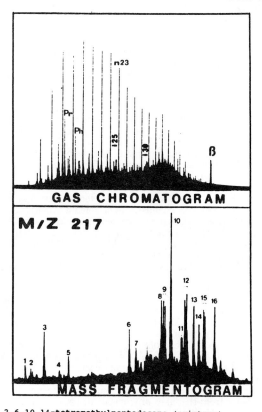

17- C_{19} tricyclic terpane
18- C_{20} tricyclic terpane
19- C_{21} tricyclic terpane
20- C_{23} tricyclic terpane
21- C_{24} tricyclic terpane
22- C_{25} tricyclic terpane
23- C_{26} tricyclic terpanes
24- C_{24} tetracyclic (Des-E)
Te- C_{24} tetracyclic (Des-A)
25- C_{28} tricyclic terpanes
26- C_{29} tricyclic terpanes
27- C_{25} tetracyclic
28- C_{27} 18α(H)-trisnorneohopane(Ts).
29- C_{30} tricyclic terpanes
T- C_{27} 25,28,30-trisnorhopane
30- C_{27} 17α(H)-trisnorhopane(Tm).
31- C_{31} tricyclic terpanes
32- 17α(H),18α(H),21β(H)-28,30-bisnorhopane(C_{28}).
N- 25-norhopane (C_{29})
33- C_{29} 17α(H),21β(H)-norhopane.
34- C_{29} 17β(H),21α(H)-norhopane.
35- C_{30} 17α(H),21β(H)-hopane.
36- C_{33} tricyclic terpanes
37- C_{30} 17β(H),21α(H)-hopane.
38- C_{34} tricyclic terpanes
39- C_{31} 17α(H),21β(H)-homohopane (22S + 22R).
40- C_{30} gammacerane.
41- C_{32} 17α(H),21β(H)-bishomohopane (22S + 22R).
42- C_{35} tricyclic terpanes
43- C_{33} 17α(H),21β(H)-trishomohopane (22S + 22R).
44- C_{34} 17α(H),21β(H)-tetrakishomohopane (22S + 22R).
45- C_{35} 17α(H),21β(H)-pentakishomohopane (22S + 22R).

Pr- 2,6,10,14-tetramethylpentadecane (pristane).
Ph- 2,6,10,14-tetramethylhexadecane (phytane).
i-C_{25} 2,6,10,14,18-pentamethyleicosane (regular).
i-C_{25} 2,6,10,15,19-pentamethyleicosane (irregular).
i-C_{30} squalane.
ß- ß-Carotane.
1- 13β(H),17α(H)-diapregnane(C_{21})
2- 5α(H),14β(H),17α(H)-pregnane(C_{21})
3- 5α(H),14β(H),17β(H) + 5α(H),14α(H),17α(H)-pregnane(C_{21})
4- 4α-methyl-5α(H),14β(H),17β(H) +
   4α-methyl-5α(H),14α(H),17α(H) homopregnane(C_{22})
5- 5α(H),14β(H),17β(H) + 5α(H),14α(H),17α(H)-
   homopregnane(C_{22}))
6- 13β(H),17α(H)-diacholestane, 20S (C_{27}-diasterane).
7- 13β(H),17α(H)-diacholestane, 20R (C_{27}-diasterane).
8- 5α(H),14α(H),17α(H), 20S (C_{27}-cholestane).
9- 5α(H),14β(H),17β(H), 20R + 20S (C_{27}-cholestane).
10- 5α(H),14α(H),17α(H), 20R (C_{27}-cholestane).
11- 5α(H),14α(H),17α(H), 20S (C_{28}-methylcholestane).
12- 5α(H),14β(H),17β(H), 20R + 20S (C_{28}-methylcholestane).
13- 5α(H),14α(H),17α(H), 20R (C_{28}-methylcholestane).
14- 5α(H),14α(H),17α(H), 20S (C_{29}-ethylcholestane).
15- 5α(H),14β(H),17β(H), 20R + 20S (C_{29}-ethylcholestane).
16- 5α(H),14α(H),17α(H), 20R (C_{29}-ethylcholestane).

APPENDIX 2

I-C_{25} + I-C_{30}: Sum of 2,6,10,14,18- and/or 2,6,10,15,19-pentamethyleicosane (I-C_{25}) and squalane (I-C_{30}) peak areas in RIC trace and normalised to added sterane standard.

ß-carotane: Peak area (ß) in RIC trace and normalised to added sterane standard.

Low molecular weight steranes: Sum of peak areas (1+2+3+5) in m/z 217 chromatogram and normalised to added sterane standard (m/z 221 chromatogram).

Sterane concentration: Sum of peak areas for 20R and 20S 5α,14α,17α(H)-cholestane (8+10) in m/z 217 chromatogram and normalised to added sterane standard (m/z 221 chromatogram).

C_{27}/C_{29} sterane: Peak area of 20R 5α,14α,17α(H)-cholestane (10) over peak area of 20R 5α,14α,17α(H)-ethylcholestane (16) in m/z 217 chromatogram.

Diasterane index: Sum of peak areas of C_{27} 20R and 20S 13ß,17α(H)-diasteranes (6+7) in m/z 217 chromatogram over sum of peak areas of C_{27} 20R and 20S 5α,14α,17α(H)-cholestane (8+10) x 100. Low < 30, Medium 30-100, High > 100.

4-Methyl sterane index: Sum of peak areas of all C_{30} 4-methyl steranes in m/z 231 chromatogram recognised using mass spectra and m/z 414 chromatogram) over sum of peak areas of C_{27} 20R and 20S 5α,14α,17α(H)-cholestane (8+10) x 100. Low < 60, Medium 60-80 High > 80.

Hopane/sterane: Peak area of C_{30} 17α,21ß(H)-hopane (35) in m/z 191 chromatogram over sum of peak areas of C_{27} 20R and 20S 5α,14α,17α(H)-cholestane (8+10) in m/z 217 chromatogram. Low < 4, Medium 4-7, High > 7.

Tricyclic index: Sum of peak areas of C_{19} to C_{29} (excluding C_{22}, C_{27}) tricyclic terpanes (18-23, 25, 26) in m/z 191 chromatogram over peak area of C_{30} 17α,21ß(H)-hopane (35) x 100. Low < 50, Medium 50-100, High > 100.

C_{34}/C_{35} Hopane: Peak areas of C_{34} 22R and 22S 17α,21ß(H)-hopanes (44) in m/z 191 chromatogram over peak areas of C_{35} counterparts (45). Low < 1, High >1.

Bisnorhopane index: Peak area of C_{28} 28,30-bisnorhopane (32) over peak area of C_{30} 17α,21ß(H)-hopane (35) x 100 in m/z 191 chromatogram. Low < 10, Medium 10-50, High 50.

Oleanane index: Peak area of 18α(H)-oleanane (X) in m/z 191 chromatogram over peak area of C_{30} 17α,21ß(H)-hopane (35) x 100 in m/z 191 chromatogram.

Ts/Tm: Peak area of 18α(H)-trisnorneohopane (Ts) (28) over peak area of 17α(H)-trisnorhopane (Tm) (30) in m/z 191 chromatogram.

Hopane concentration Peak area of 35 measured in RIC and normalised to added sterane standard.

Gammacerane index: Peak area of gammacerane (40) in m/z 191 chromatogram over peak area of 17α,21ß(H)-hopane (35) x 100. Low < 50, Medium 50-60, High > 60.

Bisnorhopane concentration: Peak area of 32 measured in RIC and normalised to added sterane.

Trisnorhopane concentration: Peak area of "T" measured in RIC and normalised to added sterane.

Tetracyclic index: Peak area of C_{24} tetracyclic (24) over peak area of C_{30} 17α,21ß(H)-hopane (35) x 100 in m/z 191 chromatogram.

Paleolimnological Signatures from Carbon and Oxygen Isotopic Ratios in Carbonates from Organic Carbon-Rich Lacustrine Sediments

M. R. Talbot
Geological Institute
University of Bergen
Bergen, Norway

K. Kelts
Swiss Federal Water Institute (EAWAG/ETH)
Dübendorf, Switzerland

Primary and early diagenetic carbonate minerals may be associated with organic carbon-rich lacustrine sediments. The stable oxygen and carbon isotopic compositions of these carbonates can yield useful paleoenvironmental information. Primary carbonates from hydrologically open lakes typically show little or only poorly correlated covariance between oxygen and carbon isotopic variations, whereas carbonates precipitated from surface waters of closed lakes display characteristic, highly correlated covariance. Oxygen isotopic ratios in early diagenetic phases reflect the isotopic composition of the waters in which the host sediments were deposited. The carbon isotopic composition of diagenetic carbonates is determined by the nature of the dominant microbiological processes involved in diagenesis of organic matter during the earliest stages of burial. In lakes poor in dissolved sulfate, bacterial methanogenesis is dominant and produces carbonates with markedly positive $\delta^{13}C$ values. Sulfate-rich lakes favor bacterial sulfate reduction that leads to precipitation of carbonates with negative $\delta^{13}C$ signatures. Using isotopic data from carbonates from Lake Bosumtwi, Ghana, as a case study, we show that characteristic relationships exist between isotopic compositions of primary and diagenetic carbonates of sulfate-poor and sulfate-rich lakes.

INTRODUCTION

Carbonate minerals are associated with many organic carbon-rich lacustrine sediments and may be the only prominent macroscopic component of otherwise superficially featureless mudstone sequences. In addition to fossil (i.e., biogenic) carbonate, primary and/or diagenetic phases also may be present. Both the mineralogy and trace element and stable isotope compositions of these carbonates can provide paleoenvironmental information. Because some organic carbon-rich lacustrine mudrocks are valuable hydrocarbon source rocks (Powell, 1986; Fleet et al., 1988), any carbonates present may provide useful insights into the conditions under which these economically important sediments accumulated. The

relationship between primary carbonate mineralogy and water chemistry has been discussed in earlier works (Müller et al., 1972; Kelts and Hsü, 1978; Eugster and Kelts, 1983). Little systematic work has been done on trace-element geochemistry of lacustrine carbonates (although Katz et al., 1977; Katz and Kolodny, 1989; and Janaway and Parnell, 1989, were notable exceptions), but a reasonable stable-isotopic database now exists. This paper reviews processes controlling stable-isotopic variations in primary and diagenetic lacustrine carbonates and shows how they may be used to reconstruct the environmental and hydrological histories of continental basins. Data from a modern tropical African lake will be presented as a case study and compared with other lake basins, ancient and modern.

CARBONATES IN LACUSTRINE MUD AND MUDSTONE

Primary Minerals

Most primary carbonates originate as precipitates from the upper water column, often in association with phytoplankton blooms, which produce chemical disequilibrium by abstracting large volumes of CO_2 from surface waters (Kelts and Hsü, 1978; Stabel, 1986). These precipitates settle through the water column as single, silt-sized crystals or crystal aggregates of fecal origin. Typically, they accumulate as carbonate mud or as discrete laminae within clastic mudstone. In some carbonate-precipitating lakes much of the carbonate mud may have been deposited in pellet form (e.g., Spencer et al., 1984; Kelts and Shahrabi, 1986).

Studies of upper Pleistocene-Holocene lake deposits show that a variety of primary carbonate minerals occur, most commonly calcite, high magnesian-calcite, and aragonite. Monohydrocalcite has been recorded from several lakes (Stoffers and Fischbeck, 1974; Krumbein, 1975), and debate continues as to whether dolomite can precipitate as a primary mineral from open waters in perennial lakes. The occurrence of apparently primary protodolomite ooids and grain coatings in upper Pleistocene sediments of Lake Albert (Uganda-Zaire; Stoffers and Singer, 1979) indicates that dolomite precursors may at least precipitate from open lake waters. Mineralogy is apparently determined by water chemistry, although the controlling factors are not known with certainty. In some lakes a correlation has been identified between mineralogy and Mg/Ca ratio, the precipitate changing from calcite to magnesian calcite and ultimately to aragonite as the Mg/Ca ratio rises (Müller et al., 1972). However, seawater experiments have shown that other factors, notably dissolved sulfate and phosphate (Walter, 1986) and pH (Burton, 1989), also influence carbonate mineralogy and probably affect carbonate precipitation in at least some continental water bodies. In ancient lacustrine sequences carbonates interpreted as primary in origin generally are calcite, although aragonite has been recorded from sediments as old as early Tertiary. All known instances of aragonite preservation in exposed, pre-upper Pleistocene deposits are from highly organic-matter-rich (OM) mudstone (Milton, 1971; Smith and Robb, 1973; Begin et al., 1974; Anadón, Cabrera et al., 1989; Bellanca et al., 1989), where hydrophobic bituminous shale has protected the aragonite from alteration by groundwater.

Diagenetic Minerals

Diagenetic carbonates have various modes of occurrence in lacustrine mudstone. They may form crusts, laminae conformable with primary bedding, nodules formed from passive, pore-filling precipitates, or displacive spherulites or nodules. They also occur as pore filling in biogenic debris. In upper Pleistocene-Holocene sediments, diagenetic calcite, high-magnesium calcite, dolomite, siderite, manganosiderite, and rhodochrosite all have been found. Dolomite typically is poorly ordered and calcium rich, but well ordered, stoichiometric dolomite also is known (Talbot and Kelts, 1986). Of all these minerals that have been identified in pre-Pleistocene lacustrine sediments, calcite, dolomite, and siderite are most common.

STABLE ISOTOPES

Although stable-isotopic analysis of biogenic carbonates can yield valuable paleoenvironmental data (Kelts and Talbot, 1990), this study examines only abiogenic primary and diagenetic phases. We first will review the main controls on carbon and oxygen isotopic composition of these two mineral assemblages and then present data from a study of Lake Bosumtwi, Ghana.

Primary Minerals

Oxygen

Oxygen isotopic composition of primary carbonates is a function of lake water composition and mineral-precipitation temperature. Water composition is determined by the composition of surficial and groundwater inflow and rainfall directly onto the lake and by the balance between inflow and outflow vs. losses due to evaporation. In lakes with short residence times where throughflow is rapid, isotopic composition will differ little from that of the inflow. As residence time increases, isotopic evolution will occur, mainly due to evaporative loss of water vapor from the lake surface. In hydrologically closed lakes with long residence times, evaporation dominates isotopic composition (Fontes and Gonfiantini, 1967; Gat, 1981; Gonfiantini, 1986; Fritz et al., 1987). The net effect is that the water body becomes enriched in ^{18}O with respect to inflow composition.

Carbon

Primary carbonates record the carbon isotopic composition of dissolved inorganic carbon (DIC) in the upper water column at the time of precipitation. Isotopic composition of lake DIC is a complex function of inflow composition, isotopic exchange with atmospheric CO_2, lake metabolism, and water residence time (McKenzie, 1985; Lee et al., 1987; Talbot and Kelts, in preparation).

Variations in carbon and oxygen isotopic compositions of primary carbonates from short-residence, hydrologically open lakes seem generally independent of one another, although carbon isotopes typically display greater variability than oxygen (Talbot, in press; Figure 1). In some open lakes carbonate $\delta^{13}C$ has been shown to be closely related to the interplay between photosynthetic fixation of surface water DIC and circulation of respired CO_2 (e.g., McKenzie, 1985). The relatively small variations in $\delta^{18}O$ reflect the limited influence that evaporatively induced isotopic enrichment has upon these water bodies. Closed-lake carbonates, on the other hand, show a strong covarying relationship between oxygen and carbon isotopes (Figure 1). Covariance reflects the effects of long residence times on isotopic evolution of closed water bodies. Oxygen isotopic variations principally are the result of changes in water balance, whereas variations in $\delta^{13}C$ result from preferential outgassing of ^{12}C-rich CO_2 from the lake surface. Changes in primary productivity, which have been proposed to explain carbon isotopic variations in some lacustrine carbonates (Stiller and Hutchinson, 1980; McKenzie, 1985; McKenzie and Eberli, 1987; Hillaire-Marcel and Casanova, 1987; Botz et al., 1988; Gasse and Fontes, 1989), seem to have little impact on the composition of lake DIC of long-residence water bodies as recorded by their primary carbonate precipitates.

Each closed lake possesses a unique covariant isotopic trend, which is a function of basin morphology, climatic and geographic setting, and hydrology. Persistence of these covariant trends over extended periods of time and through major climatic fluctuations indicates considerable long-term hydrologic stability of many closed basins. Perturbations in the covariant trend seem only to occur when basin hydrology is permanently changed.

Diagenetic Minerals

Oxygen

Oxygen isotopic composition of diagenetic carbonates reflects the composition and temperature of sediment pore waters at the time of precipitation. Minerals considered here formed at shallow burial depths—the uppermost several meters of the sediment column. Consequently, precipitation temperatures did not differ significantly from those of the hypolimnion.

Carbon

Early diagenetic reactions in OM-rich sediments are dominated by microbiological processes that, by changing pore-water pH and alkalinity, may cause carbonate dissolution or precipitation. Because many OM-rich lake sediments initially contain abundant reactive organic matter, microbial activity can be especially vigorous. Bacterially mediated early diagenetic processes are therefore of particular importance and are the principal reasons for the commonly observed association between carbonates and lacustrine oil source rocks.

Microbiological degradation of sedimentary organic matter proceeds in a sequence of redox transformations related to the relative metabolic energy efficiencies of each oxidation reaction (Figure 2), although physiological rather than strictly thermodynamic factors apparently determine the microbiological zonation (Lovley and Klug, 1986). This sequence, which depends on the availability of oxidants required for organic matter consumption, commences in the water column and continues through a series of depth-related zones as organic matter becomes buried beneath the sediment-water interface (Berner, 1981; Hanselmann, 1986; Hesse, 1986; Capone and Kiene, 1988). The relationship between these microbially mediated reactions, carbonate diagenesis, and carbon isotopic composition of carbonate minerals is now reasonably well understood, and has been discussed by various authors, mainly with reference to marine environments (Irwin et al., 1977; Pisciotto, 1981; Baker and Burns, 1985; Curtis and Coleman, 1986; Hesse, 1986; Compton, 1988; Figure 2). In most marine settings, reactions involving molecular oxygen dominate aerobic environments, and bacterial sulfate reduction is the principal form of organic-matter degradation under anaerobic conditions. OM-rich lacustrine sediments, especially those that are potential petroleum source rock precursors, mainly accumulate under anoxic conditions (Demaison and Moore, 1980; Talbot, 1988). This paper will therefore be concerned principally with anaerobic diagenetic processes.

In oxygen-poor marine systems, much of the labile organic matter usually is oxidized by sulfate-reducing bacteria (Figure 2, reaction 2). In the presence of large amounts of reactive organic matter, sulfate-reduction rates can be very high and of particular significance to other early diagenetic processes (Lyons and Gaudette, 1979; Westrich and Berner, 1984). Compared to seawater, however, many lakes have low dissolved-sulfate concentrations (Livingstone, 1963; Davison, 1988; Cerling et al., 1988). Bacterial sulfate reduction in such water bodies is almost insignificant and is probably completed within the uppermost centimeter of the sediment surface (Lovley and Klug, 1986).

The combination of low dissolved sulfate and rapid depletion of the sulfate pool in the presence of abundant organic matter allows bacterial methanogenic processes to dominate the earliest stages of diagenesis in many lacustrine sediments. Studies have documented active bacterial methanogenesis in OM-rich sediments of many lakes (e.g., Craig, 1974; Tietze et al., 1980; Whiticar et al., 1986; Cerling et al., 1988). Two principal metabolic pathways are responsible for biogenic methane production (Figure 2). In marine systems methanogenesis proceeds initially via reduction (reaction 3) of CO_2 (or HCO_3^-) liberated by decarboxylation of organic matter during sulfate reduction (Whiticar et al., 1986), but the absence of this CO_2 source in sulfate-poor interstitial water means that fermentation (reaction 4) provides the initial methanogenic pathway (Whiticar et al., 1986; Lovley and Klug, 1986; Kuivila et al., 1989). The potential impact of organic matter fermentation on carbonate solubility is

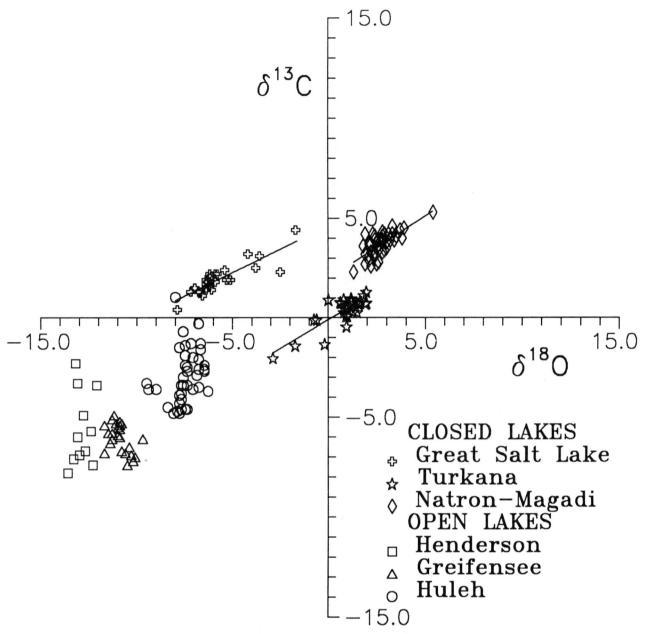

Figure 1. Typical $\delta^{13}C$ vs. $\delta^{18}O$ relationships for primary lacustrine carbonates. Great Salt Lake data replotted from Spencer et al. (1984) and McKenzie (1985); Lake Turkana (Kenya) data from Halfman et al. (1989); Lakes Natron-Magadi (Kenya-Tanzania) data replotted from Hillaire-Marcel and Casanova (1987); Henderson Lake (United States) data from Stuiver (1970); Greifensee (Switzerland) data replotted from McKenzie (1985); Lake Huleh (Israel) data replotted from Stiller and Hutchinson (1980). Carbonates from closed basins show characteristic highly correlated covariance (r normally ≥0.7).

unclear because it is, among other things, probably intimately related to concomitant reduction of unstable manganese and iron species (Kuivila and Murray, 1984; Curtis et al., 1986). In the absence of metal cation reduction it is possible that fermentation may cause carbonate dissolution, due to the fall in pore-water pH that should accompany metabolic CO_2 production. A marked rise in pore-water dissolved CO_2 concentration recently has been found where methane formation is occurring via the acetate fermentation pathway in a temperate low-sulfate lake (Kuivila et al., 1989). In some lakes dissolution due to CO_2 production probably leads to complete removal of primary carbonates, precipitation of a diagenetic phase occurring only when pH begins to rise as CO_2 reduction reactions begin to assume importance (Talbot and Kelts, 1986).

With respect to carbon isotopic compositions of the various products, the principal feature of note is that in sulfate reduction, all the carbon liberated during

Talbot and Kelts

(1) **Aerobic oxidation:**

$$CH_3COO^- + O_2 \rightarrow HCO_3^- + H_2O$$

(2) **Anaerobic sulfate reduction:**

$$CH_3COO^- + SO_4^{2-} \rightarrow 2HCO_3^- + HS^-$$

(3) **Methanogenesis - carbon dioxide reduction:**

$$HCO_3^- + 4H_2 + H^+ \rightarrow CH_4 + H_2O$$

(4) **Methanogenesis - acetate fermentation:**

$$CH_3COO^- + H_2O \rightarrow CH_4 + HCO_3^-$$

****

Carbon isotope fractionation during organic matter diagenesis:

(a)   **Reaction 2 - sulfate reduction:**

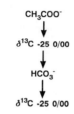

(b)   **Reaction 3 - carbon dioxide reduction:**

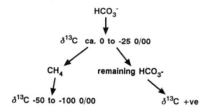

Reaction 4 - acetate fermentation:

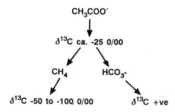

Figure 2. Principal bacterially mediated early diagenetic reactions of significance to carbonate dissolution and precipitation, in order of potential energy yield.

organic matter mineralization goes to form bicarbonate (Figure 2, reaction a), which inherits the isotopic composition of the original organic matter. In zones of intense sulfate reduction, bicarbonate of this sort dominates the DIC pool, and any resulting carbonate precipitate will therefore acquire an isotopic composition that reflects the ^{13}C-depleted character of biogenic organic matter. Methanogenesis via fermentation reactions, on the other hand, involves splitting the original OM carbon between two carbon-bearing products, methane and bicarbonate (reaction b). Because of the strong kinetic isotopic fractionation effect imposed by bacterial methanogenesis, biogenic

methane is preferentially ^{12}C enriched, which results in ^{13}C enrichment in the accompanying bicarbonate. Where methanogenesis is achieved via the carbon dioxide reduction pathway, the same kinetic fractionation effect leaves a bicarbonate pool that becomes progressively richer in ^{13}C as methanogenesis progresses. Carbonates that result from either of these two methanogenic processes will reflect this fractionation with $\delta^{13}C$ values that are considerably more positive than that of the original organic matter (Figure 2). Under conditions of high sediment-accumulation rates and intense methanogenesis, Raleigh distillation processes can lead to extreme ^{13}C enrichment in early diagenetic carbonates associated with OM-rich sediments (Claypool and Kaplan, 1974; Kelts and McKenzie, 1982; Talbot and Kelts, 1986).

CASE STUDY— LAKE BOSUMTWI, GHANA

Lake Bosumtwi, presently about 5 km in diameter with a maximum depth of 78 m, occupies a meteorite impact crater in the forest zone of Ghana, west Africa (Figure 3). This hydrologically closed lake has alkaline, mildly brackish waters with a pH of 9.1 to 9.6, total dissolved solids (TDS) of 725 mg/L, and about 0.15 meq/L dissolved sulfate. The water column is permanently anoxic below 40 m. Cores from the center of the lake are highly OM rich (5–20% TOC; Talbot, 1988) and contain laminae and micronodules of various primary and diagenetic carbonates. Former highstands are recorded by upper Pleistocene and Holocene lacustrine sediments exposed above present lake level. Prominent among these is a zone of stromatolitic and oncolitic carbonates of Late Holocene age that overlie and encrust exposed basement rocks 20–25 m above the modern shoreline (Talbot and Delibrias, 1980).

Stable Isotopes

Methods

Wherever possible, samples have been obtained by removing fragments of microlaminae using a surgical blade or handpicking individual micronodules under a binocular microscope. Samples of one or several growth laminae were drilled from stromatolite hand specimens using a 0.4-mm-diameter dental drill. All samples were digested in commercial-grade sodium hypochlorite to remove organic matter and then washed five times in distilled water prior to analysis. Where necessary, mineral compositions were checked by XRD. Isotopic analyses were carried out at the Geological Institute, ETH, Zürich, and at the National GMS Laboratory, Geological Institute, University of Bergen, using standard techniques. Analytical data from the two laboratories have been calibrated by means of a Carrara marble standard.

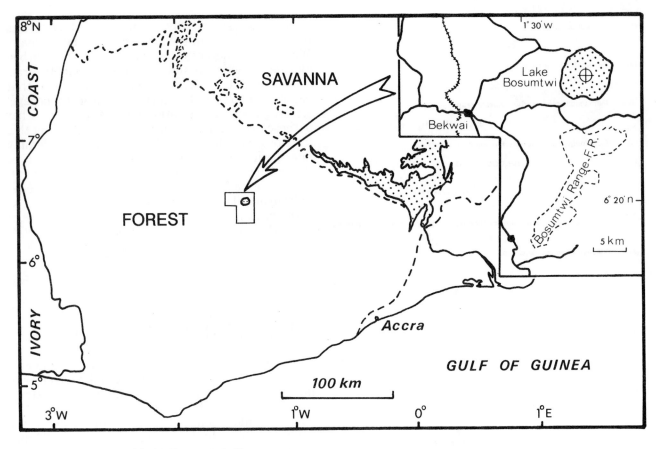

Figure 3. Location of Lake Bosumtwi, Ghana.

Results are expressed in the conventional δ notation with respect to the PDB standard.

Results

Isotopic analyses of carbonate samples from cores and algal deposits are shown in Figure 4. Although some groupings among particular minerals are apparent, the data show considerable scatter. However, when the analyses are divided into primary and diagenetic phases, it becomes apparent that primary minerals define a highly correlated linear covariant trend, and diagenetic minerals all plot on the more positive $\delta^{13}C$ side of this trend. As noted above, linear isotopic trends in primary carbonates are characteristic of hydrologically closed basins and in this case show that Lake Bosumtwi has been hydrologically closed for much of the time recorded by the cores.

The mineralogical change from calcite through magnesian calcite to aragonite is typical in continental water bodies undergoing increasing evaporative concentration (Müller et al., 1972). This is confirmed by comparing the mineralogical variations with the independently derived lake-level curve (Talbot et al., 1984; Figure 5)—aragonite forming during low-water periods and calcite or magnesian calcite at intermediate lake levels. The isotopic trend reflects these same climatically induced variations

in the lake's water balance (see Talbot, in press, for a detailed discussion).

Although less clear-cut than the primary carbonates, a similar mineralogical and $\delta^{18}O$ trend is apparent for diagenetic minerals—that is, a change from manganosiderite through calcite or magnesian calcite to dolomite with increasing $\delta^{18}O$ values (Figure 5). Comparison with the lake-level curve again confirms that both mineralogical and oxygen isotopic variations closely follow changes in the state of the lake. Manganosiderite formed in sediment deposited when the lake was particularly high and relatively dilute; calcite and magnesian calcite formed at intermediate to low levels; and dolomite characterizes sediments deposited when lake level was particularly low and probably relatively saline. The trend of manganosiderite to calcite/magnesian calcite to dolomite clearly reflects increasing pore-water salinities.

Although less clear-cut than the primary carbonates, a similar mineralogical and $\delta^{18}O$ trend is apparent for diagenetic minerals—that is, a change from manganosiderite through calcite or magnesian calcite to dolomite with increasing $\delta^{18}O$ values (Figure 5). Comparison with the lake-level curve again confirms that both mineralogical and oxygen isotopic variations closely follow changes in the state of the lake. Manganosiderite formed in sediment

Talbot and Kelts

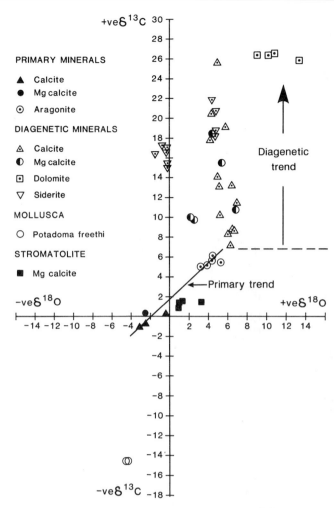

Figure 4. Isotopic composition of upper Pleistocene-Holocene carbonates from Lake Bosumtwi, Ghana. *Potadoma freethi* (gastropod) specimens were collected live from lower reaches of a stream entering the lake; stromatolite samples came from outcrops above present lake level; primary and diagenetic mineral samples were taken from cores (see Figure 5).

deposited when the lake was particularly high and relatively dilute; calcite and magnesian calcite formed at intermediate to low levels; and dolomite characterizes sediments deposited when lake level was particularly low and probably relatively saline. The trend of manganosiderite to calcite/magnesian calcite to dolomite clearly reflects increasing pore-water salinities.

Diagenetic mineral isotopic compositions follow this trend; $\delta^{18}O$ values, when projected onto the primary covariant line, indicate compositions that would be expected in primary carbonates formed at the time the host sediment was deposited (Figure 5). One can conclude therefore that although these are diagenetic minerals, their $\delta^{18}O$ compositions closely reflect the composition of the water in which the sediment accumulated. $\delta^{18}O$ values of calcites and dolomites intersect the primary trend in the zone of primary aragonites (Figures 4, 5), suggesting that

the latter could have provided a precursor mineral that was replaced by solution/reprecipitation during formation of the diagenetic minerals. No examples of partially replaced primary minerals have been observed, but distribution of the primary and diagenetic phases is mutually exclusive, and it is improbable that *no* primary carbonates were precipitated during deposition of sediments that now contain only diagenetic minerals. Primary aragonite precipitation would have been consistent with the inferred lowstand at the time sediments containing ^{18}O-rich calcite and dolomite were deposited (Talbot et al., 1984; Talbot and Kelts, 1986).

The apparent anomaly in the rather different $\delta^{18}O$ values for the two siderite groups (S1 and S2, Figure 5) can be explained by an overflow event that probably occurred in the carbonate-poor interval between 10.5 and 13.0 m (~19–23 ka). This would have had the effect of flushing the lake and resetting its bulk isotopic composition. Prior to 23 ka the lake may have had a long period of closure, which, because of the residence effect, would have caused the water body to evolve to progressively greater ^{18}O enrichment. When the $\delta^{18}O$ compositions of the S1 siderites are normalized to those of calcite, they become markedly negative (Figure 5); it is apparent that the mineral must have precipitated from lake waters that had undergone little isotopic evolution. Assuming similar precipitation temperatures, the normalized siderite ^{18}O values imply a $\delta^{18}O$ composition that is similar to that recorded by the living *Potadoma freethi* (Figure 4) collected from one of the major inflow streams, indicating that lake waters at the time of S1 precipitation had a composition close to that of the inflow. This would be possible only under conditions of rapid throughflow and thus strongly supports the previous contention (Talbot et al., 1984, and above) that S1-bearing sediments accumulated at a time when Lake Bosumtwi was hydrologically open.

Salinities at the time of precipitation of the siderite groups probably were similar (low TDS). However, isotopic evolution during periods of closure and the impact of episodic overflow on the isotopic budget of lakes are so profound that oxygen isotopic composition of neither primary nor diagenetic carbonates can be used as a reliable proxy measure of water salinity. Note, however, that the covariant trend itself does not change from one period of closure to the next. Isotopic composition of a closed lake is tightly constrained by regional rainfall pattern and basin hydrology; any variations in lake-water isotopic composition simply shifts the composition of primary carbonates along the covariant trend. Deviations from this trend occur only when major (and probably permanent) changes in basin hydrology occur, as a result of river capture or tectonically induced changes in basin morphology, for example (Talbot, in press).

Carbon isotopic composition of the diagenetic minerals is consistently shifted to more positive $\delta^{13}C$ with respect to the primary trend. As noted above,

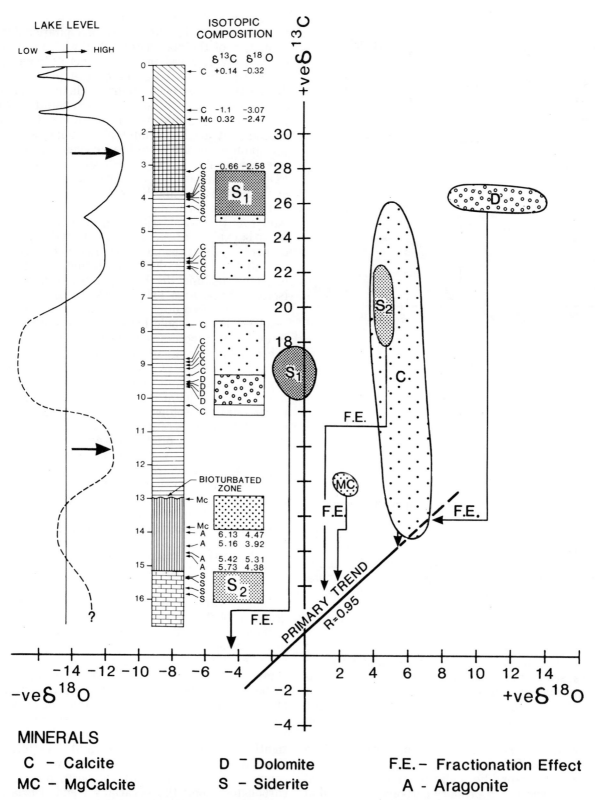

Figure 5. Generalized log of cores B6 and B7 showing stratigraphic distribution of samples (except stromatolites) and corresponding inferred lake-level changes (arrows indicate possible overflows). Vertical scale in meters. Age of base of core is about 27.5 ka. At right are covariant trend for primary carbonates and isotopic compositional fields for diagenetic minerals (from Figure 4). Patterns correspond to those of sample locations on core log. Fractionation effects (F.E.) indicate oxygen isotopic fractionations for siderite/water, magnesian calcite/water and dolomite/water with respect to calcite (siderite from Carothers et al., 1988; magnesian calcite from Tarutani et al., 1969; dolomite from Fritz and Smith, 1970).

extreme ^{13}C enrichment is characteristic of diagenetic carbonates formed in association with bacterial methanogenesis. Diagenesis in such an environment always displaces the $\delta^{13}C$ of the diagenetic phase to more positive values with respect to any corresponding primary mineral. The range of compositions shown by both groups of carbonates may therefore be simplified into two basic components—a linear trend defined by the primary minerals and a more diffuse field occupied by diagenetic minerals whose minimum $\delta^{13}C$ values are bounded by the primary carbonate trend (Figure 5).

OTHER SYSTEMS

Presently, few data are available comparable to those from Lake Bosumtwi. Lake Rukwa (Tanzania), another low-sulfate, alkaline water body, shows a similar relationship between primary and diagenetic calcites—primary minerals define a linear covariant trend (Talbot, in press), and diagenetic phases are displaced to more positive $\delta^{13}C$ values (Talbot, unpublished data).

Botz et al. (1988) obtained similar results from their study of carbonate minerals associated with OM-rich sediment in Lake Kivu. Although isotopic analysis of these minerals was complicated by difficulties in obtaining pure samples of single mineral phases, most of the pure aragonite samples, which they suggested are of primary origin, seem to define a covariant trend, and samples containing diagenetic phases all lie to the positive $\delta^{13}C$ side of this line (Figure 6). In view of possible diagenetic contamination (see below), the covariance (r = 0.68), although of lower order than is typical for most closed lakes (Talbot, submitted), suggests that the lake certainly had a long residence time and was probably a closed water body during aragonite precipitation. This interpretation is consistent with other data indicative of long hydrologic closure during the late Pleistocene and Holocene (Hecky and Degens, 1973).

Lake Kivu is notable for exceptionally high concentrations of dissolved, mainly magmatic CO_2 and biogenic CH_4 in the hypolimnion (Tietze et al., 1980). Not surprisingly, diagenetic carbonates display a typically methanic $\delta^{13}C$ signature (Figure 6). Bottom waters contain approximately 4.6 meq/L dissolved sulfate (Degens et al., 1973); some sulfate reduction is taking place within the water column (Degens et al., 1972), and pore waters likely become sulfate depleted rapidly in the presence of abundant, highly reactive organic matter (typically 5-20% TOC, Talbot, 1988; Botz et al., 1988; hydrogen index 200-560, Katz, 1988; Talbot, 1988, and unpublished).

It is noteworthy that some aragonites show extreme ^{13}C enrichment, which, according to Botz et al. (1988), reflects high rates of algal photosynthesis at the time of aragonite precipitation. However, as noted above and discussed elsewhere (Talbot and

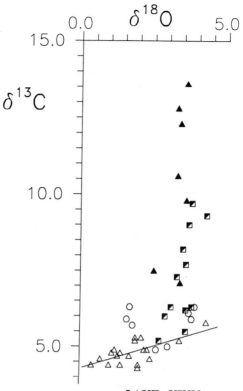

LAKE KIVU
PRIMARY AND DIAGENETIC CARBONATES

△ ARAGONITE
▲ ^{13}C–ENRICHED ARAGONITE
▨ CALCITE/DOLOMITE MIXTURES
○ ARAGONITE/CALCITE/MONOHYDROCALCITE MIXTURES

Figure 6. Isotopic composition of carbonates from upper Quaternary sediments of Lake Kivu (replotted from Botz et al., 1988). Aragonite and monohydrocalcite are considered to be primary; calcite and dolomite are diagenetic. Note similarities in primary and diagenetic phase relationships with those of Lake Bosumtwi carbonates (Figures 4 and 5). ^{13}C-enriched aragonites have $\delta^{13}C$ values that are more typical of diagenetic carbonates.

Kelts, in preparation), primary productivity apparently has no significant impact on carbon isotopic composition of carbonates from long residence-time lakes. Furthermore, such a process probably cannot cause such extreme excursions from the normally stable covariant $\delta^{18}O/\delta^{13}C$ relationship (Talbot, in press). Because these aragonites are morphologically identical to primary aragonites from Kivu and other lakes, a diagenetic origin can be dismissed (Botz et al., 1988; R. Botz, 1989, pers. comm.). A possible explanation is the release of large volumes of ^{13}C-enriched pore waters into the epilimnion at the time of aragonite precipitation. From their study of sediment and phytoplankton floras in cores from the lake, Haberyan and Hecky (1987, p.187) suggested that Lake Kivu may be subject to "extreme hydro-thermal events," such as the release of large volumes of CO_2 into the water column, and likened these to the recent explosive overturn in some Cameroonian crater lakes (Kling et al., 1987). If these events included entrainment of CO_2 generated in association with bacterial methanogenesis, they might explain the isotopically anomalous aragonites.

Primary and diagenetic carbonates in Lake Kivu sediment conform to the isotopic relationship identified in Lake Bosumtwi. We suggest, therefore, that such a compositional relationship is typical for carbonate minerals formed in closed lakes with low dissolved-sulfate contents, wherein sediments are organic-matter rich, and methanogenesis dominates early diagenesis. We are unaware of any data available for ancient lacustrine basins that convincingly demonstrate the same compositional trends, mainly because of an absence of preserved primary phases. However, Bahrig's (1989) results from a study of carbonates associated with German Cenozoic oil shale demonstrate a strong methanic $\delta^{13}C$ signature in early diagenetic siderites.

Data from other ancient basins where there are sound reasons for believing that both primary and diagenetic carbonates are preserved show rather different isotopic relationships from those seen in Lakes Bosumtwi, Rukwa, and Kivu. Examples from the Middle Devonian Orcadian basin in Scotland (Janaway and Parnell, 1989) and the Miocene Cenajo basin in Spain (Bellanca et al., 1989) are shown in Figure 7. Orcadian basin samples come from a sequence of OM-rich (1–5% TOC) carbonate laminites and dolostones. Textural and geochemical evidence suggests that some very finely laminated units preserve primary calcite deposited during high-stands. On the other hand, dolomite-bearing lithologies are believed to have formed when the Devonian lake was low (Janaway and Parnell, 1989). Cenajo basin carbonates occur in a sequence of marl, shale, diatomite, and porcellanite that accumulated in a lake that varied between shallow and saline and a deep, meromictic, relatively freshwater body. Shales from all stages are relatively rich in organic matter. Aragonite of probable primary origin occurs as fine, discrete crystals and more rarely as the dominant constituent of peloids. Diagenetic calcite and dolomite occur as pore-filling sparry cement and as microcrystalline aggregates and lenses (Bellanca et al., 1989).

As with Lake Bosumtwi, isotopic compositions of primary phases in both these Devonian and Miocene examples define a highly correlated covariant trend, whereas diagenetic minerals show considerable compositional scatter (Figure 7). A striking difference, however, is that in these cases the diagenetic field lies on the negative $\delta^{13}C$ side of the primary trend. In the discussion of microbiologically mediated diagenetic reactions, we noted that bacterial sulfate reduction produces bicarbonate with a strongly negative carbon isotopic composition that reflects the

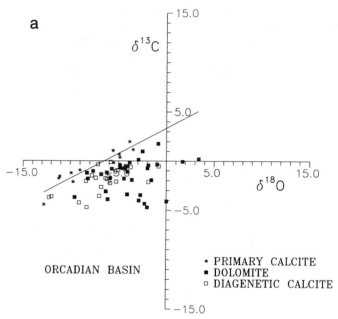

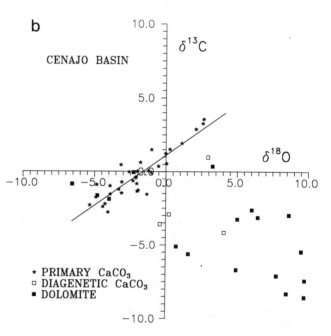

Figure 7. Primary vs. diagenetic isotopic relationships in carbonates from ancient lacustrine basins—(a) Orcadian basin (Middle Devonian), Scotland (replotted from Janaway and Parnell, 1989); (b) Cenajo basin (Miocene), Spain (replotted from Bellanca et al., 1989). Cenajo $CaCO_3$ samples consist of varying mixtures of calcite and aragonite. In both examples the primary phases show well-defined covariant trends and the diagenetic phases fall on the more negative $\delta^{13}C$ side of this line. Note also that dolomites generally have more positive $\delta^{18}O$ compositions than diagenetic $CaCO_3$.

Talbot and Kelts

[13]C-depleted nature of the parent organic matter. The distinctly negative $\delta^{13}C$ values for diagenetic minerals from the Orcadian and Cenajo basins suggest that they formed in association with active bacterial sulfate reduction. Sedimentary features associated with these carbonates independently support this notion. For example, other lacustrine deposits in the Orcadian basin contain pseudomorphs after gypsum (Parnell, 1985); studies of carbon/sulfur ratios in Orcadian sediments also suggest elevated sulfate concentrations in the lake waters (Hamilton and Trewin, 1988). In the Cenajo basin laminated gypsum units are interbedded with other lacustrine sediments, and nodules of native sulfur occur within some carbonates. These findings show that waters in both basins were relatively sulfate rich, sufficiently so for bacterial sulfate reduction to dominate the earliest stages of diagenesis. Similar carbon isotope signals occur in early diagenetic calcite nodules associated with gypsum pseudomorphs and metallic sulfides in black shale from the Culpeper basin (Upper Triassic-Lower Jurassic), Virginia (Gore, 1988; Gore and Talbot, 1988), and in carbonates that replace bedded gypsum associated with oil shale in the Libros basin (Miocene), Spain (Anadón, Rosell, and Talbot, 1989).

As in the Lake Bosumtwi example, rising $\delta^{18}O$ values in diagenetic carbonates produced in association with bacterial sulfate reduction indicate increasingly ^{18}O-enriched pore waters and suggest that carbonates formed in this diagenetic environment also preserve an isotopic record of the waters in which the host sediment was deposited. Similarly, the carbonates display the same mineralogical change, from calcite to dolomite (Figure 7), with increasing $\delta^{18}O$ that is seen in the Bosumtwi and Kivu sediments. Increasing pore-water salinity is again likely responsible for this transition.

Observations of early diagenetic dolomite associated with indicators of high dissolved-sulfate concentrations are of interest because of the debate over sulfate inhibition of dolomite precipitation (Baker and Kastner, 1981; Hardie, 1987; Burns et al., 1988). New experimental evidence supports the inhibition theory (Morrow and Ricketts, 1988). The association probably can be explained in terms of rapid depletion of the pore-water dissolved-sulfate pool due to combined rapid burial and high rates of bacterial metabolism (cf. Berner, 1978; Burns et al., 1988). Dolomite then may inherit the isotopic characteristics of bicarbonate produced in association with bacterial sulfate reduction, although it does not form in the presence of significant concentrations of dissolved sulfate. It is also noteworthy that among diagenetic carbonates, dolomites typically have the heaviest $\delta^{13}C$ values. Although this may in part reflect the composition of any precursor primary mineral, it could also indicate that dolomite precipitated at the transition into the underlying methanogenic zone, where isotopically heavy bicarbonate begins to be appear, as Burns et al. (1988)

have suggested for dolomites formed in marine OM-rich sediments.

These examples confirm that contrasting, bacterially mediated, early diagenetic processes in the OM-rich sediments of sulfate-rich and sulfate-poor lacustrine systems result in mutually exclusive carbon isotopic relationships between primary and diagenetic carbonates. Carbon isotopic compositions of the diagenetic mineral phases may therefore be used as a guide to lake-water chemistry. In low-sulfate water bodies, wherein bacterial methanogenesis dominates during earliest diagenesis, compositions are characterized by a shift to more positive $\delta^{13}C$ values, whereas early diagenetic carbonates from lakes with sufficient dissolved sulfate to allow significant bacterial sulfate reduction typically show an opposite shift, to negative $\delta^{13}C$ values (Figure 8).

CONCLUSIONS

Carbon and oxygen isotopic variations in primary lacustrine carbonates yield valuable paleohydrological information and in particular may be used to distinguish between hydrologically open and closed basins, carbonates from the latter typically displaying a highly correlated covariant relationship. Where lake sediments are rich in organic matter, intense bacterial activity favors the formation of diagenetic carbonates in the uppermost several meters of the sediment column. Although these microbiological processes may lead to dissolution of primary carbonates, the diagenetic phases themselves can yield valuable paleoenvironmental information. Water chemistry has a decisive influence upon which bacterial process is dominant during early diagenesis and leaves a clear imprint upon carbon isotopic composition of the carbonates. Sulfate-poor waters favor bacterial methanogenesis, which causes a shift to strongly positive $\delta^{13}C$ values. In relatively sulfate-rich lakes, on the other hand, bacterial sulfate reduction becomes important, producing diagenetic carbonates with markedly negative $\delta^{13}C$ compositions. Oxygen isotopic compositions are not significantly influenced by bacterial processes; in both sulfate-poor and sulfate-rich systems they preserve a record of the composition of the waters in which the host sediments were deposited. Carbonates formed in methanic and sulfidic diagenetic systems also show similar changes in mineralogy as pore-water salinity varies.

ACKNOWLEDGMENTS

The authors acknowledge support from the Royal Norwegian Research Council (NAVF). Coring operations at Lake Bosumtwi were supported by the National Science Foundation, Washington, through

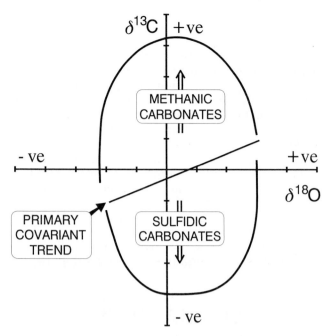

Figure 8. Summary of potential compositional relationships between primary and diagenetic carbonates from OM-rich sediments of closed-lake basins. In sulfate-poor systems, bacterially mediated methanogenesis dominates during early diagenesis processes (methanic carbonates), but in sulfate-rich water bodies bacterial sulfate reduction dominates (sulfidic carbonates).

a grant to Prof. D. A. Livingstone, Duke University. Dan Livingstone is thanked for his help and encouragement to our work on Lake Bosumtwi. Paul Baker, Barry Katz, and Judy McKenzie provided valuable refereeing. This paper was written while the senior author was on sabbatical leave at the Department of Geological Sciences, University of California at Santa Barbara. Department Chairman Mike Fuller and his colleagues are thanked for providing a friendly and stimulating environment in which to work.

REFERENCES

Anadón, P., L. Cabrera, R. Juliá, E. Roca, and L. Rosell, 1989, Lacustrine oil-shale basins in Tertiary grabens from NE Spain (western European rift system): Palaeogeography, Palaeoclimatology, Palaeoecology, v. 70, p. 7–28.

Anadón, P., L. Rosell, and M. R. Talbot, 1989, Carbonate replacement of gypsum in Upper Miocene lacustrine deposits, eastern Spain: 5th European Union of Geosciences Meeting, Strasbourg, Abstracts, p. 219–220.

Bahrig, B., 1989, Stable isotope composition of siderite as an indicator of the paleoenvironmental history of oil shale lakes: Palaeogeography, Palaeoclimatology, Palaeoecology, v. 70, p. 139–151.

Baker, P. A., and S. J. Burns, 1985, The occurrence and formation of dolomite in organic-rich continental margin sediments: AAPG Bulletin, v. 69, p. 1917–1930.

Baker, P. A., and M. Kastner, 1981, Constraints on the formation of sedimentary dolomites: Science, v. 213, p. 351–415.

Begin, Z. B., A. Ehrlich, and Y. Nathan, 1974, Lake Lisan, the Pleistocene precursor of the Dead Sea: Bulletin of the Geological Survey of Israel, v. 63, p. 1–30.

Bellanca, A., J. P. Calvo, P. Censi, E. Elizaga, and R. Neri, 1989, Evolution of lacustrine diatomite carbonate cycles of Miocene age, southeastern Spain—Petrology and isotope geochemistry: Journal of Sedimentary Petrology, v. 59, p. 45–52.

Berner, R. A., 1978, Sulfate reduction and the rate of deposition of marine sediments: Earth and Planetary Science Letters, v. 37, p. 492–498.

Berner, R. A., 1981, A new geochemical classification of sedimentary environments: Journal of Sedimentary Petrology, v. 51, p. 359–365.

Botz, R., P. Stoffers, E. Faber, and K. Tietze, 1988, Isotope geochemistry of carbonate sediments from Lake Kivu (East-Central Africa): Chemical Geology, v. 69, p. 299–308.

Burns, S. J., P. A. Baker, and W. J. Showers, 1988, The factors controlling the formation and chemistry of dolomite in organic-rich sediments—Miocene Drakes Bay Formation, California, in V. Shukla, and P. A. Baker, eds., Sedimentology and geochemistry of dolostones: SEPM Special Publication 43, p. 41–52.

Burton, E. A., 1989, The importance of fluid pH as an influence on the mineralogy of carbonate cements—Implications for early diagenesis and paleo-ocean chemistry: British Sedimentological Research Group Annual Meeting, University of Leeds, 1989, Abstracts, p. 24.

Capone, D. G., and R. P. Kiene, 1988, Comparison of microbial dynamics in marine and freshwater sediments—Contrasts in anaerobic carbon catabolism: Limnology and Oceanography, v. 33, p. 725–749.

Carothers, W. W., L. H. Adami, and K. J. Rosenbauer, 1988, Experimental oxygen isotope fractionation between siderite-water and phosphoric acid liberated CO_2-siderite: Geochimica et Cosmochimica Acta, v. 52, p. 2445–2450.

Cerling, T. E., J. R. Bowman, and J. R. O'Neil, 1988, An isotopic study of a fluvial-lacustrine sequence—The Plio-Pleistocene Koobi Fora sequence, East Africa: Palaeogeography, Palaeoclimatology, Palaeoecology, v. 62, p. 335–356.

Claypool, G. E., and I. R. Kaplan, 1974, The origin and distribution of methane in marine sediments, in I. R. Kaplan, ed., Natural gases in marine sediments: Marine Science, v. 3, p. 99–140.

Compton, J. S., 1988, Sediment composition and precipitation of dolomite and pyrite in the Neogene Monterey and Sisquoc Formations, Santa Maria basin area, California, in V. Shukla, and P. A. Baker, eds., Sedimentology and geochemistry of dolostones: SEPM Special Publication 43, p. 53–64.

Craig, H., ed., 1974, Lake Tanganyika Geochemical and Hydrographic Study—1973 Expedition: Scripps Institution of Oceanography, SIO Reference Series 75-5, p. 1–83.

Curtis, C. D., and M. L. Coleman, 1986, Controls on the precipitation of early diagenetic calcite, dolomite and siderite concretions in complex depositional sequences, in D. L. Gautier, ed., Roles of organic matter in sediment diagenesis: SEPM Special Publication 38, p. 23–33.

Curtis, C. D., M. L. Coleman, and L. G. Love, 1986, Pore water evolution during sediment burial from isotopic and mineral chemistry of calcite, dolomite and siderite concretions: Geochimica et Cosmochimica Acta, v. 50, p. 2321–2334.

Davison, W., 1988, Interaction of iron, carbon and sulfur in marine and lacustrine sediments, in A. J. Fleet, K. Kelts, and M. R. Talbot, eds., Lacustrine petroleum source rocks: Geological Society Special Publication 40, p. 131–137.

Degens, E. T., H. Okada, S. Honjo, and J. C. Hathaway, 1972, Microcrystalline sphalerite in resin globules suspended in Lake Kivu, East Africa: Mineralium Deposita, v. 7, p. 1–12.

Degens, E. T., R. P. Von Herzen, H.-K. Wong, W. G. Deuser, and H. W. Jannasch, 1973, Lake Kivu—Structure, chemistry and biology of an East African rift lake: Geologische Rundschau, v. 62, p. 245–277.

Demaison, G. J., and G. T. Moore, 1980, Anoxic environments and oil source bed genesis: Organic Geochemistry, v. 2, p. 9–31.

Eugster, H. P., and K. Kelts, 1983, Lacustrine chemical sediments, in A. S. Goudie, and K. Pye, eds., Chemical sediments and geomorphology—Precipitates and residua in the near-surface environment: London, Academic Press, p. 321–368.

Fleet, A. J., K. Kelts, and M. R. Talbot, eds., 1988, Lacustrine petroleum source rocks: Geological Society Special Publication 40, 391 p.

Fontes, J. C., and R. Gonfiantini, 1967, Comportement isotopique au cours de l'evaporation de deux bassins sahariens: Earth and Planetary Science Letters, v. 3, p. 258–266.

Fritz, P., A. V. Morgan, U. Eicher, and J. H. McAndrews, 1987, Stable isotope, fossil coleoptera and pollen stratigraphy in Late Quaternary sediments from Ontario and New York State: Palaeogeography, Palaeoclimatology, Palaeoecology, v. 58, p. 183–202.

Fritz, P., and D. C. W. Smith, 1970, The isotopic composition of secondary dolomites: Geochimica et Cosmochimica Acta, v. 34, p. 1161–1173.

Gasse, F., and J. C. Fontes, 1989, Palaeoenvironments and palaeohydrology of a tropical closed lake (Lake Asal, Djibouti) since 10000 yr B.P.: Palaeogeography, Palaeoclimatology, Palaeoecology, v. 69, p. 67–102.

Gat, J. R., 1981, Lakes, in J. R. Gat, and R. Gonfiantini, eds., Stable isotope hydrology—Deuterium and oxygen-18 in the water cycle: International Atomic Energy Agency Technical Report 210, p. 203–221.

Gonfiantini, R., 1986, Environmental isotopes in lake studies, in P. Fritz, and J. C. Fontes, eds., Handbook of environmental isotope geochemistry: New York, Elsevier, p. 113–168.

Gore, P. J. W., 1988, Lacustrine sequences in an early Mesozoic rift basin—Culpeper Basin, Virginia, USA, in A. J. Fleet, K. Kelts, and M. R. Talbot, eds., Lacustrine petroleum source rocks: Geological Society Special Publication 40, p. 247–278.

Gore, P. J. W., and M. R. Talbot, 1988, Diagenetic history of calcite concretions in lacustrine shale—Early Jurassic Midland Fishbed, Culpeper basin, Virginia (abs.): Geological Society of America Annual Meeting, Abstracts with Program, v. 20, no. 7, p. A376–A377.

Haberyan, K. A., and R. E. Hecky, 1987, The Late Pleistocene and Holocene stratigraphy and paleolimnology of Lakes Kivu and Tanganyika: Palaeogeography, Palaeoclimatology, Palaeoecology, v. 61, p. 169–197.

Halfman, J. D., T. C. Johnson, W. J. Showers, and G. S. Lister, 1989, Authigenic low-Mg calcite in Lake Turkana, Kenya: Journal of African Earth Sciences, v. 8, nos. 2–4, p. 533–540.

Hamilton, R. F. M., and N. H. Trewin, 1988, Environmental controls on fish faunas of the Middle Devonian Orcadian Basin, in N. J. McMillan, A. F. Embry, and D. J. Glass, eds., Devonian of the world: Canadian Society of Petroleum Geologists Memoir 14-3, p. 589–600.

Hanselmann, K. W., 1986, Microbially mediated processes in environmental chemistry (lake sediments as model systems): Chimia, v. 40, p. 146–159.

Hardie, L. A., 1987, Dolomitization—A critical view of some current views: Journal of Sedimentary Petrology, v. 57, p. 166–183.

Hecky, R. E., and E. T. Degens, 1973, Late Pleistocene-Holocene chemical stratigraphy and paleolimnology of the rift valley lakes of central Africa: Woods Hole Oceanographic Institution Technical Report 73-28, 93 p.

Hesse, R., 1986, Early diagenetic pore water/sediment interaction—Modern offshore basins: Geoscience Canada, v. 13, p. 165–196.

Hillaire-Marcel, C., and J. Casanova, 1987, Isotopic hydrology and paleohydrology of the Magadi (Kenya)-Natron (Tanzania) basin during the Late Quaternary: Palaeogeography, Palaeoclimatology, Palaeoecology, v. 58, p. 155–181.

Hsü, K. J., and K. Kelts, 1978, Late Neogene chemical sedimentation in the Black Sea, in A. Matter, and M. E. Tucker, eds., Modern and ancient lake sediments: International Association of Sedimentologists Special Publication 2, p. 129–145.

Irwin, H., M. L. Coleman, and C. Curtis, 1977, Isotopic evidence for the source of diagenetic carbonate during burial of organic-rich sediments: Nature, v. 269, p. 209–213.

Janaway, T. M., and J. Parnell, 1989, Carbonate production within the Orcadian basin, northern Scotland—A petrographic and geochemical study: Palaeogeography, Palaeoclimatology, Palaeoecology, v. 70, p. 89–105.

Katz, B. J., 1988, Clastic and carbonate lacustrine systems: an organic geochemical comparison (Green River Formation and East African lake sediments), in A. J. Fleet, K. Kelts, and M. R. Talbot, eds., Lacustrine petroleum source rocks: Geological Society Special Publication 40, p. 81–90.

Katz, A., and N. Kolodny, 1989, Hypersaline brine diagenesis and evolution of Dead Sea-Lake Lisan system (Israel): Geochimica et Cosmochimica Acta, v. 53, p. 59–67.

Katz, A., Y. Kolodny, and A. Nissenbaum, 1977, The geochemical

evolution of the Pleistocene Lake Lisan-Dead Sea system: Geochimica et Cosmochimica Acta, v. 41, p. 1609–1626.

Kelts, K., and K. Hsü, 1978, Freshwater carbonate sedimentation, in A. Lerman, ed., Lakes—Chemistry, geology, physics: New York, Springer-Verlag, p. 295–323.

Kelts, K., and J. A. McKenzie, 1982, Diagenetic dolomite formation in Quaternary anoxic diatomaceous muds of DSDP Leg 64, Gulf of California: Initial Reports of the Deep Sea Drilling Project, v. 64, p. 553–570.

Kelts, K., and M. Shahrabi, 1986, Holocene sedimentology of hypersaline Lake Urmia, northwestern Iran: Palaeogeography, Palaeoclimatology, Palaeoecology, v. 54, p. 105–130.

Kelts, K., and M. R. Talbot, 1990, Lacustrine carbonates as geochemical archives of environmental change and biotic-abiotic interactions, in M. M. Tilzer, and C. Serruya, eds., Ecological structure and function in large lakes: Madison, Wis., Science Tech., p. 290–317.

Kling, G. W., M. A. Clark, H. R. Compton, J. D. Devine, W. C. Evans, A. M. Humphrey, E. J. Koenigsberg, J. P. Lockwood, M. L. Tuttle, and G. N. Wagner, 1987, The 1986 Lake Nyos gas disaster in Cameroon, West Africa: Science, v. 236, p. 169–174.

Krumbein, W. E., 1975, Biogenic monohydrocalcite spherules in lake sediments of Lake Kivu (Africa) and the Solar Lake (Sinai): Sedimentology, v. 22, p. 631–634.

Kuivila, K. M., and J. W. Murray, 1984, Organic matter diagenesis in freshwater sediments—The alkalinity and total CO_2 balance and methane production in the sediments of Lake Washington: Limnology and Oceanography, v. 29, p. 1218–1230.

Kuivila, K. M., J. W. Murray, A. H. Devol, and P. C. Novelli, 1989, Methane production, sulfate reduction and competition for substrates in the sediments of Lake Washington: Geochimica et Cosmochimica Acta, v. 53, p. 409–416.

Lee, C., J. A. McKenzie, and M. Sturm, 1987, Carbon isotope fractionation and changes in flux and composition of particulate matter resulting from biological activity during a sediment trap experiment in Lake Greifen, Switzerland: Limnology and Oceanography, v. 32, p. 83–96.

Livingstone, D. A., 1963, Chemical composition of rivers and lakes: USGS Professional Paper 440-G, 64 p.

Lovley, D. R., and M. J. Klug, 1986, Model for the distribution of sulfate reduction and methanogenesis in freshwater sediments: Geochimica et Cosmochimica Acta, v. 50, p. 11–18.

Lyons, W. B., and H. E. Gaudette, 1979, Sulfate reduction and the nature of organic matter in estuarine sediments: Organic Geochemistry, v. 1, p. 151–155.

McKenzie, J. A., 1985, Carbon isotopes and productivity in the lacustrine and marine environment, in W. Stumm, ed., Chemical processes in lakes: New York, Wiley, p. 99–118.

McKenzie, J. A., and G. P. Eberli, 1987, Indications for abrupt Holocene climatic change—Late Holocene oxygen isotope stratigraphy of the Great Salt Lake, Utah, in W. H. Berger, and L. D. Labeyrie, eds., Abrupt climatic change—Evidence and implications: NATO Advanced Study Institutes Series C, p. 127–136.

Milton, C., 1971, Authigenic minerals in the Green River Formation: Wyoming University Contributions to Geology, v. 10, p. 57–63.

Morrow, D. W., and B. D. Ricketts, 1988, Experimental investigation of sulfate inhibition of dolomite and its mineral analogues, in V. Shukla, and P. A. Baker, eds., Sedimentology and geochemistry of dolostones: SEPM Special Publication 43, p. 25–38.

Müller, G., G. Irion, and U. Forstner, 1972, Formation and diagenesis of inorganic Ca-Mg carbonates in the lacustrine environment: Naturwissenchaften, v. 59, p. 158–164.

Parnell, J., 1985, Evidence for evaporites in the O.R.S. of northern Scotland—Replaced gypsum horizons in Easter Ross: Scottish Journal of Geology, v. 21, p. 377–380.

Pisciotto, K. A., 1981, Review of secondary carbonates in the Monterey Formation, California, in R. E. Garrison, and R. G. Douglas, eds., The Monterey Formation and related siliceous rocks of California: SEPM Pacific Section, Los Angeles, p. 273–283.

Powell, T. G., 1986, Petroleum geochemistry and depositional setting of lacustrine source rocks: Marine and Petroleum Geology, v. 3, p. 200–219.

Smith, J. W., and W. A. Robb, 1973, Aragonite and the genesis of carbonates in the Mahogany zone oil shales of Colorado's Green River Formation: U.S. Bureau of Mines Report of Investigations 7727, 21 p.

Spencer, R. J., M. J. Baedecker, H. P. Eugster, R. M. Forester, M. B. Goldhaber, B. F. Jones, K. Kelts, J. McKenzie, D. B. Madsen, S. L. Rettig, M. Rubin, and C. J. Bowser, 1984, Great Salt Lake, and precursors, Utah—The last 30,000 years: Contributions to Mineralogy and Petrology, v. 86, p. 321–334.

Stabel, H. H., 1986, Calcite precipitation in Lake Constance— Chemical equilibrium, sedimentation and nucleation by algae: Limnology and Oceanography, v. 31, p. 1081–1093.

Stiller, M., and G. E. Hutchinson, 1980, The waters of Merom, a study of Lake Huleh, part 1—Stable isotopic composition of carbonates of a 54 m core, paleoclimatic and paleotrophic implications: Archiv für Hydrobiologie, v. 89, p. 275–302.

Stoffers, P., and R. Fischbeck, 1974, Monohydrocalcite in the sediments of Lake Kivu (East Africa): Sedimentology, v. 21, p. 163–170.

Stoffers, P., and A. Singer, 1979, Clay minerals in Lake Mobutu Sese Seko (Lake Albert)—Their diagenetic changes as an indicator of the paleoclimate: Geologische Rundschau, v. 68, p. 1009–1024.

Stuiver, M., 1970, Oxygen and carbon isotope ratios of fresh-water carbonates as climatic indicators: Journal of Geophysical Research, v. 75, p. 5247–5257.

Talbot, M. R., 1988, The origins of lacustrine oil source rocks— Evidence from the lakes of tropical Africa, in A. J. Fleet, K. Kelts, and M. R. Talbot, eds., Lacustrine petroleum source rocks: Geological Society Special Publication 40, p. 29–43.

Talbot, M. R., (in press), The palaeohydrological interpretation of carbon and oxygen isotopic ratios in primary lacustrine carbonates: Chemical Geology, Isotope Geoscience Section.

Talbot, M. R., and G. Delibrias, 1980, A new Late Pleistocene— Holocene water-level curve for Lake Bosumtwi, Ghana: Earth and Planetary Science Letters, v. 47, p. 336–334.

Talbot, M. R., and K. Kelts, 1986, Primary and diagenetic carbonates in the anoxic sediments of Lake Bosumtwi, Ghana: Geology, v. 14, p. 912–916.

Talbot, M. R., and K. Kelts, (in preparation), Environmental controls on carbon isotopic variations in authigenic lacustrine carbonates.

Talbot, M. R., D. A. Livingstone, P. G. Palmer, J. Maley, J. M. Melack, G. Delibrias, and S. Gulliksen, 1984, Preliminary results from sediment cores from Lake Bosumtwi, Ghana: Palaeoecology of Africa, v. 16, p. 173–192.

Tarutani, T., R. N. Clayton, and T. Mayeda, 1969, The effect of polymorphism and magnesium substitution on oxygen isotope fractionation between calcium carbonate and water: Geochimica et Cosmochimica Acta, v. 33, p. 987–996.

Tietze, K., M. Geyh, H. Müller, L. Schröder, W. Stahl, and H. Wehner, 1980, The genesis of methane in Lake Kivu (Central Africa): Geologische Rundschau, v. 69, p. 452–472.

Walter, L. M., 1986, Relative efficiency of carbonate dissolution and precipitation during diagenesis—A progress report on the role of solution chemistry, in D. L. Gautier, ed., Roles of organic matter in sediment diagenesis: SEPM Special Publication 38, p. 1–11.

Westrich, J. T., and R. A. Berner, 1984, The role of sedimentary organic matter in bacterial sulfate reduction—The G model tested: Limnology and Oceanography, v. 29, p. 236–249.

Whiticar, M. J., E. Faber, and M. Schoell, 1986, Biogenic methane formation in marine and freshwater environments—CO_2 reduction vs. acetate fermentation, isotope evidence: Geochimica et Cosmochimica Acta, v. 50, p. 693–709.

Reflections on a Rift Lake

Thomas C. Johnson
Patrick Ng'ang'a*
Department of Geology
Duke University Marine Laboratory
Beaufort, North Carolina, U.S.A.

High-resolution seismic profiles, side-scan sonar records, and sediment cores collected from Lake Malawi have been analyzed to determine the nature of sedimentation in a modern rift lake. More than 4500 m of sediment have accumulated in the deepest basin in the northernmost part of the lake. If the modern sedimentation rate of 1 mm/yr is representative of most of the lake's history, then the deepest basin may, when compaction is accounted for, be on the order of 26 Ma. Although it is an open-basin lake at present, it has several times in the past been a closed-basin lake in response to drier climate. Lake level has been 100–150 m lower than present at least three times in the last 10 k.y.

The distribution of modern sediment is quite complex. Little or no deposition occurs in most regions shallower than 100 m due to storm-generated surface-wave activity. Gravitational transport of sediment by creep, debris flows, slumping, and turbidity currents is common, particularly off deltas and border faults. Diatom-rich clays occur in the deep basins far removed from major terrigenous input. These typically have organic carbon concentrations of 3–6 wt. %. Laminated sediments are common in Lake Malawi, and their frequency increases with water depth. Neither the abundance of organic carbon nor frequency of laminations increases abruptly below the depth of the chemocline. This probably results from the rise and fall of the chemocline as well as lake level in response to climatic changes. Although source rock potential of Lake Malawi as a future petroleum resource is high, the reservoir rock potential has yet to be demonstrated.

INTRODUCTION

The importance of lacustrine basins as sources and reservoirs of hydrocarbons in commercial proportions has become more apparent in recent years with the discovery of important fields in many areas, including offshore west Africa and Brazil, large basins in eastern China, and southern Sudan. This has spurred oil companies to explore other lake basins that may have the requisite size, quantity, and quality of hydrocarbons to merit exploitation.

In many of these lacustrine basins, however, the sedimentary sequences are complex and difficult to interpret in terms of depositional environments. For example, widely disparate sedimentary facies occur in close juxtaposition, evaporites occur in some sequences but not in others, signals of water depth within the sediments appear mixed, and clastic size and sorting vary rapidly and, at times, in an incomprehensible manner. Many of these complexities will be resolved only through careful analysis of sedimentation in modern large lakes. Through such studies the influence of various factors (such as tectonic and climatic setting, biological produc-

tivity, and water chemistry) on sediment composition and distribution will become apparent, and the interpretation and modeling of ancient lacustrine sequences will become more reliable.

The large lakes presently occupying the continents are of either tectonic or glacial origin and are found mostly between latitudes 0° and 20° or 40° and 70° (Johnson, 1984). Tectonic lakes are of most relevance to petroleum exploration. Glacial lakes, on the other hand, tend to be short lived and hold little potential for accumulation of significant organic matter. Among the tectonic lakes, those formed by extensional and strike-slip processes are more common than lakes formed at convergent plate margins. Large lakes of the east African rift valley perhaps are the best modern analogs of synrift lake basins. They span a wide latitudinal range, from 3°N to 14°S, and exhibit a substantial hydrological gradient, from the arid, closed basin setting of Lake Turkana (Rudolf) in northern Kenya to the relatively moist, tropical settings of Lakes Tanganyika and Malawi farther south.

The purpose of this paper is to describe the sedimentology of one of the large tropical rift lakes, Lake Malawi (Nyasa), in terms of structure, climatic setting, and limnology. Results are based primarily on high-resolution seismic-reflection profiles and supplemented by multichannel seismic data, side-

---

*Present address: National Museums of Kenya, P. O. Box 40658, Nairobi, Kenya

scan sonar, and cores. The data, acquired by Duke University's Project PROBE in 1986–1987, provide a comprehensive example of how sediment composition changes and to what degree in a rift lake in response to various environmental factors. However, we have made no attempt to develop a sedimentological model for rift lakes based on this survey. Such models can evolve only from equally exhaustive studies of other modern rift lakes. Nevertheless, the Lake Malawi results should have direct applicability to analysis and modeling of ancient lacustrine deposits.

STRUCTURAL SETTING

Tiercelin et al. (1988) attributed the Malawi trough to en echelon extension associated with dextral strike-slip motion along the Tanganyika-Rukwa-Malawi fault zone and the Zambezi lineament. The lake basin is complexly faulted and bounded by active faults that date as early as Pliocene (Crossley and Crow, 1980; Ebinger et al., 1984) and possibly early Oligocene (Tiercelin et al., 1988). It consists of a series of generally north-south–trending half-grabens of alternating polarity. The degree to which the half-grabens overlap strongly influences complexity of the resultant structure.

Rosendahl (1987) established a model of rift architecture, based largely on multichannel seismic data from Lake Tanganyika, that applies well to the northern half of the Malawi rift (Flannery, 1988). Here the border faults appear to be grossly arcuate in plan view and shift from the eastern lake shore in the northernmost basin (Livingstone) to the western shore in the next basin to the south (Usisya-Mbamba) (Figure 1). The slight degree of overlap between opposing faults gives rise to an *interference accommodation zone* (formerly termed low-relief accommodation zone). Slight bathymetric relief associated with this zone separates the Livingstone and Usisya-Mbamba basins into distinct depocenters (Figure 2). More than 4000 m of sediment underlie parts of the Livingstone basin, whereas maximum sediment thickness in the Usisya-Mbamba basin is about 3000 m (Flannery, 1988). The structure of the southern two-thirds of the Usisya-Mbamba basin is complicated by the Mbamba border fault system on the eastern shore, which along much of its length overlaps the Usisya border fault system on the western shore (Figure 1). This creates a relatively long interference accommodation zone that divides the basin into four subbasins in the subsurface structure (Flannery, 1988). Surface expression of this structure is best revealed in a northwest-trending lineament of bathymetric offset that extends from about 11°50′S, 34°45′E to 11°10′S, 34°20′E (Figure 2).

The next structural unit to the south is the Bandawe-Metangula depositional province. It includes the Likoma platform on the northeast, the

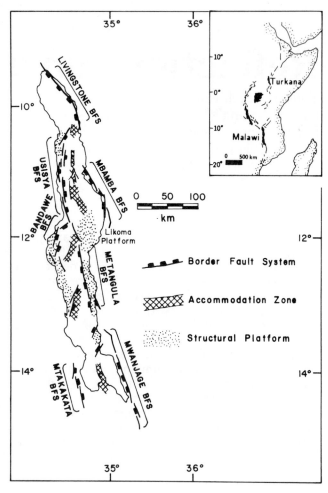

Figure 1. Structural setting of Lake Malawi (from Flannery, 1988), showing depositional basins and accommodation zones in relation to overlapping of en echelon border fault systems. Inset map shows location of Lake Malawi with respect to other lakes in the east African rift basin.

Bandawe border fault system near the western shore opposite the Likoma platform, and the Metangula border fault through the southern two-thirds of the province. Both border fault systems are unusual in terms of what Rosendahl (1987) described for the Tanganyika model. The Bandawe fault system lies within the lake and consists of several minor normal faults rather than one major fault. The linear Metangula border fault, which also lies lakeward of the eastern shoreline (Figure 1), coincides with the axial deep in this part of the lake and, as will be shown, strongly influences sedimentation in this depositional basin. Up to 2500 m of sediment underlie this province (Flannery, 1988).

Finally, the Mwanjage-Mtakataka province encompasses the lake's far southeastern and southwestern arms. The structural grain, as defined by the trend of several minor normal faults, is northeast-southwest. Relatively linear border faults of relatively minor offset trend nearly north-south, as in the basins farther north, and completely overlap to

Johnson and Ng'ang'a

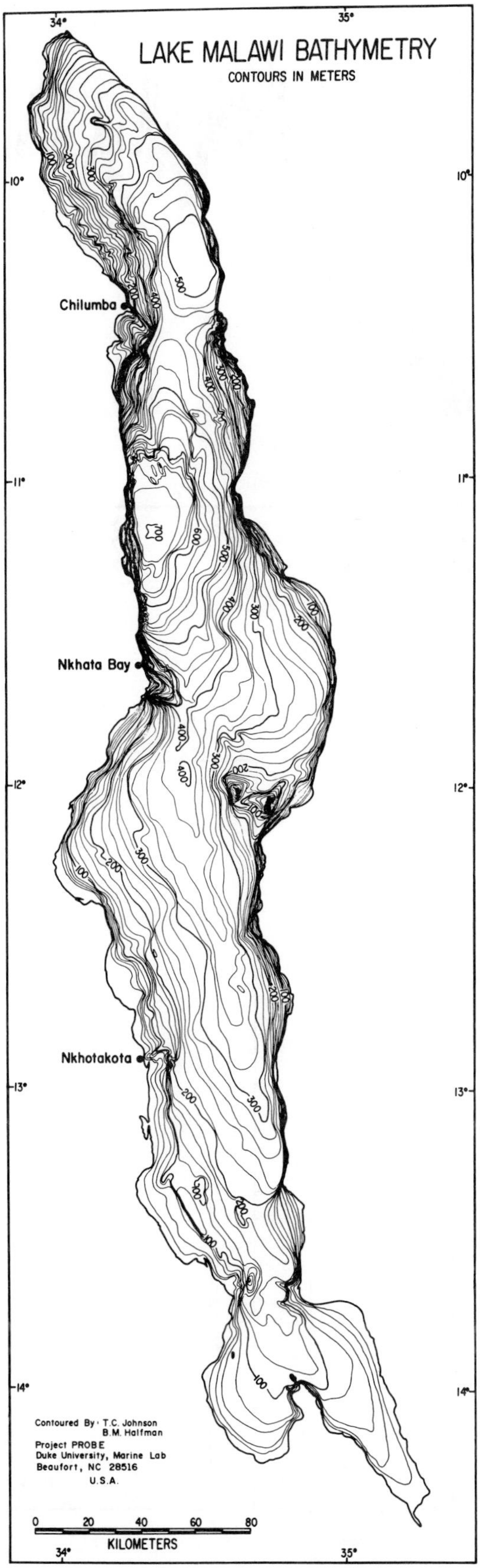

Figure 2. Bathymetric map of Lake Malawi. Contour interval 20 m.

form a nearly symmetrical, full-graben structure. Both border faults lie several kilometers inland from the lake shore. Sediment thickness in this region averages about 800 m.

The greater thickness of sediment underlying the northern basins and higher topographic relief in the northern drainage basin compared to the southern basins suggest that the age of rifting is greater in the north. Dixey (1956) estimated that rifting in the Malawi region began in the Miocene, and Crossley and Crow (1980) suggested a Miocene to Pliocene age for the modern rift. Given the maximum sediment thickness in the Livingstone basin (4500 m) and a modern sedimentation rate of 1 mm/yr, corrected for compaction (to be explained later), a maximum age for the basin is approximately 26 Ma or late Oligocene. Recent K-Ar dates on the Rungwe volcanics north of Lake Malawi loosely constrain the age of initiation of the basin at 5 to 8 Ma (C. Ebinger and A. Deino, pers. comm.); however, Tiercelin et al. (1988) published much older dates (35–40 Ma) for the "older extrusives" in the Rungwe massif.

LIMNOLOGY

Lake Malawi is about 550 km long and 50 km wide. It has a mean depth of about 200 m and a maximum depth of just over 700 m in the Usisya-Mbamba basin (Figure 2). The lake volume is 6140 km³, with a surface area of 22,490 km² and drainage-basin surface area of 65,000 km². Hydrologic balance is dominated by rainfall, river runoff, and evaporation. Outflow via the Shire River at the southern end of the lake comprises just 20% of total water loss. Rainfall and runoff add 1.73–2.5 m/yr to lake level, but evaporation averages about 1.9 m/yr; river runoff accounts for an annual loss of only 0.45 m (Pike, 1964). The large proportion of the hydrologic budget due to evaporation affects major lake-level fluctuations as a result of climatic change. The seasonal fluctuation in lake level is 0.4–1.7 m, and the interannual lake-level variability has been about 6 m during the last century (Pike and Rimmington, 1965) (Figure 3). During the lake's lowest recorded stage in 1915, a sand bar built across the Shire River outlet, stopping the outflow. By 1935 the lake level had risen 6 m, and the sand bars finally were breached, allowing outflow to resume (Beadle, 1981). This proclivity for Lake Malawi to become a closed-basin lake in response to only minor climatic change has strong implications for the evolution of its sedimentary sequences, for evidence will be presented that several times in the geologic past, lake levels dropped several hundred meters.

The thermal structure of Lake Malawi, although relatively weak, experiences distinct seasonal variability (Eccles, 1974). Surface temperatures reach 28°C in late February to early March and drop below 23°C in July and August. The thermocline that

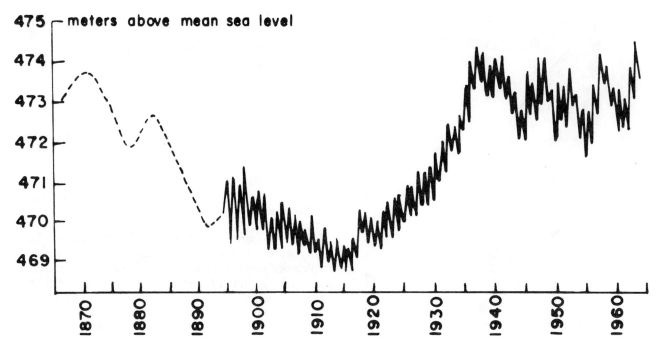

Figure 3. Lake Malawi water levels, 1865-1963 (from Beadle, 1981). Dashed line reflects measurements of early explorers prior to establishment of a monitoring system.

develops during the warm season (September to March) separates water warmer than 24°C above a depth of about 80 m from deep water at 22°–23°C (Figure 4). During the austral winter surface-water cooling in the lake's southeastern arm gives rise to density currents that flow northward along the lake floor and replace mid-depth waters in the northern basins. Seasonal mixing of the water column apparently occurs primarily in the upper 300 m, although analysis of tritium profiles suggests that as much as 20% of the hypolimnion may be replaced annually (Gonfiantini et al., 1979). Eccles (1974) reported that bottom-water temperature rose from 22.1°C in 1939 to 22.56°C in 1963; Gonfiantini et al. (1979) reported 22.70°C in 1976. This rise subsequently has been attributed to drift in calibration of the reversing thermometer that was used (J. Edmond, pers. comm.).

The salinity of Lake Malawi is very low, approximately 0.2 ppt. Conductivity ranges from 248 μmho/cm in surface waters to about 268 μmho/cm in the hypolimnion. Chemically the water is dominated by calcium, magnesium, and sodium cations and by bicarbonate anion (Table 1). Anoxia exists below a depth of 200 m. Although concentrations of most cations and anions are only slightly elevated in the hypolimnion compared to the epilimnion, density calculations based on Chen and Millero's (1977) equations indicate that salinity structure rather than thermal structure most strongly influences the stable density profile.

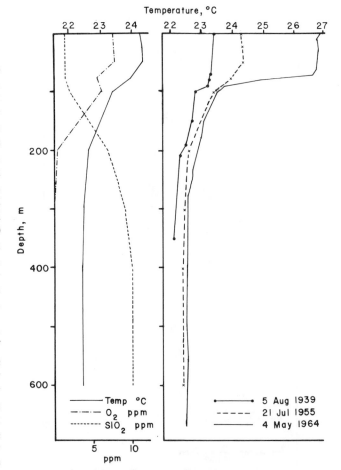

Figure 4. Profiles of temperature, oxygen, and dissolved silica off Nkhata Bay, Lake Malawi (from Eccles, 1974).

Johnson and Ng'ang'a

Table 1. Temperature, chemical, and isotopic composition of Lake Malawi at different depths at deep station (10 km east of Usisiya) on June 16, 1976 (from Gonfiantini et al., 1979)

Depth (m)	Temperature (°C)	Conductivity (µmhos/cm)	pH	O_2 (mg/L)	Ca^{+2} (meq/L)	Mg^{+2} (meq/L)	Na^+ (meq/L)	K^+ (meq/L)	HCO_3^- (meq/L)	Cl^- (meq/L)	SO_4 (meq/L)
1	25.62	248	8.0	7.90	0.90	0.61	0.88	0.16	2.36	0.14	0.13
25	25.40	247	8.1	7.90	0.94	0.62	0.87	0.16	2.32	0.14	0.19
50	25.32	248	8.1	7.82	0.90	0.61	0.87	0.16	2.33	–	0.15
75	25.09	251	8.1	7.24	0.88	0.62	0.88	0.16	2.31	–	0.15
100	23.38	254	7.75	4.72	0.94	0.63	0.90	0.16	2.40	–	0.09
125	23.12	255	7.75	3.38	0.95	0.62	0.91	0.17	2.40	–	0.09
150	22.99	259	7.75	1.96	0.94	0.63	0.91	0.17	2.40	–	–
175	22.97	260	7.7	1.51	0.96	0.62	0.90	0.16	2.42	–	–
200	22.87	260	7.7	0.36	0.98	0.63	0.90	0.17	2.41	–	–
225	22.87	262	7.35	0.00	0.98	0.63	0.90	0.17	2.40	–	–
250	22.77	265	7.4	0.00	0.98	0.63	0.91	0.17	2.42	–	–
275	22.72	261	7.45	0.00	0.99	0.63	0.89	0.17	2.45	0.20	0.15
300	22.70	265	7.5	0.00	1.01	0.63	0.89	0.17	2.44	0.21	0.13
350	22.69	264	7.4	0.00	0.97	0.63	0.90	0.17	2.44	–	0.14
400	22.72	265	7.35	0.00	0.99	0.63	0.90	0.17	2.51	–	–
450	22.69	267	7.5	0.00	0.98	0.63	0.90	0.17	2.45	–	–
500	22.76	264	7.3	0.00	0.99	0.63	0.89	0.17	2.51	–	–
520	22.66	–	–	0.00	0.98	0.63	0.90	0.17	2.48	–	–
580	22.72	–	–	0.00	0.98	0.63	0.89	0.17	2.49	–	–
600	22.75	265	7.3	0.00	1.00	0.63	0.90	0.17	2.49	0.28	0.13
640	22.70	268	7.3	0.00	0.99	0.63	0.90	0.17	2.50	0.28	0.14

The circulation in Lake Malawi is not well known. Orientation of sand spits along the shoreline and visual observations generally indicate clockwise circulation, which would be expected by the influence of the Coriolis force in the southern hemisphere. According to Eccles (1974), currents average about 0.5 km/hr. Echo soundings and bottom sampling show that erosional moats have been scoured around the bases of bedrock highs (Eccles, 1974; Scott, 1988; Johnson and Davis, 1989). Evidence will be presented for widespread erosion by bottom currents in certain areas of the lake, even in water depths of several hundred meters.

During the windy season, May to October, strong (>25 knots) southerly winds called Mweras can generate strong surface currents and waves 3–4 m high that tend to transport surface waters northward, causing depression of the pycnocline at the northern end of the lake and upwelling of nutrient-rich deep waters at the southern end. For example, the thermocline in May 1964 was 50 m deeper in northern Lake Malawi than in Monkey Bay (Eccles, 1974) (Figure 5). Relaxation of the wind stress after a Mwera gives rise to generation of seiches and internal waves. The latter can have an amplitude of about 20 m, and the turbulence associated with them effectively mixes nutrients from the metalimnion into the epilimnion (Eccles, 1974).

Biological productivity in Lake Malawi has not been measured over a sufficient period of time to be adequately quantified. Degnbol and Mapila (1982) measured values ranging between 240 and 1140 mg C/m^2/day and a lakewide average of 700 mg C/m^2/day. This is comparable to the estimated mean productivity for Lake Tanganyika (800 mg C/m^2/day) but is significantly less than that for Lakes Edward and Albert to the north (Hecky and Kling, 1987). Degnbol and Mapila's estimate for Lake Malawi is high by oceanic standards, where primary productivity ranges from about 135 mg C/m^2/day in the open ocean to about 800 mg C/m^2/day in highly productive upwelling zones (Gross, 1987). Primary productivity is believed to be higher in southern Lake Malawi than in the northern basins because of upwelling of nutrient-rich deep water in the southern half of the lake in response to the Mweras.

Lake Malawi contains between 500 and 1000 species of fish. An exact number is not possible because hundreds of new species have been identified in the past decade alone (Lewis et al., 1986). More than 90% of these species are endemic to Lake Malawi. Approximately 40,000 MT of fish are caught for local consumption each year, primarily from the lake's southeastern arm.

MODERN SEDIMENT COVER AND SEDIMENTARY PROCESSES

Distribution of sediment types in the offshore basins of Lake Malawi was mapped by 1-kHz seismic reflection profiling (Figure 6) (Johnson and Davis, 1989) and by 28-kHz echo sounding (Scott, 1988). The maps agree reasonably well with one another, despite differences in instrumentation, ship tracklines, and

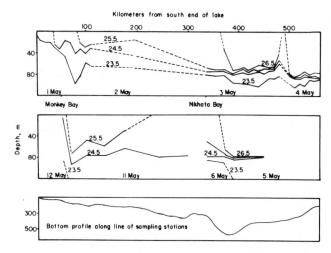

Figure 5. Isotherm depths along a north-south transect in Lake Malawi. Note the dramatic shift in the 25.5°C isotherm between May 1 and 12, 1964 (from Eccles, 1974).

criteria for defining acoustic facies. The sedimentary mosaic is complex but can be explained in terms of local tectonics, pre-existing structure, climate, and limnology (Scott, 1988; Johnson and Davis, 1989). Johnson and Davis (1989) have correlated acoustic facies with sediment type based on the 1-kHz profiles and sediment cores. Diatom ooze covers regions far removed from river deltas and border faults, where coarse-grained clastics can dilute the biogenic components. Following the northern lake shore are well developed border faults from which aprons of coarse-grained clastic sediment extend several kilometers into the lake. Higher annual rainfall in the north gives rise to greater river discharge and large river deltas, which have well-developed distributary channels that funnel turbidity currents into the deep basins offshore. That nearshore sands extend to water depths of about 100 m in most areas attests to the great depth to which surface waves can influence sedimentation in large lakes.

Among the more interesting sedimentary environments of Lake Malawi are wave-dominated nearshore regions away from major influxes of sediment, river deltas, border faults, and deep basins. The processes that affect sedimentation here vary widely. Turbidity currents, for example, are common on river deltas and border faults and eventually flow into the deep basins. Mass wasting occurs both on relatively steep slopes of lacustrine deltas and on clastic aprons shed from the border faults. Sediment scour occurs wherever lake circulation and basin physiography accelerate bottom currents to speeds exceeding critical erosion velocity; this can occur in the deep basins, in nearshore regions, or at intermediate depths along border faults.

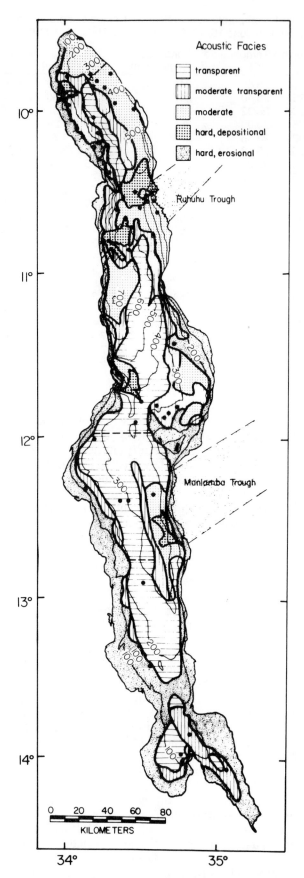

Figure 6. 1-kHz acoustic facies map (redrawn from Johnson and Davis, 1989).

118

Johnson and Ng'ang'a

Nearshore Sands

Acoustic maps of Johnson and Davis (1989) (Figure 6) and Scott (1988) show a nearshore sand facies that extends to a depth of about 100 m in many regions of the lake. This facies is characterized by high acoustic reflectivity on normal-incidence reflection profiles (Figure 7) but is difficult to sample by normal gravity or piston coring techniques. Side-scan sonar records from lake floor regions dominated by these sands indicate that they are covered by a thin, patchy veneer of mud (Figure 7). Ng'ang'a (1988) estimated that about 60% of the nearshore sands were covered by this mud veneer at a water depth of 50 m on the western side of the lake just north of Chilumba (Figure 2). The mud veneer probably is ephemeral, accumulating during relatively calm conditions and swept basinward by intermittent currents associated with storm-generated surface waves.

A 100-m-deep wave base is not unusual for a large lake. This has been observed in Lake Superior (Thomas and Dell, 1974; Johnson, 1980a) and in Lake Michigan (Lineback et al., 1971), for example. Wave size depends on wind speed, duration and fetch, and water depth. In lakes of this size, 50-knot winds can generate waves over 200 m long and 8 m high, with periods of 10 seconds (Johnson, 1980b). Waves of this size can generate orbital velocities at 100-m depths that will erode fine- to medium-grained sand (Johnson, 1980b).

Although we have not sampled the nearshore sands in detail to determine mineralogical and textural variations, they undoubtedly are immature in mineralogy and angularity because of the short transport distance and time from the source areas on the rift margin to the lake (Crossley, 1984). The nearshore and beach sands are the best sorted of the various sandy facies in the rift-lake environment because winnowing by waves, at least in very shallow waters, is very effective as a sorting mechanism.

River Deltas

Large rivers comprise the major sources of sediment to the lake. In terms of drainage basin area, the largest rivers flowing into Lake Malawi include (from south to north) Linthipe, Bua, Dwangwa, South Rukuru, Songwe, and Ruhuhu (Figure 8). Most enter the lake from the west, reflecting the influence of eastward prerift drainage (Crossley, 1984). Superimposed on this regional prerift drainage is the influence of modern rift structure—i.e., asymmetrical development of the largest drainage systems on the shoaling side of the half-graben (Le Fournier, 1980; Johnson, 1984; Rosendahl, 1987).

A pronounced north-to-south gradient has been recorded in annual rainfall on Lake Malawi, ranging from 250 cm at the northern extremity to 200 cm at Nkhata Bay to 65 cm at the southern end (Eccles, 1974). This is due in part to higher elevation in the northern drainage basin, together with prevailing southerly winds that carry moisture off the lake and into the highlands where it becomes precipitation (Eccles, 1984). Consequently, the northern large rivers have higher discharge and probably carry more sediment into the lake than do the southern rivers. This is reflected in the deep-basin sediments, which contain more detrital silt and sand in the northern basins than in the south (Johnson and Davis, 1989).

In 1987 we surveyed the North Rukuru and Ruhuhu River deltas in northern Lake Malawi with high-resolution seismic profiling, side-scan sonar, and gravity coring. North Rukuru River is of modest size and evulges across a normal fault at the shoreline (Crossley and Tweddle, 1984). Consequently, the lake floor drops rapidly to the 80-m contour but levels off thereafter. The upper section of a prodelta fan that extends about 25 km into the basin shows slump deposits and several incised turbidite channels, indicative of a high sediment-supply rate (Figure 9) (Ng'ang'a, 1988). Sediment cores from the fan lakeward of the 80-m isobath reveal numerous turbidites and sandy debris flows (Figure 10), typically 10–20 cm thick (Ng'ang'a, 1988).

The Ruhuhu deltaic fan covers an area of about 400 km², which is comparable to the North Rukuru. It is, however, different in that its upper fan is incised by several deep distributary channels that radiate from the river mouth. Side-scan sonar records show channel floors strewn with gravel; well-developed levees appear to consist of barchanlike dune features approximately 5 m high and 300 m wide with wavelengths of about 75 m (Figure 11). Slumps extend about 10 km into the basin and are replaced lakeward by sediment waves that suggest creep processes that corrugate or wrinkle the lake floor morphology (Ng'ang'a, 1988). Creep bed forms can be traced across the lake basin from the Ruhuhu fan to the deepest basin on the west adjacent to the Usisya border fault (Scott, 1988). Gravity cores recovered resemble those from North Rukuru in that they contain numerous sandy turbidites and debris flows (Figure 10). The Ruhuhu delta sediments contain more quartz and less mica than the North Rukuru deposits probably because the Ruhuhu River drains a Karroo rift-valley sequence of lacustrine and fluvial sediments, whereas the North Rukuru drainage basin is underlain primarily by Precambrian metamorphic rocks (Ng'ang'a, 1988).

Border Faults

A reconnaissance survey of the Livingstone border fault area in the northern lake basin provides insight into sedimentation patterns in a tectonically active setting. Topography along the Livingstone border fault is the most spectacular of any along the lake. The Livingstone Mountains rise more than 1000 m above lake level, and water depth plunges to 500 m within 5 km of the shoreline. Slump and debris-flow lobes and turbidite channels, which are common on the steep slope adjacent to the fault, run as far as

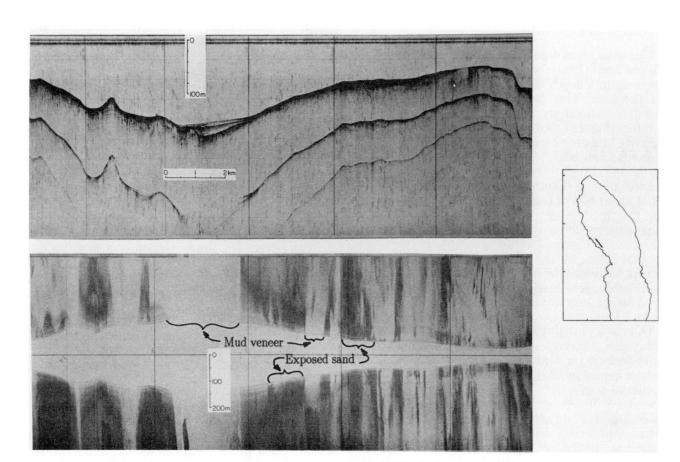

Figure 7. 1-kHz seismic reflection profile (above) and side-scan sonar record (below) offshore north of Chilumba. Note sand areas (dark patches) overlain by thin mud veneer (light areas) on side-scan sonar record. Inset map shows survey trackline.

15 km lakeward (Figure 12). Turbidity currents derived from the border fault travel to the axial deep and then apparently are diverted southward into an axially aligned turbidite channel that parallels the fault (Figure 13).

Gravity-flow features along the border fault are elongated either parallel to or normal to bathymetric contours. Contour-parallel features are mud waves that may have developed by creep, giving the lake floor a corrugated appearance. The waves, which are arcuate in plan view and concave downslope, have amplitudes of 5–7 m, crescent lengths of 200–600 m (parallel to the wave crest), and crest-to-crest wavelengths of about 80 m (Figure 12). Elongate features normal to the bathymetry include debris-flow lobes and turbidite channels. Debris-flow lobes, which can be more than 1000 m wide, typically are more reflective than the lake floor, indicating coarser sediment texture (Figure 12). Turbidite channels 100–200 m wide and 5–10 m deep are deeper and leveed farther offshore from the border fault. The axial turbidite channel that separates fault-influenced terrain from lakeward pelagic terrain 15 km offshore is about 1000 m wide and 7 m deep; its levees are about 300 m wide and rise another 7 m above the lake floor (Figure 13).

Numerous oval mounds are associated with the turbidite channels and debris-flow lobes off the Livingstone border fault. These mounds, elongate normal to the bathymetry, are up to 70 m long, 10 m wide, and 5–8 m high (Figure 13). Their origin is unknown. Subbottom reflectors beneath some of them are horizontal and continuous, whereas under others they are discontinuous and subparallel to lake floor morphology (Figure 13). Some mounds are aligned roughly parallel to nearby turbidite channels, whereas others appear as isolated features scattered over the lake floor. The mounds were not observed in other areas where side-scan surveys were taken. These features are similar to what Hovland and Judd (1988, p. 187) described as mud volcanoes, which actually are mud diapirs formed by rapid deposition on underconsolidated muds. This process could easily occur close to an active border fault.

Four gravity cores, each about 2 m long, were recovered within 20 km of the Livingstone border fault (Figure 10). All were from water depths exceeding 300 m and, because they are from the anoxic zone, are dominated by laminated diatom ooze interspersed with debris-flow sands and turbidites (Figure 10; Ng'ang'a, 1988). Cores M87–13G and M87–15G, taken within 2 km of the fault scarp, contained

Johnson and Ng'ang'a

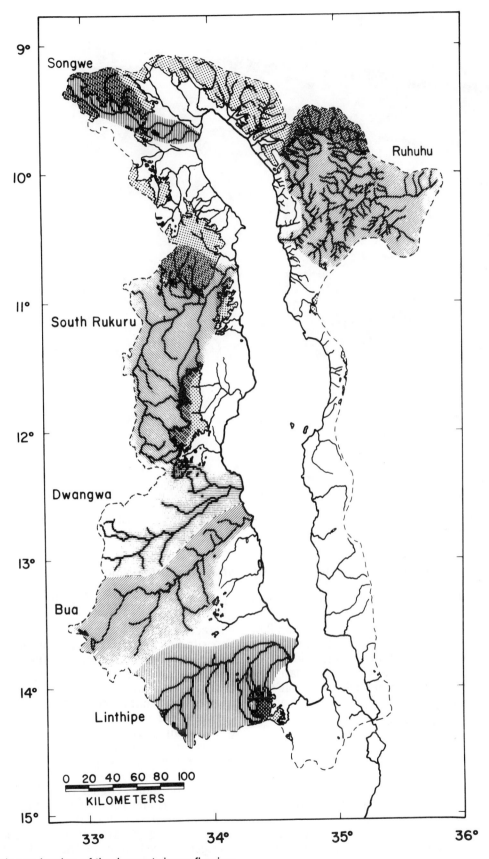

Figure 8. Drainage basins of the largest rivers flowing into Lake Malawi. Hachuring indicates areas above 1500 m elevation.

Lacustrine Basin Exploration: Case Studies and Modern Analogs

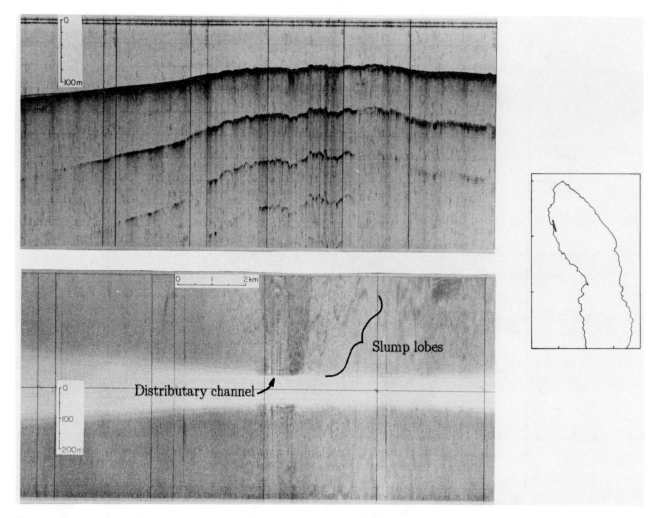

Figure 9. 1-kHz profile (above) and side-scan sonar record (below) from North Rukuru delta. Inset map shows survey trackline across the delta, parallel to the shoreline.

sandy units up to 40 cm thick that make up nearly one-half the core's length. Core M87–12G, recovered within the axial turbidite channel, contained only one sandy turbidite about 10 cm thick and two other silty, distal turbidites, comprising no more than 25% of the total core sequence. Core M87–14G, taken just east of the axial channel, showed faint evidence of two distal silty turbidites, but is more than 80% pelagic, laminated diatomaceous silt. A piston core (M86–32P) from the axial channel recovered gravel at a site nearly midway between the western and eastern shores.

Deep Basins

Most of our high-resolution seismic profiles and sediment cores were recovered from the deep basins offshore. Seismic profiles were classified according to lake-floor reflectivity—*hard*, being highly reflective with no or very weak subbottom reflectors; *moderately reflective*, often hummocky and with moderately visible subbottom reflectors; and *moder-*

ately transparent and *transparent*, with numerous subbottom reflectors, in places discernible to depths of 200 m below the lake floor (Figures 6, 14) (Johnson and Davis, 1989). Hard reflector profiles are characteristic of the nearshore sandy, erosional lag deposits but also occur in the depositional turbidite channels. Moderate reflector profiles generally denote regions where lake-floor sediments are silty, whereas acoustically transparent profiles are most common in regions of diatom ooze and clay. The correlation between surface-sediment reflectivity and texture is quite strong (Figure 15).

Because of both climatic and tectonic effects, hard or moderate reflectivity profiles are relatively more common in the northern half of the Malawi basin than in the south. Higher rainfall and greater relief in the north and the fact that major border faults are located at the shoreline rather than inland all contribute to greater influx of coarse-grained, terrigenous sediment to the north basins. In the southern basins, on the other hand, prevailing winds from the south promote upwelling and higher diatom influx.

Johnson and Ng'ang'a

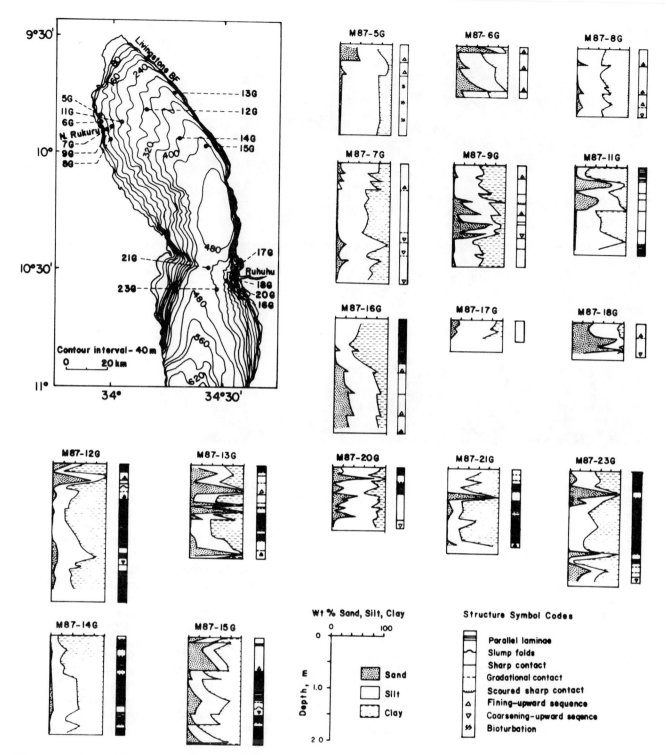

Figure 10. Sediment-core stratigraphy in northern Lake Malawi. Inset map shows bathymetry and core locations.

One region of moderate reflector profiles dramatically different from this general distribution is the elongate field along the Metangula border fault in the eastern south-central basin (Figure 6). Coarse-grained sediments here clearly were supplied from the eastern shoreline, but no major river enters there, and the drainage basin is extremely small. However, one of the small rivers drains the small Karroo Maniamba trough, whose sequences are easily eroded and contribute unusually large volumes of relatively

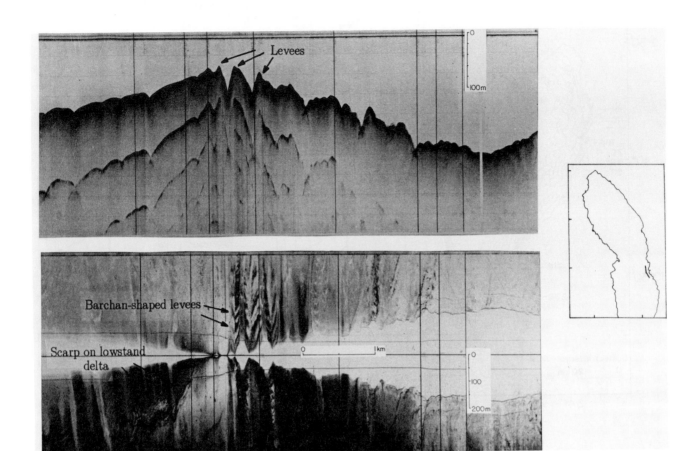

Figure 11. 1-kHz profile (above) and side-scan sonar record (below) across Ruhuhu delta. Laterally compressed side-scan sonar record clearly shows distributary channels, barchanlike dune features on levees of two main channels, and scarp of a lowstand delta. Inset map shows survey trackline.

coarse-grained material to the Malawi basin. This clearly shows the influence of pre-Cenozoic lithology on modern sediment distribution.

In some deeper regions of the lake bottom currents are strong enough to have a pronounced effect on lake-floor morphology. Erosional moats have been scoured around the bases of bedrock highs in the southern end (Figures 14, 16). Wavy bed forms also appear to be the result of bottom currents rather than mass wasting (Figure 16; Scott, 1988; Johnson and Davis, 1989). Such currents could be seasonally driven by thermal convection, wind driven, or associated with seiches. No direct measurements of this deep circulation have, however, been made in Lake Malawi.

Basinal sediments consist of biogenic, terrigenous, and authigenic components. The biogenic fraction includes diatom frustules, fish skeletal debris, terrigenous plant matter, and insect remains (Figure 17). Diatom frustules typically make up 15% of the deep basin sediment, although they can range from less than 5% to more than 50% (Johnson et al., in prep.). The diatom assemblages are dominated by *Melosira* and *Stephanodiscus* sp., with occasional horizons dominated by *Nitzschia* (Crossley and Owen, 1988; Haberyan, 1988). Other components are commonly observed but rarely exceed 2% of the silt and sand fractions. Carbonate shells are exceedingly rare except in nearshore sands, probably because deep waters are undersaturated with respect to calcium. In contrast to what Cohen and Thouin (1987) observed in Lake Tanganyika, Lake Malawi contains no recent, extensive carbonate deposits because of the difference in water chemistries. However, older lacustrine deposits, the Chiwondo Beds (Plio-Pleistocene), exposed in the basin west of the present lake, contain abundant biogenic carbonate (Thatcher, 1974; Bishop, 1987). Thus, Lake Malawi was more conducive to carbonate deposition in the past.

Total organic carbon (TOC) in Lake Malawi typically falls in the range of 3–6 wt. %, and C/N ratios of organic matter range from about 9 to 20. TOC correlates with sediment texture—finer grained sediments generally contain more TOC, as would be expected. However, no significant difference was found between the abundance or C/N ratio of TOC in sediments from the epilimnion vs. the hypolimnion, (Figure 18). Sediments in the anoxic region of the lake may have higher TOC than in oxygenated shallower regions, but the observed range of values in the two areas overlap (Figure 18). Sediment TOC values in Lake Malawi are comparable to what has

Johnson and Ng'ang'a

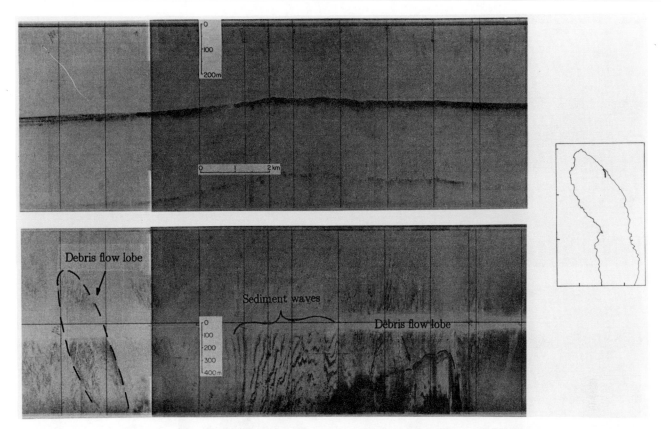

Figure 12. 1-kHz profile (above) and side-scan sonar record (below) parallel to the Livingstone border fault, showing debris-flow lobe and contour-parallel sediment waves created by downslope mass movement (creep?). Inset map shows survey trackline.

been found in most other rift lakes (Hecky and Degens, 1973; Halfman, 1987; Katz and Kelley, 1987).

One surprising aspect of high-resolution seismic stratigraphy that relates to organic-carbon content is that sediments throughout most of Lake Malawi were not found to be gaseous enough to adversely affect acoustic profiling. In Lake Turkana methane-rich sediment was a significant problem (Johnson and Davis, 1989). In view of the Lake Turkana problem, where TOC abundance is typically 0.5 wt. % and the entire water column is oxygenated, we antici-pated even greater problems in Lake Malawi, where TOC content is an order of magnitude higher and where much of the water column is anoxic. Organic-matter composition is, however, strikingly different in the two lakes. Turkana organic matter is characterized by a lower C/N ratio and higher $\delta^{15}N$ (Figure 19). Its TOC most likely is dominated by autochthonous material because the lake is in a desert setting with no significant vegetation growing within the drainage basin. In Lake Malawi sediment, on the other hand, TOC must contain a substantial amount of terrigenous organic material in addition to that derived from aquatic sources.

Terrigenous mineral components in Lake Malawi include primarily clay, silt, and sand. The sand and silt are texturally and mineralogically immature (Figure 17). Clays are dominated by smectite, some of which is authigenic nontronite (Müller and Forstner, 1973), and the remainder is detrital. Laminated muds from the northern deep basin typically contain less than 2% sand, 10–40% silt, and 50–90% clay (Ng'ang'a, 1988). Heavy-mineral analysis undoubtedly would be useful for determining provenance and mapping transport pathways, but this has not yet been attempted.

Authigenic components include nontronite, limo-nite, vivianite, opal, and calcite. Müller and Forstner (1973) reported that these minerals precipitate at the sediment-oxic water interface in water depths shallower than 250 m. They assumed that the abundance of authigenic nontronite and opal required a hydrothermal source of silica because the fluvial contribution of dissolved silica appeared to be inadequate. Johnson et al. (in prep.) found that the supply of dissolved silica by rivers nearly balances the sedimentation rate of diatom frustules. If silicate authigenesis is also an important sink for dissolved silica in Lake Malawi, then a hydrothermal source of silica, such as hot springs, is required to balance the silica budget. Hot springs are common in the Lake Malawi drainage basin, and if they occur within the

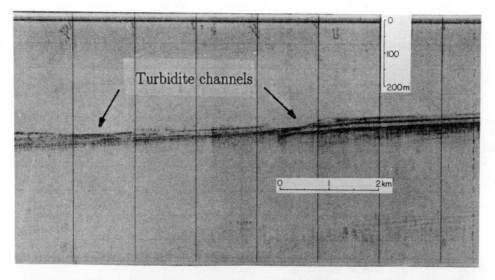

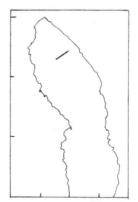

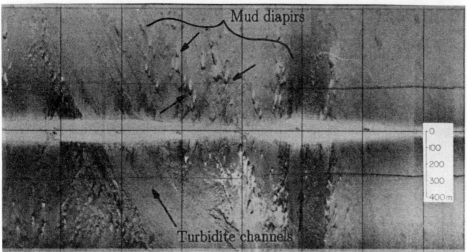

Figure 13. 1-kHz profile (above) and side-scan sonar record (below) of an axial turbidite valley, a feeder valley, and sediment mounds approximately 15 km offshore from the Livingstone border fault at latitude 9°45′S. Inset map shows survey trackline.

lake itself, hydrothermal iron sulfides may exist that are comparable to those found in Lake Tanganyika (Vaslet et al., 1987; Tiercelin et al., 1989).

Crossley and Owen (1988) reported the occurrence of euhedral calcite crystals and vivianite rosettes in several cores from Nkhata Bay at a depth of about 350 m. We also observed euhedral calcite at certain horizons in our cores and interpret its presence as an indication of earlier lowstands.

Laminations are common in the offshore sediments of Lake Malawi. Couplets of light and dark layers typically are about 1 mm thick, with the light layer dominated by diatoms and the dark layer by a combination of diatoms and terrigenous minerals and plant debris (Pilskaln and Johnson, in press). The couplets apparently reflect seasonal variability in sedimentation—diatomaceous light layers represent the highly productive, windy, dry season, and dark layers represent the calm, rainy season. This

hypothesis currently is being tested by a time-series sediment trap deployed in northern Lake Malawi. The presence or absence of laminations in a sediment core is not a simple reflection of anoxia in the water column. We have found laminated intervals in cores from water depths well above the present chemocline, as well as nonlaminated intervals from the deep anoxic basins. Although a correlation exists between water depth and the proportion of laminated core (Figure 20), it shows no sharp discontinuity at the depth of the present chemocline (Pilskaln and Johnson, in press).

Sedimentation rates in the offshore basins have been determined by ^{210}Pb and ^{14}C measurements. Linear sedimentation rates from ^{210}Pb analyses of five cores range from 0.9 to 3.6 mm/yr (Johnson et al., in prep.), corresponding to a mass-accumulation rate of about 50 mg/cm²/yr. ^{14}C dating has provided mixed results. Dates on bulk organic matter in the

Johnson and Ng'ang'a

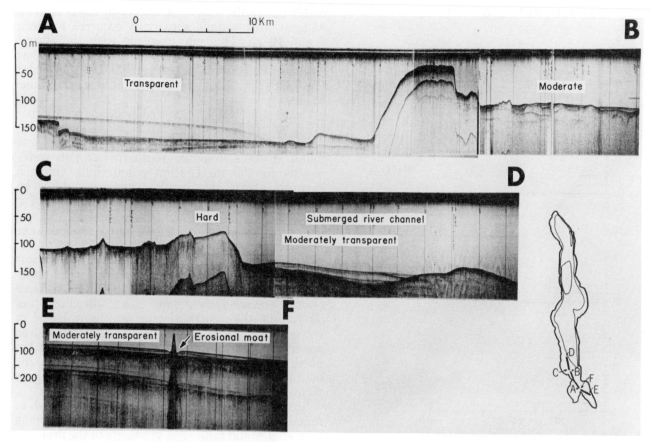

Figure 14. 1-kHz seismic-reflection profiles from southern Lake Malawi, illustrating hard, intermediate, and transparent reflectivities (top); drowned river valley filling with diatom ooze (middle); and erosional moat around a bedrock high (bottom). Inset map shows locations of profiles.

southernmost basins have been unusually old—e.g., more than 20 ka in core-top samples in cores 4P and 6P (Table 2). Other dates on bulk organic carbon appear reasonable, however, as do dates on minor amounts of authigenic calcite that have been found at certain horizons (Table 2). If we consider cores 16P and 22P representative of deep-basin sediments in the Livingstone and Metangula basins, respectively, linear sedimentation rates average 0.9 and 1.3 mm/yr, respectively, for the last 7500 yr. These rates correspond to mass-accumulation rates of 22–34 mg/cm^2/yr. The higher rates found by ^{210}Pb analysis may reflect the effects of increased land use in the drainage basin in the last century.

The age of the Lake Malawi basin, reported earlier as 26 Ma, was determined by assuming a modern sedimentation rate of 1 mm/yr and a sediment porosity of 0.9 in the upper meter of the section (Johnson et al., 1988). To calculate age with these parameters, we assume that (1) sediments compact to a porosity of 35% at 1300 m depth, based on a compaction curve for marine shale (Hamilton, 1980); and (2) the compaction curve is logarithmic from the sediment-water interface through the datum at 1300 m to the maximum burial depth of 4500 m. The resultant age-depth relationship is given by the following equation:

$$T = (0.18Z \, (\log Z) - 0.08Z) \times 10^4$$

where T is age (in years) and Z is depth (m). This yields an age of just under 26 Ma at 4500 m.

FLUCTUATIONS IN LAKE LEVEL

One of the most important factors controlling distribution of sediment facies in Lake Malawi is the dramatic fluctuations in lake level that occur in response to climatic change. Lake level has varied more than 5 m within the last century because of the near balance between freshwater input and loss by evaporation (Figure 3). When the lake was at its lowest level in historical times, outflow down the Shire River ceased. The lake achieved closed-basin status, at which point it would not have required a significantly drier climate to cause either the lake level to drop several tens of meters or the water to become substantially more saline.

Surface Reflector Strength

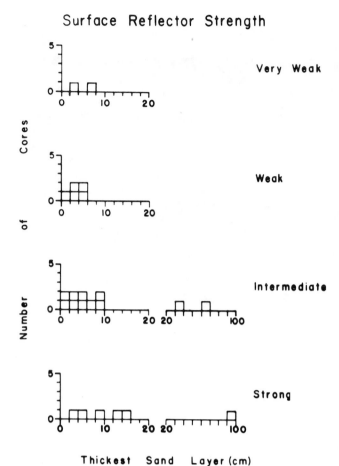

Figure 15. Maximum thickness of sand layers encountered in sediment cores from regions of varying acoustic reflector strength. Note that sand was found in only two cores from regions of very weak reflector strength. Few cores were obtained from regions of strong reflectivity because of difficulty in coring.

Scholz and Rosendahl (1988) reported seismic evidence that the lake level had been 300–400 m lower than present before about 25 ka. Although their evidence for a lowstand is indisputable, their estimate of its time is based on extrapolation of very few ^{210}Pb dates. At present the age of the major unconformity reported by Scholz and Rosendahl (1988) cannot be determined to the nearest 50 k.y., but it very likely occurred within the last 100 k.y.

We see evidence on high-resolution seismic-reflection profiles for at least two lowstands about 150–250 m below present lake level (Figure 21). The lower may coincide with Scholz and Rosendahl's (1988) unconformity. The lowstands are indicated seismically by erosional unconformities, drowned river valleys, and wave-cut scarps around the periphery of the lake basin at water depths shallower than 250 m. Farther offshore in deeper water, some of the reflectors are interpreted to correspond to periods of lowstands. Several that were cored contain nearshore benthic diatoms, desiccated mud, or authigenic calcite. We are dating these horizons to

better define the lake-level history of Lake Malawi. At least one lowstand apparently occurred between 5 and 10 ka. This may coincide with the abrupt regression at 7.5–8 ka reported for many African lakes by Street-Perrott and Roberts (1983). Dates for older lowstands have not yet been established for Lake Malawi.

Owen et al. (1990) presented evidence for another lowstand within the past 500 yr, more specifically, between 1500 and 1850 AD. An erosional hiatus was found in cores from the southern basin in water depths to 108 m. Diatom populations changed abruptly across this break, which was dated by extrapolation of ^{210}Pb dates. The presence of a lowstand at this time is substantiated by the oral history of some of the tribes living around Lake Malawi.

Lake-level fluctuations that exceed 100 m and occur frequently in response to climatic change greatly affect sedimentation patterns in a rift lake. In Lake Malawi the sandy nearshore facies will shift up to 30 km in the gently sloping southern basin in response to a 100-m drop in lake level. A 200-m drop would move the southern shoreline about 100 km north of its present position. Conversely, where the bottom slope is steeper, lateral shifts of nearshore facies would not be as dramatic.

Sedimentation rates would not be expected to remain constant as lake level rose and fell. Dropping lake level would raise river gradients in the drainage basin, thereby accelerating erosion and basinward transport of nearshore sediment. A drop also would affect authigenic mineral formation in that wind mixing would erode the upper hypolimnion, adding substantial iron and manganese to the sedimentary record. As a result of lower lake levels, higher water salinity would induce enhanced calcite precipitation and preservation. During lowstands Lake Malawi may be sufficiently saline to promote the production of nearshore carbonate facies, including stromatolites and shell-lag deposits, such as are found in Lake Tanganyika today (Cohen and Thouin, 1987). However, no evidence has yet been found for extensive nearshore carbonate deposition in Lake Malawi except in the Plio-Pleistocene deposits mentioned earlier.

IMPLICATIONS FOR OIL POTENTIAL

Sedimentation patterns in Lake Malawi are complex but understandable in terms of its tectonic, climatic, and geologic setting. Although much remains to be learned about the biological, chemical, and physical processes that control sedimentation within the lake, enough information has been obtained to show that the Malawi basin has potential as a future source of oil.

The source bed potential of Lake Malawi, like all other African rift lakes (Talbot, 1988), is high because of the widespread occurrence of deep-water muds that

Johnson and Ng'ang'a

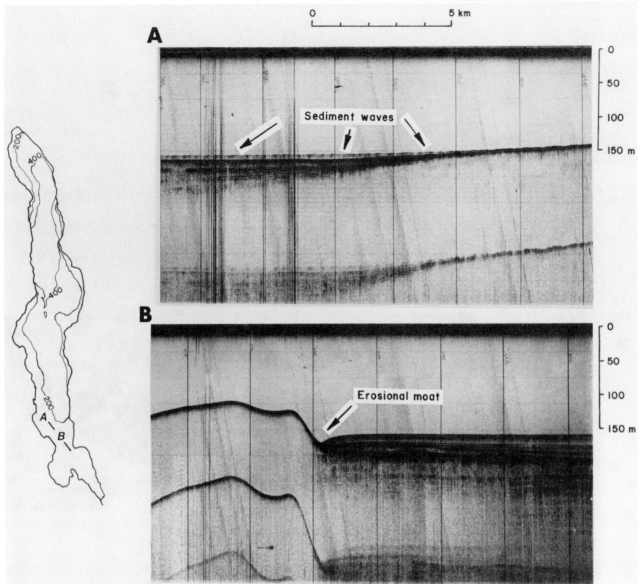

Figure 16. Seismic-reflection profiles from southern Lake Malawi, showing sediment waves (above) and erosional channel or moat along the base of a bedrock high (below). Both are evidence for bottom currents. Inset map shows location of profile.

contain about 5 wt. % organic carbon. Our analyses to characterize this organic matter suggest little difference between that which is accumulating in sediments above or below the present chemocline, which separates oxic from anoxic waters. This may be due, in part, to relatively high sedimentation rates into the lake, by which organic matter is rapidly buried in a reducing pore-water environment, regardless of the presence or absence of oxygen in the overlying water column. We see far greater influence on character of the organic matter in rift-lake sediments by climatic setting (i.e., the amount of vegetation growing in the drainage basin and whether it contributes to sedimentary organic matter) than by presence or absence of anoxia. Obviously, more sophisticated analyses of organic matter are needed to evaluate the relative importance of climatic setting vs. anoxia to source rock potential of lacustrine basins.

Perhaps a more important question regarding the oil potential of rift lakes concerns the extent and quality of reservoir beds. For example, sandy facies are associated with nearshore, deltaic, border fault, and turbidite environments. We have provided insight about the lateral extent of some of these coarse-grained deposits on the present lake floor and have presented evidence that their boundaries may shift tens of kilometers laterally as lake level fluctuates. We were, however, unable to determine thicknesses of the sand units because our coring capabilities biased the sampling toward muddy facies. Multichannel seismic data presented by

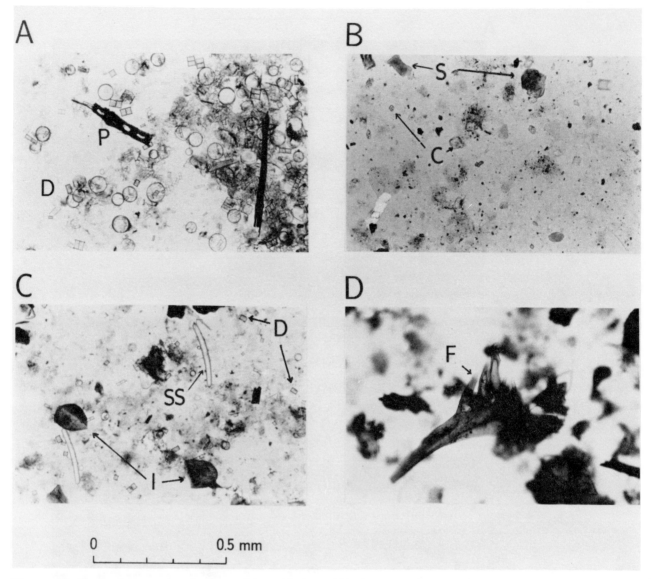

Figure 17. Photomicrographs of smear slides of Lake Malawi sediments. A, core M86-4P at 50 cm, diatom frustules (D) and plant debris (P). B, core M86-12P at 80 cm, detrital sand and silt (S) and authigenic calcite (C). C, core M87-3G at 16 cm, diatoms (D), sponge spicules (SS), and insect remains (I). D, core M86-10P at 68 cm, fish debris (F).

Flannery (1988) and Scholz and Rosendahl (1988) suggest that sand units, particularly off some of the major deltas, such as the Ruhuhu, may be on the order of 1000 m thick. Talus deposits along some of the border faults also may be of comparable thickness, but deep drilling and textural analyses will be required to evaluate these potential reservoir beds.

ACKNOWLEDGMENTS

Most of the data collection at Lake Malawi was supervised by Thomas W. Davis and John Graves, with assistance from graduate students and technicians from Duke University's Project PROBE and many able helpers from Malawi and Mozambique. We extend our sincere thanks to these people and to various officials of the governments of Malawi, Mozambique, and Tanzania, who assisted us in obtaining research permits and logistical support in the field. We thank B. Rosendahl and J. J. Tiercelin for their comments on an earlier draft of this manuscript. Financial support was provided by several oil companies (AGIP, Amoco, Exxon, Mobil, Petrofina, and Texaco) through Project PROBE, under the direction of Bruce Rosendahl, and NSF Grant ATM8816615 to TCJ.

Johnson and Ng'ang'a

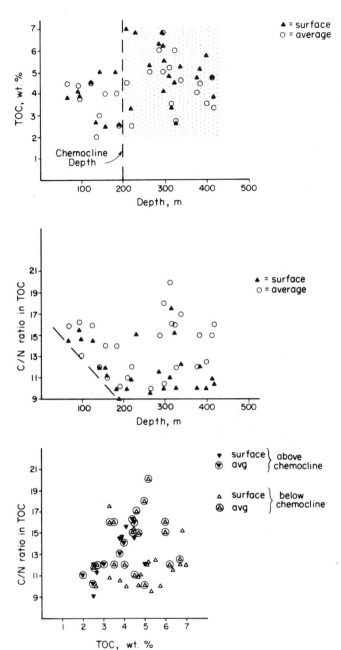

Figure 18. Total organic carbon (TOC) and C/N ratios in core-top samples and downcore averages (typically top 10 m) in Lake Malawi. Note overlap of values in samples from water depths above the chemocline where water is oxygenated and samples from anoxic deep waters.

REFERENCES

Beadle, L. C., 1981, The inland waters of tropical Africa, 2d ed.: New York, Longman, 475 p.

Bishop, M. G., 1987, Clastic depositional processes in response to rift tectonics in the Malawi rift, Malawi, Africa: M. S. Thesis, Duke University, Durham, N. C., 111 p.

Chen, C. T., and F. J. Millero, 1977, The use and misuse of pure water PVT properties for lake waters: Nature, v. 266, p. 707–708.

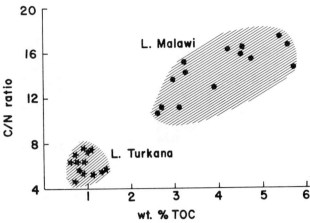

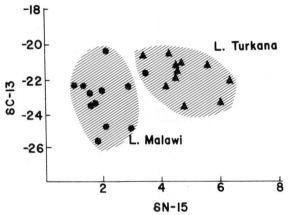

Figure 19. C/N ratio vs. TOC and δ^{13}C vs. δ^{15}N for organic matter in Lake Turkana and Lake Malawi sediments, revealing the dramatic differences attributed to the dissimilar climatic settings of the two lakes.

Cohen, A., and C. Thouin, 1987, Nearshore carbonate deposits in Lake Tanganyika: Geology, v. 15, p. 414–418.

Crossley, R., 1984, Controls of sedimentation in the Malawi rift valley, central Africa: Sedimentary Geology, v. 40, p. 33–50.

Crossley, R., and M. J. Crow, 1980, The Malawi rift, *in* A. Carrelli (president), Geodynamic evolution of Afro-Arabian rift system: Academia Nazionale dei Lincei, Proceedings no. 47, p. 78–88.

Crossley, R., and B. Owen, 1988, Sand turbidites and organic-rich diatomaceous muds from Lake Malawi, Central Africa, *in* A. J. Fleet, K. Kelts, and M. R. Talbot, eds., Lacustrine petroleum source rocks: Geological Society Special Publication 40, p. 369–374.

Crossley, R., and D. Tweddle, 1984, A note on some sedimentary features revealed by echosounding in Lake Malawi, Africa: Transactions of the Geological Society of South Africa, v. 87, p. 45–51.

Degnbol, P., and S. Mapila, 1982, Limnological observations on the pelagic zone of Lake Malawi from 1970 to 1981, *in* Biological studies on the pelagic system of Lake Malawi: Food and Agricultural Organization, FI:DP/MLW/75/019, p. 5–47.

Dixey, F., 1956, Erosion and tectonics in the East African Rift System: Quaternary Journal of the Geological Society of London, v. 95, p. 75–108.

Ebinger, C., M. J. Crow, B. R. Rosendahl, D. Livingstone, and J. LeFournier, 1984, Structural evolution of Lake Malawi: Nature, v. 308, p. 627–629.

Eccles, D. H., 1974, An outline of the physical limnology of Lake Malawi (Lake Nyasa): Limnology and Oceanography, v. 19, p. 730–742.

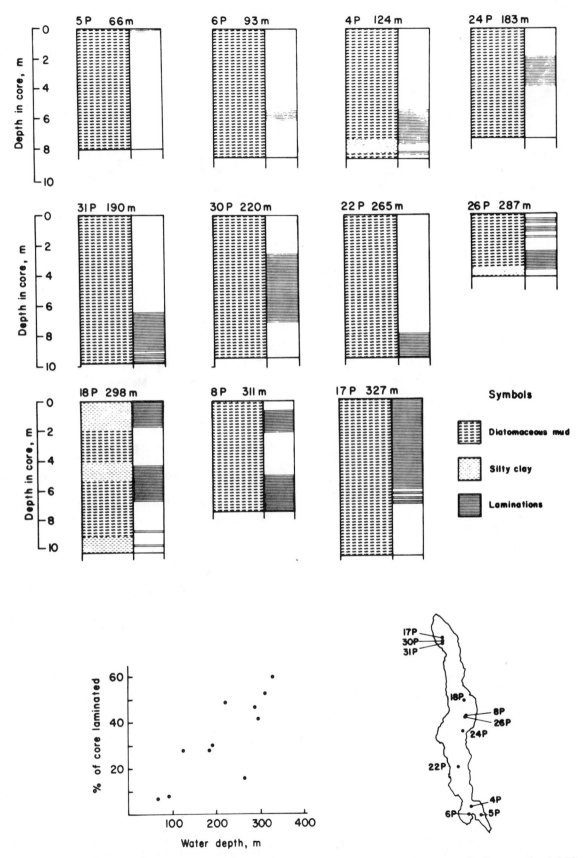

Figure 20. Lamination stratigraphy of Lake Malawi sediment cores and relationship between core site (water) depth and percentage of core laminated. Note that no major break occurs in the graph at 250 m, the present depth of the chemocline. Inset map shows core locations.

Johnson and Ng'ang'a

Table 2. Radiocarbon dates on Lake Malawi sediments

Core	Depth (cm)	Component Dated	Age (yr)	Reference No.
4P	10–30	TOC	21410±430	B18351
	230–250	TOC	21620±370	B18352
	450–468	TOC	>32700	B18353
	748–770	TOC	33780±1200	B18354
	857–883	CaCO$_3$	10740±130	B18355
6P*	10–25	TOC	27280±1350	B22867
	50–70	TOC	25170±420	B24735
	350–370	TOC	28080±880	B24736
	540–555	TOC	23790±1000	B22868
	810–823	TOC	29760±2640	B22869
12P	120–141	CaCO$_3$	7490±260	B29140
	292–312	CaCO$_3$	9140±170	B29141
13P	54–70	TOC	1240±70	B22995
	311–313	TOC	4090±220	B22997
	385–400	TOC	1290±80	B22998
16P**	189–206	TOC	3370±60	
	318–349	TOC	3750±90	
	545–569	TOC	6590±110	
	842–864	TOC	9390±110	
18P	560–600	CaCO$_3$	22050±510	B33164
22P	76–98	TOC	1950±200	B29806
	799–812	TOC	6320±240	B29807
	874–920	CaCO$_3$	7600±140	B29807
24P	197–228	CaCO$_3$	6400±100	B33165
Modern gastropods*		CaCO$_3$	107±0.7%	B24714

* Provided by K. Haberyan.

** Provided by R. B. Owen, dated by G. Cook, Scottish Universities Research and Reactor Center, Glasgow. All depths for this core had been reported by Owen as dry (shrunken) depths and have been corrected back to wet depths, assuming that shrinkage was linearly proportional to original depth. All "B" reference numbers are for dates determined by Beta Analytic, Miami, Florida.

Eccles, D. H., 1984, On the recent high levels of Lake Malawi: Suid-Afrikaanse Tyrdskrif vir Wetenskap, v. 80, p. 461–468.

Flannery, J. W., Jr., 1988, The acoustic stratigraphy of Lake Malawi, East Africa: M. S. Thesis, Duke University, Durham, N. C., 117 p.

Gonfiantini, R., G. M. Zuppi, D. H. Eccles, and W. Ferro, 1979, Isotope investigation of Lake Malawi, in Isotopes in lake studies: International Atomic Energy Agency Panel Proceedings Series STI/PUB/511, p. 195–207.

Gross, M. G., 1987, Oceanography—A view of the Earth, 4th ed.: Englewood Cliffs, N. J., Prentice-Hall, 406 p.

Haberyan, K. A., 1988, Phycology, sedimentology, and paleolimnology near Cape Maclear, Lake Malawi, Africa: Ph.D. Dissertation, Duke University, Durham, N. C., 246 p.

Halfman, J. D., 1987, High-resolution sedimentology and paleoclimatology of Lake Turkana, Kenya: Ph.D. Dissertation, Duke University, Durham, N. C., 188 p.

Hamilton, E. L., 1980, Geoacoustic modeling of the sea floor: Journal of the Acoustical Society of America, v. 68, p. 1313–1340.

Hecky, R. E., and E. T. Degens, 1973, Late Pleistocene-Holocene chemical stratigraphy and paleolimnology of the rift valley lakes of Central Africa: Woods Hole Oceanographic Institution Technical Report 73-28, 93 p.

Hecky, R. E., and H. J. Kling, 1987, Phytoplankton ecology of the great lakes in the rift valleys of central Africa, in M. Munawar, ed., Proceedings of the symposium on the phycology of large lakes of the world: Archiv für Hydrobiologie Ergebnisse der Limnologie (Supplementary Issue), v. 25, p. 197–228.

Hovland, M., and A. G. Judd, 1988, Seabed pockmarks and seepages—Impact on geology, biology, and the marine environment: Boston, Mass., Graham and Trotman Ltd., 336 p.

Johnson, T. C., 1980a, Late-glacial and post glacial sedimentation in Lake Superior based on seismic reflection profiles: Quaternary Research, v. 13, p. 380–391.

Johnson, T. C., 1980b, Sediment redistribution by waves in lakes, reservoirs and embayments, in H. G. Stefan, ed., Proceedings of the symposium on surface water impoundments: American Society of Civil Engineers, v. 2, p. 1307–1317.

Johnson, T. C., 1984, Sedimentation in large lakes: Annual Review of Earth and Planetary Sciences, v. 12, p. 179–204.

Johnson, T. C., and T. W. Davis, 1989, High resolution seismic profiles from Lake Malawi, east Africa, in B. R. Rosendahl, J. J. W. Rogers, and N. M. Rach, eds., African rifting: African Journal of Earth Sciences (and the Middle East), v. 8, nos. 2-4, p. 383–392.

Johnson, T. C., T. W. Davis, B. M. Halfman, and N. D. Vaughan, 1988, Sediment core descriptions: Malawi 86, Lake Malawi, east Africa: Beaufort, N. C., Duke University Marine Laboratory Technical Report, 101 p.

Johnson, T. C., R. E. Hecky, and D. Engstrom, (in prep.), Tributary input of dissolved silica and diatom deposition in Lake Malawi, East Africa.

Katz, B. J., and P. A. Kelley, 1987, Central Sumatra and the east African rift lake sediments—An organic geochemical comparison: Indonesian Petroleum Association 16th Annual Convention, Proceedings, p. 259–289.

LeFournier, J., 1980, Depots de preouverture de l'Atlantique Sud—Comparaison avec la sédimentation actuelle dans la branche occidentale de rifts est-Africains: Recherches Géologiques en Afrique, v. 5, p. 127–130.

Lewis, D., P. Reinthal, and J. Trendall, 1986, A guide to the fishes of Lake Malawi National Park: Gland, Switzerland, World Conservation Centre, 71 p.

Lineback, J. A., D. L. Gross, R. P. Meyer, and W. L. Unger, 1971, High-resolution seismic profiles and gravity cores in southern Lake Michigan: Illinois State Geological Survey Environmental Geology Notes 47, 41 p.

Müller, G., and U. Forstner, 1973, Recent iron ore formation in Lake Malawi, Africa: Mineralogica Deposita, v. 8, p. 278–290.

Ng'ang'a, P., 1988, Sedimentation off deltas and border faults in northern Lake Malawi—Evidence from high-resolution acoustic remote sensing and gravity cores: M. S. Thesis, Duke University, Durham, N. C., 93 p.

Owen, R. B., R. Crossley, T. C. Johnson, D. Tweddle, I. Kornfield, S. Davison, D. H. Eccles, and D. E. Engstrom, 1990, Major low levels of Lake Malawi and implications for speciation rates in cichlid fishes: Proceedings of the Royal Society of London, v. 240, no. 1299, p. 519.

Pike, J. G., 1964, The hydrology of Lake Nyasa: Journal of the Institute of Water Engineering, v. 18, p. 542–564.

Pike, J. G., and G. T. Rimmington, 1965, Malawi, a geographical study: Oxford University Press, 229 p.

Pilskaln, C. H., and T. C. Johnson, (in press), Seasonal signals in Lake Malawi sediments, east Africa: Limnology and Oceanography.

Rosendahl, B. R., 1987, Architecture of continental rifts with special reference to East Africa: Annual Review of Earth and Planetary Sciences, v. 15, p. 445–503.

Scholz, C. A., and B. R. Rosendahl, 1988, Low lake stands in Lakes Malawi and Tanganyika, east Africa, delineated with multifold seismic data: Science, v. 240, p. 1645–1648.

Scott, D. L., 1988, Modern processes in a continental rift lake—An interpretation of 28 kHz seismic profiles from Lake Malawi, east Africa: M. S. Thesis, Duke University, Durham, N. C., 82 p.

Street-Perrott, F. A., and N. Roberts, 1983, Fluctuations in closed-basin lakes as an indicator of past atmospheric circulation patterns, in F. A. Street-Perrott, M. Beran, and R. A. S. Ratcliffe, eds., Variations in the global water budget: Boston, D. Reidel Publishing Co., p. 331–345.

Talbot, M. R., 1988, The origins of lacustrine oil source rocks—Evidence from the lakes of tropical Africa, in A. J. Fleet, K. Kelts, and M. R. Talbot, eds., Lacustrine petroleum source rocks: Geological Society Special Publication 40, p. 29–43.

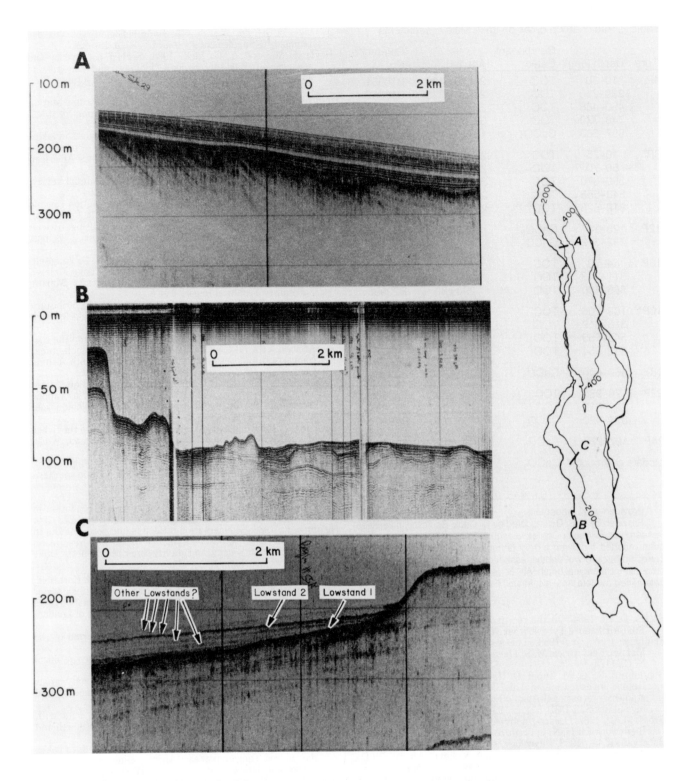

Figure 21. High-resolution seismic-reflection profiles indicative of lowstands in Lake Malawi—A, erosional unconformity at 200 m; B, submerged and buried river channels at 150 m; C, wave-cut scarp at 220 m (bottom). Inset map shows locations of profiles.

Thatcher, E. C., 1974, The geology of the Nyika area: Geological Survey Department, Malawi Ministry of Agriculture and Natural Resources, Bulletin 40, 90 p.

Thomas, R. L., and C. I. Dell, 1978, Sediments of Lake Superior, *in* J. R. Kramer, ed., Limnology of Lake Superior: Journal of Great Lakes Research, v. 4, p. 264–275.

Tiercelin, J. J., J. Chorowicz, H. Bellon, J. P. Richert, J. T. Mwanbene, and F. Walgenitz, 1988, East African rift system—Offset, age, and tectonic significance of the Tanganyika-Rukwa-Malawi intracontinental transcurrent fault zone: Tectonophysics, v. 148, p. 241–252.

Tiercelin, J. J., C. Thouin, T. Kalala, and A. Mondeguer, 1989,

Discovery of sublacustrine hydrothermal activity and associated massive sulfides and hydrocarbons in the north Tanganyika trough, East African Rift: Geology, v. 17, p. 1053–1056.

Vaslet, N., C. Thouin, Y. Fouquet, P. J. Brichard, T. Kalala, A. Mondequer, and J.-J. Tiercelin, 1987, Decouverte de sulfures massifs d'origine hydrothermale dans le Rift Est-africaine—Mineralisations sous-lacustres dans le fosse du Tanganyika: Comptes Rendus de l'Académie des Sciences de Paris, v. 305, II, p. 885–891.

Tectono-Stratigraphic Model for Sedimentation in Lake Tanganyika, Africa

Andrew S. Cohen
Department of Geosciences
University of Arizona
Tucson, Arizona, U.S.A.

Lake Tanganyika provides an excellent opportunity for understanding tectonic and climatic influences on sedimentation in a rift lake. Each tectonic setting within rift half-graben basins generates a predictable range of lithofacies architectures.

Escarpment-margin (boundary-fault) drainages are small and steep, producing small fan deltas and thick, although not broad, sublacustrine fan complexes. Most water and sediment derived from escarpments is diverted away from the rift basin, although it may reenter along an adjacent half-graben basin margin.

Drainages crossing hinge ramps or platforms are larger and well integrated. Deltaic sands on these platforms may form sheets or be channelized into older alluvial valleys. Delta positions are poorly constrained by structure, and portions of the platforms may be clastic-sediment bypassed.

Fault-bounded interbasinal ridges, termed accommodation zones, are clastic-sediment starved. They are predominantly areas of pelagic sedimentation but may become areas of littoral carbonate accumulation at appropriate lake levels.

Rift-axial streams drain moderately large areas under very low gradients. Their deltaic positions are highly constrained by rift structure, providing for abundant clastic-sediment supply across the axis. A predominance of interflows and underflows generates strong density currents across most lake margins.

Asymmetry in lithology and strata thickness is the result of lake-level fluctuations interacting with varying rates of sediment accumulation, much of which is structurally influenced. Differences in sequence geometry have implications not only for interpreting ancient rift-lake deposits but also for deposition of economically viable reservoir facies and their juxtaposition with source rocks and caprocks. Several environments deserve more attention as exploration analogs than they have previously received. Platform and axial-margin sand bodies (clastic and carbonate), accommodation-zone carbonates, and turbidites or contourites derived from platform or axial sources all have considerable potential as reservoir facies.

INTRODUCTION

Large modern rift lakes are important analogs in any attempt to interpret the facies architecture of ancient rifts. Tectonic and geophysical investigations over the last eight years have revealed the structural complexities of modern rifts and have suggested possible tectonic models for understanding their origin (Chenet and Letouzey, 1983; LeFournier et al., 1985; Rosendahl et al., 1986; Rosendahl, 1987). As sedimentological studies of rift lakes progress, it is becoming apparent that the structural asymmetry and complexity of alternating half-graben basins is mirrored by similar asymmetry and complexity of depositional environments (e.g., Frostick and Reid, 1987; Leeder and Gawthorpe, 1987). Sediment distribution within large rift lakes may be controlled locally by variations in tectonism within the basin at a given time and temporally by regional variations in climate or tectonism.

In this paper I outline a model for sediment distribution patterns in Lake Tanganyika, east Africa. The database for this model consists of direct underwater observations, analyses of surface sediment samples and gravity cores, and water chemistry analyses, in addition to published seismic and structural data. My sediment sampling has concentrated on water depths between 0 and 200 m because this interval encompasses most of the profound depth- or energy-related facies changes observed in Lake Tanganyika. The model makes predictions about facies geometries within a youthful rift lake based on observed or inferred limnological and sedimentological processes. Inasmuch as it can be tested and refined through future field observations, it can serve now as a modern analog for comparisons with ancient rift lake strata, many of which are important for hydrocarbon exploration and production (e.g., Brazil, west Africa).

Lake Tanganyika is the largest (~34,000 km²) and deepest (maximum 1,470 m) of the African rift lakes (Figure 1). The modern lake experiences a semihumid (1000–1200 mm/yr precipitation), tropical climate, and the lake is hydrologically open, although most water loss is via evaporation. Chemically the lake is mildly alkaline (6.6 meq/L CO_3^{-2} + HCO_3^-; pH = 9.0) and nonsaline (conductivity 650–700 μmho/cm, TDS). Its anomalously high Mg/Ca ratio (Mg^{+2} = 45 mg/L, Ca^{+2} = 12 mg/L), given its low salinity, is the result of upstream low-magnesium calcite precipitation in Lake Kivu, whose water enters Tanganyika via the Ruzizi River and subsequently is diluted by other (uniformly low-salinity) influents. The lake normally is supersaturated with respect to calcite (Cohen and Thouin, 1987).

Lake Tanganyika is monimolimnetic, with a transition to anoxic conditions at depths of 100–250 m, depending on season and location (Kufferath, 1952; Craig, 1974; this study). However, the monimolimnion is leaky, and deep-water upwellings are known along the northern Zaire coast and suspected elsewhere in the lake (Coulter, 1963; Cohen, unpublished data).

Most influent streams are cold and carry high suspended-sediment loads, which ensures that they enter the lake as strongly descending underflows or interflows, caused by contrasts in salinity (e.g., Ruzizi River) or temperature/suspended-load (most other influents). Late Pleistocene-Holocene sediment-accumulation rates have been measured at 0.4–0.5 mm/yr for deep-water (>500 m) basinal settings but are considerably less on structural highs (Hecky and Degens, 1973).

Seismic-reflection data show that the lake occupies a series of eight major half-graben basins whose escarpment margins (border faults) face alternating directions (Rosendahl et al., 1986; Rosendahl, 1988). More than 4000 m of sediment has accumulated in the deepest areas of some Tanganyikan basins (Rosendahl et al., 1986). The northernmost basin is probably late Miocene or early Pliocene in age, based on estimated sedimentation rates and radiometric dating of basalts in the northern Ruzizi basin (Ebinger, 1988, 1989), but the ages of other basins are unknown. Unlike other well studied African rift lakes, volcanism was of minor importance in the geologic evolution of the Tanganyikan basins.

Where Lake Tanganyika basins share common tectonic and physiographic elements, lithofacies and stratal geometries are predictable. This paper addresses these tectono-stratigraphic relationships and their implications for synrift hydrocarbon accumulation.

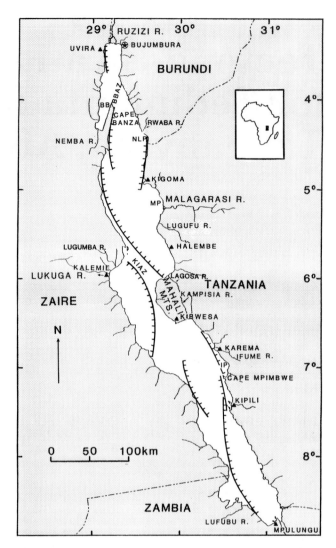

Figure 1. General geography of Lake Tanganyika and vicinity. Heavy lines denote major escarpments. Abbreviations: BB, Burton's Bay; BBAZ, Burton's Bay accommodation zone; KIAZ, Kavala Island accommodation zone; NLP, Nyanza-Lac platform; MP, Malagarasi platform; IP, Ikola platform.

TECTONO-STRATIGRAPHIC ELEMENTS OF A TANGANYIKAN BASIN

In cross section, the deeper side of an individual Lake Tanganyikan half-graben basin is bounded laterally by a steep-sided, boundary-fault escarpment (Figure 2). Axial terminations of individual basins, referred to as accommodation zones (Rosendahl et al., 1986), commonly are both structural and topographic highs. Sediment accumulating here is subject to high strain and severe faulting relative to adjacent basinal sediments. In this paper I do not distinguish separate facies models for the different

Cohen

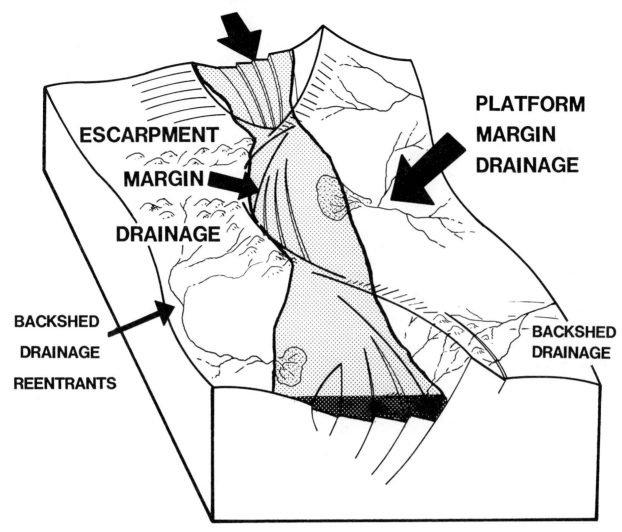

AXIAL MARGIN DRAINAGE

ESCARPMENT

MARGIN

DRAINAGE

PLATFORM MARGIN DRAINAGE

BACKSHED DRAINAGE REENTRANTS

BACKSHED DRAINAGE

Figure 2. Block diagram illustrating major rift-basin drainage classes and their relation to tectonic setting in a series of linked half-grabens (modified after Rosendahl et al., 1986).

types of accommodation zones or other variations in half-graben geometry, given the sparsity of comparative sedimentologic data. Subaerial basin-axis lake margins (hereafter referred to as axial margins) will, however, be recognized because they are minor input points for clastic sediment, unlike submerged or subaerial accommodation-zone highs.

Escarpment Lake Margins

The broad, uplifted flanks of the rift enhance the asymmetry in basin structure. An elevated escarpment consists of a steep, narrow (<10 km wide) rift-facing side and a flank that dips gently away from the rift. This geometry gives rise to backshed drainage (Figure 2), which effectively diverts most drainages on the rift shoulders away from, rather than into, the rift (Frostick and Reid, 1986; Cohen,

1989a). Drainage down the steeply faulted escarpment margin directly into the rift basin tends to be short and steep graded (Figure 3). The 79 escarpment-margin drainages entering the lake along the border faults south of Uvira, Zaire, and north of Kigoma, Tanzania, have a median maximum length of 2.6 km and a mean slope of 12°08′ ±3°31′ for all drainages longer than 1 km. That these drainages originate in the nearby uplands also dictates that they are poorly integrated and have small watersheds (median 2.7 km^2). Deep channel incision along escarpment walls renders avulsion and lateral fan-delta migration in this setting nearly impossible; thus, the positions of escarpment fan deltas are highly stable.

Clastic sedimentation patterns along youthful escarpment margins can be related directly to this pattern of input both in terms of size and spacing. Individual drainages build discrete, small fan deltas (almost always <1 km^2 in area) that rarely coalesce

Figure 3. Rift-escarpment margin in Tanzania (central Mahali Mountains). Local topographic relief may exceed 3000 m along some escarpment margins, with considerably greater structural relief.

with adjacent fan deltas. Extremely steep margins along the border faults ensure that these fan deltas do not prograde, but rather act as point sources for deep-water grain flows, grainfalls, and density flows commonly focused down sublacustrine canyons. In the immediate vicinity of most fan deltas examined, sediments are poorly sorted (sands to boulder conglomerates) and poorly stratified. Because of the small areas of the drainage hinterlands, most derived gravels tend to be lithologically homogeneous within a particular fan delta. Coarse-grained sediment derived from these inflows mantles high-angle (60–70°) bedrock escarpments immediately offshore from the delta fronts. These regions have little or no long-term sediment-accumulation potential in shallow water.

Adjacent to the small fan deltas, very coarse (up to 5-m-diameter blocks) colluvium is shed directly down the escarpments into deep water. Even on extremely steep slopes some of this material becomes rapidly lithified by algally precipitated carbonates (Figure 4). This type of cementation appears to be more common away from drainage inputs, where clastic dilution is lower (Cohen, 1989a).

Profundal sediments along escarpment margins are derived from several distinct sources. Proximal terrigenous debris derived from escarpment fan deltas and colluvial rockfall creates recognizable turbidites, debris flows, toe-of-slope talus cones, and some channelized sands that slope lakeward offshore from the escarpment walls. Their local (mostly metamorphic) basement provenance is easily recognized by comparing clasts with local bedrock lithologies. These sediments typically are poorly to very poorly sorted and often admixed with pelagic sediment. Material shed from border faults are not, however, the dominant component of basin fill in the deeply subsided areas of Lake Tanganyika. Seismic reflection data and grab samples suggest that sublacustrine coarse-clastic fan complexes normally do not prograde more than several kilometers from the border fault. Between fan deltas, deep-basinal escarpment-margin sediments consist of pebbly to bouldery mudstone formed by talus falling directly into onlapping pelagic mud.

The bulk of deep-basinal sediments within Lake Tanganyika are muds or sands derived from sources other than the escarpment itself. Pelagic diatom oozes and hemipelagic muds are strongly focused by density-flow resedimentation into areas adjacent to the escarpment margins. These sediments occur as laminites or graded beds interbedded with nonreworked pelagic microlaminae. Where they accumulate are zones of maximum subsidence within the rift basins, typically topographic lows (e.g., along the Kigoma border fault) into which pelagic and hemipelagic fines can be reworked (Rosendahl et al., 1986; Morley, 1988). Sandy contourites and mud-sand turbidites that are not derived locally apparently originate from either platform (cross-basinal) or axial sources and subsequently are transported by deep density currents into areas adjacent to escarpment margins. Their derivation and transport in Lake Tanganyika is facilitated by numerous underflows at all major deltas.

Steep escarpment margins display minimal lateral facies migration during lake-level fluctuations. During highstands, alluvial valleys may be drowned for short distances upstream, generating temporary ponding and sediment storage. When lake level recedes, most of this material then is flushed into the basin as turbidite and sublacustrine fan progradation. However, during extreme lake lowstands, water density probably would rise as salinity increased, and underflow-generated contourite deposition would decrease.

Platform or Ramp Lake Margins

Rivers entering rift basins across platforms or hinge ramps associated with structural rollovers behave very differently than those on escarpment margins. Ramped margins are uplifted but to a lesser

Figure 4. Calcite-cemented colluvial rockfall at a depth of 10 m along the Kigoma escarpment, northern Tanzania. Fish at left center is approximately 100 mm long.

degree than escarpments (Weissel and Karner, 1989), allowing downcutting to balance uplift. Because platforms serve as a basin's structural hinges, they experience relatively slow rates of subsidence compared with border-fault areas. The low-gradient surfaces on the ramp sides of half-graben basins allow large, well-integrated drainages to develop (median area 23 km²). Many platform drainage systems (e.g., Malagarasi, Lugumba, and Lagosa rivers) are backshed escarpment drainages that enter the rift via the adjacent half-graben ramp (Figure 2). Ramp drainage systems are much longer and less steep than those along escarpments. The 58 rivers entering Lake Tanganyika along the Nyanza-Lac and Malagarasi platforms have a median maximum length of 11 km and mean gradient of 2°41′ ±1°47′. During high-stands, channel stability on the platforms probably is low and avulsion common, as in other settings where local subsidence rates are low relative to sediment input (Bridge and Leeder, 1979; Blakey and Gubitosa, 1984). When lake levels drop, the channels incise and stabilize on all lake margins. Channel stability also may be reduced on platform ramps because rivers flow across the regional structural grain, thereby providing little opportunity for fault

control of drainage. Platform deltas may respond to the combination of (1) slow subsidence, (2) low gradient, (3) absence of structural control on channel position, and (4) rapid sediment input by frequently switching lobe position and reduced clastic sedimentation rates over large regions of the platforms. Sedimentologic data from the Malagarasi, Nyanza-Lac, and Ikola platforms support this view. Some areas of these platforms are being blanketed by clastic sediment, which has been carried offshore down the gently sloping ramps (Figure 5) and which may ultimately be transported into profundal, anoxic environments by density currents. In shallow water some clastic sediment may be redistributed by longshore currents, as on the platform deltas of Lake Turkana (Frostick and Reid, 1986). However, the large size and depth of Tanganyikan platforms inhibit longshore currents from redistributing sediments across their entire expanse. Large areas of the platforms are sufficiently distant from present river mouths that they do not not receive abundant clastic sediment. These are areas of either active winnowing or condensed sections.

Some regions of the platforms have been deeply incised by sublacustrine canyon systems, such as

Figure 5. Platform siliciclastic sand blanket in 3 m of water, 0.5 km offshore on the southern Malagarasi platform, Tanzania. Mounded appearance of the lake floor results from nest-building activities of cichlid fish. Larger fish are approximately 120 mm long.

offshore from the Rwaba River on the Nyanza-Lac platform or off the Lugufu River on the Malagarasi platform. These canyons (presumably of alluvial origin) are infilled with massive, sandy grain-flow deposits, muds, and some sand-mud turbidites. Where they crosscut relatively low-relief platforms, individual canyon systems may extend as far as 20 km offshore, terminating in water depths up to 700 m. Elsewhere on the platforms, sediments are dominated by greenish-gray to black muds, interbedded with both thin sands and thicker, relatively clean, channelized sands of probable contourite origin. Sediment-starved (and in shallow water, actively winnowed) zones often contain enormous (up to several hundred km² in area) coquina lags or "shell graveyards" (Cohen, 1989b). Other carbonate lithologies (mud, sand, and boundstone) also are found on platform surfaces (Figure 6) (Cohen and Thouin, 1987). Beachrock carbonate cementation is ubiquitous in areas of littoral sand deposition away from streams. In shallow littoral platform zones with considerable wave agitation, ooid shoals and coquina sands commonly develop. In more protected embayments these give way to shallow-water gastropodal, ostracodal, or *Chara*-bearing muds. Cohen (1989a) has shown that carbonate deposits on the Nyanza-Lac platform become more common along the shoreline with distance from influent streams.

Sublittoral, clastic sediment-starved platform areas are dominated by stromatolite boundstones on high-relief or hardground bottoms and by olive-gray to black mud and muddy sand on low-relief bottoms. At water depths of 20–70 m these muds typically are massive and bioturbated and commonly contain shelly-sand horizons. Below 70 m massive muds are occasionally interbedded with laminated muds. The laminae are comparatively thick (>2 mm) with diffuse contacts but become thinner with less diffuse contacts below 100 m deep. Laminated sequences make up an increasing proportion of the interbeds in water deeper than 70 m.

Stratigraphic architecture of the platform-ramp develops then through the interaction of relatively low subsidence rates, extreme lake-level fluctuations (Scholz and Rosendahl, 1987, 1988), and periodically avulsing, underflow-delta discharge. The low topographic gradient of platforms allows much broader facies belts to develop than on escarpment margins. Platform carbonate deposition is favored in areas well removed from siliciclastic input, partic-

Cohen

Figure 6. Stromatolite bank in 10 m of water, 100 m offshore on the northern Malagarasi platform, Tanzania. Fish is approximately 100 mm long.

ularly when large reaches of the platform are flooded to littoral or shallow sublittoral depths. Alluvial sand deposition may migrate basinward during moderate lowstands, and canyon cutting may occur during extreme lowstand events. Alluvial sands can be onlapped by mottled or laminated muds during moderate or extreme lake-level rises, respectively. Interfingering of mud and sand also will occur during lake highstands as delta-derived deep-water sands (carried both in incised channels and as migrating current drift deposits) pass laterally into the clastic sediment-starved areas.

Rift platforms may, in fact, act as important (if not primary) conduits for sediment transport into deeper rift-basinal environments. The distribution patterns of many large, deep-water (midlake) sand bodies suggest they have been transported tens of kilometers outward from the platforms. In some cases these sands appear to extend nearly the width of the basin (J. LeFournier, 1986, pers. comm.).

Accommodation-Zone Lake Margins and Midlake Highs

Sedimentation processes and lithofacies geometry along both the top surfaces of accommodation-zone highs and down their fault-bounded flanks are poorly understood. Even generalizations about accommodation-zone sedimentation are difficult because the areas are so highly variable, both structurally and topographically.

Comparatively few drainages originate on accommodation-zone highs because of their relatively small surface area. Where the zones intersect the modern lake shoreline, drainage basins are larger but may be moderately to extremely steep. For the eight >1-km-long drainages on the eastern (Burundi) and western (Zaire) termini of the Burton's Bay accommodation zone and at the eastern terminus of the Kavala Island accommodation zone (northern end of the Mahali Mountains, central Tanzania), the median maximum drainage length is 7.9 km with mean gradients of 5°59′ ±4°28′.

Most accommodation-zone stream systems drain laterally into adjacent basins, the streams assume overall morphologies comparable to those on platform margins. For example, drainages off the eastern lakeshore termini of the Burton's Bay and Kavala Island accommodation zones deliver sediment to adjacent platforms. Where they happen to follow an accommodation-zone axis, drainages become morphologically similar to escarpment drainages (i.e., steep and small). Under these circumstances, small fan deltas and talus rockfalls (Figures 7, 8) create sublacustrine turbidites and grainfall and debris-flow deposits similar to those found along escarpment margins.

The significant difference between accommodation-zone high and escarpment-margin clastic sedimentation styles arises from the more limited area- and relief-dependent sediment-source potential of accommodation zones. Thus, accommodation-zone margins (particularly their upper surfaces) potentially are the most clastic sediment-starved littoral and sublittoral regions of the lake. Carbonates (algal boundstone bioherms) are common in nearshore environments on accommodation-zone flanks and upper surfaces. However, because of the steep slopes, long-term accumulation of uncemented sediment is unlikely. Most shallow-water carbonates or clastics that do accumulate are likely to be transported rapidly into deeper water.

The sedimentology of submerged accommodation-zone highs is almost entirely unknown. Estimated sediment-accumulation rates for pelagic diatom oozes on these highs are among the lowest recorded for the lake (<0.07 mm/yr; Hecky and Degens, 1973). This is hardly surprising given that they form either gently inclined ridges or submerged midlake bathymetric highs. Sublacustrine transport of large quantities of debris down the narrow upper surfaces of these ridges would be extremely difficult. Thus, when lake level only slightly submerges significant lengths of a given interbasinal high, an ideal situation arises for development of a large (by lake standards) carbonate bank. Hypothetically, carbonate banks could develop as completely submerged banks and shoals or as rims around small islands. Because the flanks of accommodation zones may be quite steep, shallow-water carbonates could easily be reworked into adjacent deeper water environments by slumping, debris flows, or grain flows. Although such deposits have not been documented in Lake Tanganyika, the appropriate environments have yet to be investigated (for example, the lakeward extension of the Cape Banza–Burton's Bay accommodation-zone ridge).

Axial Lake Margins

Axial rift drainage and its associated deltas have received more attention than other types of rift drainage and are commonly thought to play a dominant role in rift-lake infilling (LeFournier, 1980; Lambiase and Rodgers, 1988). In Lake Tanganyika the Ruzizi River (draining the Ruzizi basin and Lake Kivu to the north) is, by far, the largest axial drainage. Smaller axial rivers occur at the northern end of the lake, and additional major streams such as the Nemba River (south of Burton's Bay) drain other axial lake margins.

Axial streams around Lake Tanganyika have very low gradients and moderate to large drainage basins compared with the other types of lake margins. Most axial basins have well-integrated drainage systems dominated by a single major trunk stream. For all axial streams longer than 1 km in the Ruzizi and Nemba basins, the median drainage length is 8.25 km, and mean stream gradient is 0°43′ ±1°03′. The positions of axial rivers and their deltas are constrained by basin structure (specifically, geometry of adjacent border-fault systems) to a greater extent than those rivers that enter the lake laterally. Most axial drainage is confined to narrow zones of rapid subsidence, which limits large-scale channel avulsion to a much greater degree than on platform margins.

The Ruzizi River enters Lake Tanganyika as a dense underflow (Figure 9), the result of salinity contrasts (Hecky and Degens, 1973). River-mouth bar sands are strongly reworked by wave action and transported both on and offshore. The combination of underflow current, a narrowly confined zone of clastic sedimentation, and rapid subsidence has created a thick, prograding sandy-delta system. The Ruzizi delta is relatively steep fronted (0°43′ slope for 20 km offshore) in comparison to rift-axial deltas that exhibit hypopycnal flow in other lakes (0°04′ for 20 km offshore of the Omo River at the north end of Lake Turkana). However, all deltas forming along rift axes in Lake Tanganyika are gently inclined compared with deltas formed in other tectonic settings. This is because platform and escarpment deltas form across major rift faults, whereas axial deltas form parallel to the major fault systems and are rarely cut by faults perpendicular to the rift axis.

Oversteepening on the underflow-dominated Ruzizi delta front frequently leads to soft-sediment deformation and generation of turbidity currents. Massive sand blankets occur far offshore from the river mouth, although their mode of origin is unknown. On the Ruzizi delta margin, clastic sediment is being distributed across most of the axial margins littoral-sublittoral zone. This is evident from the uniformly high suspension and traction loads across the northern end of the lake. This type of continuous sand transport across a specific shoreline interval contrasts markedly to both platform and escarpment-margin sedimentation, where shallow clastic deposition is much more discontinuous.

Elsewhere in Lake Tanganyika, deep-water axial sand transport by near-continuous density currents has been well documented for the rift-axial Nemba River delta (Mondeguer, 1984; Mondeguer et al.,

Cohen

Figure 7. Fan delta of the Gatorongoro River on the northern flank of the eastern (Burundian) extension of the Burton's Bay accommodation zone. Total above-water fan extent is 0.01 km², typical of the extremely small fan deltas observed on both accommodation zone and escarpment margins.

1986). In Burton's Bay, Mondeguer showed that deep-water sediment transport is directed axially along the deepest available channels (i.e., those adjacent to the rift escarpment).

At present the relative contribution of platform vs. axial sources of clastic sediment (sand in particular) into Lake Tanganyika is unknown. The fate of these contributions is, however, much clearer. Lake Tanganyikan axial margins are long-term sediment storage zones, whereas platform margins act more like conveyor belts, delivering large quantities of sediment to deep water but storing relatively little.

DISCUSSION

Sedimentation patterns in large rift lakes are complex—a function of variable interactions of climate, tectonism, erosion and, in some lakes, volcanism. In Lake Tanganyika, as in other rift lakes, climate affects sedimentation principally by altering influent sediment and water discharge, influent and lake water chemistry and, perhaps most importantly,

lake level. Tectonism affects the positioning of pathways and barriers to sediment input and the loci of ultimate sediment storage, both along the rift axis and perpendicular to rift structures.

Figure 10 is an attempt to portray these interactions through hypothetical stratigraphic sequences resulting from two cycles of lake-level fluctuation. The transfer of sequence-stratigraphic concepts from marine basins (Abbott and Branson, 1979) to lakes is by no means straightforward, and lakes ultimately may require their own terminology and sequence theory. Seismic-reflection data collected by Ebinger et al. (1984), Rosendahl et al. (1986), and Scholz and Rosendahl (1987, 1988), however, demonstrate that rift-lake strata are organized into discrete sequences commonly bounded by unconformities that are linked to major lake-level fluctuations. In the block diagram, stratal geometry is displayed in cross section between escarpment and platform lake margins and in plan view across two adjacent half-grabens. Sedimentary sequences A and B illustrate the changes in stratigraphy as basin hydrology changes from closed to open conditions during two episodes of water-level rise.

Figure 8. Colluvial, conglomeratic sand in 15 m of water on an interfluve of the northeastern (Burundian) margin of the Burton's Bay accommodation zone. Eel is approximately 700 mm long.

In the cross-section on the left, a platform drains both the interior of its own structural basin and the shoulders of the adjacent rift escarpment. At the base of each sequence, lowstand alluvium and littoral sand have been deposited over erosional surfaces. The degree of erosion represented by these surfaces decreases lakeward as they pass into their correlative conformities below the lowstand base level. If any of these surfaces developed as deeply incised canyons during lowstands, they could act as sublacustrine sediment conduits during the next cycle of lake-level rise. Although the lakeward extent of the alluvial/littoral sands may be considerable on the platform side, slower subsidence rates there limit total thickness of the platform facies. The rapid fluctuations in water level that can occur in rift lakes are reflected in the abrupt transition to onlapping deep-water mud (first bioturbated clastic mud and then laminated diatom oozes).

At highstand, sand-body geometry in any given section will depend on the position of platform deltas during that interval. Because platform deltas are freer to avulse than either escarpment fan deltas or axial deltas, a given locality may be either coarse-clastic inundated (sequence A) or clastic bypassed (sequence B) during a given time interval. Where clastic input is high, deltas may prograde during a lake highstand (approximating a stillstand in an open-basin rift lake). On the other hand low rates of clastic input will permit carbonate precipitation and deposition, if the water chemistry is appropriate. Dissolved calcium supply and calcium carbonate production in Lake Tanganyika during the Holocene has been almost entirely dependent on overflow of the Ruzizi and is therefore highly correlatable with highstands (Hecky and Degens, 1973; Haberyan and Hecky, 1987). If water chemistry is inappropriate for carbonate precipitation, condensed stratigraphic sections will form. Active winnowing at times of low lake level can lead to formation of lag coquinas.

Offshore transport of platform-derived sediment may involve both sheetlike and channelized turbidite and contourite sands. These may be concentrated in deep-water environments adjacent to an escarpment or reworked and carried parallel to the shoreline.

In the section at the lower right of Figure 10, the rapidly subsiding escarpment margin is illustrated during the same two cycles. During initial lowstand conditions in sequence A, alluvium is flushed from

Figure 9. The Ruzizi River delta at the northern end of Lake Tanganyika. Note the sharp demarcation between open lake water and the descending Ruzizi underflow.

the escarpment drainage basins and fan deltas onto the basin floor by turbidity currents, debris flows, and grain flows. Deposits so formed may be very thick but typically are areally restricted to within several kilometers of shore.

As lake level begins rising, the remaining alluvium on the fan deltas is progressively drowned and onlapped by massive muds and then by pelagic diatom oozes. At extreme highstands, pelagic mud will be deposited very close to the lake margin. Pelagic oozes and subordinate interbedded siliciclastics (mud and sand) comprise most of the deep-basin fill, although little of this material is actually derived from the adjacent escarpment margin. Most of the sediment has been concentrated into these bathymetric troughs from one of three sources—pelagic diatoms, ramps or platforms across the basin, or adjacent platform or axial-margin sources, via axial troughs. Occasional increases in subsidence rates along the border faults may be accompanied by temporary cessation of local clastic input and onlap of pelagic sediments onto the border faults (sequence B). In Mesozoic rift basins in Mexico Blair (1987) has

documented similar facies relationships associated with increased rates of tectonism. Pebbly mudstone and escarpment-margin carbonate also will occur sporadically through the onlapped section.

IMPLICATIONS FOR SYNRIFT RESERVOIR FACIES

The petroleum industry commonly minimizes the probability that economically viable synrift reservoir lithologies form in lakes, despite numerous examples to the contrary (South Atlantic and China in particular). However, sedimentologic studies in Lake Tanganyika suggest several possible modern analogs for stratigraphic traps of adequate size and appropriate structural setting that would facilitate development of petroleum reservoirs. Furthermore, the studies also shed light on why other environmental and tectonic settings are less favorable to reservoir formation.

The most obvious potential reservoirs are the large sand bodies associated with axial deltas, such as at the Ruzizi River. These are volumetrically the largest sand bodies within the lake, and in many rift basins, axial deltas have proven to be important reservoirs (J. Lambiase, 1989, pers. comm.). The Ruzizi delta, however, illustrates some of the potential problems associated with axial-delta reservoirs. First, the updip terminus of the Ruzizi delta has no effective synrift caprock facies. Second, sediments on the delta plain are predominantly sand, and even under conditions of highstand, the structural confinement of an axial delta position produces an overwhelming influx of sand, even in relatively deep water. Third, because of clastic mud dilution of organic material offshore, potential source rocks adjacent to an axial delta are less attractive than midlake basin muds.

Platform deltas may be superior to axial deltas in terms of potential source-reservoir-seal geometry. The areal extent of the largest platform delta (Malagarasi River) may be comparable to, or larger than, the axial deltas, although their sands more likely are thinner. Furthermore, platform ramps are sufficiently steep and channelized to provide numerous pathways for isolated sand bodies into deep water. The importance of large platform-deltaic sources for deep-water sand reservoir facies in rift lakes is illustrated by the Candeias Formation turbidites of the Lower Cretaceous Reconcavo rift basin, Brazil (Medeiros and Ponte, 1981; Petrobrás, 1988). At Itaparica Island and Ponta da Sapoca, deep-water turbidites occur close to the escarpment (only 3 km from the border fault). Paleocurrent data show, however, that they are consistently derived from the basin's platform margin 40–50 km away.

Escarpment-derived sands probably have low reservoir potential. Escarpment fan deltas usually are extremely small, and their locations along a border fault are difficult to predict. Most sublacus-

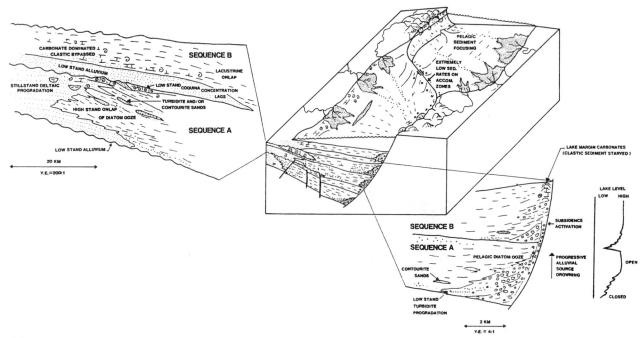

Figure 10. A sequence-stratigraphic model for Lake Tanganyika. The model illustrates the effects of two lake-level fluctuation cycles (curve at far right) on stratigraphic geometry along both platform (left) and escarpment (right) lake margins. Multiple sequences are illustrated to demonstrate the effect of abundant clastic sedimentation (sequence A) and sediment starvation (sequence B) on platform stratal geometry. Structural geometry modified from Morley (1988).

trine fans associated with escarpment margins in Lake Tanganyika examined to date are extremely muddy and probably would prove too tight, if lithified, to hold adequate reservoir rock potential. This problem is well illustrated by escarpment-derived clastics adjacent to the major (eastern) border fault of the Reconcavo basin (Medeiros and Ponte, 1981; Milani, 1987; Petrobrás, 1988). In the Salvador Formation both boulder conglomerates (associated with fan deltas and sublacustrine fans) and diamictites occur. The diamictites, which clearly are of rift-talus (and not glacial) origin, contain soft-sediment deformed muds associated with "dropstones" (presumably bounced off steep escarpment walls), many of which are striated! Similar problems could be expected with turbidite and debris-flow sands shed from accommodation-zone highs.

However, as seen at Itaparica Island, not all sands that accumulate along the bases of escarpments are derived from those escarpments. Some may be concentrated near escarpments or accommodation zones by contour or turbidity current activity. Where this has occurred, either from axial flow or cross-basinal platforms, sands as close as several kilometers to either a border fault or an accommodation-zone structure likely will be much cleaner than the local coarse-grained clastics shed off these highs.

Another class of exploration targets, suggested by Lake Tanganyika facies geometry and South Atlantic Cretaceous rifts, includes carbonate shoals on platforms (clastic sediment-bypassed) and accommodation-zone highs (clastic sediment-starved). Both coquinas and carbonate grainstones on structural highs have considerable potential as hydrocarbon reservoirs, as demonstrated by the Lagoa Feia Formation in the Campos basin, Brazil (Bertani and Carozzi, 1984; Abrahao, 1987) and the Lower Cretaceous Toca carbonates of Cabinda, Angola (McHargue, 1988). Accumulating adequate thicknesses of carbonate sand to form viable exploration targets may be difficult, but coquinas present fewer problems. Cohen (1989b) has shown how lacustrine coquinas might become stacked into relatively thick sequences, which form through active winnowing when lake level drops to expose platforms or other bathymetric highs to shallow-water conditions. Platform and accommodation-zone-high carbonates are both easily onlapped and encased by organic-rich mudrock during subsequent highstands.

At present, we understand the major controls on facies architecture in Lake Tanganyika, as discussed here. As an analog for exploration strategies, however, the model needs refinement. Predictive details of the relationship between lithofacies architecture (e.g., precise size and shape of each important facies element) and reservoir occurrence are still unknown. Because other African rift lakes possess strikingly different facies associations and geometries, the exploration geologist must remember that no one modern lake will provide a precise analog for a given ancient lake (Cohen, 1989a). Future analog studies of modern facies at Lake Tanganyika and other large rift lakes will reveal the spectrum of variation in lithofacies and sequence geometries that the explorationist may encounter in the stratigraphic record.

ACKNOWLEDGMENTS

This research was conducted under the auspices of the Université du Burundi and the National Research Council of Tanzania. I thank Pontien Ndabaneze, Gaspard Ntakimazi, Laurent Ntahuga, Andre Mondeguer, Chris Scholz, and Jacques LeFournier for their assistance and valuable discussion on this project and Ian Reid, Cindy Ebinger, Joe Lambiase, and an anonymous reviewer for valuable suggestions on earlier versions of this paper. This work was partially funded by NSF No. BSR 8415289.

REFERENCES

Abbott, W. O., and D. L. Branson, 1979, Atlas of seismic stratigraphy: Houston, Texas, Shell Oil Co., v. 1, 44 p.

Abrahao, D., 1987, Lacustrine and associated deposits in a rifted continental-margin—The Lagoa Feia Formation, Lower Cretaceous, Campos basin, offshore Brazil: M. S. Thesis 3318, Colorado School of Mines, Golden, Colo., 193 p.

Bertani, R. T., and A. V. Carozzi, 1984, Microfacies, depositional models and diagenesis of the Lagoa Feia Fm., Campos Basin: Petrobrás Ciencias Técnica Petroleo, no. 14, 104 p.

Blair, T., 1987, Tectonic and hydrologic controls on cyclic alluvial fan, fluvial, and lacustrine rift-basin sedimentation, Jurassic-Lowermost Cretaceous Todos Santos Formation, Chiapas, Mexico: Journal of Sedimentary Petrology, v. 57, p. 845-862.

Blakey, R., and R. Gubitosa, 1984, Controls of sandstone geometry and architecture in the Chinle Formation (Upper Triassic), Colorado Plateau: Sedimentary Geology, v. 38, p. 51-86.

Bridge, J., and M. Leeder, 1979, A simulation model of alluvial stratigraphy: Sedimentology, v. 26, p. 617-644.

Chenet, P.Y., and J. Letouzey, 1983, Tectonique de la zone compris entre Abudurba et Gebel Mezzazat (Sinai, Egypte) dans le contexte de l'evolution du rift du Suez: Bulletin des Centres de Recherche, Exploration-Production Elf-Aquitaine, v. 7, p. 201-215.

Cohen, A., 1989a, Facies relationships and sedimentation in large rift lakes and implications for hydrocarbon exploration—Example from Lakes Turkana and Tanganyika: Palaeogeography, Palaeoclimatology, Palaeoecology, v. 70, p. 65-80.

Cohen, A., 1989b, The taphonomy of gastropod shell accumulations in large lakes—An example from Lake Tanganyika, Africa: Paleobiology, v. 15, p. 26-44.

Cohen, A., and C. Thouin, 1987, Nearshore carbonate deposits in Lake Tanganyika: Geology, v. 15, p. 414-418.

Coulter, G. W., 1963, Hydrological changes in relation to biological productivity in southern Lake Tanganyika: Limnology and Oceanography, v. 8, p. 463-477.

Craig, H., ed., 1974, Lake Tanganyika Geochemical and Hydrographic Study—1973 Expedition: Scripps Institution of Oceanography, SIO Reference Series 75-5.

Ebinger, C., 1988, Thermal and mechanical development of the East African Rift System: Ph. D. Dissertation, Massachusetts Institute of Technology, Cambridge, 260 p.

Ebinger, C., 1989, Tectonic development of the western branch of the East African rift system: GSA Bull., v. 101, p. 885-903.

Ebinger, C., M. J. Crow, B. R. Rosendahl, D. Livingstone, and J. LeFournier, 1984, Structural evolution of Lake Malawi, Africa: Nature, v. 308, p. 627-629.

Frostick, L. E., and I. Reid, 1986, Evolution and sedimentary character of lake deltas fed by ephemeral rivers in the Turkana basin, northern Kenya, in L. E. Frostick, R. W. Renaut, I. Reid, and J. J. Tiercelin, eds., Sedimentation in the African rifts: Geological Society Special Publication 25, p. 113-126.

Frostick, L. E., and I. Reid, 1987, Tectonic control of desert sediment in rift basins ancient and modern, in L. E. Frostick, and I. Reid, eds., Desert sediments—Ancient and modern: Geological Society Special Publication 35, p. 53-68.

Haberyan, K. A., and R. E. Hecky, 1987, The Late Pleistocene and Holocene stratigraphy and paleolimnology of Lakes Kivu and Tanganyika: Palaeogeography, Palaeoclimatology, Palaeoecology, v. 61, p. 169-197.

Hecky, R. E., and E. T. Degens, 1973, Late Pleistocene-Holocene chemical stratigraphy and paleolimnology of the rift valley lakes of Central Africa: Woods Hole Oceanographic Institution Technical Report 73-28, 93 p.

Kufferath, J., 1952, Le milieu biochemique, exploration hydrobiologique du Lac Tanganika—Resultats scientifiques: Institut Royale des Sciences Naturelles de Belgique, v. 1, p. 29-47.

Lambiase, J., and M. Rodgers, 1988, A model for tectonic control of lacustrine stratigraphic sequences in continental rift basins, in Lacustrine exploration—Case studies and modern analogues (abs.): AAPG Research Conference Abstract with Program, p. 1.

Leeder, M., and R. Gawthorpe, 1987, Sedimentary models for extensional tilt-block/half-graben basins, in M. P. Coward, J. F. Dewey, and P. L. Hancock, eds., Continental extensional tectonics: Geological Society Special Publication 28, p. 139-152.

LeFournier, J., 1980, Depots de preouverture de l'Atlantique Sud—Comparaison avec la sedimentation actuelle dans la branche occidentale des rifts Est-Africains: Recherches Geologiques en Afrique, v. 5, p. 127-130.

LeFournier, J., J. Chorowicz, C. Thouin, F. Balzer, P. Chenet, J. Henriet, J. Masson, A. Mondeguer, B. Rosendahl, F. Spy-Anderson, and J. Tiercelin, 1985, Le bassin du Lac Tanganyika—Evolution tectonique et sedimentaire: Comptes Rendus de l'Académie des Sciences de Paris, v. 301, p. 1053-1058.

McHargue, T., 1988, Tectonostratigraphic development of Proto-Atlantic rifting in Cabinda, Angola—A petroliferous lake basin, in Lacustrine exploration—Case studies and modern analogues (abs.): AAPG Research Conference Abstracts with Program, insert.

Medeiros, R., and F. Ponte, 1981, Roteiro geologico da Bacia do Reconcavo: Petrobrás, Setor de ensino da Bahia, Salvador, Bahia, Brazil, 63 p.

Milani, E., 1987, Aspectos da evolucao tectonica das Bacias do Reconcavo e Tucano Sul, Bahia, Brasil: Petrobrás, Centro de Pesquisas e Desenvolvimento, Secao Exploracao de Petroleo, no. 18, 61 p.

Mondeguer, A., 1984, La Baie de Burton—Approche sedimentologique et structurale: Thèse, Université Bretagne Occidentale, Brest, France, 90 p.

Mondeguer, A., J. J. Tiercelin, M. Hoffert, P. Larque, J. LeFournier, and P. Tucholka, 1986, Sedimentation actuelle et Recente dans un petit bassin en contexte extensif et decrochant—La Baie de Burton Fosse Nord Tanganyika, Rift Est-Africain: Bulletin des Centres de Recherche, Exploration-Production Elf-Aquitaine, v. 10, p. 229-247.

Morley, C., 1988, Variable extension in Lake Tanganyika: Tectonics, v. 7, p. 785-801.

Petrobrás, 1988, Roteiro para visita de afloramentos na Bacia do Reconcavo: Petrobrás, 25 p.

Rosendahl, B. R., 1987, Architecture of continental rifts with special reference to East Africa: Annual Review of Earth and

Planetary Sciences, v. 15, p. 445–503.

Rosendahl, B. R., ed., 1988, Seismic atlas of Lake Tanganyika, East Africa: Duke University Project PROBE Geophysical Atlas Series, v. 1, 82 p.

Rosendahl, B. R., D. J. Reynolds, P. M. Lorber, C. F. Burgess, J. McGill, D. Scott, J. J. Lambiase, and S. J. Derksen, 1986, Structural expressions of rifting—Lessons from Lake Tanganyika, Africa, *in* L. E. Frostick, R. W. Renaut, I. Reid, and J. J. Tiercelin, eds., Sedimentation in African rifts: Geological Society Special Publication 25, p. 29-43.

Scholz, C. A., and B. R. Rosendahl, 1987, Late Pleistocene low lake levels in two African rift lakes (abs.): GSA Annual Meeting Abstracts with Program, v. 19, p. 834.

Scholz, C. A., and B. R. Rosendahl, 1988, Low lake stands in Lakes Malawi and Tanganyika, east Africa, delineated with multifold seismic data: Science, v. 240, p. 1645-1648.

Weissel, J. K., and G. D. Karner, 1989, Flexural uplift of rift flanks due to mechanical unloading of the lithosphere during extension: Journal of Geophysical Research v. 94, p. 13919-13950.

Coarse-Clastic Facies and Stratigraphic Sequence Models from Lakes Malawi and Tanganyika, East Africa

Christopher A. Scholz
Department of Geology
Duke University
Durham, North Carolina, U.S.A.

Bruce R. Rosendahl
Rosenstiel School of Marine and Atmospheric Science
Miami, Florida, U.S.A.

Seismic-reflection data from Lakes Malawi and Tanganyika in the western branch of the east African rift system reveal a variety of coarse-grained depositional facies. These facies include fan deltas and slope aprons adjacent to border faults, deep-water sublacustrine fans and channel systems, lowstand deltas, and an array of clastic and carbonate littoral deposits. Each is located in specific areas within half-grabens and develops at specific times within the cycle of lake-level variation. Rift lakes in tropical settings are highly sensitive to level fluctuations. High-amplitude and high-frequency lake-level variations may cause the resulting depositional sequence and facies architecture to be more complex than on passive margins. Controls on sequence development, such as sediment supply and lake-level variation, may be more closely coupled than on passive margins. Progradational clinoform depositional packages are uncommon in these basins probably because of the small size of catchments relative to lake surface areas and because of high slope gradients on the basin margins. Erosional truncation surfaces are more readily observed in these seismic data than are downlap surfaces.

INTRODUCTION

The large east African rift-valley lakes commonly are considered the best modern analogs of lacustrine basins in rifted terranes. Recent studies of Lakes Malawi (Nyasa) and Tanganyika have suggested that they are among the oldest lakes in the world (Abrahão and Warme, this volume; Burgess et al., 1988; Flannery, 1988). Rift-basin lakes and tectonically formed lake basins in general are long-lived geological features compared to the world's large glacial lakes (Johnson, 1984). Organic-rich shale accumulation in lacustrine basins commonly is associated with tropical lakes prone to permanent anoxia at depth (Demaison and Moore, 1980). Furthermore, these basins have been found to be prolific producers of oil and gas (e.g., Xijiang, 1988; Abrahão and Warme, 1988). One difficulty in exploring for hydrocarbons in lacustrine basins is predicting the location of coarse-grained reservoir rocks within the synrift section that are closely juxtaposed to hydrocarbon source rocks.

We present herein examples of multifold seismic-reflection data (CDP) and high-resolution reflection data from Lakes Malawi and Tanganyika in the east African rift system (EARS) (Figure 1). These examples illustrate coarse-grained deposits that likely occur in those large lakes and suggest possible analogs for reservoir facies in ancient lacustrine rift basins. The seismic data collected by Project PROBE, formerly at Duke University and now at the University of Miami, represent some of the most detailed acoustic images obtained from a lacustrine rift environment.

These acoustic examples are placed in a sequence stratigraphic framework not unlike that commonly described for passive-margin basins (e.g., Vail, 1987). We hope that such models can be used to interpret ancient rift basins and successfully predict the location of coarse-grained facies with hydrocarbon

151

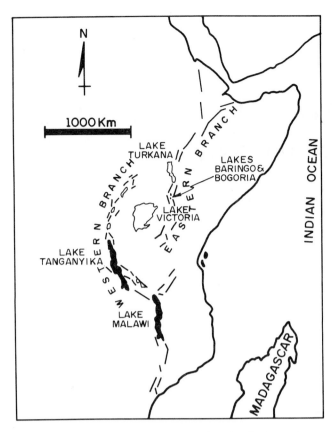

Figure 1. Large lakes and major border-fault systems of the east Africa rift system.

reservoir potential. The models presented here are built around the half-graben framework and keyed to specific stages of the cycle of lake-level variation.

METHODS

Seismic-reflection data were acquired aboard the R/V NYANJA, an 11-m portable seismic vessel designed for and dedicated to African rift lake studies. The seismic source was an array of one to four airguns with a total output of 0.65–2.3 L. All data are 24-fold and were collected using a 960-m, 48-channel cable with a maximum offset of about 1450 m and recorded on a Texas Instruments DFS-V recording system. Navigation was controlled by TRANSIT satellites, coupled to speed logs and radar, and intermittently by GPS satellite data.

Data underwent routine multifold processing at Duke University and the University of Miami using Digicon DISCO processing software. The principal steps included trace editing, velocity analysis, NMO correction, prestack deconvolution, and CDP stack. All data presented are 24-fold stacked, unmigrated time sections unless otherwise noted. Approximately 3500 km of multifold data exist for Lake Malawi and about 1500 km for Lake Tanganyika (Figure 2).

Seismic line spacing was about 8 km on Lake Malawi and about 15 km on Lake Tanganyika. Because this grid has a reconnaissance line spacing, it is difficult to accurately interpret the true three-dimensional external geometry of many facies examples presented herein; consequently, we have relied upon two-dimensional profile images for most of our interpretation. Fortunately we have extensive high-resolution single-channel coverage of Lake Malawi (>10000 km), and those data are used to augment and in some cases more clearly image the shallow facies seen also on multifold data. The high-resolution data were collected with a Furuno 28-kHz echosounder and a 1-kHz Geopulse profiling system.

REGIONAL FRAMEWORK

Studies of the EARS have shown distinct hierarchical scales of rifting, from that of separate branches (western vs. eastern), to zones (e.g., the Tanganyika vs. Malawi rifts zones), to half-graben units, and to blocks or subbasins within half-graben units (Rosendahl, 1987). Perhaps of greatest importance with respect to depositional facies architecture and sedimentary fill are the half-graben units. They typically measure about 110 km by 70 km and have major depocenters just offshore and adjacent to bounding border faults (Reynolds and Rosendahl, 1984; Rosendahl et al., 1986). Onshore next to the border faults are high rift mountains towering 2000 m or more above the lakes. The lakes are often many hundreds of meters deep just offshore of the border faults. Most well-developed half-graben units have slopes of 6° or more on the border-fault side, a basin plain with slopes less than about 1:300 (0.02°), a gentle concave-upward profile at the base of the shoaling margin, and a convex-upward profile near the upper edge of the shoaling margin (Figure 3). On the shoaling margin, the lake-bottom gradient is gentler, with slopes typically less than 1.5°. This shoaling margin geometry is analogous to the "rollover" margin of Frostick and Reid (1987). A thin veneer of sediment commonly covers crystalline basement rocks in shallower water on the shoaling sides over broad "platform" areas (Rosendahl, 1987; Burgess et al., 1988).

Regions known as accommodation zones link adjacent half-graben units (Rosendahl et al., 1986; Rosendahl, 1987). These features have a variety of morphologies, which may have contrasting kinematic development (e.g., Specht and Rosendahl, 1989). For example, they can be characterized as high-relief structures or horsts in some localities, broad antiforms with subdued relief in others, zones of intermeshing faults representing either oblique extension or oblique compression, or zones of simple basement warping. Rosendahl (1987) described in detail the structures and their relationships to various half-graben linking modes. Such features not

Scholz and Rosendahl

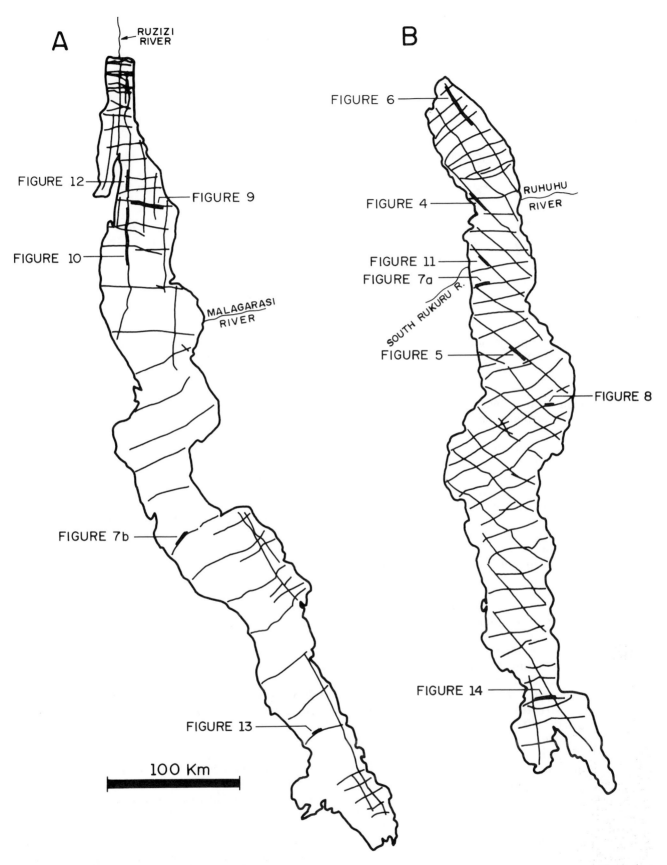

Figure 2. CDP seismic-reflection tracklines for (A) Lake Tanganyika and (B) Lake Malawi. Heavy lines indicate location of sections shown in Figures 4–14.

Lacustrine Basin Exploration: Case Studies and Modern Analogs

153

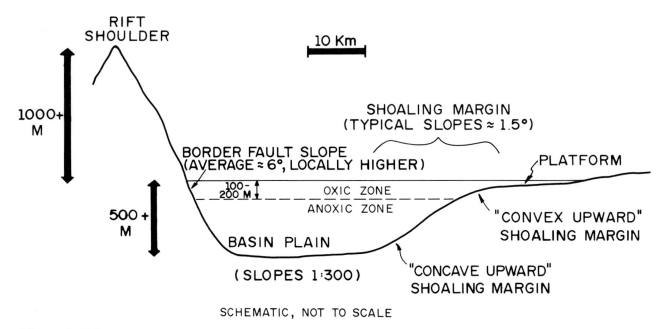

RIFT
SHOULDER

10 Km

SHOALING MARGIN
(TYPICAL SLOPES ≈ 1.5°)

1000+
M

BORDER FAULT SLOPE
(AVERAGE ≈ 6°, LOCALLY HIGHER)

PLATFORM

100-
200 M   OXIC ZONE
ANOXIC ZONE

500 +
M

"CONVEX UPWARD"
SHOALING MARGIN

BASIN PLAIN

(SLOPES 1:300)

"CONCAVE UPWARD"
SHOALING MARGIN

SCHEMATIC, NOT TO SCALE

Figure 3. Schematic profile of a half-graben showing the major physiographic provinces of a typical well-developed half-graben from the western branch of the east African rift system.

only are fundamental to understanding the kinematics of rifting but also are important in the development of facies architecture in rift basins. Accommodation zones and transfer faults (their passive-margin equivalents) have been associated with hydrocarbon reservoirs in some ancient rift systems (Etheridge et al., 1984, 1988) in large part because they greatly influence depositional pathways between half-graben units.

Preexisting structural and tectonic features also exert important controls on rift development (Versfelt and Rosendahl, 1989) and facies patterns in individual half-graben units. Older Mesozoic rifts that intersect the Cenozoic rift, such as the Ruhuhu system in Lake Malawi (Frostick and Reid, 1989), contain large volumes of easily erodible sedimentary rock. These old rifts may contribute inordinate amounts of clastic material into the lakes compared to that from surrounding Precambrian terrane. Major river systems, which themselves may predate the young Cenozoic rifts (Crossley, 1984), commonly follow prerift structural trends. Clastic material accumulates in the rifts where these older lineaments intersect the modern lakes. Intersecting preexisting structures can greatly influence subsequent fault-block patterns and subbasin development in half-graben units, which further alter depositional pathways.

Volcanic centers are uncommon in the western branch of the EARS and in general are neither significant sediment sources nor impediments to sediment transport, with only a few exceptions. In the eastern branch, however, volcanic edifices dot the landscape and are the sources of much of the primary and secondary sedimentary fill within the

rift valleys. The volcanic structures themselves can control depositional pathways (Cohen et al., 1986; Watkins, 1986).

Detailed studies of lacustrine deposits have revealed evidence for paleoclimatic fluctuations widespread in space and time over east Africa during the late Pleistocene (e.g., Butzer et al., 1972). The large western branch lakes are located within a tropical to subtropical climatic setting dominated mainly by savannah and woodland. Many paleoclimate studies agree that temperatures were considerably lower in the late Pleistocene, prior to about 12 ka, and that much of east Africa was drier as well. Unfortunately, paleoclimatic data for the time period prior to about 25 ka are limited, and no consensus exists as to conditions then.

Considerable climatic variation is evident over east Africa even today—rainfall in the eastern branch (Lake Turkana region) averages 15 cm/yr (East African Meteorological Department, 1975, from Halfman, 1987) and more than 200 cm/yr (Eccles, 1974) in the western branch (northern Lake Malawi). Present conditions in a basin such as Lake Turkana may, in fact, resemble to some degree conditions operative in the past in a western branch lake, such as Malawi. Distinctive acoustic facies observed at depth in the seismic records from Lakes Malawi and Tanganyika may have modern analogs in shallow deposits in Lakes Baringo and Turkana in the eastern branch of the EARS.

Drainage patterns into the African rift lakes tend to be repetitive on the scale of half-graben units (Frostick and Reid, 1987). Where relief is greatest—that is, adjacent to the rift shoulder on the border-fault sides of rift mountains—drainage areas are very

Scholz and Rosendahl

small and steep, and streams are often ephemeral. Major river systems adjacent to the rift mountains in fact drain away from the rift (Frostick and Reid, 1987). On the shoaling sides, drainage areas are substantially larger. Axial streams that enter the "end" or "terminal" half-graben in a rift zone in some cases also have substantial discharge, such as the case of the Omo River in Lake Turkana (Ferguson and Harbott, 1982). In general, drainage patterns are important controls on the ultimate depositional facies architecture within these basins.

Where several rift lakes are connected by a single drainage system, oversized axial river systems may profoundly influence deposition and lake hydrochemistry (Hecky and Degens, 1973; Haberyan and Hecky, 1987). Such is the case of the Ruzizi River, which is both the outlet for Lake Kivu and a primary input for Lake Tanganyika (Figure 2). In the cases of Lakes Malawi and Tanganyika, their outlets apparently have been tectonically stable in the recent geologic past. However, minor changes in the precipitation/evaporation balance can rapidly lower lake levels below the outlet thresholds, thereby transforming the lakes into hydrologically closed basins. This has reportedly occurred in both lakes during historical times (Beadle, 1981).

EVIDENCE FOR MAJOR LAKE LOWSTANDS

It has long been recognized that lakes in the arid and semiarid regions of eastern Africa are sensitive to fluctuations in level. Annual fluctuations in some rift-system lakes are one meter or more (Beadle, 1981). Paleolimnological studies in Lake Tanganyika have suggested lake-level drops of hundreds of meters (Livingstone, 1965; Degens et al., 1971; Haberyan and Hecky, 1987; Tiercelin et al., 1988). The recognition of seismic and sedimentary sequence boundaries and associated unconformities in the sections of large lakes (Johnson et al., 1987; Burgess et al., 1988; Flannery, 1988; Dunkelman et al., 1988) is further evidence for dramatic changes in water level in these basins.

Scholz and Rosendahl (1988) provided evidence for prolonged lowstands in the Malawi and Tanganyika basins in the form of widespread erosional surfaces beneath shallow sequence boundaries. They estimated that late Pleistocene lake levels were more than 350 m below present levels. In Figure 4 a seismic line illustrates this major erosional surface in Lake Malawi. Shallow sequence boundaries in both lakes are overlain by thin sequences of similar thickness (<100 m). The sequence boundaries are interpreted to have developed at approximately the same time in both lakes. Simultaneous lowstand events associated with these boundaries are thus interpreted as being climatically controlled. Recent sedimentation rates (~1 mm/yr) of the overlying sequences

together with thickness estimates suggest a minimum age of 75 ka or more for this unconformity. Because no core or well-bore data exists for dating the materials above or below the unconformity, extensive sampling will be required before age relationships are precisely understood.

Another section from central Lake Malawi (Figure 5) shows sequence boundaries (erosional surfaces) at various stratigraphic levels. It is uncertain whether or not all these boundaries formed as a result of tectonic or paleoclimatic events. However, water fluctuations, such as evidenced here, clearly will be a major control on depositional architecture in these basins. Such pronounced changes in water level exceed analogous fluctuations described for the major ocean basins, wherein minor sea-level fluctuations in epeiric seas can cause drastic lateral shifts in depositional facies. Presumably, the effects of higher amplitude water-level fluctuations on facies development in rift lakes also would be extreme.

CLASTIC FACIES AND DEPOSITIONAL ENVIRONMENTS

A variety of depositional facies have been interpreted to lie juxtaposed in modern environments of large rift lakes. Whereas these systems are dominated by fine-grained hemipelagic deposits, many contain significant amounts of coarse-grained material. Adjacent to border faults are broad slope aprons composed of small fan deltas near breaks in the rift shoulders, subaqueous fans in deep water off these point sources, and talus and debris-flow deposits along the base of the rift shoulders (Ng'ang'a, 1988). Large sublacustrine fans with channel and levee complexes that are observed over large areas of the half-graben basin floors appear to be fed mainly by large river systems (Scott et al., in press). Small "sediment wave" bed forms, perhaps the result of contour-current activity, have been described in southern Lake Malawi (Scott, 1988). Large sand sheets have been reported in association with old, subaerially exposed rift deposits onshore (Johnson and Davis, 1989; Owen and Crossley, 1989). Lag deposits are inferred in "erosional moats" around islands that perhaps were subjected to contour- or bottom-current activity (Scott, 1988). In addition, various carbonate (Cohen and Thouin, 1987) and clastic deposits are present in the littoral zones (Owen and Crossley, 1989). Coarse-grained eolian and beach deposits are important proximal facies in littoral zones as well as in lagoons (Bishop, 1987).

Modern deposits adjacent to the border faults are mainly subaqueous in origin primarily because little subaerially exposed surface exists along the rift escarpment. Only small surface areas of the fan deltas that cling to the steep slopes have alluvial fan

NW   3 KM   SE

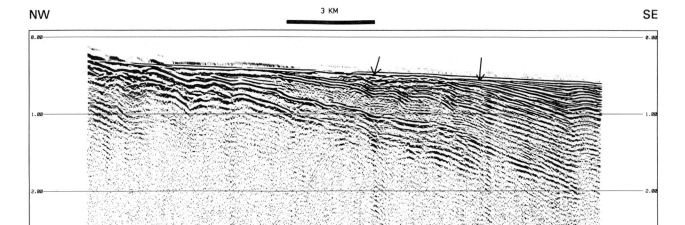

Figure 4. Seismic section from northeastern Lake Malawi, showing shallow, profound unconformity beneath a thin depositional sequence. Numbers on vertical axes on all seismic sections represent two-way travel time in seconds.

2 KM

NW   SE

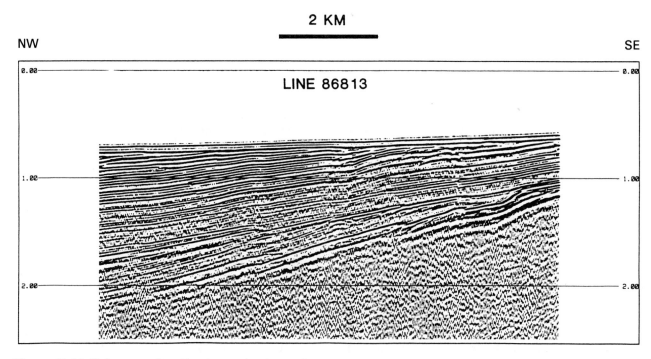

LINE 86813

Figure 5. Multiple unconformities in a seismic section from central Lake Malawi.

characteristics. In general, downlapping reflections resembling Gilbert delta megaforesets are not recognized in any CDP data associated with these deltas probably because of the steep gradients upon which these features are constructed; such deltas are not unlike the fan deltas in pull-apart basins (Dunne and Hempton, 1984). Studies of ancient rifts have shown that fan deltas have a high preservation potential (Wescott and Etheridge, 1983; McPherson et al., 1987) and that they can become major hydrocarbon reservoirs (Harms et al., 1981; Stow et al., 1982; Harms and McMichael, 1983). The CDP seismic section in Figure 6 shows what we interpret as a series of fan deltas offshore of the Livingstone Mountains border fault, Lake Malawi. Although lacustrine fan deltas have been examined in seismic

Scholz and Rosendahl

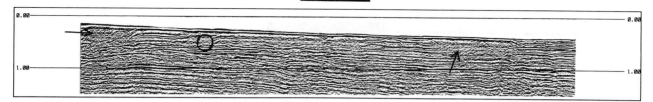

Figure 6. Seismic section along strike of the Livingstone Mountains rift shoulder, Lake Malawi. Reflection-free zones (arrows and circle) are interpreted as coarse- grained fan-delta deposits shed off steep slopes of the rift shoulder.

sections elsewhere (e.g., Xijiang, 1988), criteria for distinguishing between subaerial and subaqueous components have not been well defined. The chaotic, reflection-free zones are interpreted to be proximal coarse-grained material deposited in an alluvial fan setting during the lowstand discussed previously.

Deep-water sublacustrine fans with channel and levee complexes have been described in the recent environments in Lake Malawi (Scott 1988; Scott et al., in press). The largest subaqueous fans in Lake Malawi and Lake Tanganyika are fed by the largest rivers, the Ruhuhu and Malagarasi, respectively (Figure 2). Both rivers drain large areas of sedimentary terrane and are therefore able to deliver sizable quantities of erodible material to the lakes. Channels up to 3 km across and more than 100 m deep have been recognized in Lakes Malawi and Tanganyika (Flannery 1988; Scholz et al., 1990). A CDP record from central Lake Malawi (Figure 7A) illustrates several such channels. Buried beneath the active channels is a mounded seismic facies, similar in form to the modern levees, and interpreted to represent an older channel complex that is part of the same or similar sublacustrine fan system. Figure 7B is a similar occurrence in Lake Tanganyika. Echosounder records across such fans show prolonged reflections in the deep-water channel bases, which suggest the presence of coarse-grained material. This implies that the channels are active conduits for turbidity flows. Both the extent and the external and internal geometries of these features are poorly constrained.

Large "braid-deltas" (terminology of McPherson et al., 1987) have developed adjacent to several EARS lakes where axial rivers enter the end half-graben units in a rift zone. Rivers such as the Omo in Lake Turkana and the Ruzizi in Lake Tanganyika deposit material that is typically finer grained than that originating along border-fault fan deltas (Frostick and Reid, 1986). The Omo River, for example, has developed a classic bird-foot morphology (Galloway, 1975) that is characteristic of very fine-grained, river-dominated delta systems. Braid-deltas develop where rivers enter the lakes onto slopes at gradients significantly less than adjacent to border faults.

Numerous small, shallow channels have been recognized on the shoaling sides of half-graben units. These are typically a maximum several hundred meters across and a few meters to several tens of meters deep from the channel bases to the tops of the adjoining levees. The example in Figure 8 is interpreted to represent a fluvial channel submerged since the last major lowstand and now buried by tens of meters of fine-grained lacustrine mud.

Analogs for some deeply buried seismic facies in large EARS lakes cannot always be found in the Holocene deep-lake environments described by Johnson and Ng'ang'a (this volume), Scott (1988), Ng'ang'a (1988), Johnson and Davis (1989), Cohen and Thouin (1987), or Owen and Crossley (1989). One must look at seismic data sets from other parts of the world and other geologic settings to reasonably interpret some of these features. Lakes in the eastern branch of the EARS may even provide more appropriate modern analogs for some of the deposits associated with lowstand events in the large western branch lakes.

In an example from northern Lake Tanganyika, Figure 9 shows a 100- to 200-ms-thick transparent, reflection-free acoustic facies that is widespread through the lake at this stratigraphic level. The feature is interpreted to be a time-transgressive, coarse-grained unit that was emplaced successively higher up the slope of the shoaling margin during a lacustrine transgression following a major lowstand. Its distinctive character makes it a useful acoustic marker for correlations between half-graben units that may have no direct stratigraphic continuity. Thickness of the unit varies laterally, depending on location within the half-graben. It commonly is thicker on the downthrown sides of major faults (Figure 9) and thus may in part be composed of both reworked beach and littoral deposits as well as fan deltas. This unit also is located atop a large antiform in the central Kigoma basin and could represent a hydrocarbon exploration prospect (Figure 10).

Clinoform (downlap) reflection geometries associated with prograding deltas are scarce in Malawi and Tanganyika CDP data. The few localities where they are observed are interpreted to be associated with lowstand deposits. Figure 11 shows a series of downlapping reflections that developed when the proto-South Rukuru River flowed down the steep slopes of the northern Lake Malawi basin and onto the basin plain. In northern Lake Tanganyika (Figure 12) a lowstand delta deposit has prograded over a

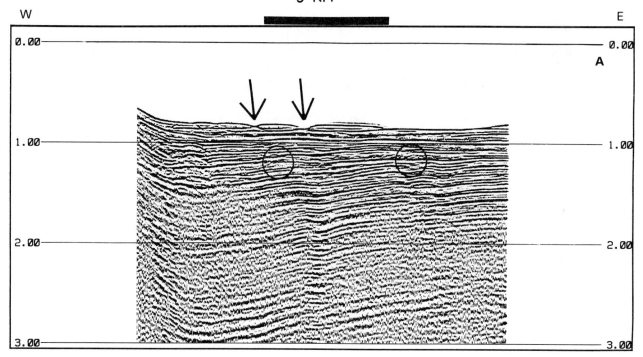

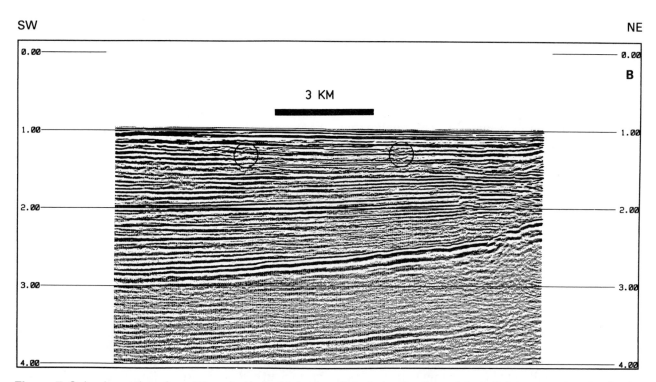

Figure 7. Seismic sections from (A) central Lake Malawi and (B) central Lake Tanganyika, showing deep-water channels (arrows in A) and mounded, wavy, and distorted reflections (circled) interpreted to be buried channels and levees.

Scholz and Rosendahl

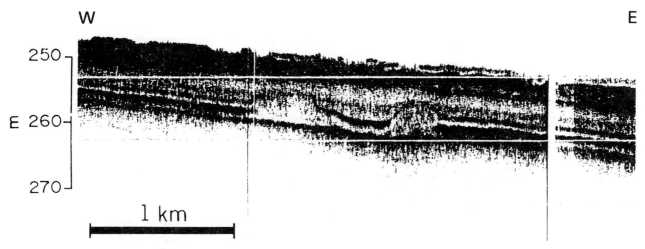

Figure 8. High-resolution, 28-kHz echosounder profile of a small buried channel and accompanying levees on shoaling margin in central Lake Malawi.

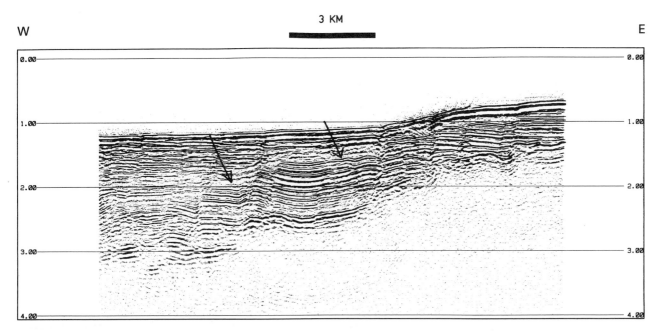

Figure 9. Seismic section from northern Lake Tanganyika, showing 100- to 200-ms-thick seismic facies overlying a sequence boundary (arrows). This acoustic facies is interpreted to be a transgressive package deposited after a major lake lowstand.

large area from what may have been the proto-Ruzizi River into a lowstand lake that occupied the present Kigoma basin. In both examples, hemipelagic mud and pelagic diatom oozes that are being deposited on the lake floor reflect the current lake highstand.

The limited occurrence of progradational clinoform deposits associated with deltas is attributable to several factors. These include the steep slope morphology and small size of the catchment areas relative to the lakes themselves. Most rivers enter the lake where the slopes are relatively steep, generally greater than 1°. Even on shoaling margins, gradients are substantial. Rarely does vertical accretion reach a "depositional base level" or threshold such that sediment bypassing occurs. Thus major delta foresets (downlapping reflections) fail to develop on a scale that is resolvable with conventional CDP data. The high basinal subsidence rate relative to sediment input enhances this situation over geologic time, and the classic clinoform geometry

S   N

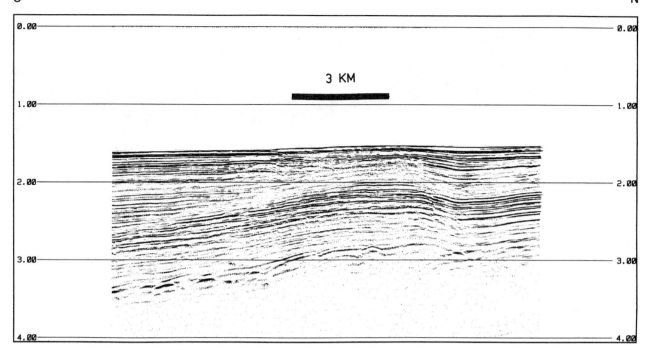

Figure 10. Rift-axial seismic section through a broad antiform in central Lake Tanganyika. Transparent seismic facies similar to that in Figure 9 can be recognized atop the antiform at about 2.0 s.

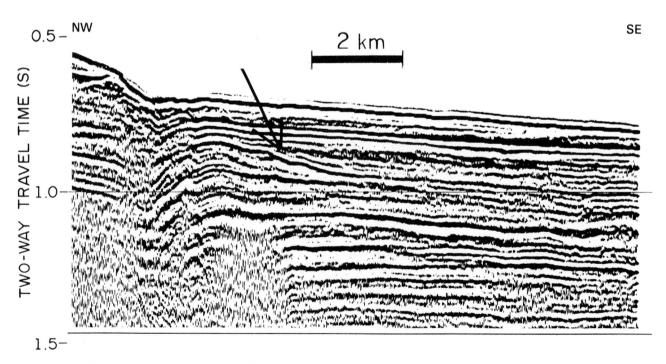

Figure 11. Seismic section from central Lake Malawi showing downlapping reflections (arrow at about 0.8 s) interpreted as a delta deposited during a lacustrine lowstand.

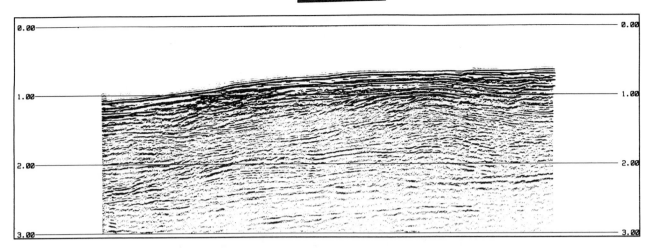

Figure 12. Seismic section from north-central Lake Tanganyika, showing a series of inclined reflections interpreted to be a progradational package deposited during a lacustrine lowstand.

(Rich, 1951; Scruton, 1960; Mitchum et al., 1977), with corresponding toplap and downlap reflection terminations, cannot develop.

Proximal shoreline sand deposits in the form of beach and shoreface sands, and eolian deposits are important in the modern setting. However, they generally are of insufficient thickness and distribution to be recognized easily in CDP data. Bishop (1987) has observed fine-grained deposits in Lake Malawi lagoons bounded by tilted fault blocks, and similar occurrences are interpreted in seismic data from Lake Tanganyika (Figure 13). In the Malawi and Tanganyika rift basins, tilted fault blocks have resulted in the development of wedge-shaped subbasins that may have been hydrologically isolated during times of drastically low lake level. Sand sheets also are documented in the littoral zone and atop horst blocks and ramps in Lake Malawi (Owen and Crossley, 1989).

Sediment waves and abyssal bed-form deposits formed by contour currents are common in many regions of the deep sea. These features also have been described in large lake systems such as Lake Superior (Johnson et al., 1980). Deposition controlled by contour-current flow has been interpreted from CDP and echosounder data from southern Lake Malawi (Figure 14). At either end of the echosounder record are downlapping reflections that grade into prolonged reflections interpreted to be coarse-grained lag deposits formed by accelerated nearshore contour currents. Similar geometries are interpreted at the top of the CDP section and may also be evident at depth (line drawing, Figure 14). Although such deposits likely are volumetrically insignificant constituents of the basin fill in these lakes, this is clearly a unique phenomenon and important in some localities of Lake Malawi.

APPLICATION OF SEQUENCE STRATIGRAPHY

From burgeoning research activities in east Africa have come several models for rift-valley sedimentation (e.g., Crossley, 1984; Frostick and Reid, 1987; Tiercelin et al., 1987; Lambiase, 1988; Cohen, 1989). In addition, fan-delta and alluvial fan models are emplaced within a rift context (e.g., McPherson et al., 1987) or within a strike-slip or pull-apart basin framework (Dunne and Hempton, 1984; Manspeizer, 1985). Strike-slip basins share similar tectonic and geomorphological characteristics with east African basins—marked basin asymmetry, high aspect-ratio cross sections, and limited, tectonically controlled catchment areas.

Seismic stratigraphic studies of passive margins and other basins, conducted mainly by the oil and gas industry, have rekindled the sequence-analysis approach (Sloss et al., 1949) to the study of sedimentary basins (e.g., Vail et al., 1977; Posamentier et al., 1988). Controls on sequence development along passive margins are disputed, some regard eustatic control as primary (e.g., Vail et al., 1977, 1984; Haq et al., 1987), whereas others maintain that sequences develop as an equal consequence of eustasy, tectonism, and sediment supply (Hubbard, 1988; Galloway, 1989).

The recognition of extreme water-level fluctuations, as described above, presents an opportunity to apply models of lacustrine rift-system facies development in the context of sequence stratigraphy. To do this, one first must identify discrete depositional packages separated by unconformities and their correlative conformities in seismic data. The groundwork for this in the large EARS lakes was

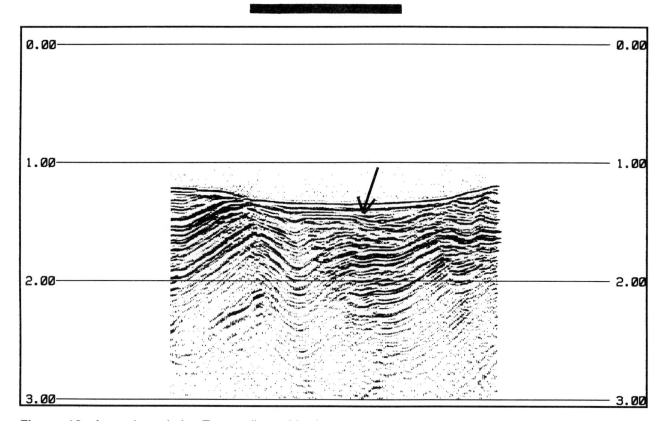

W   3 KM   E

Figure 13. A southern Lake Tanganyika subbasin (arrow) that may have been an isolated water body during a major lowstand.

established by Dunkelman et al. (1988), Burgess et al. (1988), and Flannery (1988). Most published sequence-stratigraphy and systems-tract models have been developed for passive-margin marine systems. It is therefore worthwhile to examine the similarities and differences between these depositional systems and lacustrine rift basins.

The major difference is physical scale. Systems tracts (Vail et al., 1984) in the marine setting, especially on a shelf/slope/rise type of passive margin, commonly are measured in hundreds of kilometers. Individual facies components within them may encompass tens to hundreds of kilometers. This is in marked contrast to lacustrine rift systems, wherein entire drainage basins are commonly less than 100 km wide and only several hundred kilometers long. Facies changes here typically are observed occurring within tens of kilometers or less (Rosendahl and Livingstone, 1983). The high cross-sectional aspect ratio of the asymmetric basins leads to rapid vertical facies changes as well.

Both the amplitude and frequency of lake-level changes are greater than sea level changes. Annual fluctuations of one meter or more are common in Lake Malawi, and fluctuations of tens to hundreds of meters are believed to have occurred in many African lakes during the last 15,000 yr (e.g., Butzer et al., 1972; Livingstone, 1965; Haberyan and Hecky, 1987; Halfman, 1987; Tiercelin et al., 1988). Even during extreme glacial oscillations in the Pleistocene, sea level did not fluctuate more than a few hundred meters. Furthermore, those changes occurred over a period of tens of thousands of years. A consequence of the higher frequency lacustrine fluctuations presumably would be stratigraphic sequences thinner than those typically preserved on passive margins.

Regional climate can be perturbed either by external factors such as Indian Ocean monsoons or by local factors such as those induced by rift-shoulder uplift. Any such change probably will have some effect on lake water level as well as potentially significant effects on sediment supply to the basins. How lake-level changes and sediment-supply controls influence generation of depositional sequences is quite unlike the conditions on passive-margin systems. In the marine setting, eustasy and local sediment supply are, for the most part, unrelated, but in a rift lake system, these controls may be coupled. For instance, during times of lacustrine

                                             Scholz and Rosendahl

BOTTOM CURRENT-CONTROLLED DEPOSITION
(ECHO-SOUNDER AND MCS PROFILES AT SAME HORIZONTAL SCALE)

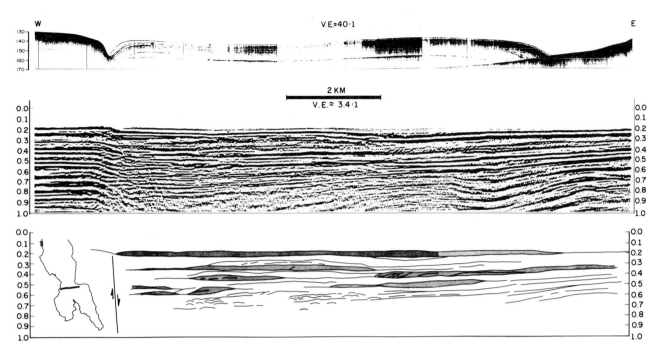

Figure 14. Echosounder record (top), multifold seismic section (center), and line drawing (bottom) at same horizontal scale from southern Lake Malawi. Prolonged acoustically opaque reflections may represent lag deposits caused by accelerated contour-current flow through the narrow part of the lake (see inset map).

transgression, the onset of wetter climatic conditions responsible for lake-level rise also would be directly responsible for increased river discharge and possibly anomalously high clastic input.

The measurement of tectonic and thermal subsidence is critical for understanding basin development. However, the lack of borehole lithologic and biostratigraphic data precludes quantitative analysis and evaluation of subsidence in EARS lake basins. Dense fault patterns (Sander, 1986; Rosendahl et al., 1986; Specht and Rosendahl, 1989) and divergent reflection patterns seen in the seismic data do suggest severe tectonic effects. Measurement and dating of tectonically induced variations in basin subsidence are difficult to evaluate for all except the youngest depositional sequences recognized in the rift lakes (e.g., Scholz and Rosendahl, 1988). It appears that climatically induced fluctuations occur more frequently than tectonically induced lake-level changes.

SYSTEMS-TRACT MODELS

Another initial step in the application of sequence stratigraphy to lacustrine rifts is to develop models to represent end-member stages in the lake-level cycle. Such systems-tract models, when placed within the half-graben framework, ideally show three-dimensional geometry and internal structure of depositional facies components. Figure 15 is an idealized representation of lacustrine highstand and lowstand systems tracts within a framework of half-graben units that alternate polarity along rift axes. The highstand model (Figure 15A) illustrates major facies components identified or interpreted from CDP and echosounder data from the larger EARS lakes. The lowstand model (Figure 15B) is more difficult to conceptualize because few or no high-resolution seismic data are available to document such deposits in Lake Malawi and Lake Tanganyika. However, smaller lakes of the eastern branch of the EARS may provide analogs and facies examples for some lowstand features interpreted from CDP data. It should be noted here that these models best approximate the tectonic conditions in the western branch of the EARS, where rifts have undergone considerable subsidence but limited amounts of extension. Parts of the eastern branch have experienced greater extension (Mohr, 1987) and may better approximate synrift sections of rifted continental margins.

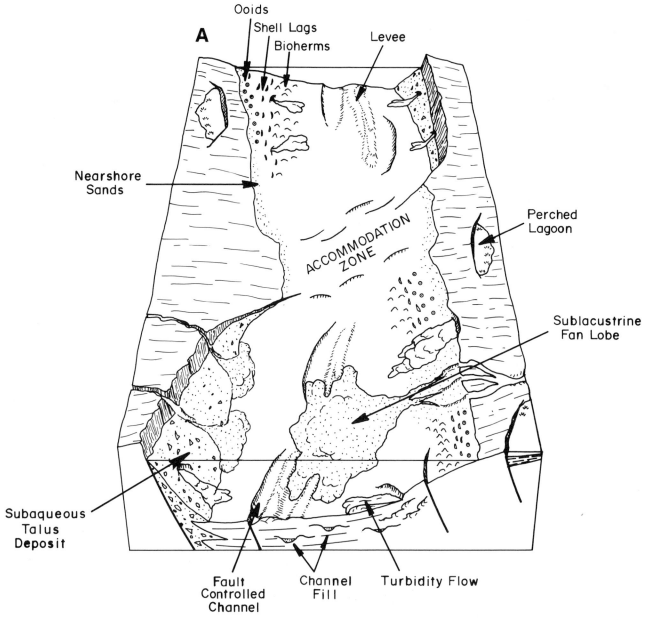

Figure 15. Idealized systems-tract models (not to scale) of rift-basin lacustrine facies and structure in an alternating half-graben framework. A, highstand model; B, lowstand model. Important coarse-grained facies may be developed within lowstand deltas and fan-delta systems.

EXPLORATION SIGNIFICANCE

The proximity of organic-rich sediments (mainly profundal diatomaceous oozes) to the coarse-grained deposits described here seemingly would signify numerous structural and stratigraphic exploration targets. Over geologic time, however, subaerial exposure and erosion of half-graben shoaling margins likely will lead to exceedingly poor preservation of shoaling-margin highstand systems tracts, as seen in Figure 4. Although various littoral deposits, both clastic and carbonate, ultimately could become excellent reservoir rocks, enhanced erosion along the basin rims may preclude these facies types from constituting a major percentage of the stratigraphic section.

The best preserved coarse-grained deposits likely will be found in the deeper parts of the basins—namely, subaqueous fans formed during highstands and lowstand deltaic deposits. Although lacustrine turbidity deposits may not have an historical record of significant hydrocarbon production, lowstand

Scholz and Rosendahl

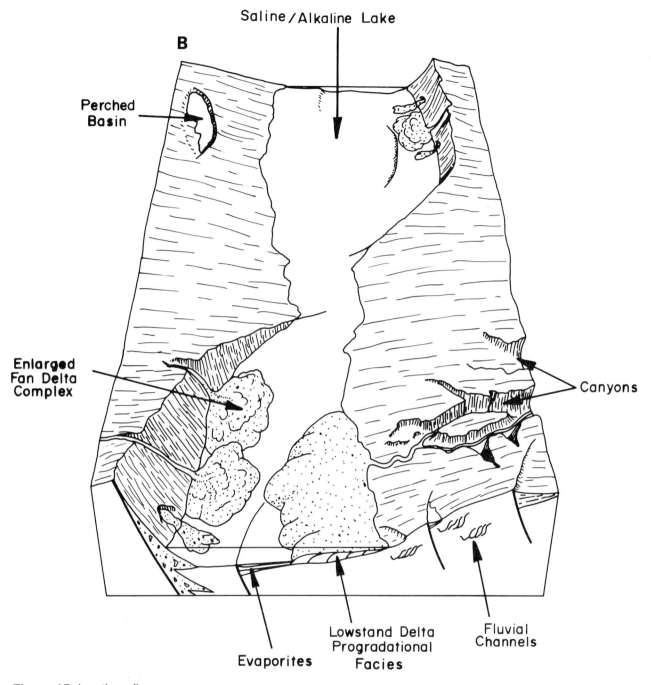

Saline/Alkaline Lake

B

Perched
Basin

Enlarged
Fan Delta
Complex

Canyons

Evaporites

Lowstand Delta
Progradational
Facies

Fluvial
Channels

Figure 15. (continued)

deltas and highstand subaqueous fans, both associated with the largest shoaling-margin river
systems, have the potential to accumulate clean,
coarse-grained material, given the appropriate source
material. Whereas border fault-related deposits are
likely to be very coarse grained, we would anticipate
those bodies to be smaller than shoaling-margin
deposits and more contaminated with fine-grained
matrix material.

DISCUSSION AND
CONCLUSIONS

Fluctuations in lake levels in the east African rift
system over geologic time probably are as much a
function of climatic change as of tectonism. The
transition from a single large basin into several
discrete, shallow lakes simply by lowering lake level

probably can be accomplished over several tens to hundreds of thousands of years. The various stages between the end-member highstand and lowstand models also may have distinct depositional signatures. For instance, changes in the level of the hypolimnion have been observed in ancient lacustrine sections (Anadón et al., 1988) and in large lakes (B. Finney, pers. comm.).

The large lakes in the western branch of the EARS are located within high-relief terrain but today are dominated by fine-grained deposits. Large progradational deposits are noticeably absent within these systems because of various factors, including large ratios of lake area to drainage area and the steep slopes upon which deltas develop. Because of their greater termination angles, sequence boundaries (erosional surfaces) are more easily recognizable in CDP data in these lakes than are downlap surfaces and condensed sections.

The systems-tract models presented here are necessarily simplified and based primarily on seismic images rather than detailed sedimentological studies. However, they are a first step in the use of sequence stratigraphic concepts in such landlocked basins. Many questions must be answered before the models can be considered complete, including discriminating between climatically vs. tectonically produced sequence boundaries. Important driving elements of basin development, such as eustasy, subsidence, uplift, catchment development, and sediment supply are more localized in lacustrine basins and more closely coupled to the resulting depositional sequences than in marine settings. EARS lakes could yield vital information on late Mesozoic and Cenozoic climatic changes and tectonic development of east Africa. Furthermore, they represent ideal laboratories for studying other interrelated sedimentation and tectonics problems, such as development of tectonic cyclothems and quantitative analysis of unconformities. Additional geophysical data acquired in conjunction with more ground data should prove highly fruitful in studies of these basins.

ACKNOWLEDGMENTS

We thank the countries of Burundi, Malawi, Mozambique, Tanzania, Zaire, and Zambia for their permission to conduct this research. We appreciate the comments of A. S. Cohen, H. Dypvik, L. Liro, D. Schunk, and W. H. Wheeler on earlier versions of this manuscript. This work was supported by Project PROBE's industrial consortium made up of the following companies: Agip, Amoco, ARCO, Esso/Exxon, Mobil, Pecten, Petrofina, Placid, Shell, Texaco, and the World Bank. Acknowledgment is made to the Donors of The Petroleum Research Fund, administered by the American Chemical Society, for the partial support of this research.

REFERENCES CITED

Abrahão, D., and J. E. Warme, 1988, Lacustrine and associated deposits in a rifted continental margin—The Lagoa Formation, Campos basin, offshore Brazil, in Lacustrine exploration—Case studies and modern analogues (abs.): AAPG Research Conference Abstract with Program.

Abrahão, D., and J. E. Warme, Lacustrine and associated deposits in a rifted continental margin—Lower Cretaceous Lagoa Feia Formation, Campos basin, offshore Brazil, this volume.

Anadón, P., L. Cabrera, and R. Julia, 1988, Anoxic-oxic cyclical sedimentation in the Miocene Rubielos de Mora Basin, Spain, in A. J. Fleet, K. Kelts, and M. R. Talbot, eds., Lacustrine petroleum source rocks: Geological Society Special Publication 40, p. 353–367.

Beadle, L. C., 1981, The inland waters of tropical Africa, 2d ed.: New York, Longman, 475 p.

Bishop, M. G., 1987, Clastic depositional processes in response to rift tectonics in the Malawi rift, Malawi, Africa: M. S. Thesis, Duke University, Durham, N. C., 111 p.

Burgess, C. F., B. R. Rosendahl, S. Sander, C. A. Burgess, J. Lambiase, S. Derksen, N. Meader, 1988, The structural and stratigraphic evolution of Lake Tanganyika—A case study of continental rifting, in W. Manspeizer, ed., Triassic-Jurassic rifting, continental breakup and the origin of the Atlantic Ocean and passive margins: New York, Elsevier, v. 2, p. 859–881.

Butzer, K., G. Isaac, J. Richardson, and C. Washbourn-Kamau, 1972, Radiocarbon dating of East African lake levels: Science, v. 175, p. 1069–1076.

Cohen, A., 1989, Facies relationships and sedimentation in large rift lakes and implications for hydrocarbon exploration—Example from Lakes Turkana and Tanganyika: Palaeogeography, Palaeoclimatology, Palaeoecology, v. 70, p. 65–80.

Cohen, A., D. S. Ferguson, P. M. Gram, S. L Hubler, and K. W. Sims, 1986, The distribution of coarse-grained sediments in modern Lake Turkana, Kenya—Implications for clastic sedimentation models of rift lakes, in L. E. Frostick, R. W. Renaut, I. Reid, and J. J. Tiercelin, eds., Sedimentation in the African rifts: Geological Society Special Publication 25, p. 127–139.

Cohen, A., and C. Thouin, 1987, Nearshore carbonate deposits in Lake Tanganyika: Geology, v. 15, p. 414–418.

Crossley, R., 1984, Controls on sedimentation in the Malawi rift valley, central Africa: Sedimentary Geology, v. 40, p. 33–50.

Degens, E. T., R. P. Von Herzen, and H. K. Wong, 1971, Lake Tanganyika—Water chemistry, sediments, geological structure: Naturwissenschaften, v. 58, p. 229–241.

Demaison, G. J. and G. T. Moore, 1980, Anoxic environments and oil source bed genesis: AAPG Bulletin, v. 64, p. 1179–1209.

Dunkelman, T. J., J. A. Karson, and B. R. Rosendahl, 1988, Structural style of the Turkana rift, Kenya: Geology, v. 16, p. 258–261.

Dunne, L. A., and M. R. Hempton, 1984, Deltaic sedimentation in the Lake Hazar pull-apart basin, south-eastern Turkey: Sedimentology, v. 31, p. 401–412.

East African Meteorological Department, 1975, Climatological records for East Africa: Nairobi, Kenya, East Africa Community, 92 p.

Eccles, D. H., 1974, An outline of the physical limnology of Lake Malawi (Lake Nyasa): Limnology and Oceanography, v. 19, p. 730–742.

Etheridge, M. A., J. C. Branson, D. A. Falvey, K. L. Lockwood, P. G. Stuart-Smith, and A. S. Scherl, 1984, Basin-forming structures and their relevance to hydrocarbon exploration in Bass Basin, southeastern Australia: BMR Journal of Australian Geology and Geophysics, v. 9, p. 197–206.

Etheridge, M. A., P. A. Symonds, T. G. Powell, and G. S. Lister, 1988, Application of the detachment model for continental exploration in extension basins: Australian Petroleum Exploration Association Journal, p. 167–187.

Ferguson, A. J. D., and B. J. Harbott, 1982, Geographical, physical and chemical aspects of Lake Turkana, in A. J. Hopson, ed., Lake Turkana—A report on the findings of the Lake Turkana Project, 1972–1975: London, Overseas Development Administration, p. 1–107.

Flannery, J. W., Jr., 1988, The acoustic stratigraphy of Lake Malawi, East Africa: M. S. Thesis, Duke University, Durham,

Scholz and Rosendahl

N. C., 117 p.

Frostick, L. E., and I. Reid, 1986, Evolution and sedimentary character of lake deltas fed by ephemeral rivers in the Turkana basin, northern Kenya, *in* L. E. Frostick, R. W. Renaut, I. Reid, and J. J. Tiercelin, eds., Sedimentation in the African rifts: Geological Society Special Publication 25, p. 113-126.

Frostick, L. E., and I. Reid, 1987, Tectonic control of desert sediment in rift basins ancient and modern, *in* L. E. Frostick, and I. Reid, eds., Desert sediments—Ancient and modern: Geological Society Special Publication 35, p. 53-68.

Frostick, L. E., and I. Reid, 1989, Is structure the main control of river drainage and sedimentation in rifts?: Journal of African Earth Sciences (and the Middle East), v. 8, p. 165-182.

Galloway, W. E., 1975, Process framework for describing the morphologic and stratigraphic evolution of deltaic depositional systems, *in* M. L. Broussard, ed., Deltas: Houston Geological Society, p. 87-98.

Galloway, W. E., 1989, Genetic stratigraphic sequences in basin analysis I: Architecture and genesis of flooding-surface bounded depositional units: AAPG Bulletin, v. 73, p. 125-142.

Haberyan, K. A., and R. E. Hecky, 1987, The Late Pleistocene and Holocene stratigraphy and paleolimnology of Lakes Kivu and Tanganyika: Palaeogeography, Palaeoclimatology, Palaeoecology, v. 61, p. 169-197.

Halfman, J. D., 1987, High-resolution sedimentology and paleoclimatology of Lake Turkana, Kenya: Ph.D. Dissertation, Duke University, Durham, N. C., 188 p.

Haq, B. U., J. Hardenbol, and P. R. Vail, 1987, Chronology of fluctuating sea levels since the Triassic: Science, v. 235, p. 1156-1167.

Harms, J. C., and W. J. McMichael, 1983, Sedimentology of the Brae Oilfield area, North Sea—Comments: Journal of Petroleum Geology, v. 5, p. 437-439.

Harms, J. C., P. Tackenberg, E. Pickles, and R. E. Pollock, 1981, The Brae oilfield area, *in* L. V. Illing, and G. D. Hobson, eds., Petroleum geology of the continental shelf of north-west Europe: Philadelphia, Heyden and Son, p. 352-357.

Hecky, R. E. and E. T. Degens, 1973, Late Pleistocene-Holocene chemical stratigraphy and paleolimnology of the rift valley lakes of Central Africa: Woods Hole Oceanographic Institution Technical Report 73-28, 93 p.

Hubbard, R. J., 1988, Age and significance of sequence boundaries on Jurassic and early Cretaceous rifted continental margins: AAPG Bulletin, v. 72, p. 49-72.

Johnson, T. C., 1984, Sedimentation in large lakes: Annual Review of Earth and Planetary Sciences, v. 12, p. 179-204.

Johnson, T. C., T. Carlson, and J. E. Evans, 1980, Contourites in Lake Superior: Geology, v. 8, p. 437-441.

Johnson, T. C., and T. W. Davis, 1989, High resolution seismic profiles from Lake Malawi, east Africa, *in* B. R. Rosendahl, J. J. W. Rogers, and N. M. Rach, eds., African rifting: African Journal of Earth Sciences (and the Middle East), v. 8, nos. 2-4, p. 383-392.

Johnson, T. C., J. D. Halfman, B. R. Rosendahl, and G. Lister, 1987, Climatic and tectonic effects on sedimentation in a rift-valley lake—Evidence from high-resolution seismic profiles, Lake Turkana, Kenya: GSA Bulletin, v. 98, p. 439-447.

Johnson T. C., and P. Ng'ang'a, Reflections on a rift lake, this volume.

Lambiase, J., and M. Rodgers, 1988, A model for tectonic control of lacustrine stratigraphic sequences in continental rift basins, *in* Lacustrine exploration—Case studies and modern analogues (abs.): AAPG Research Conference Abstract with Program, p. 1.

Livingstone, D. A., 1965, Sedimentation and the history of water level change in Lake Tanganyika: Limnology and Oceanography, v. 10, p. 607-610.

Manspeizer, W., 1985, The Dead Sea rift—Impact of climate and tectonism on Pleistocene and Holocene sedimentation, *in* K. T. Biddle, and N. Christie-Blick, eds., Strike-slip deformation, basin formation, and sedimentation: SEPM Special Publication 37, p. 143-158.

McPherson, J. G., G. Shanmugam, and R. J. Moiola, 1987, Fan-deltas and braid deltas: varieties of coarse-grained deltas: GSA Bulletin, v. 99, p. 331-340.

Mitchum, R. M., Jr., P. R. Vail, and J. B. Sangree, 1977, Stratigraphic interpretation of seismic reflection patterns in depositional sequences, *in* C. E. Payton, ed., Seismic stratigraphy—Applications to hydrocarbon exploration,

AAPG Memoir 26, p. 117-133.

Mohr, P., 1987, Structural style of continental rifting in Ethiopia—Reverse décollements: EOS, v. 68, p. 721-730.

Ng'ang'a, P., 1988, Sedimentation off deltas and border faults in northern Lake Malawi—Evidence from high-resolution acoustic remote sensing and gravity cores: M. S. Thesis, Duke University, Durham, N. C., 93 p.

Owen, R. B., and R. Crossley, 1989, Rift structures and facies in Lake Malawi: Journal African Earth Science, v. 8, p. 415-427.

Posamentier, H. W., M. T. Jervey, and P. R. Vail, 1988, Eustatic controls on clastic deposition, *in* C. K. Wilgus, and others, eds., Sea-level changes—An integrated approach: SEPM Special Publication 42, p. 109-124.

Reynolds, D. J., and B. R. Rosendahl, 1984, Tectonic expressions of continental rifting (abs.): EOS, v. 65, p. 1116.

Rich, J. L., 1951, Three critical environments of deposition, and criteria for recognition of rocks deposited in each of them: GSA Bulletin, v. 62, p 1-20.

Rosendahl, B. R., 1987, Architecture of continental rifts with special reference to East Africa: Annual Review of Earth and Planetary Sciences, v. 15, p. 445-503.

Rosendahl, B. R., and D. A. Livingstone, 1983, Rift lakes of East Africa—New seismic data and implications for future research: Episodes, v. 1983, no. 1, p. 14-19.

Rosendahl, B. R., D. J. Reynolds, P. M. Lorber, C. F. Burgess, J. McGill, D. Scott, J. J. Lambiase, and S. J. Derksen, 1986, Structural expressions of rifting—Lessons from Lake Tanganyika, Africa, *in* L. E. Frostick, R. W. Renaut, I. Reid, and J. J. Tiercelin, eds., Sedimentation in African rifts: Geological Society Special Publication 25, p. 29-43.

Sander, S., 1986, The geometry of rifting in Lake Tanganyika, East Africa: M.S. Thesis, Duke University, Durham, N. C., 46 p.

Scholz, C. A., and B. R. Rosendahl, 1988, Low lake stands in Lakes Malawi and Tanganyika, east Africa, delineated with multi-fold seismic data: Science, v. 240, p. 1645-1648.

Scholz, C. A., B. R. Rosendahl, and D. L Scott, 1990, Development of coarse-grained facies in lacustrine rift basins—Examples from East Africa: Geology, v. 18, p. 140-144.

Scott, D. L., 1988, Modern processes in a continental rift lake—An interpretation of 28 kHz seismic profiles from Lake Malawi, east Africa: M. S. Thesis, Duke University, Durham, N. C., 82 p.

Scott, D. L., P. Ng'ang'a, T. C. Johnson, and B. R. Rosendahl, (in press), High-resolution acoustic character of Lake Malawi (Nyasa), and its relationship to sedimentary processes: Sedimentology.

Scruton, P. C., 1960, Delta building and the deltaic sequence, *in* F. P. Shepard, F. B. Phleger, and T. J. H. van Andel, eds., Recent sediments, northwest Gulf of Mexico: AAPG, p. 82-102.

Sloss, L. L., W. C. Krumbein, and E. C. Dapples, 1949, Integrated facies analysis, *in* C. R. Longwell, (chairman), Sedimentary facies in geological history: GSA Memoir 39, p. 91-124.

Specht, T. D., and B. R. Rosendahl, 1989, Architecture of the Malawi Rift, East Africa: Journal of African Earth Sciences (and the Middle East), v. 8, p. 355-382.

Stow, D. A., C. D. Bishop, and S. J. Mills, 1982, Sedimentology of the Brae oilfield, North Sea—Fan models and controls: Journal of Petroleum Geology, v. 5, p. 129-148.

Tiercelin, J. J., and others, 1987, The Baringo-Bogoria half-graben, Gregory Rift, Kenya—30,000 years of hydrological and sedimentary history: Bulletin des Centres de Recherche, Exploration-Production Elf-Aquitaine, v. 11, no. 2, p. 249-540.

Tiercelin, J. J., A. Mondeguer, F. Gasse, C. Hillaire-Marcel, M. Hoffert, P. Larque, V. Ledee, P. Marestang, C. Ravenne, J. F. Raynaud, N. Thouveny, A. Vincens, and D. Williamson, 1988, 25,000 years of hydrological and sedimentary history of Lake Tanganyika, East African rift: Comptes Rendus de l'Académie des Sciences de Paris, v. 307, p. 1375-1382.

Vail, P. R., 1987, Seismic stratigraphy interpretation procedure, *in* A. W Bally, ed., Atlas of seismic stratigraphy: AAPG Studies in Geology 27, v. 1, p. 1-10.

Vail, P. R., J. Hardenbol, and R. G. Todd, 1984, Jurassic unconformities, chronostratigraphy, and sea level change from seismic stratigraphy and biostratigraphy, *in* J. S. Schlee, ed., Interregional unconformities and hydrocarbon exploration: AAPG Memoir 36, p. 129-144.

Vail, P. R., R. M. Mitchum, Jr., R. G. Todd, J. M. Widmier, S. Thompson, III, J. B. Sangree, J. N. Bubb, and W. G. Hatlelid, 1977, Seismic stratigraphy and global changes of sea level, *in* C. E. Payton, ed., Seismic stratigraphy—Applications to hydrocarbon exploration: AAPG Memoir 26, p. 49–212 (various authored papers).

Versfelt, J. W., and B. R. Rosendahl, 1989, Relationships between pre-rift structure and rift architecture in Lakes Tanganyika and Malawi, East Africa: Nature, v. 337, p. 354–357.

Watkins, R. T., 1986, Volcano-tectonic control on sedimentation in the Koobi Fora sedimentary basin, Lake Turkana, *in* L. E. Frostick, R. W. Renaut, I. Reid, and J. J. Tiercelin, eds., Sedimentation in the African rifts: Geological Society Special Publication 25, p. 113–125.

Wescott, W. A., and F. G. Etheridge, 1983, Eocene fan delta-submarine fan deposition in the Wagwater trough, east-central Jamaica: Sedimentology, v. 30, p. 235–247.

Xijiang, J., 1988, The seismic reflection characteristic and oil-gas-bearing condition of main sedimentary bodies in the Eocene faulting-Lake basin in Bohai Bay, *in* A. W. Bally, ed., Atlas of seismic stratigraphy: AAPG Studies in Geology 27, v. 2, p. 22–33.

Northern Lake Tanganyika— An Example of Organic Sedimentation in an Anoxic Rift Lake

A. Y. Huc
Institut Francais du Pétrole
Rueil-Malmaison, France

J. Le Fournier
Société Nationale Elf-Aquitaine
St. Martory, France

M. Vandenbroucke
G. Bessereau
Institut Francais du Pétrole
Rueil-Malmaison, France

Lake Tanganyika is a well-known example of a large anoxic lake that acts as a sink for organic matter. More than 400 grab samples and 25 cores were collected to study distribution of organic facies in the Bujumbura and Rumonge subbasins of the northern lake area. The overall organic richness of the lake sediments (up to 12% TOC) is favored by anoxic conditions, which prevail below about 100 m. This situation is the combined result of (1) organic productivity, (2) lake geometry (deep and narrow), which is controlled by rift architecture and hinders water circulation, and (3) equable warm tropical climate, promoting stable water stratification.

The distribution of organic facies in bottom sediments shows considerable and rather complex lateral variability. Sedimentological and organic geochemical evidence (Rock-Eval, kerogen δ^{13}C, hydrocarbon composition) suggests that organic variability is controlled mainly by depositional processes—gravity-flow deposits, containing organic matter that reflects (in terms of concentration and properties) source area environment, interfinger with autochthonous organic-rich pelagic/hemipelagic muds. The importance of redepositional processes at the basin scale probably is a response to the lake bottom's rugged topography, which is controlled by rift tectonism. Such a situation, with its implications in terms of organic heterogeneity, must be kept in mind when considering source rocks deposited in similar continental rift settings.

INTRODUCTION

Mechanisms of petroleum generation and migration have been investigated to such an extent that geoscientists now describe them with physical and chemical laws. These laws are used to construct deterministic numerical models that simulate and quantify hydrocarbon formation and movement, an important part of evaluating petroleum potential of sedimentary basins in modern exploration (Tissot and Welte, 1984).

In a realistic scenario, even the most sophisticated numerical models require pertinent descriptive information about the geological systems under consideration. In petroleum-oriented basin modeling special attention must be directed to the source rocks, in which occur two determinant processes—cracking of the kerogen and primary migration of generated hydrocarbons.

In this respect, according to current knowledge, one of the most sensitive input data is source rock organic-matter distribution, including organic richness and organic nature (Huc, 1988c). These parameters obviously are the fundamental factors that control the quantity and composition of hydrocarbons generated in a given source rock.

Moreover, they probably are critical in primary migration behavior because the amount of hydrocarbons locally generated directly influence local hydrocarbon saturation and pressure in the source rock pore system. From this standpoint it is clear that vertical or lateral variations in source beds, including sedimentologic fabric, organic richness, and quality, likely influence the oil-sourcing patterns in a sedimentary basin. Accordingly, organic-facies distribution in source horizons has become an important area of research in the prediction of source rock behavior.

Modern sedimentary basins where significant organic matter is accumulating probably are the most favorable settings in which to understand the factors controlling source bed formation and the relationships between organic-matter distribution and sedimentary facies, from which conceptual models can be developed and applied in exploration as guides for source rock evaluation.

Lake Tanganyika is, in this respect, an attractive site because of its organic-rich deposits and rift character, especially when one considers the numerous examples of prolific source rocks in rift settings (Huff, 1980; Perrodon, 1983)—for example, Gulf of Suez, Gulf of Khambhat in India, the Paleozoic Dniepr-Donetz basins in the Soviet Union (Maksimov et al., 1977; Brosse et al., 1986), sedimentary basins of interior Sudan (Schull, 1984), and the Paleocene Pematang Group in Sumatra (Williams et al., 1985). According to a survey by Klemme (1980), 10% of the world's oil reserves has been discovered in rift contexts, which themselves account for only 5% of the area of sedimentary basins.

More specifically, the east Africa rift is commonly proposed as a modern analog of the Cretaceous early rift system of the South Atlantic, in which organic-rich presalt lacustrine sediments have sourced much of the oil found along the Atlantic margins of Africa and South America (Brice et al., 1980; Mello et al., 1988; Abrahão and Warme, this volume).

This work, which is a part of the GEORIFT program supported by Elf-Aquitaine, is devoted to study of organic-matter distribution in northern Lake Tanganyika where an extensive sampling program, including grab samples and cores, has been undertaken.

TECTONISM, TOPOGRAPHY, AND SEDIMENTARY FACIES

Lake Tanganyika, located in east equatorial Africa, is the longest and deepest (1500 m maximum water depth) of several lakes situated along a system of rift valleys that extends from the Zambezi River in Mozambique north to the Red Sea. The rift system results from separation of the African and Somalian plates, and Lake Tanganyika occupies a northwest-southeast–trending extensional component (Figure 1). Within this general tectonic framework, northern Lake Tanganyika can be considered a pull-apart basin in which the direction of opening is controlled by a system of ancestral Precambrian structures oriented northwest and north-northeast (Chorowicz and Mukonki, 1980).

Vertical movements along the older directions create geomorphic features such as tilted blocks and horsts separating north-south elongated basins. The basic architecture of Lake Tanganyika has been described in terms of half-grabens and accommodation zones as a succession of subbasins separated by sills or shallow sublacustrine ridges (Rosendahl, 1987). Cape Banza and its sublacustrine extension is an example of an accommodation zone; Burton Bay and Bujumbura basin are examples of elongated half-graben basins (Figure 2).

As a consequence of this tectonic style, present lake topography is characterized by steep hypsographic gradients. The half-grabens impart an asymmetrical cross-sectional basin morphology—a steep side associated with the major border-fault system (e.g., western Rumonge basin) and a ramping side (eastern Rumonge basin).

This tectonic topography is important in several ways in controlling distribution of sublacustrine sedimentary facies (Le Fournier et al., 1985). First, adjacent to major border faults, coarse-grained clastic debris, including gravel and brecchialike material, is carried into deep water by sliding and other mass-transport mechanisms. Second, on the ramping side of the half-grabens, lateral deltaic systems tend to form a continuous shallow littoral shelf (water depth <50 m) extending 3–5 km offshore. This shelf progrades lakeward as a steep slope dissected by numerous canyons through which sediment is transported and redeposited as deep lake fans.

Third, because of their large drainage basins and sediment supply, rivers such as the Ruzizi River entering the lake have constructed longitudinal or rift-axial deltas, which are constrained laterally by the narrowness of the lake basins and thus progressively infill the entire width of the basins. Their large sedimentary discharges tend to offset the negative sedimentation-subsidence balance prevailing in the lake.

Fourth, the deep basin floors are occupied by fine-grained sediment corresponding to distal turbidites and hemipelagic muds, which are interbedded with relatively thick layers of diatom ooze.

Finally, the topographically high sublacustrine ridges or sills are beyond the influence of turbidity flows, and accumulated sediment consists almost entirely of laminated diatom ooze and clay probably resulting from deposition of suspended material.

ORGANIC PRODUCTION

Diatoms are a major component of the modern phytoplankton biomass in Lake Tanganyika, a

Huc et al.

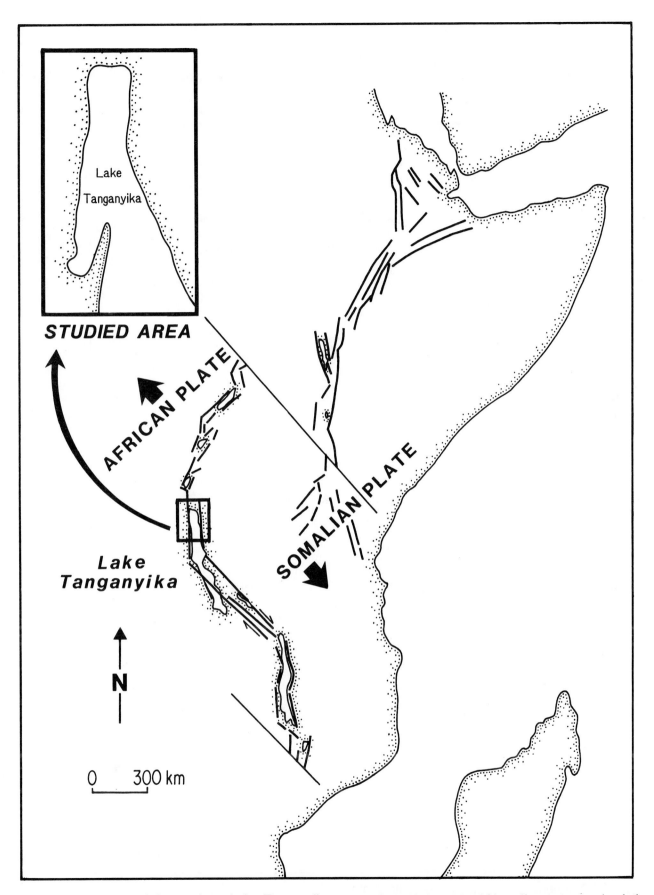

Figure 1. Location of the northern Lake Tanganyika study area with respect to the eastern and western branches of the east Africa rift system and relative motion of the African and Somalian plates.

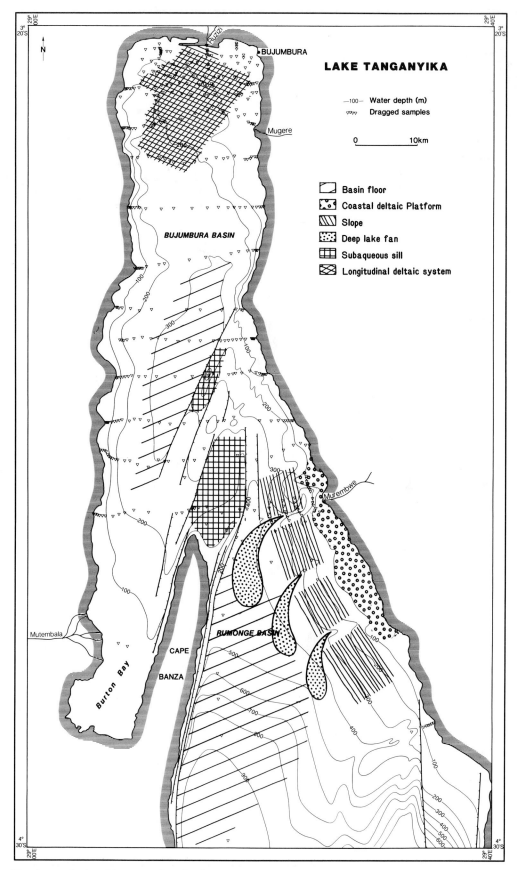

Figure 2. Northern Lake Tanganyika bathymetry (contour interval 100 m), grab-sample locations, and generalized distribution of principal sedimentary facies.

172   Huc et al.

situation reflected in the fine-grained basin sediments that are often composed entirely of the algal skeletal remains (Degens et al., 1971). Besides diatoms, other primary producers include various species of cyanophyta, chlorophyta, chrysophyta, and cryptophyta (Hecky and Kling, 1981). Although the phytoplanktonic species composition is fairly well known, the actual level of primary productivity is, unfortunately, not well assessed because of few measurements and the complexity of the productivity pattern, including lateral and temporal variability.

Year-round monitoring of primary production (Coulter, pers. comm.) has shown a seasonal variation characterized by maximum algal biomass between May and August in the south basin and between September and November in the north. This increase in algal growth is associated with strong, consistent winds during the relative cool, dry season. Evaporative cooling and wind stress induce active circulation down to several hundred meters, and upwelling of deep water brings to the surface nutrients that stimulate increased primary productivity. The warm, wet season from November to April is a period of stable stratification, which prevents replenishment of nutrients in surface waters. Because river discharge into Lake Tanganyika is low—19 km³/yr (Hecky, 1978)—riverine flux of nutrients is subordinate, and organic productivity during this period remains limited.

Large differences in productivity that have been recognized among the northern, central, and southern basins, include algal growth rate and shift of the blooming periods (Hecky et al., 1981). Moreover, the density of planktonic algae has been reported to be generally higher in inshore waters and shallow bays (Coulter, 1963). It is noteworthy that spatio-temporal variations in organic primary productivity are associated with significant changes in phytoplankton composition. For example, diatoms are abundant only during periods of major water mixing; blue-green algae locally dominate during subsequent rapid surface warming; and chlorophytae are prominent during the phase of stable stratification (Hecky and Kling, 1981).

Accordingly, the available primary productivity data should be looked upon as only indicative. Published estimates range from 400 to 500 g C/m²/yr (Hecky and Kling, 1981; Kelts, 1988). These values are not very high compared to other tropical lakes, which can reach up to 2500 g C/m²/yr (Likens, 1975). For reference, data from other east African rift lakes are as follows: Lake Victoria, 370-1460 g C/m²/yr; Lake Kivu, 450-770 g C/m²/yr; Lake Turkana, 300-1500 g C/m²/yr; Lake Nakuru, 900-1200 g C/m²/yr; and Lake Natron, 900-1200 g C/m²/yr (Kelts, 1988). In terms of global productivity patterns, however, algal productivity in Lake Tanganyika is significantly higher than mean values of 50 and 100 g C/m²/yr reported, respectively, for the open ocean and the oceanic coastal zone; productivity of the most prolific oceanic upwelling areas averages 300 g C/m²/yr (Bunt, 1975).

In addition to phytoplanktonic organisms, various species of protozoa contribute substantially to the pelagic biomass of Lake Tanganyika. Note that from mid-February through the end of April, the protozoan biomass equals or exceeds the phytoplankton biomass. Moreover, bacterial counts in the euphotic zone reveal that the bacterial biomass frequently equals or even exceeds the combined biomass of algae and protozoans (Hecky and Kling, 1981).

In Lake Tanganyika, where the stand of living pelagic fishes is high, the mean annual fish biomass is estimated as high as 8.6 g C/m² (Johannesson, 1974). When compared to the annual mean of the algal standing crops (0.6 g C/m²), the lake clearly has a high carbon efficiency transfer from primary production to fish production (Hecky and Kling, 1981). Consequently, one can infer that most of the organic matter incorporated into the sediment probably has undergone transition through the trophic chain, including bacterial reworking, and that only a minute amount of phytoplanktonic biomass likely is a direct precursor of sedimentary organic matter.

Besides the biomass generated in the lake itself, allochthonous biomass produced on the surrounding dry land can be considered another potential source of sedimentary organic matter. Because the lake's tectonic setting limits drainage basin size and because climatic conditions limit land productivity (mainly savannah), the source of terrestrial organic matter generally is confined to swampy areas within coastal deltaic environments. This limited input of higher plant-derived organic matter is recognizable microscopically as a subordinate fraction of woody particles.

WATER STRATIFICATION

Equatorial climate is characterized by the absence of significant seasonal variation in air temperature. As a result warm, low-density surface water of the lake permanently overlies a cooler, higher density deep water, thereby inducing stable water stratification. Such stratification is apparent in Figure 3, where water temperature in Lake Tanganyika decreases to a depth of about 150 m, the position of the thermocline. Actually, the situation is more complex because the depth of the thermocline fluctuates annually—during stable stratification it is at 40-50 m, but during mixing periods it deepens to 75-100 m in the north basin and deteriorates in the south basin into a slight thermal discontinuity at 200-300 m, sufficient to maintain stratification (Coulter, pers. comm.).

As a result of this stratification, the oxygen, which is consumed by organic detritus showering from the surface, cannot be replenished at a sufficient rate in sluggish and stagnant deep water. This situation promotes development of an anoxic bottom-water mass as revealed by the oxygen profile in Figure 3.

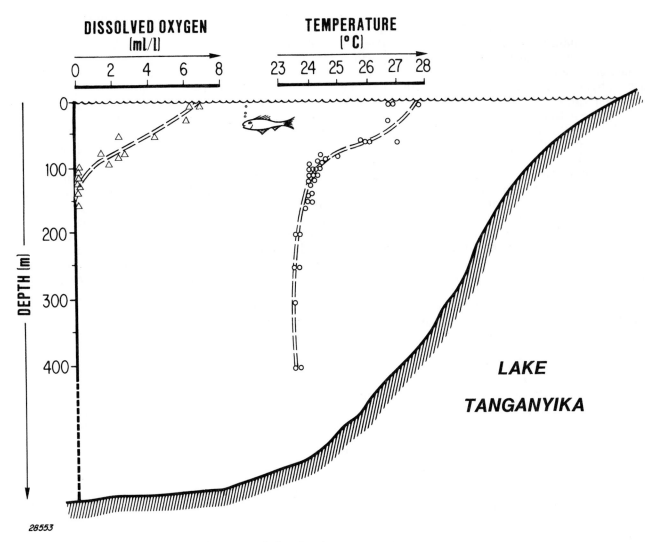

Figure 3. Depth profiles of temperature and dissolved oxygen in Lake Tanganyika (data from Degens et al., 1971).

Water beneath the thermocline is devoid of oxygen. As a result of water depth and the lake's steep topography, more than 90% of the lake-bottom area is permanently covered by anoxic water.

DISTRIBUTION OF SEDIMENTARY ORGANIC MATTER

Methodology and Results

Based on the information above, it can be stated that in Lake Tanganyika occur what are considered to be the basic ingredients for significant organic accumulation in sediment—high productivity and anoxia

(Demaison and Moore, 1980; Huc, 1988b). In fact, total organic carbon (TOC) content of up to 12% has been reported by Degens et al. (1971) in a 2.5-m sediment core recovered in deep water (1400 m) of the central lake basin.

To evaluate sedimentary organic content on a firmer statistical basis, more than 400 surface grab samples collected under the GEORIFT program were analyzed by means of a Rock-Eval apparatus equipped with a carbon module. Representative sampling was conducted along regularly spaced transects across the northern lake basin (Figure 2). TOC values, which have been confirmed by LECO and CARMHOGRAPH analysis, are summarized in Figure 4a. The resulting histogram is bimodal, including a mode centered on low values (<1% TOC) and a mode between 3 and 5% TOC. On the whole, as expected, sedimentary organic content is fairly high, averaging 3% TOC with maximum values reaching 10% TOC.

Huc et al.

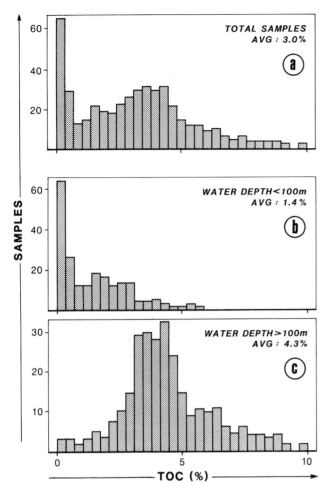

Figure 4. Histograms of organic-carbon content of surface sediments in Lake Tanganyika. a, all samples; b, samples from water depths <100 m; c, samples from water depths >100 m.

Interpretation of Organic Distribution Patterns

On the assumption that the grab samples are representative of the surface sediments, a map of organic-carbon distribution was constructed (Figure 5) of relatively great accuracy due to the high sampling density. Although extreme spatial heterogeneity is apparent, several trends can be discerned from the complex patterns.

Lake Margin

The lake periphery is characterized by organic-poor sediments (<1% TOC). If we exclude the northeastern Bujumbura basin, organic-carbon content increases abruptly, however, from less than 1% to more than 2%, within a narrow area often associated with continuous, shallow deltaic platforms, on which sediments are limited by the basinward edges. As a rule, this rapid transition to organic richer sediment occurs at a depth of 50 m, approximately the higher

stand of the thermocline during the phase of stable stratification (Figure 5).

Two explanations can be proposed for these observations. The first is implied by the occurrence of dissolved oxygen in the water mass above the thermocline. Oxygen allows benthic organisms to live and consume organic matter descending from the surface. In addition, mixing of surface sediments by burrowing activity, which usually occurs under such conditions, increases the exposure time of organic matter to decomposition processes and significantly decreases organic-matter content (Demaison and Moore, 1980; Pratt, 1984; Huc et al., in press).

The second explanation relates to hydrodynamic movements in the water layer above the thermocline. An important but often overlooked property of organic matter is its low density—1.1 to 1.7 g/cm³ (Nwachukwu and Baker, 1985). Water agitation within the epilimnion inhibits settling, and most of the organic matter tends to remain in suspension until ultimately it is carried into more quiescent waters, that is, below the thermocline. Consequently, bottom sediments that lay permanently within the epilimnion (water depth <50 m, the higher stand of the thermocline) will be organic poor. This hydraulic control on the shallow littoral platform is clearly reflected by coarse sediment grain sizes and thanatocoenoses of marsh-dwelling mollusks. In contrast, sediment below 50 m generally is fine grained and massive or laminated with clay- and silica-rich layers, which may reflect seasonal variations in productivity blooms and water circulation.

Sublacustrine Slopes and Ridges

A second pattern of organic-carbon distribution—gradual basinward increase—occurs along slopes to the lake floor (Figures 2, 5). This pattern cannot be explained by differential degradation of organic matter because all the sediment there is deposited in anoxic water. Moreover, Rock-Eval pyrolysis data show no changes in organic quality along these gradients; hydrogen indices of kerogens remain in the range of 400–600, with no relationship to organic content (slope and lake-floor sediments, Figure 6). Sorting of the low-density organic matter along decreasing hydraulic energy gradients most likely explains this pattern (Huc, 1988a). Besides surface water agitation, which may be responsible for sweeping organic matter from the narrow littoral platform, other hydraulic movements have been recognized in the water masses of Lake Tanganyika. Sediment as deep as 100 m is still subjected to direct stirring induced by wind stress at the surface and is affected by seasonal fluctuations of the thermocline (Coulter, 1963). These hydrodynamic conditions are inferred to result in limited accumulation of organic matter to 100 m depth.

In this respect, note that samples shallower than 100 m, including those from the littoral platform and uppermost slope, account for nearly the entire low-TOC mode shown in the histogram in Figure 4a.

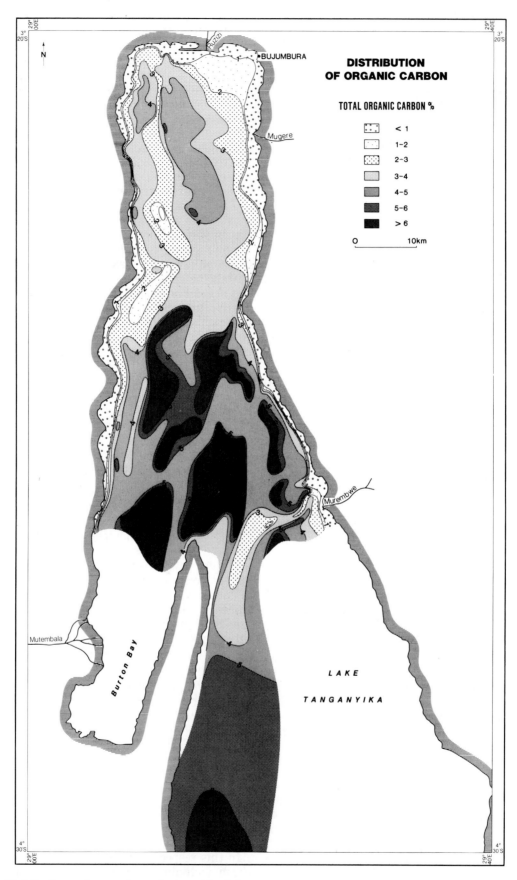

Figure 5. Distribution of organic-carbon content in surface sediments of northern Lake Tanganyika.

Huc et al.

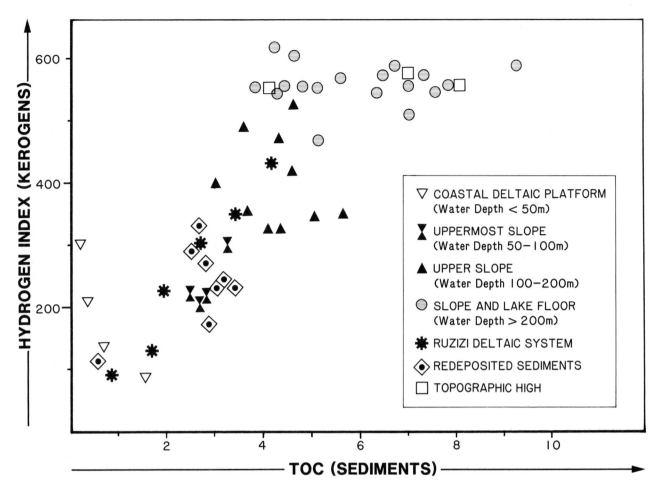

Figure 6. Sediment TOC vs. kerogen hydrogen index of organic facies samples from northern Lake Tanganyika.

When the two modes are differentiated on the basis of water depth above or below 100 m, the resulting means are 1.4% TOC for shallow samples and 4.3% TOC for those sediments permanently situated below the thermocline and beyond the effects of wave action (Figures 4b, 4c).

At greater depths water movements indeed are greatly limited. However, the enormous volume of water below the thermocline is not entirely stagnant because, as Coulter (1963) documented, it is still affected by hydrodynamic phenomena, such as rare and temporary deep stirring caused by unusually violent storms, and large-amplitude internal standing waves induced by seasonal changes in wind stress.

Consequently, one can infer that organic matter tends to be progressively winnowed toward the most quiescent settings of the lake bottom, beginning from continually agitated water in shallow areas, through depths subjected to occasional stirring, sluggish seasonal water circulation, and finally into the virtually stagnant, deepest waters.

This hydrodynamic model not only satisfactorily explains the organic poverty of the littoral platform

and progressive increase of organic carbon content downslope but also explains why topographic highs such as the sublacustrine extension of Cape Banza, as soon as they descend into deep, quiescent waters, retain high quantities of pelagic organic matter (Figures 5, 6).

Bujumbura Basin

The northeastern basin end exhibits a progressive basinward increase in organic carbon content. The anomalously wide zone of 1-2% TOC (Figure 5) just offshore from Bujumbura is associated with the dominant sedimentary feature of this part of the basin—the axial deltaic system at the mouth of the Ruzizi River. Overall southward progradation of this complex, which supplies relatively large amounts of detrital sediment compared to lateral deltas, is infilling the Bujumbura basin. By embedding the bottom topography, this infilling results, at least in nonactive areas, in a generally gentle slope from the deltaic platform to the basin floor.

In this framework the decreasing influence of incoming detritus containing relatively little allochthonous organic matter likely contributes to the basinward increase in organic content. The allochthonous organic matter, in part derived from higher plants, probably dilutes the autochthonous organic input of the lacustrine biomass. Rock-Eval pyrolysis of kerogens supports this interpretation by showing variations in hydrogen index (Figure 7) that can be related to differential mixing of different types of organic matter. Allochthonous organic input of the river clearly dominates the vicinity of the Ruzizi mouth, where low-TOC sediments exhibit HI values as low as 94. As the influence of river sediments progressively decreases, both organic content and hydrogen index steadily increase (Figure 6 and labeled values in Figure 7).

Rumonge Basin

The last organic-matter distribution pattern relates to elongated areas of organic-poor sediment in several deep parts of the basin, the most characteristic of which occurs in Rumonge basin around the Murembwe River mouth (Figure 5).

According to echosounding surveys, the lake bottom in this area exhibits subaqueous canyon morphologies. The organic-poor area, which was delineated by high-density sampling, follows in its narrowest part a canyon feature and thus provides a clue to its origin—a deep lake fan—implying redepositional processes. This is confirmed by turbiditic structures observed in a core recovered from the center of the area.

Organic geochemical studies of surface sediments show that hydrogen index values (Figure 8a) of kerogens from the organic-poor area, corresponding to the proximal location in the fan system, are consistent (200–400) and clearly distinctive from HI values of kerogens in the surrounding slope and basinal sediments (400–600) but similar to those from sediment within the shallower deltaic platform under aerobic to dysaerobic conditions. Similar comparisons can be made from elemental analyses (H/C and O/C ratios) and ^{13}C isotopic signatures (Figures 8b–8d).

Eleven chloroform-extracted samples were studied for hydrocarbon composition. The first notable trend is revealed by the total yield of extractable bitumen, which parallels the kerogen properties—i.e., relatively low yields (10–27 mg HC/g TOC) from sediment redeposited as proximal turbidites and from shallow settings in contrast to significantly higher yields (58–140 mg HC/g TOC) for surrounding slope and basinal sediments.

The saturated hydrocarbons fraction in all samples shows a strong odd/even predominance for n-alkanes dominated by nC_{29} and nC_{31} (Figure 9). That this pattern is a classical feature of most young sediments can be explained by the sensitivity to n-alkanes derived from cuticular waxes of higher plants, even when these plants are a subordinate contributor to sedimentary organic matter (Tissot and Welte, 1984).

Abundant hopanes that occur in the branched and cyclic fractions of the alkanes (Figure 9) are dominated by the $17\beta(H)$, $21\beta(H)$ series, which exhibits the biological configuration of hopanes, pointing to a genuine origin in these young, immature sediments. The relative importance of hopanoids, often considered to be specific makers of prokaryote cellular membranes (Ourisson et al., 1979), probably can be related to our earlier observation about bacterial abundance in lake biomass.

On the other hand, the presence of subordinate $17\alpha(H)$, $21\beta(H)$ hopanes, including extended homologs for which the isomerization at position 22 is completed, suggests that some hydrocarbons in these sediments could have undergone significant thermal maturation, implying earlier deep burial or contamination by fossil hydrocarbons. Besides possible anthropogenic pollution, which cannot be disregarded, this feature also might reflect the occurrence of oil seeps, whose surface manifestations, including blocks of asphaltic material, have been reported elsewhere in the lake—for example, along Cape Kalamba and in Burton Bay (Mondeguer et al., 1986; Tiercelin et al., 1989).

In addition to hopanoid abundance, another remarkable compositional aspect of the alkanes is the occurrence of a major peak identified by comparison with published mass spectra (Corbet, 1980) as the de-A-10β(H) lupane (Figure 9).

The unsaturated hydrocarbons fraction (Figure 10) shows significant amount of n-alkenes dominated by nC_{27} and nC_{29}. Although cyclic alkenes consist mainly of diploptene and hopenes components, the most salient feature is the prominent peak tentatively assigned to a de-A-arborene structure (published mass spectra, Kimble, 1972; Albrecht, pers. comm.).

On gas-chromatography (GC) traces of the total fraction of alkanes plus alkenes, the two peaks assigned to de-A-10β(H) lupane and de-A-arborene are easily recognized. By comparing chromatographic patterns for various samples in the Murembwe area, we observe an interesting picture (Figure 11). Irrespective of their origin or fate, we do note that, when compared to de-A-arborene, de-A-10β(H) lupane is dominant in the deltaic platform and in most transported sediments, except for a sample from the most distal part.

On the other hand, de-A-arborene is largely dominant in the surrounding slope and basinal sediments, where in most GC traces it is the major peak. These observations strongly suggest shallow-water origin of most organic matter associated with mass-transported sediment. However, note that a relative and progressive increase in de-A-arborene can be followed away from sediment-source areas into distal areas, indicating that increased mixing probably occurs with organic matter that settles more directly onto the basin floor.

Consequently, it appears that gravity-transport mechanisms in Lake Tanganyika clearly are responsible for conveying sediments and their characteristic organic content along specific pathways. This results in deposition of what can be called "allochthonous organic facies."

Huc et al.

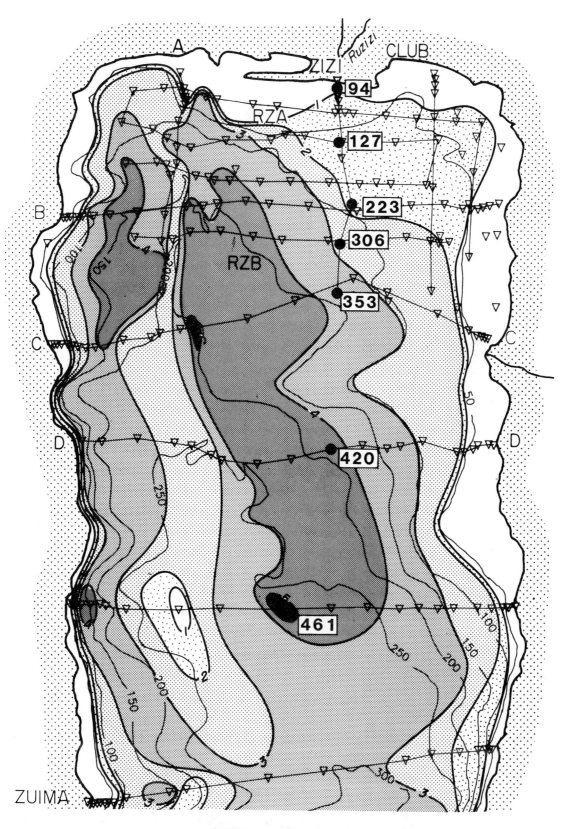

● 420 **HYDROGEN INDEX**

Figure 7. Hydrogen index of kerogens and TOC (see key in Figure 5) of organic facies samples from northern Lake Tanganyika. Labeled values along the transect from the Ruzizi River inlet correspond to Ruzizi samples plotted in Figure 6.

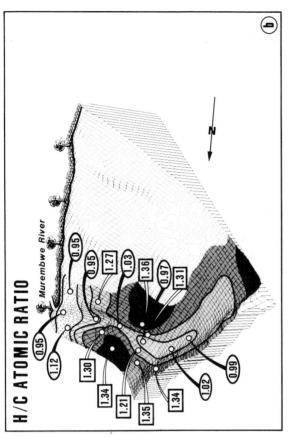

HYDROGEN INDEX (mgHC/gORG.C)

TOTAL ORG. C
[%]

< 1
1-2
2-3
3-4
4-5
> 5

0   10km

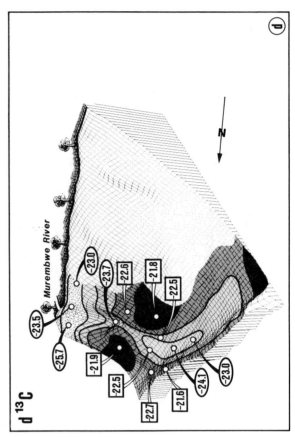

H/C ATOMIC RATIO

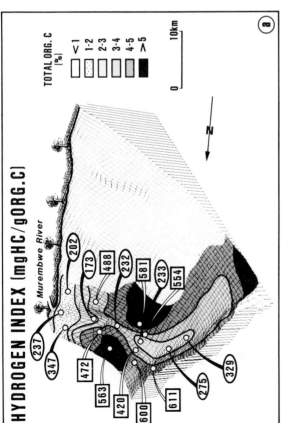

O/C ATOMIC RATIO (X 100)

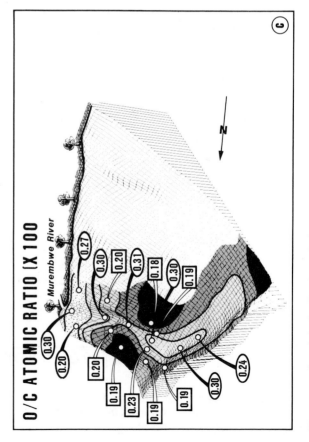

d ¹³C

Figure 8. Spatial distribution of chemical properties of kerogens in Murembwe basin. a, hydrogen index; b, H/C atomic ratio; c, O/C atomic ratio; d, kerogen $\delta^{13}C$.

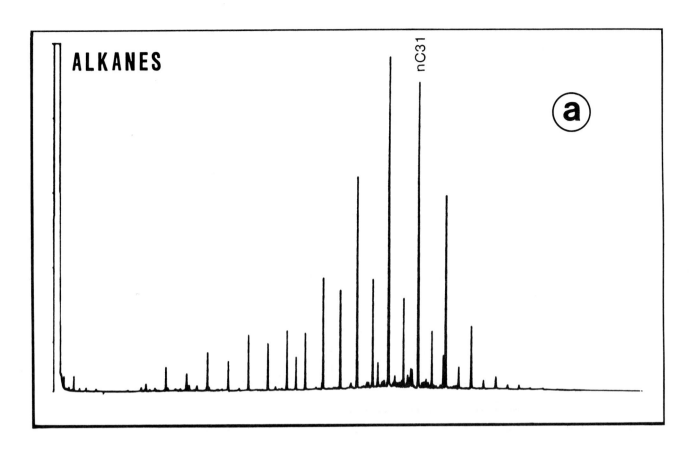

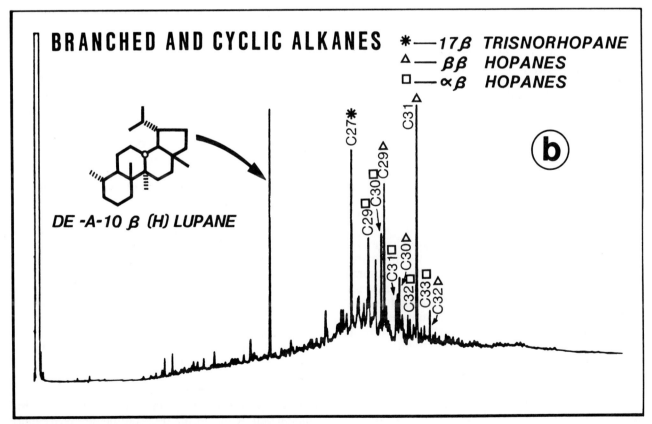

Figure 9. Chromatograms of (a) total alkanes and (b) branched/cyclic alkanes. Note prominent peak identified as de-A-10β(H) lupane.

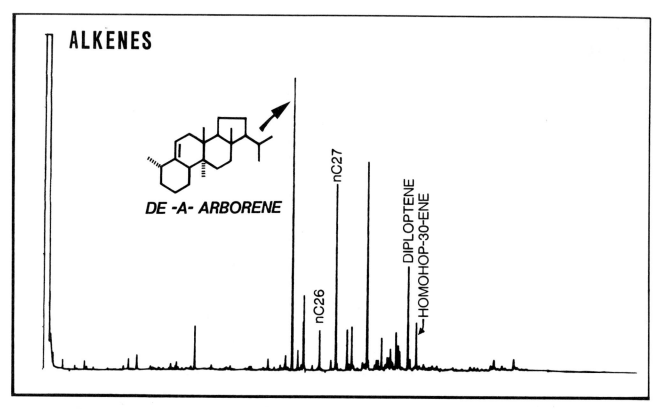

Figure 10. Chromatogram of alkenes, showing prominent peak tentatively assigned to de-A-arborene.

DISCUSSION

In the considered case, organic matter that was previously deposited and altered in aerobic to dysaerobic shallow-water conditions ultimately accumulates in basinal, strictly anaerobic conditions.

If the high level of biomass productivity associated with a stable meromictic regime can be invoked to account for the significant accumulation of organic matter in bottom sediments of Lake Tanganyika, then the extreme spatial heterogeneity of the organic facies, a salient feature of the surface sediments, must be rationalized in terms of transport processes.

Low-density organic matter cannot accumulate to any degree in sediment deposited in agitated waters, but rather it tends to be incorporated into the permanent suspension load. Consequently, coastal deltaic platforms within the oxygenated epilimnion on the ramping sides of half-grabens are characterized by low concentrations of altered organic matter. Organic matter from the agitated water is progressively concentrated toward the more quiescent areas of the lake bottom, including the basin floor and the slope, where organic content increases as a function of waning hydrodynamic conditions.

Because of steep slope instability, some of the organic-bearing sediment slumps and is redeposited as fine-grained turbidites on the basin floor. Due to episodic destruction of the organic-poor coastal platform sediments by channel avulsion and abandonment of constructive deltas, periodic changes in water level, or uplifting of rift shoulders, however, this organic facies is transitory and unlikely to be preserved in the sedimentary record. Instead, it is reworked and redeposited as deeper delta fans. As a result, in basinal settings redeposited organic facies coexist with organic-rich pelagic/hemipelagic muds and with organic-rich, slumping-induced distal turbidites.

Sublacustrine topographic highs, or accommodation zones, are beyond turbiditic influence. When they are deep enough to be protected from surface water agitation, they represent a specific organic-rich setting in which organic matter settles from the suspension load.

The supply of organic-poor detritus that builds longitudinal deltas locally dilutes autochthonous organic-rich lacustrine sedimentary input, resulting in deposits of moderate organic content. Note that, in contrast to the transitory character of the lateral coastal deltaic platform, longitudinal deltas are more likely to be preserved in the sedimentary record.

One of lakes' main characteristics is that they respond dynamically and rapidly to environmental

Huc et al.

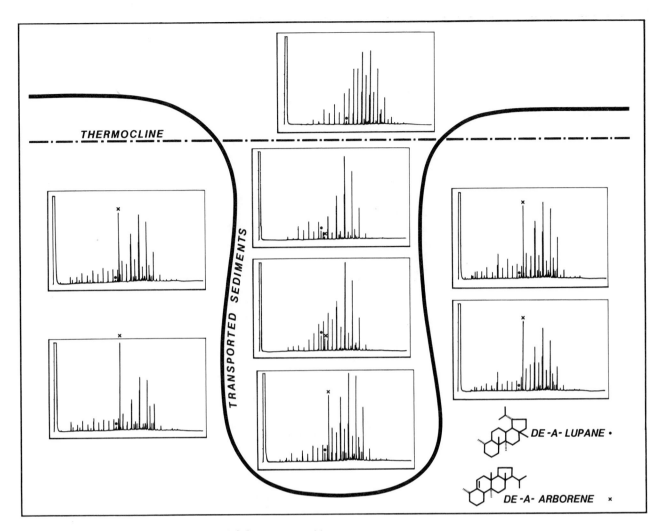

Figure 11. Generalized distribution of GC patterns with respect to the thermocline and area of transported organic-poor sediment in the Murembwe basin.

fluctuations. Consequently, the pattern of present-day organic facies is an instantaneous picture that must be considered in view of short-term changes—climate or tectonic—that may occur. For example, lakes typically experience rapid water-level changes, phenomena documented in the geologically recent history of Lake Tanganyika. Before about 100 ka water level was more than 600 m below the present lake level (Scholz and Rosendahl, 1988, Tiercelin et al., in press). During the last 25 k.y. two highstands separated by a regressive episode have been recognized and correlated with climatic phases identified in east Africa (Tiercelin et al., 1988).

Water-level fluctuations also are known to control the occurrence and extent of facies redeposited on basin floors. In the case of moderate water-level change, one can postulate that during lowstands, coastal deltaic platforms are subject to active subaerial erosion, resulting in redeposition of material, mostly through turbiditic processes.

Therefore, one can infer that, as a response to increased turbidity current activity, deep lake fans become common sedimentary features, interfingering and diluting the pelagic and hemipelagic organic oozes in the residual lake. In contrast, during highstands coastal deltaic platforms are likely to develop allowing deposition of more extensive organic-rich layers in the basinal setting, as is now the case.

Moreover, it can be anticipated that under semiarid climatic conditions, dramatic lowering of lake level would result in a totally different sedimentary situation—huge, subaerial alluvial fan assemblages on the ramping sides of half-grabens grading laterally into evaporitic mud flats and playa deposits, sporadically enriched in organic matter.

Another example of the influence of short-term events on organic facies is the outbreak of the Ruzizi River into Lake Tanganyika. A sudden overflow of Lake Kivu into the Ruzizi River occurred between

6 and 2.5 ka, during the Holocene humid period (Vincens, in press). As a result, a 1-dm-thick layer of organic-poor gray clay (<1% TOC) was deposited over a wide area. This layer, recognized in most cores, appears to mark the base of the present constructive episode of the Ruzizi River's longitudinal deltaic system. This situation implies significant time-related heterogeneity of the organic facies in addition to lateral changes.

The complex model of organic sedimentation, strongly controlled by climate and tectonic setting, proposed for modern Lake Tanganyika under tropical conditions accounts for only a specific phase of rift evolution. Le Fournier et al. (1985), Mondeguer et al. (1987), and Lambiase (this volume) have proposed that during the initial spreading phase, intracratonic rift areas are affected by a dense network of fault systems of small vertical displacement. Under tropical conditions the resulting depressional settings likely will be dominated by marshy wetlands, such as the modern Okavango basin in Botswana. In the sedimentary record this environment can be preserved as fluviodeltaic facies, a facies often recognized in early rift formations (Williams et al., 1985) in which type III organic matter (land-plant-derived organic matter of moderate hydrocarbon potential), including coal, eventually occurs. In subsequent rift phases, only specifically oriented faults become active, resulting in large vertical displacements associated with block tilting.

These tectonic processes progressively build a topographic architecture similar to that displayed in the Tanganyika graben system. At this stage uplifted rift shoulders reduce the size of drainage basins, thereby preventing significant input of land-plant-derived organic matter and dilution of autochthonous lacustrine organic matter by large detrital input. In such a situation sediments accumulating on the lake floor will be characterized by high content of organic matter, derived mainly from algal and bacterial biomass, which can be related to type I and/or type II.

of buried organic matter originating mainly from lacustrine biomass. Third, the complexity of the spatial distribution of organic matter in surface sediment reflects contrasting depositional processes—pelagic/hemipelagic deposition vs. gravity transport (turbidites, slumps), the latter's relative importance being a response mainly to rift structure (basin shape and bathymetric gradient).

All these factors affecting organic facies distribution and that are solely related to the basin's rift character likely will be valid for most other source rocks deposited in similar tectonic settings, provided that environmental conditions, such as climate, are the same.

Consequently, besides the general character of the organic facies (organic richness, high petroleum potential), which can be used for prognosis, the heterogeneity of such source rocks must be carefully considered when assessing the petroleum potential of a given volume of source rock based on analysis of sparse cuttings and when discussing their properties in terms of hydrocarbon migration.

ACKNOWLEDGMENTS

This work has been conducted as part of the GEORIFT project. The authors are grateful to SNEA(P) for allowing publication and express their thanks to M. Bernon, B. Buton, M. Da Silva, M. Fabre, G. Pichaud, and C. Schwartz for their technical contributions. The paper has been improved by the constructive remarks of Dr. F. Baltzer (University of Orsay, France), Dr. Coulter (New-Zealand), Dr. J. J. Tiercelin (University of Bretagne, France), and Dr. A. Vincens (University of Marseille, France). The manuscript also benefited from critical reviews by Dr. G. Demaison (Chevron Oil Co.) and Dr. B. J. Katz (Texaco Inc.).

CONCLUSIONS

Besides biological factors, including high productivity, and climatic conditions, the organic sedimentation pattern prevailing in Lake Tanganyika is strongly influenced by three aspects of the basin's rift morphology. First, the great depth of the lake and basin shape hamper efficient water mixing, which, hence, favors stability of water stratification and development of an enormous oxygen-depleted water mass. Second, uplifted rift shoulders are responsible for the sediment-starved situation of the lake, which favors the organic richness of sediment undiluted by large detrital input. Furthermore, it is responsible for the low contribution of land-plant-derived organic matter in the lake's organic stock and, consequently, to the high petroleum potential

REFERENCES

Abrahão, D., and J. E. Warme, Lacustrine and associated deposits in a rifted continental margin—Lower Cretaceous Lagoa Feia Formation, Campos basin, offshore Brazil, this volume.
Brice, S. E., K. R. Kelts, and M. A. Arthur, 1980, Lower Cretaceous lacustrine source beds from early rifting phase of South Atlantic (abs.): AAPG Bulletin, v. 64, p. 680–681.
Brosse, E., G. Deroo, J. Roucache, and T. A. Botneva, 1986, Organic geochemistry as a test of validity for results of a modelisation in the Pripiat basin (Bielorussia), in J. Burrus, ed., Thermal modeling in sedimentary basins: Paris, Éditions Technip, p. 497–516.
Bunt, J. S., 1975, Primary productivity of marine ecosystems, in H. Lieth, and R. H. Whittaker, eds., Primary productivity of the biosphere: New York, Springer-Verlag, p. 169–183.
Chorowicz, J., and M. N. B. Mukonki, 1980, Linéaments anciens, zones transformantes récentes et géotectonique des fosses de l'est Africain, d'après la télédétection et la microtectonique: Tervuren Belgium, Musee Royal de l'Afrique Centrale, Departement de Geologie et de Mineralogie, Rapport Annuel, v. 1979, p. 143–166.

Corbet, B., 1980, Origine et transformation des triterpènes dans des sédiments récents: Ph.D. Thèse, Université de Strasbourg, 148 p.

Coulter, G. W., 1963, Hydrological changes in relation to biological productivity in southern Lake Tanganyika: Limnology and Oceanography, v. 8, p. 463–477.

Degens, E. T., Von Herzen, R. P., and H. K., Wong, 1971, Lake Tanganyika—Water chemistry, sediments, geological structure: Naturwissenschaften, v. 58, p. 229–240.

Demaison, G. J., and G. T. Moore, 1980, Anoxic environments and oil source bed genesis: AAPG Bulletin, v. 64, p. 1179–1209.

Durand, B., and G. Nicaise, 1980, Procedures for kerogen isolation, in B. Durand, ed., Kerogen—Insoluble organic matter from sedimentary rocks: Paris, Éditions Technip, p. 35–56.

Hecky, R. E., 1978, The Kivu-Tanganyika basin—The last 14000 years: Polskie Archiwum Hydrobiologii, v. 25, p. 159–165.

Hecky, R. E., and H. J. Kling, 1981, The phytoplankton and protozooplankton of the euphotic zone of lake Tanganyika—Species composition, biomass, chlorophyll content and spatio-temporal distribution: Limnology and Oceanography, v. 26, p. 548–564.

Huc, A. Y., 1988a, Aspects of depositional processes of organic matter in sedimentary basins, in L. Mattavelli, and L. Novelli, eds., Advances in organic geochemistry 1987: New York, Pergamon Press, p. 263–272.

Huc, A. Y., 1988b, Sedimentology of organic matter, in F. H. Frimmel, and R. F. Christman, eds., Report of the Dahlem workshop on humic substances and their role in the environment: New York, John Wiley, p. 215–243.

Huc, A. Y., 1988c, Understanding organic facies—A key to improve quantitative petroleum evaluation of sedimentary basins (abs.): AAPG Bulletin, v. 72, p. 1008.

Huc, A. Y., E. Lallier-Verges, P. Bertrand, and B. Carpentier, (in press), Organic matter response to change of depositional environment in Kimmeridgian shales, UK, in J. K. Whelan, and J. Farrington eds., Productivity, accumulation and preservation of organic matter in recent and ancient sediments: Columbia University Press.

Huff, K. F., 1980, Facts and principles of world petroleum occurrences: frontiers of world exploration: Bulletin of Canadian Petroleum Geology, v. 6, p. 343–362.

Johannesson, K. A., 1974, Preliminary quantitative estimates of pelagic fish stocks in Lake Tanganyika by use of echo integration techniques: Aviemore, Scotland, EIFAC Symposium Paper 54 (UNFAO Publication), 16 p.

Kelts, K., 1988, Environments of deposition of lacustrine petroleum source rocks—An introduction, in A. J. Fleet, K. Kelts, and M. R. Talbot, eds., Lacustrine petroleum source rocks: Geological Society Special Publication 40, p. 3–26.

Kimble, B. J., 1972, The geochemistry of triterpenoids hydrocarbons: Ph.D. Dissertation, University of Bristol, 301 p.

Klemme, H. D. 1980, Petroleum basins—Classification and characteristics: Journal of Petroleum Geology, v. 3, no. 2, p. 187–207.

Lambiase, J. J., A model for tectonic control of lacustrine stratigraphic sequences in continental rift basins, this volume.

Le Fournier, J., J. Chorowicz, C. Thouin, F. Balzer, P. Y. Chenet, J. P. Henriet, D. G. Masson, A. Mondeguer, B. R. Rosendahl, F. L. Spy-Anderson, and J. J. Tiercelin, 1985, Le bassin du lac Tanganyika—Évolution tectonique et sédimentaire: Comptes Rendus de l'Académie des Sciences de Paris, v. 301, II, p. 1053–1058.

Likens, G. E., 1975, Primary production of inland aquatic ecosystems, in H. Lieth, and R. H. Whittaker, eds., Primary productivity of the biosphere: New York, Springer-Verlag, p. 185–202.

Maksimov, S. P., P. V. Antsupov, and B. D. Goncharenko, 1977,

Ways for increasing effectiveness of geological exploration for oil and gas in the Dnieper-Donets depression: Geologiya Nefti i Gaza, v. 4, p. 9–14.

Mello, M. R., N. Telnaes, P. C. Gaglianone, M. I. Chicarelli, S. C. Brassell, and J. R. Maxwell, 1988, Organic geochemical characterisation of depositional palaeoenvironments of source rocks and oils in Brazilian marginal basins, in L. Mattavelli, and L. Novelli, eds., Advances in organic geochemistry 1987: New York, Pergamon Press, p. 31–45.

Mondeguer, A., J. J. Tiercelin, M. Hoffert, P. Larque, J. Le Fournier, and P. Tucholka, 1986, Sedimentation actuelle et Recente dans un petit bassin en contexte extensif et decrochant—La Baie de Burton Fosse Nord Tanganyika, Rift Est-Africain: Bulletin des Centres de Recherche, Exploration-Production Elf-Aquitaine, v. 10, p. 229–247.

Mondeguer, A., J. J. Tiercelin, and J. Le Fournier, 1987, Neotectonique et sedimentation associee—La terminaison sud du fosse du lac Tanganyika (Rift est-africain): Comptes Rendus de l'Académie des Sciences de Paris, v. 304, II, p. 371–376.

Nwachukwu, J. I., and C. Baker, 1985, Variation in kerogen densities of sediments from the Orinoco delta, Venezuela: Chemical Geology, v. 51, p. 193–198.

Ourisson, G., P. Albrecht, and M. Rohmer, 1979, The hopanoids—Paleochemistry and biochemistry of a group of natural products: Pure and Applied Chemistry, v. 51, p. 709–729.

Perrodon, A., 1983, Rifts et ressources energetiques fossiles: Bulletin des Centres de Recherche, Exploration-Production Elf-Aquitaine, v. 7, no. 1, p. 129–135.

Pratt, L. M., 1984, Influence of paleoenvironment factors on the preservation of organic matter in Middle Cretaceous Greenhorn Formation, Pueblo, Colorado: AAPG Bulletin, v. 68, p. 1146–1159.

Rosendahl, B. R., 1987, Architecture of continental rifts with special reference to East Africa: Annual Review of Earth and Planetary Sciences, v. 15, p. 445–503.

Scholz, C. A., and B. R. Rosendahl, 1988, Low lake stands in Lakes Malawi and Tanganyika, east Africa, delineated with multi-fold seismic data: Science, v. 240, p. 1645–1648.

Schull, T. J., 1984, Oil exploration in nonmarine rift basins of interior Sudan (abs.): AAPG Bulletin, v. 68, p. 526.

Tiercelin, J. J., A. Mondeguer, F. Gasse, C. Hillaire-Marcel, M. Hoffert, P. Larque, V. Ledee, P. Marestang, C. Ravenne, J. F. Raynaud, N. Thouveny, A. Vincens, and D. Williamson, 1988, 25,000 years of hydrological and sedimentary history of Lake Tanganyika, East African rift: Comptes Rendus de l'Académie des Sciences de Paris, v. 307, p. 1375–1382.

Tiercelin, J. J., C. Thouin, T. Kalala, and A. Mondeguer, 1989, Discovery of sublacustrine hydrothermal activity and associated massive sulfides and hydrocarbons in the north Tanganyika trough, East African Rift: Geology, v. 17, p. 1053–1056.

Tiercelin, J. J., A. Scholz, A. Mondeguer, B. R. Rosendahl, and C. Ravenne, (in press), Discontinuités sismiques et sedimentaires dans la série de remplissage du fossé du Tanganyika, Rift Est-Africain: Comptes Rendus de l'Académie des Sciences de Paris.

Tissot, B. P., and D. H. Welte, 1984, Petroleum formation and occurrence, 2d ed.: New York, Springer-Verlag, 699 p.

Vincens, A., (in press), Paleoenvironments du bassin nord Tanganyika (Zaire, Burundi, Tanzanie) au cours des 13 derniers mille ans—Apport de la palynologie: Review of Palaeobotany and Palynology, v. 61.

Williams, H. H., P. A. Kelley, J. S. Janks, and R. M. Christensen, 1985, The Paleogene rift basin source rocks of central Sumatra, in The past, the present, the future: Indonesian Petroleum Association 14th Annual Convention, Proceedings, v. 2, p. 57–90.

Cyclical Sedimentation in Lake Turkana, Kenya

John D. Halfman
Department of Earth Sciences
University of Notre Dame
Notre Dame, Indiana, U.S.A.

Paul J. Hearty
Department of Geology
Duke University Marine Laboratory
Beaufort, North Carolina, U.S.A.

An 11-m piston core recovered from Lake Turkana contains a strongly cyclical record of carbonate abundance and lamination thickness on a time scale of decades to centuries. Additional carbonate, radiocarbon dates and thickness data together with preliminary magnetic-susceptibility profiles are presented from other cores recovered from the north basin and one core from the south basin to investigate basinwide significance of the cyclical record. Parallel fluctuations of relative carbonate content from north basin cores establish a lithostratigraphy. Lamina thickness and magnetic-susceptibility profiles provide independent correlations of these cores, and both are consistent with carbonate lithostratigraphy. Magnetic susceptibility of the sediment is interpreted to reflect input of fluvial and eolian ferromagnetic materials with localized, episodic input of volcanic ash. Based on plots of carbonate and magnetic susceptibility vs. radiocarbon age, an extension of north basin stratigraphies into the south basin is not possible.

INTRODUCTION

Rhythmic variability of sedimentary processes occasionally is preserved in modern sediments and ancient stratigraphic sequences. Two stratigraphic scales have dominated recent investigations in the literature. On the scale of decimeters to meters, repetitive bedding of marine sequences frequently is associated with Milankovitch orbital perturbations (Hays et al., 1976; Imbrie et al., 1984; Arthur and Garrison, 1986; Olsen, 1986; Schwarzacher, 1987). On the scale of submillimeters to centimeters, laminae and thin beds observed in lacustrine environments commonly are associated with annual events or varves (O'Sullivan, 1983; Anderson, 1986; Fischer, 1986).

This paper focuses on the cyclical sedimentary record preserved in sediments recovered by 12-m piston cores from Lake Turkana, Kenya. Previous results were based on analysis of a single, 11-m piston core (LT84–8P) recovered from the north lake basin (Halfman and Johnson, 1988). Time-series analyses of bulk-carbonate concentrations and thickness of light/dark couplets of laminae reveal consistent cycles on a decimeter to meter scale. Carbonate concentrations in the sediment are influenced primarily by dilution of autochthonous carbonates

with a variable influx of siliceous detritus, dependent on discharge of the Omo River. Laminae are differentiated by carbonate content and are interpreted to reflect interannual (about 4 yr) events. Variability in sedimentation is on a time scale of decades to centuries calculated from best-fit linear interpolation of five radiocarbon dates vs. depth in the earlier core.

Sediments recovered in this core were assumed to represent the last 4 k.y. of deposition for the entire lake. Although bulk-carbonate profiles provided a preliminary correlation of the basin's upper stratigraphy, the lithostratigraphy is open to reinterpretation. Resolution of carbonate analyses in core LT84–8P was approximately every 10 cm; the other cores lacked this detail, with samples every 30–40 cm downcore.

The goals of this paper are to (1) substantiate carbonate lithostratigraphy by increasing the downcore resolution of the carbonate data; (2) develop independent stratigraphic methods with additional couplet-thickness profiles, radiocarbon dates, and preliminary magnetic-susceptibility profiles; and (3) determine if the cyclical variability in sedimentation is consistent throughout modern sediments in the lake.

LAKE TURKANA HYDROLOGY AND MODERN SEDIMENTS

Lake Turkana (formerly Lake Rudolf) is the largest closed-basin lake in the east African rift system (Figure 1). The modern lake is divided into two bathymetric basins, north and south of a structural high adjacent to the Turkwel and Kerio Rivers. Maximum water depths are 80 m in the north basin and 100 m in the south basin. The lake is moderately saline (2.5°/∘∘), alkaline (20 meq/L, pH 9.2), and well mixed by strong, diurnal winds (Yuretich and Cerling, 1983). The perennial Omo River provides about 90% of the fresh water flowing into the lake (Ferguson and Harbott, 1982). Headwaters are to the north draining the Tertiary volcanic terrane of the Ethiopian Plateau. Seasonal flows in the Turkwel and Kerio rivers, which drain highlands of Precambrian gneiss to the southwest, contribute most of the remaining fluvial input. The local climate is hot and dry with a mean annual temperature of 30°C and annual rainfall less than 200 mm. This lake is an "arid-region end member" of large rift-valley lakes and an important modern analog for ancient lacustrine sequences in Africa and elsewhere.

Siliceous detritus dominates the modern lake sediments (Yuretich, 1979, 1986). Nearshore, sandy silts and sands are influenced by high wave energy and the presence of volcanic headlands (Cohen et al., 1986). Offshore, muds accumulate within the north and south basins by infilling from river point sources and occasionally are offset by recent high-angle faulting (Johnson et al., 1987). The relative abundance of autochthonous components (authigenic carbonate and biogenic material) increases with distance from the Omo River—i.e., from rare, poorly preserved diatom clay in the north basin to well-preserved ostracode-diatom silty clay in the south basin. This trend is interpreted to reflect decreased dilution of the autochthonous fraction by detrital input with distance from the Omo delta (Halfman, 1987). Multifold seismic profiles reveal that clastics in the north basin blanket four half-graben structural units, which suggests that clastic deposition from the Omo River masks structural features that earlier may have controlled deposition (Dunkelman et al., 1988).

Carbonate is the next most abundant mineral component in the recent sediments (Yuretich, 1986; Halfman, 1987). Bulk-carbonate concentrations range from 1–28% (Halfman, 1987) and average 10% ($CaCO_3$, dry weight). Offshore carbonates have two main components—sand-sized ostracode carapaces and fine silt-sized, euhedral, semiacicular crystals of low-magnesium calcite. Fine-grained carbonate, the bulk of the fraction, is autochthonous and probably precipitates in superheated (35°C) surface waters of the lake (Halfman et al., 1989). Appreciable influx of detrital carbonate is lacking (Cerling et al., 1988).

Faint laminae are preserved in the offshore sediments despite well-oxygenated bottom waters (Yuretich and Cerling, 1983; Cohen, 1984). The

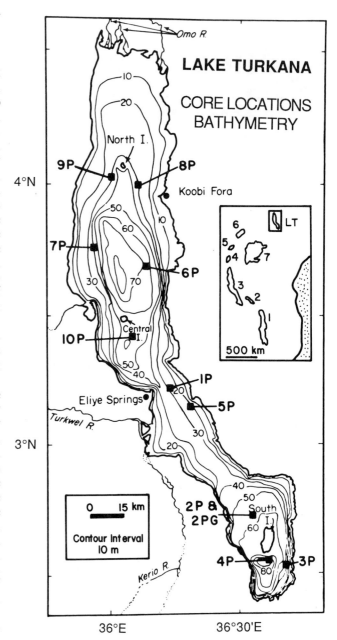

Figure 1. Core location and bathymetric map of Lake Turkana, Kenya. North basin extends from the Omo River inlet south to Eliye Springs. South basin is approximated by 40-m contour that surrounds South Island. Inset shows location of Lake Turkana (LT) in relation to other large east African lakes (1, Malawi; 2, Rukwa; 3, Tanganyika; 4, Kivu; 5, Idi Amin Dada; 6, Mobutu Sese Seko; 7, Victoria).

benthic standing crop, which could potentially homogenize the sediment, consists primarily of epibenthic detritovores (e.g., ostracodes), whose abundance declines from nearshore to offshore environments presumably because of a parallel decline in edible detritus (Cohen, 1984).

Halfman and Hearty

METHODOLOGY

Ten piston cores, each approximately 12 m long, were collected in November 1984 (Figure 1). Four radiocarbon dates were obtained from the bulk-carbonate fractions of 20-cm-thick intervals of core LT84-2P (Beta Analytic nos. 16446–16449). Dates are reported as radiocarbon years before A.D. 1950, using a half-life of 5.568 k.y. and errors (one standard deviation) from the counting statistics. Weight percentage of bulk carbonate ($CaCO_3$) in the dry sediment was determined by a vacuum-gasometric technique (Jones and Kaiters, 1983), with a reproducibility of ±0.25 wt. % from duplicate measurements. Lamina thickness (±0.1 mm) was measured with a dissecting microscope directly from the archived section of core LT84-9P. X-radiography decreased the resolution of lamina contacts.

Magnetic susceptibility (MS) was measured on archived sections of cores LT84-2P, 7P, and 8P. (For simplicity, the prefix "LT84" will be omitted from subsequent core references.) Duplicate MS values were measured every 5 cm downcore with a Bartington Instruments MS 1 meter, using a low-field, 7-cm-diameter, 2-cm-wide loop and a $MnCO_3$ calibration standard provided by the manufacturer. MS values are reported as volume susceptibility in 10^{-4} SI units (Payne, 1981). Duplicate MS measurements were within 1.0% of the respective mean value. For consistency, cyclic variability was tested following a procedure outlined by Halfman and Johnson (1988).

RESULTS

Radiocarbon Geochronology

Radiocarbon dating of bulk carbonate usually results in significant errors. This is not the case in Lake Turkana, however. Five dates determined earlier from core 8P (Table 1) were shown to be reliable (Halfman and Johnson, 1988) for the following reasons. First, a modern age was reported for the core-top sample. Second, carbonate in the profundal sediments is autochthonous (Halfman et al., 1989). Third, the lake's high alkalinity promotes rapid equilibrium of CO_2 with the atmosphere (Peng and Broecker, 1980). Plotting the four radiocarbon ages vs. core depth in 2P reveals a highly linear trend ($r^2 = 0.99$) and yields an average sedimentation rate of 4.9 mm/yr. This rate is similar to other estimates by different methods at different sites (Yuretich, 1979; Yuretich and Cerling, 1983; Cerling, 1986; Barton and Torgerson, 1988). The variability can be accounted for by proximity to sediment source or sediment focusing.

Interpolation of age vs. depth for core 2P indicated a core-top age of approximately 1.3 ka. That its age should have been modern led us to believe that the

Table 1. Radiocarbon ages for cores LT84-2P and LT84-8P from Lake Turkana

Core	Depth (cm)	Age (yr B.P.)
LT84-8P	20	Modern
	290	850 ± 180
	447	1930 ± 80
	725	3110 ± 760
	1090	3820 ± 200
LT84-2P	15	1400 ± 160
	332	2200 ± 80
	735	2820 ± 190
	1178	3840 ± 220

core top must have overpenetrated the sediment-water interface by at least 5 m. Evidence for this overpenetration includes the following: (1) upon core recovery, the core head obviously was buried below the sediment-water interface; (2) water contents of 95% in the 1-m-long trigger core indicated fluid and extremely penetrable sediments; (3) ^{210}Pb activity profiles from the trigger core revealed background activities (Halfman, 1987), which suggests that the trigger core overpenetrated the interface by at least 0.5 m; and (4) offsets in the downcore water-content gradient (Halfman, 1987) and pore-water geochemical data (Cerling et al., 1985) between the trigger core and the top of the piston core indicate overpenetration of the piston core head by at least 4 m beyond the base of the trigger core.

Carbonate Profiles

Carbonate profiles in north basin cores reveal a general trend of increasing carbonate content with core depth and with distance from the Omo River inlet (left to right in Figure 2). However, within the general trend, carbonate content varies considerably—amplitudes of fluctuations exceed detection limits. This variability is significant even in the smoothed profiles (1-m running averages) because an n-point running average (1-m or three-point) retains the signal if n does not exceed the wavelength of the oscillation in question. Close examination of the smoothed-profile oscillations reveals as many as five intervals of relative carbonate concentration. Within these intervals are higher frequency oscillations. Correlation of the intervals and the more distinctive higher frequency oscillations from core to core define a lithostratigraphy for the north basin in Figure 2. Core 10P recovered all five intervals. Cores 7P and 8P recovered four intervals, and 9P recovered three intervals. When the profiles are tied to 8P, the 9P profile corresponds to the upper 7.5 m of 8P and the upper 9 m of 7P and upper 6 m of 10P correspond to the complete section recovered by 8P.

The profile of carbonate content from south basin core 2P is inconsistent with north basin stratigraphy (Figure 3a). Although its mean carbonate content

CARBONATE (wt%)

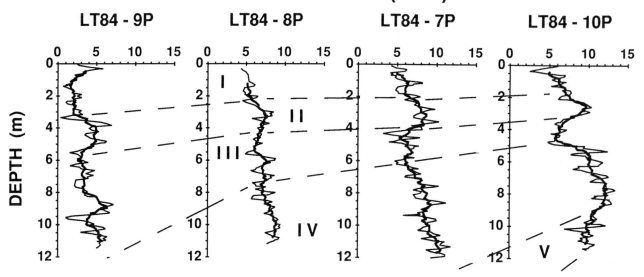

Figure 2. Dry-sediment carbonate content (wt. % CaCO₃) vs. core depth in 7P, 8P, 9P, and 10P. Profiles are arranged with increasing distance (to the right) from Omo River delta. Thin line is a three-point running average of original data (about 10 analyses/m). Heavy line is a 1-m running average of original data. Intervals I, III, and V denote relatively lower carbonate concentrations than intervals II and IV. Inferred correlations (dashed lines) thus delineate general lithostratigraphy.

(17%) is much higher than in north basin cores (5-9%), concentration generally decreases with core depth, opposite to the trend in north basin cores. Variability between oscillations is larger in the south basin core as well. The differences between basins are most clearly revealed in Figure 3a, carbonate content vs. radiocarbon age for cores 2P and 8P (assuming linear sedimentation rates), wherein cross-correlation based on carbonate concentrations for similar ages (within 20 yr) is weak and negative ($r^2 = 0.23$).

Couplet Thickness

Profiles of couplet thickness from the upper 3 m of cores 8P and 9P, located 15 km apart and equidistant from the Omo River, reveal noisy but similar patterns (Figure 4). An interval of thicker couplets occurs in the upper 2 m of each core. A smaller number of relatively thinner couplets divide these intervals at a core depth of about 1 m.

Magnetic Susceptibility

Magnetic susceptibility (MS) down two north basin cores alternates about their respective means—32.7 in 7P and 30.0 in 8P (Figure 5). Both profiles exhibit lower MS in the upper 3 m and higher values through most of the remainder of core, as observed on the smoothed (1-m running average) profiles. Variability in both cores is significant (i.e., greater than detection limits), and relative noise between the two also differs. Within the higher MS intervals prominent lows are observed at about 5, 7.5, and 11 m in core 7P and at 6 and 8 m in core 8P. On the other hand, various peaks appear in 7P (e.g., 450, 535, 570, 670, 780, and 935 cm) but not in 8P. The largest peaks in 7P occur at 450 and 935 cm and are four to five times larger than the core's mean value. These peaks are unique because they are defined by only one data point.

In contrast, MS in south basin core 2P alternates about a mean value of 20.2 (Figure 3b), which is less than in north basin cores. A peak at 110 cm in its profile corresponds to a single depth. Plotting MS vs. radiocarbon age for cores 2P and 8P reveals additional inconsistency between the two basins. Fluctuations in MS are nonparallel between the profiles except for the interval of high MS at approximately 2.6 ka. No correlation was found by cross-correlation of MS data at similar radiocarbon ages ($r^2 = 0.00$).

DISCUSSION

North Basin Sedimentation

Temporal and spatial differences in sedimentation rates can account for variations among carbonate profiles in the north basin. For example, consider core 9P, which recovered the smallest fraction of the lithostratigraphy. During the flood season, a turbid

Halfman and Hearty

CARBONATE (wt%)

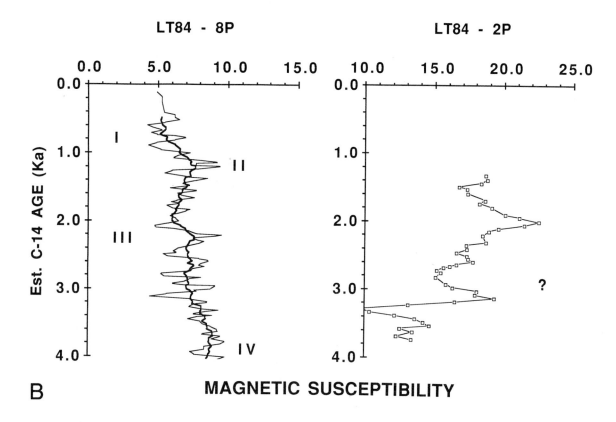

MAGNETIC SUSCEPTIBILITY

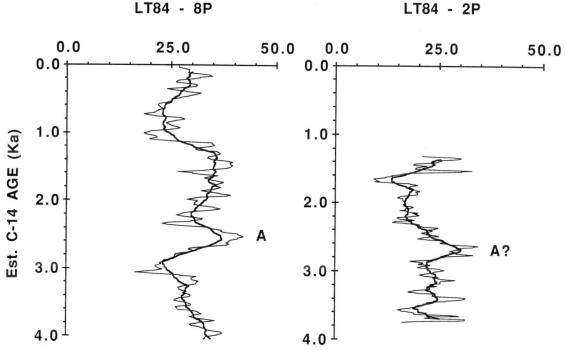

Figure 3. (A) Carbonate content (wt. %) and (B) magnetic susceptibility vs. estimated radiocarbon age in cores LT84-2P and LT84-8P from the south and north basins, respectively. A linear best-fit interpolation of radiocarbon age vs. depth in each core (four dates in 2P and five dates in 8P) was used to construct the time axis. Peak A in magnetic profiles is the only depth that may correlate between profiles. Other depths are divergent. Roman numerals in the carbonate profile 8P correspond to north basin lithostratigraphy from Figure 2.

Lacustrine Basin Exploration: Case Studies and Modern Analogs

COUPLET THICKNESS (mm)

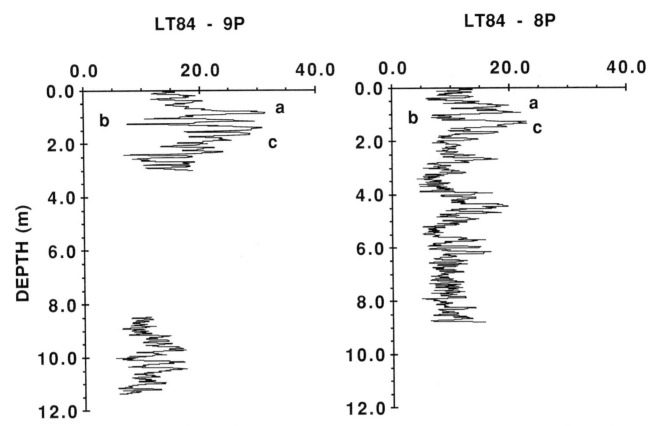

Figure 4. Thickness of successive light/dark couplets vs. core depth in LT84-8P and LT84-9P. Each plot is a five-point running average of original data. Points a and c mark correlative zones of relatively thick couplets separated by thinner couplets (b). Continuous sections of distinct laminae are absent in middle of core LT84-9P and in other cores.

plume of suspended sediment typically extends down the western lake shore from the Omo River delta to the area just beyond site 9P (Figure 1) (Yuretich, 1979; Ferguson and Harbott, 1982). This increased supply of detritus here is consistent with younger sediment at the base of the core. Site 10P, situated on an isolated bathymetric high near the basin's south end, is farthest from the delta. Implied lower sedimentation rates there result in recovery of older sediment. Based on the trends in Figure 2, both mean carbonate content and amplitude of carbonate variability generally increase with increasing distance from the Omo River inlet. Carbonate lithostratigraphy then supports our previously stated hypothesis that carbonate concentrations reflect dilution by variable influx of siliceous detritus.

Similar arguments explain major discrepancies between couplet-thickness profiles. For example, the mean thickness (17 mm) of couplets in the upper 3 m of core 9P is larger than the equivalent section of 8P (10 mm). Juxtaposition of the Omo River turbid plume to site 9P is consistent with observed thicker couplets there. Localized changes in sedimentation through time may, of course, account for minor

differences between the profiles. Compaction is assumed minimal and uniform between the two sites because the downcore decrease in porosity is linear and small (0.91 to 0.87; Halfman, 1987). Correlation of couplet thicknesses between the upper sections of 8P and 9P is consistent with the carbonate lithostratigraphy.

Faint laminae occur in the other cores recovered from the north basin. Besides the measured sections, however, no other cores revealed more than 3 m of continuous laminae. Differences in preservation could have resulted from sediment homogenization by wind-generated waves or by bioturbation. First, fetch length of prevailing southeasterly winds is shorter at site 8P (20 km) than at sites 7P (150 km), 9P (50 km), or 10P (120 km) and thus may offer protection from wave homogenization. Water depth is, however, shallower at site 8P (29 m) compared with 7P (40 m), 6P (70 m), and 10P (53 m), and lake level for the last 4 k.y. is inferred to be lower than the present (Halfman and Johnson, 1988). Second, Cohen (1984) concluded that the dearth of edible detritus precluded a benthic standing crop in the profundal deposits. The mean organic carbon content

Halfman and Hearty

MAGNETIC SUSCEPTIBILITY

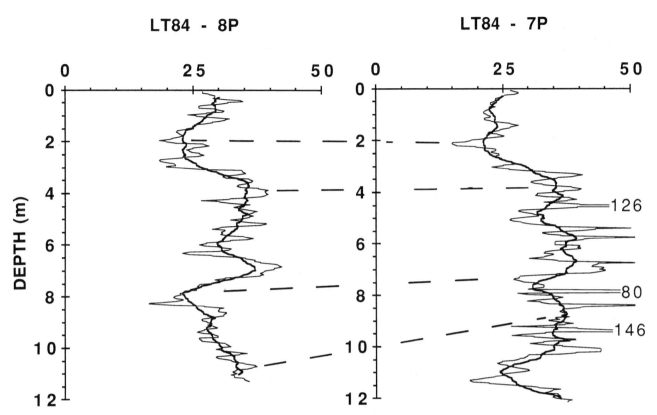

Figure 5. Magnetic susceptibility vs. core depth in LT84-7P and LT84-8P. Thin line is a three-point running average of original data (about 20 analyses/m). Heavy line is a 1-m running average of original data. Peaks at 450, 780, and 935 cm downcore LT84-7P are off the scale and excluded from 1-m running average. Dotted lines delineate tentative magnetostratigraphy.

in north basin cores is lowest in 8P—0.7 wt. % TOC compared with 0.8 wt. % in 7P and 0.9 wt. % in 9P, 10P, and 6P (Halfman, 1987). Organic carbon concentrations possibly are low enough at 8P to prevent benthic colonization but high enough for colonization at the other sites. X-radiography and visual observations revealed some bioturbation in all cores.

Magnetic Susceptibility

Magnetic susceptibility (MS) primarily is a function of the volume of ferromagnetic minerals. It is complicated by the additional influences of grain size and mineralogy (Banerjee et al., 1979; Creer, 1982; King et al., 1983; Thompson and Oldfield, 1986). Ferromagnetic materials in Lake Turkana sediments are believed to be fluvial, eolian, and volcanic in origin. Mafic volcanics and gneisses crop out in the lake's drainage basin. Sand grains in the modern sediment probably represent an eolian contribution derived from exposed metamorphic terrane south and east of the lake (Yuretich, 1979). Finally, all three

islands (Curtis, 1987) and the southern end of the lake (Truckle, 1976) show evidence for recent volcanism. The closest site downwind from Central Island is 7P, and 2P is nearest South Island. Both cores 7P and 2P recovered black, submillimeter-thick volcanic ash (Halfman, 1987). Ash at core depths of 450 and 933 cm in 7P corresponds to the largest MS values (Figure 5). Ash at core depth 111 cm in 2P corresponds to another isolated peak in MS. Episodic volcanic activity in Lake Turkana therefore appears to overprint the baseline input of magnetic material from fluvial and eolian sources.

Localized volcanism could influence site 7P but not site 8P. According to the tentative magnetostratigraphy in Figure 5, the MS profile from core 8P corresponds to the upper 9 m of that from 7P after considering the baseline magnetic signature and ignoring isolated peaks in 7P. The correlation of magnetic data is consistent with the carbonate lithostratigraphy. However, MS profiles do not mimic the respective carbonate profiles. This suggests that relative contributions by fluvial, eolian, and volcanic sources to the total detrital mass, which impact the carbonate profiles, is unique with respect to their relative contributions to the total volume of ferromagnetic material, which impact the MS profiles.

South Basin Sedimentation

Plots of carbonate content and MS vs. radiocarbon age from core 2P are not parallel to their corresponding profiles from 8P (Figure 3). This suggests that the north and south basins were depositionally isolated for at least the last 4 k.y., assuming that carbonate profiles from both basins reflect carbonate dilution by detrital influx and that MS profiles primarily reflect input of ferromagnetic minerals from fluvial, eolian, and volcanic sources. Therefore the Omo River's influence on lake sedimentology must not extend southward across the bathymetric high offshore from the Turkwel and Kerio Rivers. If the radiocarbon dates are not valid, then other stratigraphic correlations between the two basins are possible.

Variations in climate, basin morphometry, wind stress, and productivity may have contributed to the sedimentological isolation of the south basin. The Turkana region certainly is warmer (mean annual temperature of 30°C vs. 18°C) and drier (annual rainfall of 200 mm vs. 1200 mm) than the Omo drainage basin (East African Meteorological Department, 1975; Shahin, 1985). In fact, precipitation data reveal a unique trend for the Turkana region compared with the rest of east Africa (Rodhe and Virji, 1976), and a tentative paleohydrologic reconstruction from Lake Bogoria, located in the Kenyan rift 200 km south of Lake Turkana, is inconsistent with parallel fluctuations in climate recorded in Ethiopia (Tiercelin et al., 1981).

Shoreline morphology and wind stress also may directly impact the relative contributions of fluvial and eolian sediment into each basin. Extensive cliff-face shorelines and small drainage basins dominate the south basin, in contrast to low-lying, sandy or muddy shorelines to the north (Cohen et al., 1986). Prevailing winds over the south basin average 33 km/hr, and nighttime winds extend through the daylight hours. On the other hand, north basin winds are slower and shorter in duration with average velocities of 12 km/hr west of Central Island and 7 km/hr in the Omo delta; nighttime winds rarely persist past midmorning (Butzer, 1971; Ferguson and Harbott, 1982). These differences suggest that less weathered and possibly coarser grained detrital material, with a greater contribution from eolian sources, enters the south basin than the north basin.

Indirectly, both shoreline morphology and wind stress may impact primary productivity and formation of autochthonous material in each basin. North basin productivity is strongly coupled to the input of nutrient-rich floodwaters from the Omo River (Harbott, 1982). Algal blooms are reported during calm afternoons in the northwestern section of the lake (Harbott, 1982). In contrast, upwelling of nutrient-rich water may be more important to productivity in the southern sector. Although productivity data are lacking to determine precise relationships, south basin cores recovered higher concentrations of autochthonous materials, such as diatoms, ostracodes, organic matter, and autochthonous carbonate precipitates (Halfman, 1987), than in the north. Furthermore, lower mean MS values at site 2P (20.2) than in north basin cores (32.7 in 7P and 30.0 in 8P) reflect, in part, the higher concentrations of autochthonous and diamagnetic materials.

Cyclical Sedimentation in North Basin

The cyclical nature of carbonate and couplet-thickness profiles from the north basin was tested by Fast Fourier Transformation (FFT). For each carbonate data series, which was smoothed with a three-point running average, a value was interpolated every centimeter with a clamped, cubic-spline routine. The linear trend was removed from each series by a least-squares linear regression prior to analysis. A uniform sedimentation rate was assumed for each site, i.e., each centimeter of carbonate and each couplet were deposited in identical time increments. The resulting periods are expressed in cm/cycle (carbonate) or couplets/cycle. The 95% confidence limit for each power spectrum was determined by a chi-square test.

As seen in Table 2, several significant periods occur at each site. Carbonate periods cannot, however, be directly compared from profile to profile because each site experienced a different sedimentation rate. Intercore consistencies are found after normalizing each site's sedimentation rate to that at 8P based on the carbonate lithostratigraphy. Each identified period in 8P was found in the other cores, although not always in all cores. The differences among the cores may arise from aliasing of smoothed and

Table 2. Time-series analysis of core carbonate data

LT84-8P cm/cycle	LT84-7P cm/cycle	LT84-9P cm/cycle	LT84-10P cm/cycle	Average yr/cycle
93.1	104.3 ± 9	87.3 ± 12	93.9 ± 5	350
73.1	69.5 ± 4	69.8 ± 8	72.7 ± 3	260
53.9	52.0 ± 2	53.7 ± 4	53.6 ± 2	200
44.5	—	46.6 ± 3	49.4 ± 1.5	175
39.4	—	38.8 ± 2	40.8 ± 1*	145
29.3	—	30.4 ± 1	—	110

Reported periods (cm/cycle) are significant at the 95% confidence limit. The errors in periods of LT84-7P, LT84-9P, and LT84-10P are based on the interval between successive harmonics for the respective core. Average periodicity in years is calculated by averaging cm/cycle data in each row and using a linear sedimentation rate of 2.7 mm/yr.

*Prominent but not significant at the 95% confidence level.

Halfman and Hearty

interpolated data, or the assumption of linear sedimentation does not truly reflect the geology. The uniformity indicates cyclic sedimentation throughout the modern sediments of the north basin in Lake Turkana.

Thicknesses of groups of laminae also appear to oscillate periodically. FFT analysis of the 8P and two sections of 9P thickness data (172 couplets in the top 3 m and 272 couplets in the bottom 3 m of core) reveals several significant periods with intercore consistency (Table 3). Both data sets from 9P are too short to mathematically detect periods longer than about 40 couplets/cycle. They also dictate a smaller signal-to-noise ratio than the longer 8P data series; consequently, some reported periods from the 9P data are definite but not above the 95% confidence limit. Although thickness data are not as extensive as carbonate analyses, they support the hypothesis of cyclic sedimentation in the north basin on a time scale of decades to centuries.

Table 3. Time-series analysis of couplet-thickness data

LT84-8P couplet/cycle	LT84-9P (top) couplet/cycle	LT84-9P (bottom) couplet/cycle	Average yr/cycle
73.1 (75.8)	—	—	280
51.2 (53.1)	—	—	195
41.0 (42.5)	36.6 ± 6*	—	150
25.6 (26.5)	—	25.6 ± 2.8	100
20.1 (20.9)	18.3 ± 1.5	—	75
11.5 (11.9)	13.5 ± 0.4	12.1 ± 0.6*	45
8.1 (8.4)	8.1 ± 0.3	8.3 ± 0.3*	31
6.6 (6.8)	—	6.4 ± 0.3*	25

Reported periods (couplets/cycle) are significant at the 95% confidence limit. The error in LT84-9P analyses is based on the interval between successive harmonics. Values in parentheses for LT84-8P periods are recalculated in units of the carbonate FFT analysis (cm/cycle) using a linear conversion of 903 couplets within 935 cm of core. Average periodicity in years is calculated by averaging couplet/cycle data in each row and using a linear sedimentation rate of 2.7 mm/yr.

*Periods are prominent but not at the 95% confidence level.

CONCLUSIONS

1. A lithostratigraphy is defined by relative concentrations of bulk carbonate in cores recovered from the north basin of Lake Turkana. The record reflects long-term variation in the dilution of autochthonous carbonates by variable input of siliceous detritus from the Omo River.

2. Profiles of couplet thickness provide additional support for the carbonate lithostratigraphy. Unfortunately, only two cores recovered more than 3 m of distinct laminae, which probably reflects sporadic distribution of sediment mixing by benthic communities through space and time.

3. Magnetic susceptibility of the sediments reflects the input of ferromagnetic materials from fluvial, eolian, and episodic volcanic sources. MS profiles define a tentative magnetostratigraphy in the north basin that, upon removing localized volcanic events, is consistent with the lithostratigraphy.

4. Additional data will be required to definitively extend north basin lithostratigraphy and magnetostratigraphy into the south basin. Preliminary results suggest that the two basins are depositionally dissimilar.

5. Carbonate and couplet-thickness data reveal a number of significant periods with intracore consistencies. North basin deposition is cyclical and related to variable input of detritus from the Omo River.

ACKNOWLEDGMENTS

The writers thank the Kenyan Government for permission to work on Lake Turkana, G. Lister for assistance with coring, and the Captain and crew of the M/V Halcyon for their assistance on the lake. Project PROBE at Duke University provided financial support for field work; and NSF Grant ATM–8903649 and the University of Notre Dame provided support for laboratory analyses. Special thanks are extended to M. A. Perlmutter, R. Y. Anderson, and B. J. Katz for their helpful review of an earlier draft of the manuscript.

REFERENCES

Anderson, R. Y., 1986, The varve microcosm—Propagator of cyclic bedding: Paleoceanography, v. 1, p. 373–382.

Arthur, M. A., and R. E. Garrison, 1986, Cyclicity in the Milankovitch band through geologic time—An introduction: Paleoceanography, v. 1, p. 369–372.

Banerjee, S. K., S. P. Lund, and S. Levi, 1979, Geomagnetic records in Minnesota lake sediments—Absence of the Gothenberg and Erieau excursions: Geology, v. 7, p. 588–591.

Barton, C. E., and T. Torgerson, 1988, Paleomagnetic and ^{210}Pb estimates of sedimentation in Lake Turkana, East Africa: Palaeogeography, Palaeoclimatology, Palaeoecology, v. 68, p. 53–59.

Butzer, K. W., 1971, Recent history of an Ethiopian delta: University of Chicago Press, Geography Department Research Paper 136, p. 184.

Cerling, T. E., 1986, A mass-balance approach to basin sedimentation: Constraints on the recent history of the Turkana Basin: Palaeogeography, Palaeoclimatology, Palaeoecology, v. 43, p. 129–151.

Cerling, T. E., J. R. Bowman, and J. R. O'Neil, 1988, An isotopic study of a fluvial-lacustrine sequence—The Plio-Pleistocene Koobi Fora sequence, East Africa: Palaeogeography, Palaeoclimatology, Palaeoecology, v. 63, p. 335–356.

Cerling, T. E., T. C. Johnson, and J. D. Halfman, 1985, Pore water chemistry of an alkaline rift lake—Lake Turkana, Kenya (abs.): GSA Abstracts with Program, v. 17, no. 7, p. 541.

Cohen, A. S., 1984, Effects of zoobenthic standing crop on laminae preservation in tropical lake sediment, Lake Turkana, East Africa: Journal of Paleontology, v. 58, p. 499–510.

Cohen, A., D. S. Ferguson, P. M. Gram, S. L Hubler, and K. W. Sims, 1986, The distribution of coarse-grained sediments in modern Lake Turkana, Kenya—Implications for clastic sedimentation models of rift lakes, *in* L. E. Frostick, R. W. Renaut, I. Reid, and J. J. Tiercelin, eds., Sedimentation in the African rifts: Geological Society Special Publication 25, p.127-139.

Creer, K. M., 1982, Lake sediments as recorders of geomagnetic field variations—Applications to dating post-glacial sediments: Hydrobiologia, v. 92, p. 587-596.

Curtis, P. K., 1987, The structure and petrology of Central Island, Lake Turkana, Kenya: M.S. Thesis, Duke University, Durham, North Carolina, 170 p.

Dunkelman, T. J., J. A. Karson, and B. R. Rosendahl, 1988, Structural style of the Turkana rift, Kenya: Geology, v. 16, p. 258-261.

East African Meteorological Department, 1975, Climatological records for East Africa, part I—Kenya: Nairobi, Kenya, East Africa Community, p. 22.

Ferguson, A. D. J., and B. J. Harbott, 1982, Geographical, physical and chemical aspects of Lake Turkana, *in* A. J. Hopson, ed., Lake Turkana—A report on the findings of the Lake Turkana Project, 1972-1975: London, Overseas Development Administration, p. 1-107.

Fischer, A. G., 1986, Climatic rhythms recorded in strata: Annual Reviews of Earth and Planetary Sciences, v. 14, p. 351-376.

Halfman, J. D., 1987, High-resolution sedimentology and paleoclimatology of Lake Turkana, Kenya: Ph.D. Dissertation, Duke University, Durham, North Carolina, 188 p.

Halfman, J. D., and T. C. Johnson, 1988, High-resolution record of cyclic climatic change during the last 4 Ka from Lake Turkana, Kenya: Geology, v. 16, p. 496-500.

Halfman, J. D., T. C. Johnson, W. J. Showers, and G. S. Lister, 1989, Authigenic low-Mg calcite in Lake Turkana, Kenya: Journal of African Earth Sciences, v. 8, nos. 2-4, p. 533-540.

Harbott, B. J., 1982, Studies on algal dynamics and primary productivity in Lake Turkana, *in* A. J. Hopson, ed., Lake Turkana—A report on the findings of the Lake Turkana Project, 1972-1975: London, Overseas Development Office, p. 108-161.

Hays, J. D., J. Imbrie, and N. J. Shackleton, 1976, Variations in the earth's orbit—Pacemaker of the ice ages: Science, v. 194, p. 1121-1132.

Imbrie, J., J. D. Hays, D. G. Martinson, A. McIntyre, A. C. Mix, J. J. Morey, N. G. Pisias, W. L. Prell, and N. J. Shackleton, 1984, The orbital theory of Pleistocene climate—Support from a revised chronology of the marine ^{18}O record, *in* A. Berger, J. Imbrie, J. Hays, G. Kukla, and B. Saltzman, eds., Milankovitch and climate—Understanding the response to astronomical forcing: Boston, D. Reidel Publishing Co., p. 269-306.

Johnson, T. C., J. D. Halfman, B. R. Rosendahl, and G. Lister, 1987, Climatic and tectonic effects on sedimentation in a rift-valley lake—Evidence from high-resolution seismic profiles, Lake Turkana, Kenya: GSA Bulletin, v. 98, p. 439-447.

Jones, G. A., and P. Kaiters, 1983, A vacuum-gasometric technique for rapid and precise analysis of calcium carbonate in sediments and soils: Journal of Sedimentary Petrology, v. 53, p. 104-115.

King, J. W., S. K. Banerjee, J. Marvin, and S. P. Lund, 1983, Use of small-amplitude paleomagnetic fluctuations for correlation and dating of continental climate changes: Palaeogeography, Palaeoclimatology, Palaeoecology, v. 42, p. 167-183.

O'Sullivan, P. E., 1983, Annually-laminated lake sediments and the study of Quaternary environmental change—A review: Quaternary Science Reviews, v. 1, p. 245-313.

Olsen, P. E., 1986, A 40-million-year lake record of Early Mesozoic orbital climatic forcing: Science, v. 234, p. 842-848.

Payne, M. A., 1981, SI and Gaussian CGS units, conversions and equations for use in geomagnetism: Physics of the Earth and Planetary Interiors, v. 26, p. 10-16.

Peng, T., and W. Broecker, 1980, Gas exchange rates for three closed-basin lakes: Limnology and Oceanography, v. 25, p. 789-796.

Rodhe, H., and H. Virji, 1976, Trends and periodicities in east African rainfall data: Monthly Weather Review, v. 111, p. 517-528.

Schwarzacher, W., 1987, The analysis and interpretation of stratification cycles: Paleoceanography, v. 2, p. 79-95.

Shahin, M., 1985, Hydrology of the Nile basin: New York, Elsevier, Developments in Water Science 21, 575 p.

Thompson, R., and F. Oldfield, 1986, Environmental magnetism: Boston, Allen and Unwin, 227 p.

Tiercelin, J. J., R. W. Renault, G. Delibrias, J. Le Fournier, and S. Bideda, 1981, Late Pleistocene and Holocene lake level fluctuations in Lake Bogoria basin, northern Kenya rift valley: Palaeoecology of Africa, v. 13, p. 105-120.

Truckle, P. H., 1976, Geology and Late Cenozoic lake sediments of the Suguta trough, Kenya: Nature, v. 263, p. 380-383.

Yuretich, R. F., 1979, Modern sediments and sedimentary processes in Lake Rudolf (Lake Turkana), eastern rift valley, Kenya: Sedimentology, v. 26, p. 313-331.

Yuretich, R. F., 1986, Controls on the composition of modern sediments, Lake Turkana, Kenya, *in* L. E. Frostick, R. W. Renaut, I. Reid, and J. J. Tiercelin, eds., Sedimentation in the African rifts: Geological Society Special Publication 25, p. 141-152.

Yuretich, R. F., and T. E. Cerling, 1983, Hydrogeochemistry of Lake Turkana, Kenya—Mass balance and mineral reactions in an alkaline lake: Geochimica et Cosmochimica Acta, v. 47, p. 1099-1109.

Halfman and Hearty

Organic Geochemistry and Sedimentology of Middle Proterozoic Nonesuch Formation—Hydrocarbon Source Rock Assessment of a Lacustrine Rift Deposit

Scott W. Imbus
Michael H. Engel
R. Douglas Elmore
School of Geology and Geophysics
The University of Oklahoma
Norman, Oklahoma, U.S.A.

The middle Proterozoic Nonesuch Formation is part of a transgressive-regressive sequence that fills the Keweenawan trough in northern Michigan (Upper Peninsula) and northern Wisconsin. The Nonesuch is conformable with the underlying Copper Harbor Conglomerate (alluvial) and overlying Freda Sandstone (fluvial). Based on integration of outcrop and core data, three genetic facies assemblages have been recognized. A marginal-lacustrine assemblage, characterized by interbedded sandstone, siltstone, mudstone, and sandstone/shale couplets, represents deposition on a sandflat/mudflat complex. A lacustrine assemblage is characterized by massive to well-laminated, dark shaly siltstone, carbonate laminites, shale, siltstone, and mudstone. These sediments were deposited in a progressively shallowing perennial lake that periodically may have been thermally stratified. A gradual transition from a lacustrine environment to a fluvial environment is represented by red, horizontally laminated and rippled, fine-grained sandstone and siltstone of the fluvial-lacustrine assemblage. Interactions among subsidence rates, sedimentation rates, lake-level fluctuations, and possible climatic changes have resulted in variable vertical facies sequences.

Total organic carbon analyses show a strong correlation between organic richness and the shale facies (lacustrine assemblage) and parts of the marginal-lacustrine assemblage. Quantitative assessment of organic-carbon levels for the shale facies reveals the presence of organic-prone lithologies (average >0.50% TOC) comprising at least 50% of five of the eight core sections considered. Organic petrographic and geochemical analyses of selected samples, including incident white light and reflected blue-light fluorescent microscopy, pyrolysis-flame ionization detection, Rock-Eval pyrolysis, and pyrolysis-gas chromatography-mass spectrometry, indicate that most kerogens may be classified as type I and/or type II. A type III designation for some specimens is suggested based on pyrolysis results. Distinct petrographic and geochemical characteristics among these samples, viewed in terms of geographic and stratigraphic distribution, may be interpreted as the result of differential preservation of similar source organic materials rather than differential incorporation of source materials or varying thermal maturation histories within the basin.

Consideration of all petrographic and geochemical data suggests that limited intervals of the Nonesuch Formation qualify as moderate to good hydrocarbon source rocks that have experienced a mild thermal history (i.e., "oil window" thermal regime). Successful hydrocarbon exploration efforts in the Mid-Continent rift system will depend on how well one can correlate the presence of better source rocks with identification of suitable reservoir rocks and trapping mechanisms. Given the antiquity of these rocks, however, details about timing of hydrocarbon migration and accumulation and preservation of reservoirs also must be addressed.

INTRODUCTION

The potential for discovering oil and gas generated from middle Proterozoic* source rocks (~1100-Ma Oronto Group, Nonesuch Formation, and stratigraphic equivalents) arguably makes the Mid-Continent rift system one of the most intriguing North American exploration targets of the decade. To date, however, reported occurrences of indigenous oil in the region have been restricted to seeps at the White Pine Mine (Ontonagon County, Michigan; Barghoorn et al., 1965) and to evidence of migrated hydrocarbons from old mining cores in the Bayfield basin, Wisconsin (Seglund, 1989; Imbus, unpub. data). Seglund (1989) cited gas and condensate occurrences, possibly sourced by the Nonesuch Formation, in the Prairie de Chien Formation (Middle Ordovician) of Michigan's Lower Peninsula.

Farther southwest along the rift, drilling to pre-Phanerozoic strata in Washington County, northeastern Kansas (Texaco, March 1985), not only resulted in a dry hole but no evidence of hydrocarbon shows or organic-rich rock were found in the well cuttings (Ritter, 1988). Preliminary results from an Amoco deep test in Carroll County, Iowa, indicate that (1) the Keweenawan sequence in this section of the rift is analogous to that in the Lake Superior region; (2) lithologic features indicate, as in the Nonesuch Formation, that an organic-rich lacustrine sequence is present; (3) gas shows (methane and ethane), although barely detectable, persist through much of the shaly interval penetrated; and (4) the kerogens present likely are thermally postmature with respect to oil generation (R. M. McKay, Iowa Geological Survey, 1990, pers. comm.).

The high risks of exploring frontier basins are well known. Careful evaluation of all available geological, geophysical, and geochemical data is therefore essential in targeting the most promising area(s) of a prospective basin. This report seeks to present and synthesize sedimentologic and organic petrographic/geochemical data from our study of the Nonesuch Formation in Wisconsin and Michigan. Fundamental to these studies is the relationship between organic-carbon content and distribution of defined facies and facies assemblages established for the Nonesuch Formation. Additional studies have been directed toward establishing kerogen type and maturity and how these factors might bear on hydrocarbon products possibly generated from the Nonesuch Formation. Because detailed assessments have been published elsewhere (Imbus et al., 1988; Elmore et al., 1989), the sedimentologic and organic petrographic/geochemical findings are only summarized here. The quantitative assessment of organic-carbon distribution and presentation of Rock-Eval pyrolysis data are, however, original to this report.

---

* The middle Proterozoic Erathem typically is considered to represent the time interval between 1600 and 900 Ma.

GEOLOGY AND SEDIMENTOLOGY

The tectonic and sedimentologic evolution of the Oronto Group has been examined by Chase and Gilmer (1973), Fowler and Kuenzi (1978), Elmore (1981, 1984), Dickas (1986), and Elmore et al. (1989). The Oronto Group is a thick, volcanic and clastic basin-fill sequence occupying the Keweenawan trough, a northern extension of the Mid-Continent rift system, which began as a proto-oceanic rift approximately 1100 Ma. The rift system, which extends approximately 800 mi (1300 km) from northeastern Kansas to the Lake Superior region (and probably south into the Lower Peninsula of Michigan), eventually failed but not before 56 mi (90 km) of extension was achieved in the Lake Superior region.

The Nonesuch Formation, whose maximum thickness exceeds 650 ft (200 m), is located near the middle of the Oronto Group and is conformable with both the underlying Copper Harbor Conglomerate and overlying Freda Sandstone (Figure 1). The formation consists primarily of euxinic silt and shale that likely accumulated in a lacustrine setting. By synthesizing outcrop (Elmore, 1981) and core (Milavec, 1986) sedimentologic data (see Figure 2 for core and outcrop locations), we recognize nine facies in the Nonesuch Formation. Seven facies have each been assigned to one of three genetic facies assemblages (Elmore et al., 1989; Table 1). The basal, marginal-lacustrine facies assemblage consists of interbedded lithic sandstone, siltstone, mudstone, and sandstone/shale couplets and is interpreted as deposition on a sandflat/mudflat complex. The overlying lacustrine facies assemblage likely accumulated in a progressively shallowing perennial lake where redox conditions evolved from anoxic to oxic; the rocks consist primarily of massive to well-laminated, dark shaly siltstone, carbonate laminites, shale, siltstone, and mudstone. Overlying this assemblage and grading into the Freda Sandstone Formation is the fluvial-lacustrine facies assemblage. This unit, which reflects a transition from a dominantly lacustrine to a dominantly fluvial environment, consists principally of red to brown, massive, horizontally laminated and rippled, fine-grained sandstone, siltstone, and mudstone. According to Elmore et al. (1989), interactions among subsidence rates, sedimentation rates, lake-level fluctuations, possible climatic changes, and differences in tectonic setting controlled sedimentation in the Nonesuch paleolake, resulting in variable vertical facies sequences. Other researchers (Imbus et al., 1988; Elmore et al., 1989) have sought to establish what depositional control prevailed to produce observed organic-carbon levels as well as kerogen type.

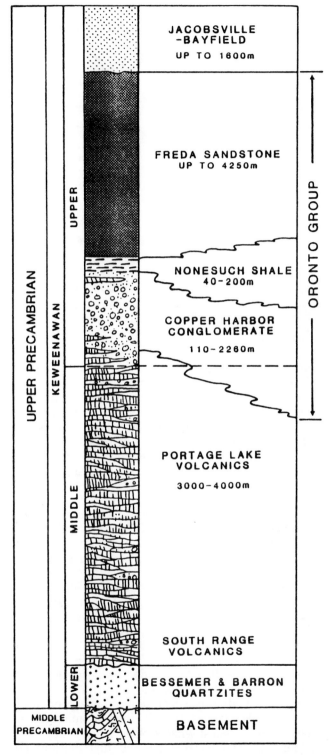

Figure 1. Generalized geologic column of Keweenawan rocks in northern Wisconsin and Upper Peninsula of Michigan. (From Elmore et al., 1989; reprinted by permission of Elsevier Science Publishers.)

ORGANIC CARBON— QUANTITY AND DISTRIBUTION

Total organic carbon (TOC) was determined in 183 Nonesuch Formation specimens collected from among eight complete cores and four complete outcrop sections. A range of essentially zero to nearly 3.0% TOC was observed. Figure 3 depicts the distribution of TOC values with respect to facies and facies assemblage assignment and geographic location. Organic-rich rocks in all core sections are concentrated in the shale facies of the lacustrine assemblage. However, marked geographic variations in organic-carbon distribution also occur in nonshaly facies. More numerous occurrences of organic-rich sediments are seen in marginal-lacustrine assemblage and nonshale facies lacustrine assemblage in the eastern (Michigan) part of the basin than in the west (Wisconsin).

If only core sections are considered (106 analyses), it is possible, given the sampling procedure, tentatively to estimate the average TOC for individual facies. During sample collection, we delineated core intervals by general lithologies (e.g., gray to black mudstone, structureless and laminated muddy siltstone, and siltstone and carbonate laminites) and took specimens we believed were representative of these lithologies. Based on mean TOC values of the samples representing each lithology and on thicknesses of the respective lithologic units, we calculated the approximate percentage of shale-facies "organic-prone" lithologies hosted by each core section. (Organic-prone lithologies are defined arbitrarily as those that on average exceed 0.5% TOC.) Table 2 summarizes these calculations together with generalized thicknesses and TOC distribution of the organic-prone lithologies, which comprise a considerable portion of the shale facies in core sections WC–25, WC–3, PI–2, WPB–8, and WPB–1. The remaining sections each contain less than 50% organic-prone lithologies.

It is difficult to explain why variations in organic-prone lithology exist among cores from the western basin or especially why cores in close proximity (WPB series) show such a large range in percentage organic-prone lithology. In part, the discrepancy may be attributed to rapid, local facies changes or to location relative to the basin edge. Another explanation may involve, at least in the case of eastern sections, underestimation of organic-rich intervals due to interfingering with nonshale facies units (neither of these units was included in the shale facies).

In summary, the method applied to establish average TOC values for core section intervals is best described as tentative. An inherent flaw exists in some cases in that average TOC values for specific lithologies from all core intervals may not be appropriately applied to individual cores. A more

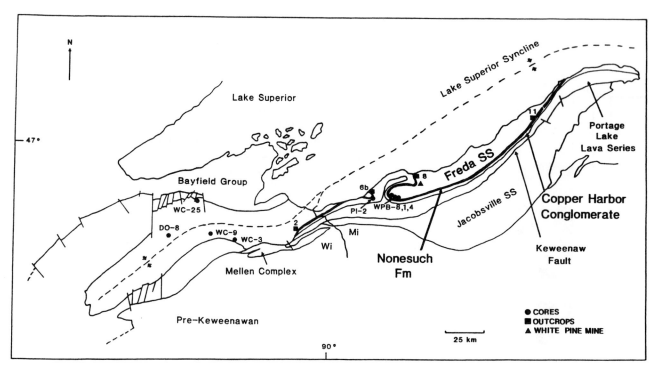

Figure 2. Geologic map of core and outcrop section locations sampled for organic petrographic and geochemical analysis. Eastern and western regions discussed in the text are located in Michigan (Mi) and Wisconsin (Wi), respectively. (From Imbus et al., 1988; reprinted by permission of Pergamon Press PLC.)

accurate assessment would, of course, entail analysis of an impractically large number of samples.

A TOC content of 0.5% normally is considered a minimum value for a viable, clastic source rock. Values higher than 1.0% are, however, considered threshold levels by many workers. Of the 106 core samples analyzed, 18 samples (or 17%) exceeded 1.0% TOC. These samples are principally from the middle shale facies (especially WPB–4 and WPB–8), the mudstone facies of the marginal-lacustrine assemblage (eastern cores only), and from the lacustrine assemblage (shale and nonshale facies) in core PI–2. Many of these values represent intervals of 10 ft (3 m) or more. From this assessment it appears that hydrocarbons most likely have been generated and expelled from the Presque Isle River (PI–2, Gogebic County, Michigan) and White Pine Mine areas (WPB series, Ontonagon County, Michigan). Such a statement relies, of course, on considerations of organic type and maturity, which are discussed below.

ORGANIC TYPE AND MATURITY

Petrographic and pyrolytic techniques were used to assess the kerogen type (i.e., chemical types I, II, or III; Tissot et al., 1974) and relative thermal maturities of selected Nonesuch Formation specimens. Rock-Eval pyrolysis (Espitalié et al., 1977) is an industry standard that allows characterization of kerogen chemical type as well as estimation of kerogen maturity and genetic potential with respect to hydrocarbon generation. An analogous technique, pyrolysis-flame ionization detection (PY/FID) using a fused-quartz pyrolysis apparatus (Ruska Laboratories, Houston, Texas), is helpful in establishing relative thermal-maturity levels via the parameter T_{max}, the temperature at which maximum kerogen pyrolysate evolution occurs. Zumberge et al. (1988) have summarized the relationships between Rock-Eval and fused-quartz-pyrolysis T_{max} values. Organic petrographic analysis in incident white light and blue-light fluorescence allows visual characterization of organic remains with respect to organoclast distribution, kerogen type, and relative maturity (Mukhopadhyay et al., 1985; Teichmüller, 1986). Finally, high-resolution pyrolysis-gas chromatography-mass spectrometry (PY/GC/MS) permits detection of specific compounds, the ratios of which signify organic-matter type and maturity.

Aspects of all these techniques have been integrated to (1) discuss the findings of microscopic analyses and suggest what geologic factors may have influenced organoclast type and distribution; (2) suggest whether or not differences exist in kerogen thermal maturity throughout the basin; (3) determine if kerogen type varies with respect to lithology, facies, or location in the basin; and (4) based on these

200

Imbus et al.

Table 1. Results of sedimentologic analysis of of the Nonesuch Formation (Milavec, 1986; Elmore et al., 1989)
Interpretations based on lithology, bedding, sedimentary structures, and other features

Facies Succession	Facies Assemblage	Facies	Interpretation
A	Marginal-lacustrine	Sandstone	Sandflat environment, located at the top of an alluvial fan, large sheet floods and/or channel migration resulted in sand deposition.
		Mudstone	Deposition on a mudflat that experienced periodic exposure, fluctuating water levels, and variations in salinity; laminated and graded beds present were result of (1) turbidity currents that entered the expanded lake, or (2) simple settling in the expanded lake.
	Lacustrine	Shale	Deposition from suspension in quiet water of perennial lake, possible anoxic conditions; siltstones are the result of bottom reworking during more oxic times.
		Siltstone-mudstone	Deposition in standing bodies of water; silts were deposited by bottom currents; silts with flutes were deposited by turbidity currents; mudstone resulted from settling in standing water.
		Thick-bedded mudstone	Deposition from suspension in quiet water below wave base in anoxic environment; siltstones were result of bottom reworking.
		Siltstone	Wave and current deposition in a nearshore, shallow-water environment; mudstones are result of settling farther offshore in a perennial lake.
	Fluvial-lacustrine	Fluvial-lacustrine	Gradual transition from a lake-dominated environment; fluvial floodplain setting.
	?	Carbonate	Probably a calcitized evaporite bed. Occurs only in core WC-25.
	?	Sandstone-shale	Graded sequences were deposited by waning sheet floods; fine-grained material settled from suspension in an expanded lake; thicker sands are probably either channel or sheet-flood deposits.

findings, suggest what, if any, hydrocarbon products might have been generated from organic-rich lithologies of the Nonesuch Formation and what area(s) of the basin may be most prospective as exploration targets.

Organic Petrography

Petrographic analysis (incident white light and reflected blue-light fluorescence) of 55 Nonesuch Formation samples from throughout the study area revealed a complex array of organic constituents from which two "organic petrographies" (OP-1 and OP-2) were defined (Imbus et al., 1988). The major distinction between the two is the relatively diminished fluorescent qualities of filamentous (algal?) organoclasts and the apparent presence of bituminite (a possible algal/bacterial degradation product cited by Teichmüller, 1982) in OP-2 samples. As petrographic data were assessed with respect to sample distribution, a geographical segregation became at once apparent. OP-1 occurs primarily in the eastern (Michigan) region, whereas OP-2 is prevalent in the west (Wisconsin).

Several explanations may be offered for the distinction between the two organic petrographies, including differential input of source material, thermal maturity differences, and differential preservation of the same source material. Based on geochemical findings (PY/FID and PY/GC/MS) and consideration of geologic and sedimentologic data, Imbus et al. (1988) favored the differential-preservation explanation. We will address this conclusion in more detail.

Rock-Eval Pyrolysis

Rock-Eval pyrolysis was performed on 25 samples representing at least one specimen from each of eight cores and four outcrop sections of the Nonesuch Formation (Table 3). As a consequence of the relatively small number of samples analyzed and the desirability of representing all sections, the specimens chosen do not necessarily represent the most

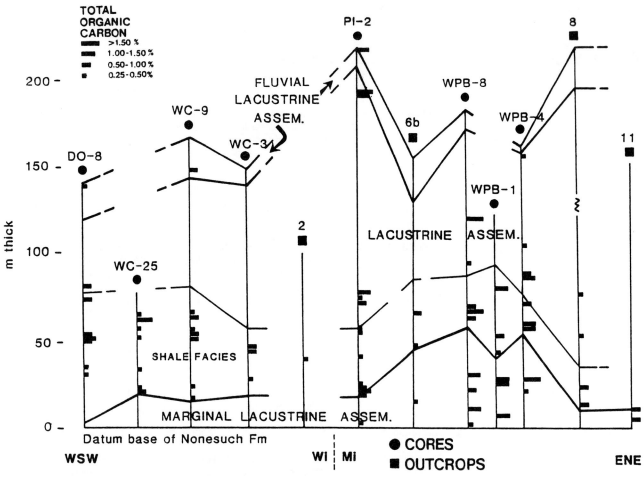

Figure 3. Fence diagram showing total organic-carbon value ranges for core and outcrop sections of the Nonesuch Formation, Michigan and Wisconsin. Depositional environment interpretations after Milavec (1986). Although sections are arranged from west-southwest to east-northeast, no horizontal scale is implied.

organic-rich samples. Figure 4 is a modified Van Krevelen diagram (Espitalié et al., 1977) of Rock-Eval analyses plotted relative to the three major "coalification" tracks—oil prone (type I), oil/gas prone (type II), and gas prone (type III).

Substantial scatter among the data suggests differences in kerogen thermal maturity and/or kerogen type among the samples analyzed. Most data points (18) plot in the area of low hydrogen and oxygen indices (Group A, Figure 4), which suggests either highly thermally evolved type I or II kerogens or a failure of accurate appraisal due to low pyrolysate yields. The latter explanation indeed may be the case for some samples because none of the 25 samples analyzed yielded T_{max} values greater than 450° (i.e., all are indicative of an "oil window" or lower thermal regime).

Two samples (Group B) are characterized by the highest oxygen indices, >25 mg/g (mg HC/g organic carbon), and relatively low hydrogen indices, 200-250 mg/g. If these samples were Phanerozoic in age, they likely would be interpreted to represent a mixture of marine and terrestrially derived organic matter because they plot between the oil/gas and gas-prone coalification tracks. Type III kerogen, representing preserved land-derived organic debris, is quite unlikely for any pre-Silurian rocks (Gray and Boucot, 1978), although kerogens exhibiting type III chemistry have been reported in Cambrian and Proterozoic rocks. McKirdy et al. (1980) reported type III kerogens in Proterozoic stromatolites. This material, considered primary organic material, was interpreted as a derivative of cyanophyte mucilage (mucopolysaccharide). If our two Group B samples indeed have a type III component—i.e., input of primary oxygen-rich organic material or the result of in situ oxidation of type I and/or type II organic matter—it appears to be the first such reported occurrence in middle Proterozoic clastic rocks. As will be discussed later, PY/GC/MS analysis of kerogen

Imbus et al.

Table 2. Summary of shale facies organic quantity and distribution (core sections only)

Rock Unit	Core Section							
	DO-8	WC-25	WC-9	WC-3	PI-2	WPB-8	WPB-1	WPB-4
Nonesuch Formation								
Total thickness, ft (m)	417 (127)	232 (71)	500 (152)	444 (135)	653 (199)	549 (167)	361 (110)	650 (198)
TOC range, % (no. analyses)	0.1–1.1 (11)	0.0–1.8 (17)	0.0–1.0 (15)	0.0–0.8 (12)	0.1–1.6 (12)	0.1–3.0 (19)	0.1–1.4 (12)	0.1–1.6 (10)
Shale facies (Lacustrine assemblage)								
Total thickness, ft (m)	78 (24)	65 (20)	85 (26)	77 (24)	107 (33)	66 (20)	85 (26)	74 (23)
TOC range, % (no. analyses)	0.1–1.1 (5)	0.1–0.3 (5)	0.0–0.9 (6)	0.2–0.8 (5)	0.1–1.3 (7)	0.4–2.7 (4)	0.9–1.1 (2)	0.1–1.0 (4)
Organic-prone lithologies: Distribution	discontinuous, up to 10 ft (3.1)	continuous-discontinuous, grading with organic-lean lithofacies	continuous-discontinuous, up to 14 ft (4.3 m)	mostly continuous, up to 19 ft (5.8 m)	mostly continuous, up to 55 ft (16.8 m)	continuous-discontinuous, units up to 18 ft (5.5 m)	continuous-discontinuous, grading with organic-lean lithofacies	continuous-discontinuous, grading with organic-lean lithofacies
Percent[a]	45	72	23	77	71	77	56	36

[a] Percent organic-prone lithologies comprising the shale facies of lacustrine-facies assemblage.

isolates (from other specimens) has confirmed the presence of significant quantities of oxygen-containing pyrolysate compounds in some samples.

A cluster of three samples (Group C), which may be closely associated with Group A, are singled out because of their absence of S_3 peaks (oxygen index = 0). Again, this phenomenon may result from low pyrolysis yields or may signify the presence of true type I kerogens. If the latter explanation is valid, deposition in an extremely anoxic, lacustrine environment would be indicated (Jones, 1987).

The last two samples (Group D) may be described as having relatively high hydrogen indices (>400 mg/g) but low oxygen indices (<20 mg/g). In addition, they appear to be only moderately, thermally evolved and representative of type I and/or type II kerogens.

Only a tentative assessment of kerogen type is possible because of low pyrolysate yields and the relatively small number of samples analyzed (Table 3). However, most samples appear to be either moderately thermally evolved (oil-window thermal regime) or, less likely, anomalously highly thermally evolved (compared to T_{max} data discussed below). All samples analyzed, except possibly two, may be readily classified as either type I or II or type I/II mixed kerogens. Several samples apparently are comprised exclusively of type I kerogen (Group C and some Group A samples).

T_{max} values characteristic of kerogen transformation (>400°C) could be determined only for nine of the 25 samples (Table 3). Again, this may be a function of low pyrolysate yield or alternatively, the presence of unusual (e.g., bituminite) or adventitious organic constituents. The nine T_{max} values determined ranged from 433 to 449°C ($\mu = 442 \pm 5$°C), which is characteristic of a thermal regime in the center of the oil window as suggested by Tissot and Welte (1984).

Another commonly applied Rock-Eval parameter is *genetic potential* (i.e., Rock-Eval $S_1 + S_2$ peaks or thermally extractable hydrocarbons + hydrocarbons generated from pyrolytic destruction of kerogen). According to Tissot and Welte's (1984) criteria, only two rock samples analyzed may be considered moderate to good source rocks based on this parameter. Sample 6b–0 (shale facies, Gogebic County, Michigan) is considered a moderately good source rock (2–6 mg/g), whereas WPB-8 M (shale facies, Ontonagon County, Michigan) is classified as a good source rock (>6 mg/g). Eight other samples have marginal source quality (1–2 mg/g)—no potential for oil but some potential for gas. Although results of the Rock-Eval study would appear pessimistic in terms of the ability of Nonesuch Formation rocks to source oil and gas, remember that sample selection was designed to properly represent each core and not necessarily the most organic-rich facies.

It is interesting to compare the results of our Rock-Eval analysis of the Nonesuch Formation to those of Hatch and Morey (1985) on the Solor Church Formation, a probable stratigraphic equivalent of the

Table 3. Results of Rock-Eval analysis for 25 Nonesuch Formation and Freda Sandstone rock samples (pyrolysis parameters defined by Espitalié et al., 1977). See Figure 4 for graphical representation of hydrogen and oxygen indices

Sample ID	Facies Assemblage[a]	Hydrogen Index[b]	Oxygen Index[c]	Genetic Potential[d]	T_{max} (°C)
D0–8 K	Lacustrine (shale facies)	107	3	0.46	–[e]
D0–8 L	Lacustrine (shale facies)	183	3	1.02	–[e]
WC–25 F	Lacustrine (shale facies)	179	15	0.79	–[e]
WC–25 M	Lacustrine	123	9	1.75	449
WC–9 I	Lacustrine (shale facies)	147	6	0.91	–[e]
WC–9 L	Lacustrine (shale facies)	159	0	0.82	–[e]
WC–3 K	Lacustrine (shale facies)	115	15	0.76	–[e]
WC–3 L	Lacustrine (shale facies)	70	0	0.47	–[e]
2–J	Facies succession B	58	16	0.17	–[e]
2–K	Facies succession B	17	13	0.06	–[e]
PI–2 R	Lacustrine	190	0	0.49	–[e]
PI–2 CC	Freda Sandstone	105	0	0.54	–[e]
6b–0	Lacustrine (shale facies)	415	11	3.50	433
6b–P	Lacustrine (shale facies)	313	0	1.82	447
WPB–8 G	Marginal-lacustrine	89	21	0.48	–[e]
WPB–8 M	Lacustrine (shale facies)	454	15	7.82	445
WPB–I G	Marginal-lacustrine	245	46	1.92	444
WPB–4 B	Marginal-lacustrine	238	30	1.62	444
WPB–4 I	Lacustrine (shale facies)	295	0	1.38	440
WPB–4 L	Lacustrine	257	0	1.10	435
WPB–4 Q	Lacustrine	188	24	0.99	–[e]
8–F	Marginal-lacustrine	18	7	0.39	–[e]
8–G	Lacustrine (shale facies)	31	20	0.42	–[e]
11–F	Marginal-lacustrine	126	17	0.97	–[e]
11–H	Lacustrine (shale facies)	196	5	1.98	439

[a] See text for explanation; succession B is unique to outcrop section 2 (see Elmore et al , 1989).
[b] Mg hydrocarbon/g organic carbon (S_2/TOC).
[c] Mg CO_2/g organic carbon (S_3/TOC).
[d] Sum of S_1 and S_2 peaks/g rock (mg hydrocarbon/g rock).
[e] T_{max} values less than 400°C are not representative of kerogen transformation.

Nonesuch Formation in the subsurface of southeastern Minnesota. Of the 25 Solor Church Formation samples they analyzed, only one exceeded 1% TOC. Hydrogen indices all fell below 50 mg/g, whereas oxygen indices ranged up to 62 mg/g. They considered only one of the six T_{max} values obtained as reliable because of poorly defined S_2 peaks. This value, 494°C, indicates a considerably greater level of thermal evolution than T_{max} values we determined for the Nonesuch Formation. Hatch and Morey concluded that Solor Church Formation kerogens are thermally postmature and placed them near the transition between the wet-gas phase of catagenesis and the dry-gas zone of metagenesis.

PY/FID and PY/GC/MS Analyses

Nine Nonesuch Formation samples were selected for kerogen isolation and subsequent analysis by PY/FID and PY/GC/MS (Imbus et al., 1988). Samples were selected to represent shale-facies and closely associated marginal lacustrine assemblage lithologies in both the eastern and western regions, marginal-lacustrine and lacustrine (nonshale facies) assemblages in the east, and the fluvial-lacustrine assemblage in the west. Six samples are representative of OP-1, and three represent OP-2. Included among these were two samples that appeared to be exceptions to the general geographic distributions of OP-1 and OP-2 (see discussion under Organic Petrography). PY/FID analysis was performed on each of the nine kerogen isolates primarily to obtain a T_{max} value (proportional to but not numerically identical to Rock-Eval:T_{max}) that would indicate any obvious differences in thermal maturity among the samples. Analysis by PY/GC/MS provides specific information with respect to chemical composition of kerogen pyrolysates. Ratios derived from quantitative assessments of the compounds that evolved are indicative of kerogen type and maturity, although few standard parameters have been established.

Table 4 summarizes selected parameters derived from the PY/FID and PY/GC/MS analyses. The

Imbus et al.

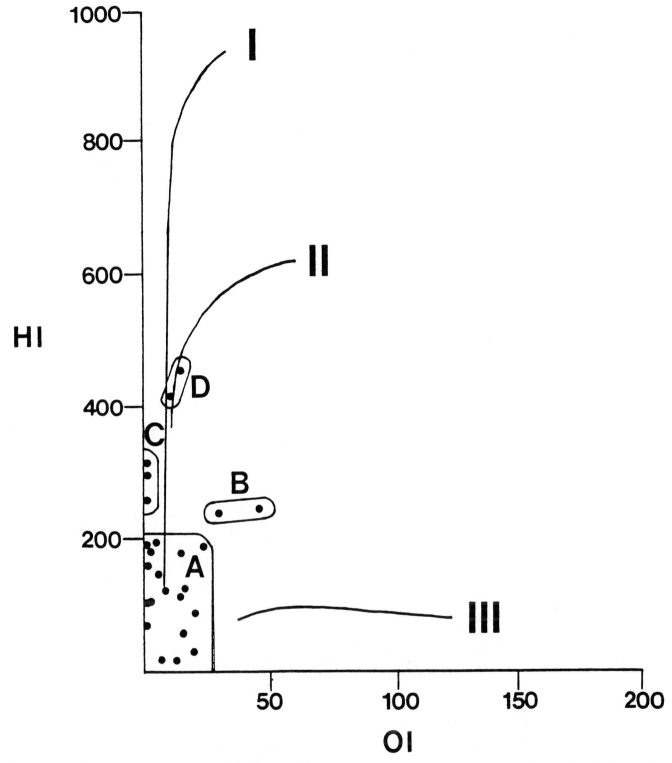

Figure 4. Modified van Krevelen diagram of hydrogen index vs. oxygen index, as derived from Rock-Eval pyrolysis. Apparent groupings of samples, designated by A, B, C, and D, are discussed in the text.

parameters listed are divided into two categories: (1) maturity-related and (2) source-related. In addition, results of stable carbon-isotope ($\delta^{13}C$) analysis are listed under the source parameters. Apparent from consideration of the alkene + alkane:n-C$_8$/n-C$_{20}$ values (maturity-related) and normalized abundances of n-octene, m + p xylenes, and phenol (source-related) parameters is a distinct correlation between values listed and designation of samples as OP-1 or OP-2. A similar tentative correlation is made between the PYRAN Level I:T_{max} (maturity-related) and $\delta^{13}C$ (source-related) parameters. OP-1 samples have, in general, the following characteristics relative to OP-2 samples: (1) slightly higher T_{max}, (2) lower alkene + alkane:n-C_8/n-C_{20} ratios, (3) slightly more depleted $\delta^{13}C$ values, and (4) distinctly higher normalized n-octene values. In terms of independent maturity parameters (PYRAN Level I:T_{max} and alkene + alkane:n-C_8/n-C_{20} values), the data are conflicting, and thus neither organic petrography can be said to be more mature than the other based on this assessment. With respect to stable carbon-isotope determinations, the values differ little, and this parameter is not in any case directly indicative of kerogen type. Normalized abundances of n-octene, m + p xylenes, and phenol, however, clearly show that OP-1 kerogens range from aliphatic to aliphatic/aromatic (i.e., indicative of type I-type II), whereas OP-2 kerogens are predominantly aromatic and/or phenolic (i.e., indicative of type II-type III), which suggests more highly oxidized organic matter.

It was suggested previously that differential preservation of the same source material might be responsible for the distinction between OP-1 and OP-2 rather than fundamentally different source inputs or thermal maturity. Considering the data from Table 4 and the geographic and stratigraphic distribution of the samples analyzed, this conclusion appears to be valid. Substantial thermal maturity differences can easily be eliminated as a factor because, in at least one case each for the eastern and western regions, both organic petrographies were found in the same core. Differential source input as an explanation is tenuous because sedimentologic features of rocks in both study areas have been determined to be essentially equivalent (Milavec, 1986). An alternative explanation invokes the possibility of differential preservation (with OP-1 remains preserved better than OP-2 remains) due either to varying residence time of organic remains passing through the water column, or to more or less favorable bottom conditions in different areas of the basin owing to the magnitude of currents or the status of the epilimnion-hypolimnion boundary (Elmore et al., 1989).

SUMMARY AND CONCLUSIONS— HYDROCARBON POTENTIAL

Three major criteria are used to assess source rock quality and to predict reservoired hydrocarbon products—organic carbon quantity (TOC), kerogen type, and kerogen maturity. Using a variety of petrographic and geochemical analytical techniques, such an assessment has been made for selected Nonesuch Formation samples. The data generated are useful in assessing hydrocarbon prospects of the Nonesuch Formation in general and in suggesting what areas of the Upper Peninsula and Wisconsin parts of the rift most likely experienced significant hydrocarbon generation.

Based on the organic-carbon assessment, it appears that several intervals of the Nonesuch Formation, particularly those of the lacustrine assemblage shale facies, have sufficient organic-carbon levels to qualify as moderate to good source rocks. Intervals of organic-rich rocks are particularly prevalent in the vicinity of the White Pine Mine and the Presque Isle River (Ontonagon and Gogebic Counties, Michigan, respectively).

Petrographic analysis in incident white light and reflected blue-light fluorescence reveals the existence of two organic petrographies, OP-1 and OP-2. Their distinction is not readily evident from Rock-Eval hydrogen and oxygen indices; however, geochemical and sedimentologic data suggest that their dissimilarity originates by differential preservation of organic remains rather than different organic source materials or varying thermal histories. Firm evidence that most Nonesuch Formation samples analyzed consist of type I and/or type II kerogens was, however, established from these data.

Parameters used to assess thermal maturity gave conflicting results for many of the rocks and kerogens analyzed (i.e., Rock-Eval T_{max} vs. PY/FID:T_{max} and PY/GC/MS compound ratios, Table 3). At this time no conclusion therefore can be made with respect to thermal maturity status of one part of the basin relative to another. Relative fluorescent intensities of filamentous organoclasts, present in samples from throughout the study area, and moderate to low Rock-Eval and PY/FID T_{max} values clearly indicate, however, that the thermal history of Nonesuch Formation rocks was indeed mild, probably not exceeding oil-window thermal regimes (R_{0max} = 1.0%, equivalent to high-volatile A bituminous coal rank according to correlation of organic-maturity parameters by Bustin et al., 1985).

Petrographic and geochemical data generated from analysis of Nonesuch Formation samples indicate overall poor to moderate source rock potential for oil with limited good source intervals in areas of Michigan. Moderate to good gas potential is likely for several intervals throughout the basin. Regardless of our estimates of source rock quality, geologic factors affecting hydrocarbon generation, migration, accumulation, and preservation must be considered in evaluating the economic potential of this basin. With respect to hydrocarbon prospects of the Mid-Continent rift, we can make no firm conclusions at this time. Other reports (Fritz, 1985; Dickas, 1984, 1986; Mudrey, 1986; Seglund, 1989) offer geological information pertinent to targeting specific regions of the rift as hosting reservoir rocks and trapping mechanisms. For example, Seglund's (1989) assessment that the Jacobsville basin (just east of our study area) is a prime area for petroleum exploration was based on deeper burial of the Nonesuch Formation, proximity to the White Pine Mine oil seeps, and

Table 4. Selected maturity and source-related parameters for nine Nonesuch Formation kerogen isolates analysed by PY/FID, PY/GC/MS and stable carbon-isotope mass spectrometry (adopted from Elmore et al., 1989)

Kerogen ID[a]	Basin Region[b]	Facies Assemblage	Lithology	OP[b]	%TOC	%CaCO$_3$	Pyran Level I:T_{max}[d]	R_c[e]	n-C$_8$/n-C$_{20}$[g]	δ^{13}C PDB kerogen (°/oo)	n-Octene	m+p Xylenes	Phenol
WPB-8 M	Eastern	Lacustrine[c]	Laminated siltstone	1	2.75	11.3	463	–[f]	0.06	-33.45	49.4	28.7	21.9
WPB-4 B	Eastern	Marginal-lacustrine	Carbonate laminite	1	1.55	35.4	449	–[f]	0.09	-33.66	38.9	37.6	23.6
WPB-4 Q	Eastern	Lacustrine[c]	Muddy siltstone	1	1.24	10.8	451	0.67	0.14	-32.68	29.7	40.2	30.1
PI-2 E	Eastern	Lacustrine[c]	Laminated siltstone	1	0.70	16.0	452	0.69	0.11	-33.44	27.0	34.8	33.9
6b P	Eastern	Lacustrine[c]	Laminated siltstone	1	0.57	11.0	450	–[f]	0.04	-33.29	65.0	33.3	1.7
WC-25 B	Western	Marginal-Lacustrine	Carbonate laminite	1	0.98	42.8	460	0.72	0.07	-34.08	44.7	39.1	16.2
PI-2 AA	Eastern	Lacustrine	Silty mudstone	2	1.58	9.4	~410	0.98	0.64	-32.84	15.8	42.1	42.1
WC-9 T	Western	Fluvial-lacustrine	Mudstone	2	0.98	8.4	448	0.67	0.67	-30.83	13.4	43.9	42.7
WC-25 M	Western	Lacustrine	Laminated siltstone	2	1.84	8.6	446	0.65	0.39	-33.12	15.8	26.0	58.2

a See Figure 2 for sample locations.
b See text for explanation.
c Shale facies of lacustrine assemblage.
d Analogous to Rock-Eval:T_{max} (e.g., Zumberge et al., 1988). See Table 3 to compare Rock-Eval and Pyran Level I (PY/FID) T_{max} parameters for samples WPB-8 M, WPB-4 R, WPB-4 Q, 6b-0, and WC-25 M.
e Calculated vitrinite reflectance equivalent (Radke et al., 1982; Radke and Welte, 1983).
f All constituents necessary for determination not detected.
g Sum of alkene + alkene n-C$_8$/n-C$_{20}$.
h After Larter (1984).

anticlinal folding associated with the Keweenaw reverse fault (i.e., reservoir and trapping opportunities). It is clear, however, regardless of the results of our source rock assessment or other geological and geophysical studies of the rift system, explorationists involved with the rift system will face uncertainties that are inherent to frontier basin exploration.

ACKNOWLEDGMENTS

The authors thank Texaco Inc. for support of this project and for performing Rock-Eval analysis. M. H. Engel acknowledges the National Science Foundation's Division of Earth Sciences (Grant EAR-8352055) and the U.S. Department of Energy's Office of Basic Energy Sciences (Grant DE-FG05-89ER14075) for partial support of this research. We thank G. H. Milavec and J. E. Zumberge for assistance with analyses and M. Starr for typing the manuscript. We also thank M. Smith and C. Robison for their helpful comments.

REFERENCES

Barghoorn, E. S., W. G. Meinschein, and J. W. Schopf, 1965, Paleobiology of a Precambrian shale: Science, v. 148, p. 461–472.

Bustin, R. M, M. A. Barnes, and W. C. Barnes, 1985, Diagenesis 10—Quantification and modelling of organic diagenesis: Geoscience Canada, v. 12, p. 4–21.

Chase, C. G., and T. H. Gilmer, 1973, Precambrian plate tectonics—The Midcontinent gravity high: Earth and Planetary Science Letters, v. 21, p. 70–78.

Dickas, A. B., 1984, Midcontinent rift system—Precambrian hydrocarbon target: Oil & Gas Journal, v. 82, p. 151–159.

Dickas, A. B., 1986, Comparative Precambrian stratigraphy and structure along the Mid-Continent rift: AAPG Bulletin, v. 70, p. 225–238.

Elmore, R. D., 1981, The Copper Harbor Conglomerate and Nonesuch Shale—Sedimentation in a Precambrian Intracontinental Rift, Upper Michigan: Ph.D. Dissertation, University of Michigan, Ann Arbor, Michigan, 200 p.

Elmore, R. D., 1984, The Copper Harbor Conglomerate—Late Precambrian fining-upward alluvial fan sequence in northern Michigan: GSA Bulletin, v. 95, p. 610–617.

Elmore, R. D., G. J. Milavec, S. W. Imbus, and M. H. Engel, 1989, The Precambrian Nonesuch Formation of the North American Mid-Continent Rift, sedimentology and organic geochemical aspects of lacustrine deposition: Precambrian Research, v. 43, p. 191–213.

Espitalié, J., J. L. Laporte, M. Madec, F. Marquis, P. Leplat, J. Paulet, and A. Boutefeu, 1977, Méthode rapide de charactérisation des roches meres, et de leur potentiel pétrolier et de leur degre d'evolution: Revue de l'Institut Francais du Pétrole, v. 32, p. 23–42.

Fowler, J. H., and W. D. Kuenzi, 1978, Keweenawan turbidites in Michigan (deep borehole red beds)—A foundered basin sequence developed during evolution of a protoceanic rift system: Journal of Geophysical Research, v. 83, p. 5833–5843.

Fritz, M., 1985, Mid-Continent exploration continues—Rift remains an intriguing frontier: AAPG Explorer, v. 6, no. 7, p. 18–20.

Gray, J., and A. J. Boucot, 1978, The advent of plant life: Geology, v. 6, p. 489–492.

Hatch, J. R., and G. B. Morey, 1985, Hydrocarbon source rock evaluation of middle Proterozoic Solor Church Formation, North American Mid-Continent rift system, Rice County, Minnesota: AAPG Bulletin, v. 69, p. 1208–1216.

Imbus, S. W., M. H. Engel, R. D. Elmore, and J. E. Zumberge, 1988, The origin distribution and hydrocarbon generation potential of organic-rich facies in the Nonesuch Formation, central North American rift system—A regional study: Organic Geochemistry, v. 13, p. 207–219.

Jones, R. W., 1987, Organic facies: Advances in Petroleum Geochemistry, v. 2, p. 1–90.

Larter, S. R., 1984, Application of analytical pyrolysis techniques to kerogen characterization and fossil fuel exploration/exploitation, in K. J. Voorhees, ed., Analytical pyrolysis—Techniques and explorations: Boston, Butterworth, p. 212–275.

McKirdy, D. M., D. J. McHugh, and J. W. Tardif, 1980, Comparative analysis of stromatolitic and other microbial kerogens by pyrolysis-hydrogenation-gas chromatography (PHGC), in P. A. Trudinger, M. R. Walter, and B. J. Ralph, eds., Biogeochemistry of ancient and modern environments: New York, Springer-Verlag, p. 187–200.

Milavec, G. J., 1986, The Nonesuch Formation—Precambrian sedimentation in an intracratonic rift: M.S. thesis, University of Oklahoma, Norman, Oklahoma, 142 p.

Mudrey, M. G., Jr., ed., 1986, Precambrian petroleum potential along the Midcontinent Trend: Geoscience Wisconsin, v. 11, 85 p.

Mukhopadhyay, P. K., H. W. Hagemann, and J. R. Gormly, 1985, Characterization of kerogens as seen under the aspect of maturation and hydrocarbon generation: Erdöl und Kohle, v. 38, p. 7–18.

Radke, M., and D. H. Welte, 1983, The Methylphenanthrene Index (MPI)—A maturity parameter based on aromatic hydrocarbons, in M. Bjorøy, and others, eds., Advances in organic geochemistry 1981: New York, Wiley, p. 504–512.

Radke, M., D. H. Welte, and H. Willsch, 1982, Geochemical study on a well in the Western Canada Basin—Relation of aromatic distribution patterns to maturity of organic matter: Geochimica et Cosmochimica Acta, v. 46, p. 1–10.

Ritter, D. M., 1988, Details released on Texaco's Midcontinent Rift test: Shale Shaker, v. 39, no. 2, p. 43.

Seglund, J. A., 1989, Midcontinent rift continues to show promise as petroleum prospect: Oil & Gas Journal, v. 87, p. 55–58.

Teichmüller, M., 1982, Origin of the petrographic constituents of coal, in E. Stach, M. T. Mackowsky, M. Teichmüller, G. H. Taylor, D. Chandra, and R. Teichmüller, eds., Stach's textbook of coal petrology: Berlin, Gebruder Borntraeger, p. 291–294.

Teichmüller, M., 1986, Organic petrology of source rocks, history and state of the art: Organic Geochemistry, v. 10, p. 581–599.

Tissot, B., B. Durand, J. Espitalié, and A. Combaz, 1974, Influence of nature and diagenesis of organic matter in formation of petroleum: AAPG Bulletin, v. 58, p. 499–506.

Tissot, B. P., and D. H. Welte, 1984, Petroleum formation and occurrence, 2d ed.: New York, Springer-Verlag, 699 p.

Zumberge, J. E., C. Sutton, S. J. Martin, and R. D. Worden, 1988, Determining oil generation kinetic parameters by using a fused-quartz pyrolysis system: Energy and Fuels, v. 2, p. 264–266.

Tectonic, Climatic, and Biotic Modulation of Lacustrine Ecosystems—Examples from Newark Supergroup of Eastern North America

Paul E. Olsen
Lamont-Doherty Geological Observatory
Columbia University
Palisades, New York, U.S.A.

Rift-related early Mesozoic lacustrine strata of the Newark Supergroup provide a background for exploration of general concepts of lacustrine paleoecology and stratigraphic architecture. The large-scale tripartite sequence of depositional environments (i.e., fluvial basal part, deep lacustrine middle part, and shallow lacustrine or fluvial upper part), commonly seen in lacustrine deposits, including the Newark, can be quantitatively modeled as the result of relatively simple interaction between basin filling and subsidence. Whereas tectonic processes produced the rifts and the maximum depths of the lakes they contain, high-frequency fluctuations in the depths of rift lakes are largely controlled by climate. Milankovitch-type climatic cycles caused by variations in the Earth's axis and orbit produce lake-level cycles with periods of 21, 41, 100, and 400 k.y. The magnitude and mode of these lake-level changes are governed by position within the climate system and by orography.

Three major classes of lacustrine facies complexes are recognized in the Newark Supergroup. These are, in order of increasing overall dryness, the Richmond, Newark, and Fundy types. Each is characterized by different suites of highstand and lowstand deposits, different sedimentary cycle types, and different amounts of organic carbon-rich rocks. Organic-carbon content of the strata is largely a function of ecosystem efficiency, which, in turn, responds to lake depth. Finally, the long-term trends in evolution of bioturbators must be taken into account because they affect not only the carbon cycle and oxygen state within lakes but also our ability to interpret the metabolic state of ancient lake systems.

INTRODUCTION

Triassic and Lower Jurassic lacustrine rocks of the Newark Supergroup of eastern North America (Figure 1) were deposited in a long chain of rift valleys for about 30 m.y. prior to the breakup of Pangea. They comprise one of the world's largest examples of fossil lacustrine systems and provide a rich environment for examining macroscopic controls on lacustrine ecosystem development and evolution. This paper outlines the major themes of Newark Supergroup lacustrine ecosystems and their relation to lake systems in general.

GEOLOGICAL CONTEXT

Newark Supergroup deposits (Olsen, 1978; Froelich and Olsen, 1984) occupy a series of half-grabens exposed along the eastern Appalachian orogene from Nova Scotia to South Carolina. In cross section each half-graben is bounded on one side by a series of major basinward-dipping faults (the border fault system) and on the other by a gently dipping, more diffusely deformed floor of hanging-wall basement rocks with some synthetic as well as antithetic faults (Figure 2). In longitudinal section the basins comprise gentle synforms. Basins tend to be linked together into

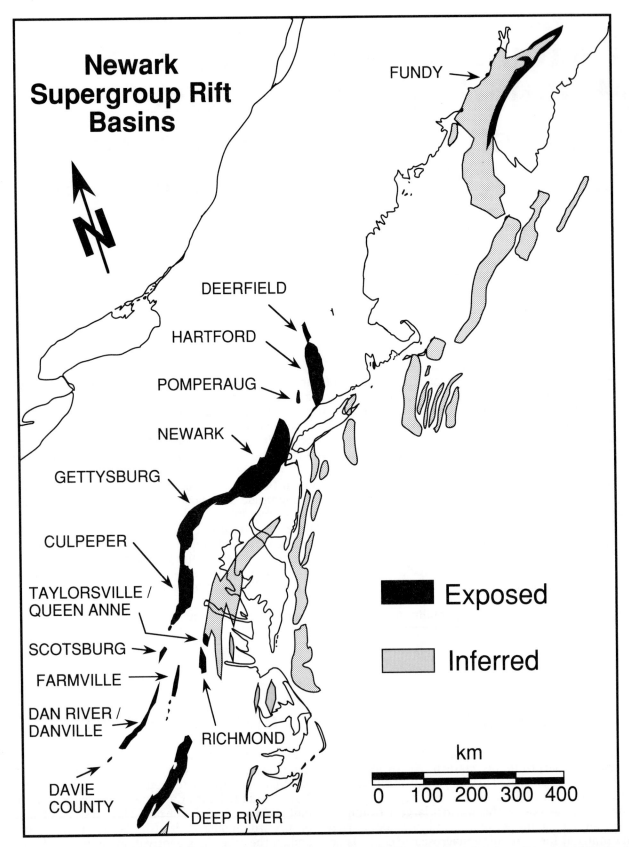

Figure 1. The Newark Supergroup of eastern North America. Adapted from Schlische and Olsen (1990). Note that the Fundy, Newark, Gettysburg, Culpeper, Dan River/Danville, Davie County, Farmville, Richmond, and Taylorsville/Queen Anne basins all have border faults on the west sides of the basins, and the Deerfield, Hartford, Pomperaug, and Deep River basins have border faults on the east sides of the basins.

Olsen

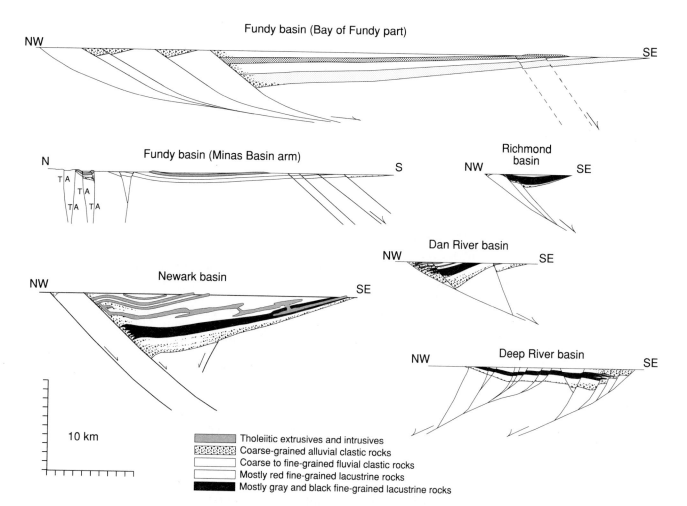

Figure 2. Cross sections of selected Newark Super-group basins (see Figure 1 for locations). Note half-graben shape of all basins except the Minas basin arm of the Fundy basin, which lies along the largely strike-slip part of the basin where it trends east-west. Cross sections adapted from the following: Fundy basin (Olsen and Schlische, 1990); Newark basin (Schlische and Olsen, 1990); Dan River and Deep River basins (Olsen et al., in press); Richmond basin, original.

basin-subbasin complexes with their border fault systems all on the same side (e.g., the Newark-Gettysburg-Culpeper system) (Figure 1). These fault systems appear to be discrete, reactivated Paleozoic compressional faults (Ratcliffe and Burton, 1985; Ratcliffe et al., 1986). Thus, Newark Supergroup rifts differ substantially from east African rifts, whose subbasins tend to link with opposing border-fault systems that do not closely follow preexisting structures (Reynolds and Schlische, 1989).

Strata within the half-graben dip toward the border fault system and generally thicken toward it. Available and proprietary seismic profiles and outcrop data (Anderson et al., 1983; Olsen et al., 1989; Schlische and Olsen, 1990) suggest onlap of successively younger strata onto hanging-wall basement both transversely and longitudinally (Figure 2). In addition, most basins are cut by major internal faults, and transverse folds are present with their greatest amplitudes expressed in the hanging walls adjacent to most major faults (Schlische and Olsen, 1988). At least some internal faulting and folding appears to be synchronous with deposition (Schlische and Olsen, 1988; Olsen et al., 1989). All the exposed basins are deeply eroded, and major sections of basin fill have been removed since the Middle Jurassic.

STRATIGRAPHY AND MAJOR DEPOSITIONAL PATTERNS

Newark Supergroup basins show two fundamental depositional patterns (Schlische and Olsen, 1990). The first is fluvial, formed in a hydrologically open basin and characterized by (1) basinwide channel systems; (2) large- to small-scale lenticular bedding; (3) conglomeratic intervals that can span the basin; (4) paleocurrent patterns that are often axial or that are dominated by one direction across the basin, indicating through-going drainage; and (5) absence of evidence for large lakes, although some pond or paludal deposits have been found. The second pattern is lacustrine and fits accepted criteria for a hydrologically closed-basin model. According to Smoot (1985), these criteria include (1) systematic increase in grain size toward all basin boundaries, (2) paleocurrent patterns away from the basin's borders, (3) local provenance of coarse-grained strata near basin margins, (4) presence of evaporites or evaporite crystal casts, and (5) cyclicity of fine-grained sedimentary rocks in the central basin. Cyclical lacustrine beds have considerable lateral continuity (Olsen, 1988), and basin-marginal areas can be dominated by deposits of lake-margin fluvial and deltaic sequences and alluvial fans. Within a basin these two generalized depositional systems have considerable persistence through time and are the largest scale stratigraphic elements (Figure 3).

Given these two depositional modes, Newark Supergroup basins tend to display a gross tripartite stratigraphy (Figure 3), consisting of a basal fluvial interval overlain by a deeper water lacustrine interval in which inferred lake depth generally decreases upsection, which in turn is overlain by a shallower water lacustrine to fluvial interval. This is a common pattern in many nonmarine basins (Lambiase, this volume). Southern Newark Supergroup basins (Deep River, Dan River, Farmville, Richmond, and Taylorsville; Figures 1, 3) contain only Triassic deposits. The northern Newark basins (Culpeper, Gettysburg, Newark, Pomperaug, Hartford, Deerfield, and Fundy; Figures 1, 3) also contain Lower Jurassic sequences that show an abrupt return to deeper water lake environments followed by a slow shoaling upward. These Lower Jurassic sequences are interbedded with thick, tholeiitic lava flows. Lambiase (this volume) refers to the northern Newark basins as dual-cycle basins.

BASIN EVOLUTION

Lakes exist for two reasons—(1) a deficit between the sediment supply to a basin and the volume created by subsidence, leading to a basin with a perched outlet; and (2) a surplus in supply of water to the basin over losses through evaporation and outflow. The balance among supplied sediment, basin growth, and outlet erosion determines the maximum possible depth and area of the lake basin (Figure 4). According to Schlische and Olsen's (1990) mass-balance models, which are based on uniform rift-basin subsidence and sediment supply with no outlet erosion, if the tectonic subsidence rate is slow enough or the rate of sediment supply is high enough, an extensional basin initially will fill with fluvial sediment, and excess sediment and water will leave the basin. As the basin continues to subside, the same sediment volume will spread over an increasingly larger surface area. At some

point the volume of sediment will just fill the basin; thereafter, a sediment-supply deficit occurs, and a lake basin forms. Sedimentation rate then will show a hyperbolic decrease through time, and the depth of the lake that could fill the basin (maximum possible lake depth) correspondingly will increase (Figure 4). After a brief period of exponential increase in maximum possible lake depth, the realized lake depth will slowly and hyperbolically decrease through time because the volume of water entering the basin is finite and reaches equilibrium with evaporation and seepage. If sediment supply is relatively low or subsidence rate relatively high, the basin will begin as lacustrine and remain so, with an ever increasing disparity between outlet position and sediment surface. Under conditions of uniform subsidence and sediment supply, this process is sufficient to produce the tripartite divisions of the Newark Supergroup and other extensional basins.

The aforementioned internal onlap pattern seen in Newark Supergroup basins, as well as other extensional basins (see compendium in Schlische and Olsen, 1990), shows that depositional surface area did increase through time, even during times of fluvial (open basin) deposition. This growth suggests progressive collapse of the hanging wall and migration of the hanging-wall hinge line away from the hanging wall cutoff. According to widely used volume- or area-balancing models of extension over a detachment (Gibbs, 1983; Groshong, 1989), this increasing area of subsidence should be balanced by decreasing rate of subsidence over the basin as a whole to conserve volume. Under these assumptions and uniform extension, subsidence rate would decrease in time, and Schlische and Olsen's basin-filling model could not produce the change from fluvial (open) to lacustrine (closed basin) conditions, although a decease in sedimentation rate still would be observed. Because this basin-filling model is based on the simplest geometrical assumptions, either the assumptions of the balancing models are incorrect, or extension rates changed during development of the basin. If extension rates increased early in basin history and volume-balancing models are correct, then the subsidence rate in the basin could have been approximately constant or could even have increased through time; thus, the change from fluvial to lacustrine deposition is still predicted by Schlische and Olsen's basin-filling model.

An abrupt increase in subsidence rate should increase basin asymmetry and temporarily decrease depositional and lake surface areas, thus increasing sedimentation rate and maximum lake depth (Schlische and Olsen, 1990). Accordingly, an earliest Jurassic increase in subsidence rate due to an increase in extension rate could have led to the stratigraphic patterns seen in northern Newark basins, as well as the extrusion of voluminous lava flows (Schlische and Olsen,1990; Olsen et al., 1989).

The along-strike linkage of several Newark Supergroup basins suggests the possibility of filling relays along a chain of subbasins, as described by

Olsen

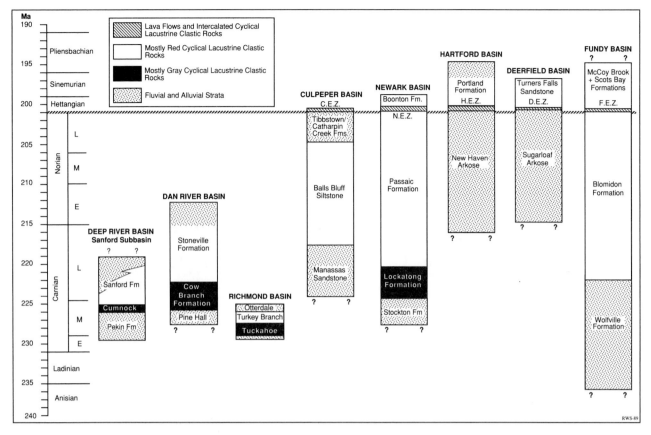

Figure 3. Correlation and facies types in Newark Supergroup basins (from Olsen et al., 1989). Abbreviations: CEZ, Culpeper basin extrusive zone; NEZ, Newark basin extrusive zone; HEZ, Hartford basin extrusive zone; DEZ, Deerfield basin extrusive zone; FEZ, Fundy basin extrusive zone.

Lambiase (this volume). Such a relationship is, in fact, supported by differences in timing between the onset and cessation of lacustrine deposition in the subbasins of the Deep River basin (described in Olsen et al., in press) and the Newark-Gettysburg-Culpeper basin system (Olsen et al., 1989). These possibilities have yet to be explored in detail, however.

CONTROL OF WATER IN THE LAKE

Water is needed in a basin to produce a lake; thus, whether or not a lake exists depends as much on climate and groundwater as on basin morphology. Recent studies of Quaternary climate and global climate models show that climate is not static, even in the tropics (Hays et al., 1976; Rossignol-Strick, 1983; Street-Perrott, 1986; Kutzbach, 1987). The history of a single lake is thus decoupled, in part, from the history of its basin.

Three broad types of lacustrine facies complexes can be recognized in the Newark Supergroup, each of which is characterized by magnitude and frequency of lake-level change. These I call the Newark-type, Richmond-type and Fundy-type lacustrine facies complexes. The most common is the Newark-type (Figure 5), in which water inflow was closely balanced by outflow. This type of facies complex occurs in the Dan River/Danville, Culpeper, Gettysburg, Newark, Pomperaug, Hartford, and Deerfield basins. Changes in precipitation resulted in dramatic changes in lake level from perhaps 200 m or more in depth to complete exposure, producing repetitive sequences of sedimentary cycles called Van Houten cycles (Figure 6) (Olsen, 1984b, 1986). These cycles range from about 1.5 to 35 m thick, and Milankovitch-type climate changes appear to have been their cause. Based on Fourier analyses of dozens of sections from several basins (one example in Figure 7), Van Houten cycles appear to have been under the control of the precession cycle of about 21 k.y., and clusters of Van Houten cycles make up larger cycles under the control of the eccentricity cycles of about 100, 400, and 2000 k.y. (Olsen, 1986; Olsen et al., 1989). A small effect of the obliquity cycle (41 k.y.) also is apparent.

Drastic changes in lake level apparently inhibited the buildup of high-relief sedimentary features (other than alluvial fans) within the basin both by wave action during transgression and regression and by the brief time the water was deep. Consequently, lacustrine strata are characterized by extreme lateral

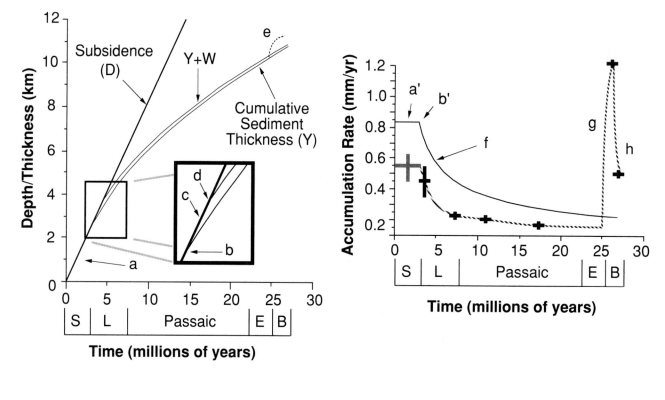

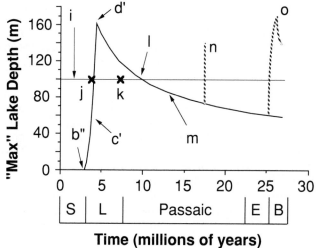

Figure 4. Predictions of the Schlische and Olsen (1990) model of stratigraphic development, assuming constant sediment input and constant subsidence (solid lines), compared to data from the Newark basin (dotted lines and crosses). Model is based on full-graben geometry but is parameterized for Newark basin. Y + W curve represents lake height above basin floor. Abbreviations: a, filling equals subsidence, resulting in fluvial sedimentation, with accumulation rate (a′) equal to subsidence rate; b, b′, and b″, onset of lacustrine deposition and sedimentologic closure; c and c′, rapid increase in lake depth as depositional surface subsides below hydrologic outlet; d and d′, deepest model-predicted lake and onset of hydrologic closure; e, major deviation from cumulative thickness curve during Early Jurassic; f, model-predicted hyperbolic decrease in accumulation rate; solid crosses represent data on accumulation rates from cyclical lacustrine strata from scattered outcrops; shaded cross represents estimated accumulation rate in fluvial Stockton Formation; g, anomalously high accumulation rate in extrusive zone, a deviation from model predictions; h, decreasing accumulation rate in Boonton Formation; i, 100-m minimum depth for formation of microlaminated sediment; j, actual first occurrence of microlaminated sediment; k, actual last occurrence of microlaminated sediment in lower Passaic Formation; l, predicted last occurrence of microlaminated sediment; m, slow decrease in lake depth; n, anomalous "superwet" climatic anomaly; o, anomalously deep Jurassic lakes. S, Stockton Formation; L, Lockatong Formation; E, Newark basin extrusive zone; B, Boonton Formation.

Olsen

RICHMOND-TYPE LACUSTRINE FACIES COMPLEX

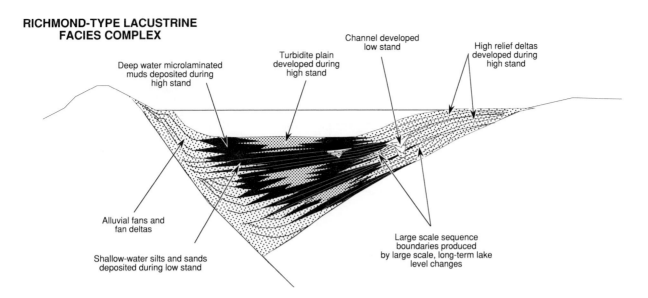

Deep water microlaminated muds deposited during high stand

Turbidite plain developed during high stand

Channel developed low stand

High relief deltas developed during high stand

Alluvial fans and fan deltas

Shallow-water silts and sands deposited during low stand

Large scale sequence boundaries produced by large scale, long-term lake level changes

NEWARK-TYPE LACUSTRINE FACIES COMPLEX

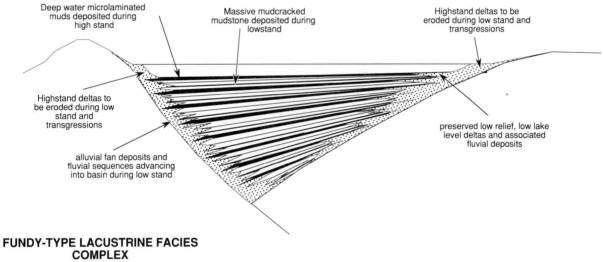

Deep water microlaminated muds deposited during high stand

Massive mudcracked mudstone deposited during lowstand

Highstand deltas to be eroded during low stand and transgressions

Highstand deltas to be eroded during low stand and transgressions

alluvial fan deposits and fluvial sequences advancing into basin during low stand

preserved low relief, low lake level deltas and associated fluvial deposits

FUNDY-TYPE LACUSTRINE FACIES COMPLEX

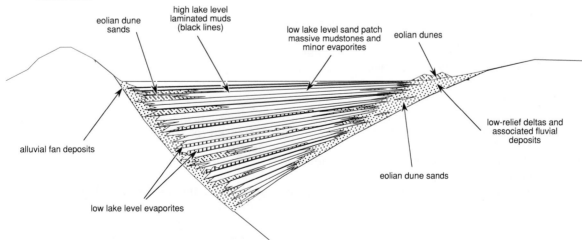

eolian dune sands

high lake level laminated muds (black lines)

low lake level sand patch massive mudstones and minor evaporites

eolian dunes

alluvial fan deposits

low lake level evaporites

eolian dune sands

low-relief deltas and associated fluvial deposits

Figure 5. Idealized types of lacustrine facies complexes in Newark Supergroup. Note that no historical trend through the basin history is shown.

Lacustrine Basin Exploration: Case Studies and Modern Analogs

215

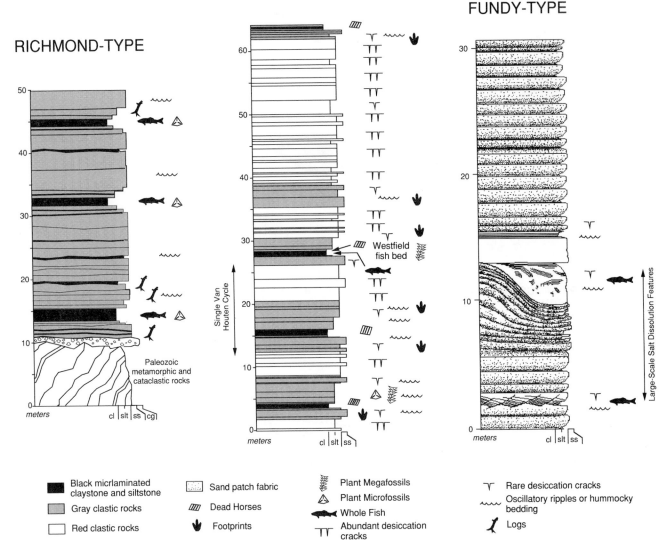

Figure 6. Comparison of sequences representative of the Richmond, Newark, and Fundy types of lacustrine facies complexes, illustrating repetitive sedimentary cycles (Van Houten cycles). Measured sections are from the following locations: Richmond type, Vinita Beds of Richmond basin (early Carnian), exposed at Boscobel Quarry, Manakin, Virginia; Newark type, middle East Berlin Formation (Hettangian) of Hartford basin, intersection of Routes 15 and 72, East Berlin, Connecticut (adapted from Olsen et al., 1989); Fundy type, lower Blomidon Formation, cliff outcrop at Blomidon, Nova Scotia (adapted from Olsen et al., 1989).

continuity and by a tendency for coarse-grained sediment to be absent from deeper water facies and restricted to basin margins (Olsen, 1985; Olsen et al., 1989). Large-scale sequence boundaries and large deltas apparently are absent from the main basin fill. In addition, although many Van Houten cycles contain thick, organic-rich deep-water units, the ratio of organic-rich to organic-poor units is very low. No close modern analogs to the exaggerated lake-level fluctuations seen in the Newark-type lacustrine facies complex apparently are known. However, apart from dramatic structural differences, the sediments of Lake Turkana (e.g., Cohen, 1989), among all east

African rift lakes, appear to be most similar to the Newark-type complex during its intermediate water depths.

The Richmond-type lacustrine facies complex is less common and is known for certain only in the Richmond and Taylorsville basins (Figures 5, 6). To date, this lacustrine facies complex is relatively poorly known, and its description is somewhat speculative. These basins contain significant coals and highly bioturbated shallow-water and fluvial sequences, suggesting more persistently humid conditions. The basins also are characterized by considerable thickness (>50 m) of microlaminated

Olsen

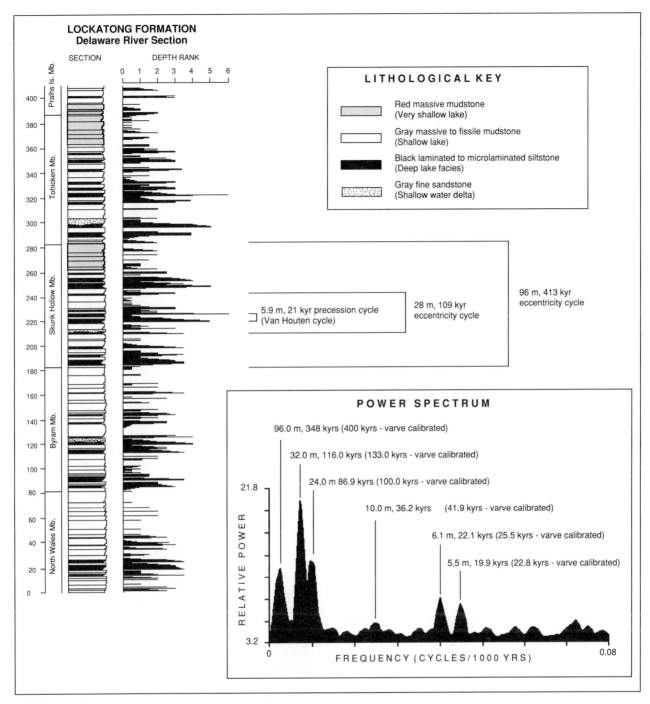

LITHOLOGICAL KEY

Red massive mudstone
(Very shallow lake)

Gray massive to fissile mudstone
(Shallow lake)

Black laminated to microlaminated siltstone
(Deep lake facies)

Gray fine sandstone
(Shallow water delta)

5.9 m, 21 kyr precession cycle
(Van Houten cycle)

28 m, 109 kyr
eccentricity cycle

96 m, 413 kyr
eccentricity cycle

POWER SPECTRUM

96.0 m, 348 kyrs (400 kyrs - varve calibrated)

32.0 m, 116.0 kyrs (133.0 kyrs - varve calibrated)

24,0 m 86.9 kyrs (100.0 kyrs - varve calibrated)

10.0 m, 36.2 kyrs   (41.9 kyrs - varve calibrated)

6.1 m, 22.1 kyrs (25.5 kyrs - varve calibrated)

5.5 m, 19.9 kyrs (22.8 kyrs - varve calibrated)

RELATIVE POWER

FREQUENCY (CYCLES/1000 YRS)

Figure 7. Section of Newark-type lacustrine facies complex in middle Lockatong Formation, Newark basin, and Fourier analysis of the Delaware river section, using the method described by Olsen (1986). Note that depth rank is a ranking of sedimentary fabrics in order of increasing inferred water depth.

siltstone interbedded with sandstone with no evidence of subaerial exposure (Figure 6). Relative to Newark-type basins, water inflow most often was greater than evaporation, and complete desiccation was rare. Consequently, the lake may have remained deep long enough that, on average, high-relief sedimentary features could have been built. These should include large prograding deltas, fan deltas, submarine channels, and turbidite fans. During the phase of the basin's history when deep lakes were possible because of the sedimentation-subsidence deficit, abrupt lateral changes in grain size of contemporaneous strata should be present, and large amounts of sand could be deposited virtually anywhere in the basin. When lake levels were low during longer-term climate cycles, such as the driest phases of 400-k.y. eccentricity cycles or longer cycles, high-relief features could be deeply eroded to produce large-scale unconformities and perhaps subaerial and subaqueous canyons. Large-scale depositional sequences bounded by sequence boundaries should characterize a major part of the basin fill. The ratio of organic-rich to organic-poor units would be high at least locally. Lakes Tanganyika and Malawi (not the specific structure) of the east African rift system appear to be reasonable modern analogs (Cohen, 1989) of the deep-lake phases of the Richmond-type lacustrine facies complex.

The Deep River basin of North Carolina and the Farmville and associated basins in Virginia appear to be of the Richmond type, except that microlaminated strata apparently are absent, suggesting that although desiccation was rare, the lakes never became as deep as in the Richmond or Taylorsville basins, perhaps because of lower outlet elevations (Gore, 1989; Olsen et al., 1989).

Fundy-type lacustrine facies complexes are characterized by a cyclicity consisting mostly of what are termed sand-patch cycles (Figures 5, 6) (Smoot and Olsen, 1985, 1988). These represent alternations between shallow perennial lakes and playas with well developed efflorescent salt crusts, perhaps representing the 100-k.y. eccentricity cycle (Olsen et al., 1989). These cycles are very thin (~1.5 m where exposed) compared with lake-level cycles in the more southern basins. Associated deposits in the Fundy basin include abundant gypsum nodules, salt-collapse structures (Olsen et al., 1989), and eolian dunes (Hubert and Mertz, 1984). Water inflow rarely matched evaporation, and deep-lake deposits are exceedingly uncommon. Consequently, as in Newark-type sequences, depositional relief within the basin would be very low, and lateral continuity of units would be extensive.. Sand should be uncommon in the basin center except for that of eolian origin in the sand-patch fabric and wave-reworked material from sand-patch cycles and eolian dunes. Because of the extreme rarity of deep lakes, virtually no organic carbon-rich intervals should be present. Saline Valley, California (Smoot and Castens-Seidel, 1982; Smoot and Olsen, 1988) seems a reasonable modern analog for the Fundy-type lacustrine facies complex.

The "wettest" looking basins with Richmond-type facies complexes are found in the southern Newark Supergroup, whereas the "driest" looking basins with Fundy-type facies complexes occur in the northernmost well-exposed basin (Figure 1). This apparent gradient could reflect a regional climatic gradient, altitudinal differences, or orographic effects. The fact that a Fundy-type facies complex occurs in the Argana basin of Morocco (Smoot and Olsen, 1988), which was at the same paleolatitude as the Newark basin, suggests that the apparent gradient has at least a strong altitudinal or orographic component, as suggested by Manspeizer (1982). It is critical to observe that climatic regime probably is more important to the style of lacustrine deposition in Newark basins than is tectonic environment, once the tectonic environment permitted a lake to develop.

The Newark Supergroup was located between latitudes –3°S and 8°N during the early Mesozoic, and qualitative as well as numerical models for this period suggest a strong monsoonal climate for the region (Robinson, 1973; Manspeizer, 1982; Kutzbach and Gallimore, 1989; Chandler, in preparation). The supercontinent of Pangea would produce supermonsoons perhaps accounting for the rather extreme expression of cyclicity (Figure 6) in Newark-type facies complexes. Whereas maximum lake depths were controlled by subsidence-sedimentation balance, their realized depths were controlled by climate.

The growth of an extensional basin through time also has a major direct effect on lake depths (Figure 4) (Schlische and Olsen, 1990). As described above, a given volume of water supplied by precipitation and balanced by evaporation might fill a deep basin when its area is small, but would be incapable of filling the basin as it widened. That is why I expect lacustrine extensional basins to show a slow decrease in apparent lake depth (both for humid and arid phases of climate cycles) after some maximum is reached early in the basin's history, even though the elevational difference between the outlet position and the sediment surface still might be increasing. Evidence for the continued presence of a high outlet are the rare deep-lake episodes in younger Triassic (Carnian) sections of the Newark Supergroup (Figure 4). These represent rare "superwet" events, which supplied much larger volumes of water than during "normal wet" phases of climate cycles. These events could produce a deep lake only because of the presence of a high outlet.

High-frequency lake-level oscillations are not at all unique to the Newark Supergroup or the early Mesozoic. Perhaps the most striking example is the Devonian Caithness Flagstones of the Orcadian basin in Scotland (Figure 8), which has cycles similar to the Van Houten cycles of Newark-type complexes (Donovan et al., 1974; Donovan, 1980; Janaway and Parnell, 1989). It has long been known that large lake basins of the Basin and Range province of the western United States have experienced large-scale lake-level cycles controlled by cyclical climate (Gilbert, 1890), and the great lakes of east Africa apparently have experienced even larger lake-level oscillations (Johnson et al., 1987; Scholz and Rosendahl, 1988), although not at such a high frequency.

LAKE DEPTH CONTROL OF MATERIAL DISTRIBUTION

In large tropical lakes, depth is the main control on material distribution. This is because the main sources of energy for vertical and horizontal material transport are turbulence and currents driven by the wind, the work of which is transmitted through the surface of the lake. Depth of wave base, one measure

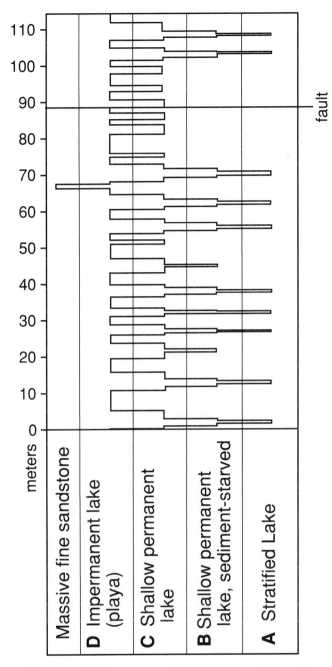

meters

D Impermanent lake (playa)
C Shallow permanent lake
B Shallow permanent lake, sediment-starved
A Stratified Lake

Massive fine sandstone

Figure 8. Sedimentary cycles in Caithness Flagstones of the Orcadian basin, Scotland (adapted from Donovan, 1980). Lacustrine facies types ranked in order of increasing inferred water depth in a manner analogous to Olsen (1986).

of this work, is a function of the distance over which the lake is exposed to wind (fetch) and the speed and duration of the wind itself. All other things equal, a lake covering a larger surface area has a deeper wave base than a lake covering a smaller area. For two lakes of equal area, the deeper lake will have a smaller proportion of its water column affected by wave mixing than the shallower one.

Manspeizer and Olsen (1981) and Olsen (1984a) have developed a relationship among fetch, wind speed, and wave base (Figure 9) using the semiempirical equations of Bretscheider (1952) and Smith and Sinclair (1972). Depth of wave base calculated for wind speeds of 10–40 m/sec correlates well ($r^2 = 0.95$ for a 30-m/sec wind) with depth of the chemocline in nonsaline, tropical stratified lakes (Figures 9 and 10). This is because the chemocline is maintained by an active, turbulent, wind-mixed surface layer that is dependent on area in a way similar to wave base. Observations of the chemocline in stratified lakes such as Tanganyika and Malawi show it deepening during windy periods and shallowing during calm times (Beadle, 1974). The reason these lakes become chemically stratified is that, below the turbulent mixed layer, oxygen is transported from the surface too slowly to allow aerobic breakdown of all organic matter produced within and washed into the lake. Thus, the deeper areas of the lake become anaerobic. Stratification in turbulence exists because the lake is deep compared to its possible fetch. That chemical stratification follows the boundaries of turbulent stratification is a consequence mostly of high organic productivity.

Lakes of similar dimensions as tropical stratified lakes but with low organic production rates do not become chemically stratified, even though turbulent stratification is present. Lake Baikal has nearly the same dimensions as Lake Tanganyika and definitely exhibits turbulent stratification (Carmack et al., 1989; Figure 10). However, primary production values are relatively low, 25–100 g C/m²/yr (Likens, 1975; Weiss et al., 1989), compared with Lake Tanganyika's 328 g C/m²/yr (Hecky and Fee, 1981). Hence, although turbulent stratification exists in Baikal, primary productivity is insufficient to develop perennial anoxia in the water mass below the turbulent layer. The same relationship between possible fetch and wave base suggests that lakes shallower than their predicted wave bases should not become perennially stratified. This seems to be borne out by observation (Figure 10). It must be noted, however, that this is mostly an empirical relationship, not a hydrodynamic one, and the exact relationship among chemocline position, wind, and lake dimension has yet to determined. Gorham and Boyce (1986) reported a similar empirical relationship between thermocline depth and fetch in temperate lakes, although thermocline depth is generally much shallower.

Nonetheless, the empirical relationships in Figures 9 and 10 permit some understanding of how deep some Newark Supergroup lakes were during their highstands. For example, individual microlaminated shales within single Van Houten cycles of the Lockatong Formation have been traced about 180 km across the Newark basin (Olsen, 1984a, 1988). Within these strata are submillimeter-thick laminae, traceable over large areas and often containing articulated fish and reptile fossils. Pinch-and-swell laminae are rare within the microlaminated inter-

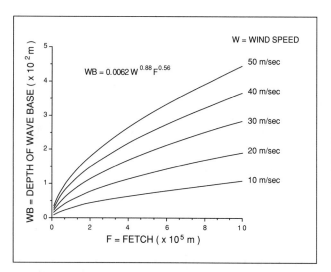

Figure 9. Calculated relationships between maximum potential fetch of a lake and predicted wave base for winds of various speeds, based on equations in Olsen (1984a) and Manspeizer and Olsen (1981).

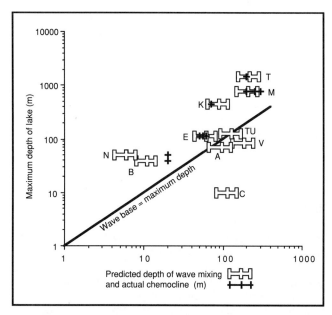

Figure 10. Relationship between predicted depth of wave base and actual depth of chemocline for several east African lakes. Abbreviations: A, Lake Albert; B, Lake Bunyuni; C, Lake Chad; E, Lake Edward; K, Lake Kivu; M, Lake Malawi; N, Lake Nhugute; T, Lake Tanganyika; TU, Lake Turkana; V, Lake Victoria. Note that for lakes in which maximum depths are less than predicted depths of wave mixing, no chemocline exists, and oxygenated waters reach the lake bottoms. Note also that Lake Biakal would plot almost coincidentally with Lake Tanganyika (T) in predicted depth of wave mixing, lake depth, and depth to the base of the measured turbulent layer; it has no chemocline, and oxygen reaches the bottom. Calculated depth of wave mixing is based on relationships in Figure 9 and data in Beadle (1974).

vals. If we reasonably assume that these microlaminated strata were not exposed to wave base during their deposition, then the longest dimension of the microlaminated unit can be used to calculate a minimum lake depth using the unit's length as the maximum potential fetch (Manspeizer and Olsen, 1981; Olsen, 1984a). For the Lockatong Formation this method yields depths of about 70, 100, and 130 m for wind speeds of 20, 30, and 40 m/sec, respectively (Olsen 1984a). The absence of bioturbation in these layers together with the preservation of whole fish and reptiles suggest an anoxic bottom. Therefore, these estimates represent minimum depths to the chemocline, and the lakes could have been much deeper.

ECOSYSTEM PRODUCTION, CONSUMPTION, AND EFFICIENCY

The main sources of energy for lacustrine ecosystems are primary production and inflow of allochthonous organic material. The balance between production and consumption of this organic material is mediated by the relationship between the lake's upper mixed layer and its depth.

The amount of organic material that accumulates in sediment is a function of ecosystem efficiency (ratio of ecosystem respiration to production $\times$ 100), which is highest in shallow water and lowest in deep water. Most aquatic ecosystems are extraordinarily efficient (approaching 100%), and even those that do produce organic-rich sediment only lose a small fraction to the sediments. The Black Sea has an ecosystem efficiency of about 96%, losing only about 4% of primary production to the sediment (Degens and Ross, 1974).

In the turbulent layer of a lake the residence time of an organic particle in the water column is relatively long, and the diffusion rate of oxygen is high. Numerous detritovores further mill down large particles, and bioturbators recycle sedimented material. Below the turbulent layer, in tropical lakes, residence time is relatively short, and the diffusion rate of oxygen is much less. Bacterial activity tends to consume available oxygen at a rate greater than can be supplied, and the water column becomes anoxic. Although researchers still debate whether aerobic or anaerobic organisms are more efficient consumers, the key to the slower consumption rates in the anaerobic zone is probably the much reduced range of organisms there and shorter residence time of organic particles in the water column. Consequently, the anoxic water mass (hypolimnion) is less efficient at consuming organic matter than the upper mixed layer.

The depths of lakes thus have a powerful control on the efficiency of their ecosystems. In the Newark Supergroup black microlaminated units were produced only in the deepest, probably perennially chemically stratified lakes, with depressed ecosystem efficiencies and little or no bioturbation. In contrast, shallow lakes of high ecosystem efficiency produced mostly red or gray massive mudstone with little or no preserved organic material, even though they probably had as high or higher primary productivity. The fluctuations in organic-carbon content characterizing Van Houten cycles reflect changing ecosystem efficiency much more than changing productivity. The same control probably is responsible for the remarkably different sedimentary carbon contents in Lakes Turkana and Tanganyika, both of which are fairly productive (Hecky and Fee, 1981; Cohen, 1984). However, Turkana has low carbon values associated with a shallow lake (maximum depth 125 m), while Tanganyika has high carbon values and greater depth (1400 m).

HISTORICAL TRENDS

Both production and consumption rates depend not only on proximal causes, such as relative depth of the water body and nutrients levels, but also on ultimate causes such as long-term evolution of new consumer groups with new innovations.

The effects of long-term historical trends are most apparent in the bioturbation of lacustrine sediments, just as in marine sediments (Thayer, 1979). Freshwater systems have very low diversities of bioturbators compared to marine systems (Lopez, 1988), and many of the groups responsible for lacustrine bioturbation appear to be of relatively recent origin. Apparently through the Phanerozoic there is a trend of increasing bioturbation of sediments deposited under low-oxygen conditions following the evolution of bioturbators tolerant of low-oxygen environments (Figure 11).

In modern lakes the important macrobioturbators in low-oxygen conditions are tubificid worms, ostracodes, chironomid and chaoborid fly larvae, and sphaerid and unionid clams (Lopez, 1988) (Figure 11). However, no definite tubificid worms nor any definitive oligochaetes are known in the fossil record (Conway et al., 1982), except for some possible Carboniferous tubificid fossils (Gray, 1988). Modern darwinulid ostracodes, which can tolerate brief periods of low or no oxygen, are known from putatively freshwater deposits as old as Ordovician (Gray, 1988). The oldest known true flies are Triassic in age (Olsen et al., 1978; Wootton, 1988), and although they belong to the larger group containing chaoborid and chironomid flies (Nematocerata), chironomids are not known before the Jurassic (Kalugina, 1980; Gray, 1988), and chaoborids are not

known before the Cretaceous (Gray, 1988). Clams of the order Unionoida are known from the Devonian (Gray, 1988), often from organic-rich rocks. However, members of the heterodont family Sphaeridae, including the important anoxia-resistant *Pisidium* (Gray, 1988), are not known until Late Jurassic. Of course, clams leave obvious fossils, and their presence usually is easily recognized.

Based on first appearances of groups, entire suites of critical modern bioturbators are absent from progressively more ancient lakes. We can therefore expect that a wider range of ancient, low-oxygen lacustrine environments should have excluded more bioturbators than at present. This means that many ancient lakes with only seasonally anoxic or low-oxygen bottom waters could have accumulated sediment devoid of bioturbation. In addition, because bioturbators under low oxygen levels are ventilators of the sediment and consumers of organic material, their absence would therefore reduce the lake's ecosystem efficiency and encourage more complete bottom anoxia under lower levels of primary production than obtained today. A higher proportion of older than younger lacustrine strata should therefore be microlaminated.

An extreme example of this trend can be seen by comparing a hypothetical early Proterozoic lake with a modern one. The Proterozoic lake produced microlaminated sediment in all situations of slow, fine-grained, suspension-dominated deposition below wave base, regardless of bottom-water oxygen levels or levels of primary production. In contrast, a modern lake with moderately high productivity has bioturbators in all environments except those perennially devoid of oxygen. Lakes of intermediate age should be bioturbated to intermediate extents. Therefore, the absence of macroscopic bioturbation in the deepest water Newark Supergroup lakes cannot be assumed necessarily to reflect perennial anoxic bottom waters. Simple analogy to modern lakes is inappropriate because the biological context has changed.

Newark Supergroup lakes probably experienced some degree of bioturbation in low-oxygen environments by ostracodes, possibly tubificid worms, and perhaps sphaerid and unionid clams. However, on the whole, a considerably wider range of oxygenation conditions probably resulted in the production of microlaminated sediment because of the absence of key bioturbators, such as chironomids. In addition, the frequency with which the lakes became chemically stratified probably was somewhat greater than we would expect from our knowledge of modern lakes.

These processes and relationships are only some of the factors important to lake system ontogeny and lacustrine ecosystem evolution. Not discussed here is *in situ* evolution of lacustrine organisms that occurs during the life cycle of an individual lake, for example, during one 21-k.y. climate cycle (McCune et al., 1984). Important successional processes occur on even faster time scales (Lopez, 1988).

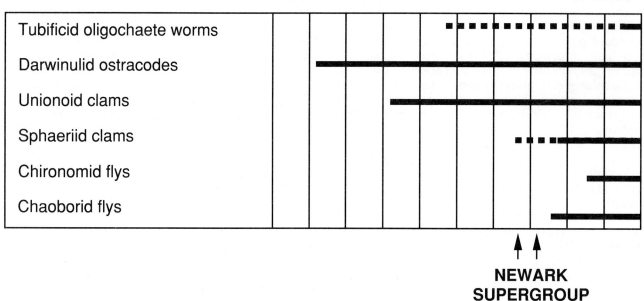

Figure 11. Times of appearances of major mud bioturbators in modern lakes. Based on data from Gray (1988) and Wootton (1988).

CONCLUSIONS

Tectonism produces a subsiding basin and controls its overall geometry, but basin filling by sediment interactively alters the geometry of the depositional surface in predictable ways. The distinctive tripartite stratigraphic pattern (fluvial, deep lake, shallow lake or fluvial) seen in extensional lacustrine basins may be due to basin filling during a long-lived pulse of extension.

Lake systems owe as much to climatic milieu as to tectonic and basin-filling processes, however. The balance between water inflow and outflow (principally by evaporation) is the critical factor governing the depths and duration of individual lakes (as opposed to the lake basin). Relatively high-frequency climatic cycles induce lake-level cycles that follow the 21-, 41-, 100-, and 400-k.y. Milankovitch cycles caused by variations in the Earth's orbit. The magnitude of these lake-level changes is governed by position within the climate system and by orography. Thus, even in a region as limited as eastern North America, one finds major differences among basins.

The frequency and magnitude of lake-level changes have major effects on the types of lacustrine sedimentary environments. Within the Newark Supergroup three types of lacustrine facies complexes are recognized (in order of increasing severity and frequency of low lake levels)—Richmond, Newark, and Fundy types.

Lake level controls ecosystem efficiency, which in turn is responsible for the loss of organic carbon-rich material to lake sediments. All other things equal, lakes that are deep compared to their areas tend to be more inefficient than others. Hence, lake-level cycles are reflected in sedimentary cycles wherein shallow-water strata are organic carbon poor, and deep-lake strata are organic carbon rich. Ultimately, control of lake metabolism is a function of the evolution of diverse organisms capable of surviving the range of lacustrine environments. The long-term trend in the degree of bioturbation of lacustrine sediments through the Phanerozoic probably reflects the proliferation of organisms capable of surviving low-oxygen environments. This long-term trend must be taken into account in paleoenvironmental interpretations of ancient lacustrine systems because it affects not only the range of sediments that escape bioturbation but also the frequency with which stratified lakes can develop.

ACKNOWLEDGMENTS

I thank Andrew Cohen, Bruce Cornet, Joseph Lambiase, Martin Link, David Reynolds, Elenora Robbins, Roy Schlische, Joseph Smoot, and Martha Withjack for many enlightening discussions about lakes, rift basins, and the Newark Supergroup. Cynthia Banach, Lloyd Burkel, Matthew Matthews, and Roy Schlische are thanked for reading the manuscript and suggesting numerous changes that substantially improved it. Support for this research has been provided by grants from the Donors of the American Chemical Society administered by the Petroleum Research Fund (PRF 19878-G2) and the National Science Foundation (BSR 87 17707), for which I am very grateful. This paper is Lamont-Doherty Geological Observatory contribution number 4568.

REFERENCES

Anderson, R. E., M. L. Zoback, and G. A. Thomson, 1983, Implications of selected subsurface data on the structural form and evolution of some basins in the northern Basin and Range province, Nevada and Utah: GSA Bulletin, v. 94, p. 1055-1072.

Beadle, L. C., 1974, The inland waters of tropical Africa: New York, Longman, 475 p.

Bretscheider, C. L., 1952, Revised wave forecasting relationships, in Proceedings, 2d Conference on Coastal Engineering: New York, Engineering Foundation, Council on Wave Research, p. 28-45.

Carmack, E. C., R. F. Weiss, and V. M. Koropalov, 1989, Physical studies of mixing processes in Lake Baikal: EOS, v. 70, p. 1136.

Chandler, M. A., (in preparation), The climate of the Early Jurassic as simulated by the GISS General Circulation Model.

Cohen, A., 1984, Effects of zoobenthic standing crop on laminae preservation in tropical lake sediment, Lake Turkana, East Africa: Journal of Paleontology, v. 58, p. 499-510.

Cohen, A., 1989, Facies relationships and sedimentation in large rift lakes and implications for hydrocarbon exploration—Examples from Lakes Turkana and Tanganyika: Palaeogeography, Palaeoclimatology, Palaeoecology, v. 70, p. 65-80.

Conway M. S., R. K. Pickerill, and T. L. Harland, 1982, A possible annelid from the Trenton Limestone (Ordovician) of Quebec, with a review of fossil oligochaetes and other annulate worms: Canadian Journal of Earth Science, v. 19, p. 2150-2157.

Degens, E. T., and D. A. Ross, 1974, The Black Sea—Geology, chemistry, and biota. AAPG Memoir 20, 633 p.

Donovan, R. N., 1980, Lacustrine cycles, fish ecology, and stratigraphic zonation in the middle Devonian of Caithness: Scottish Journal of Geology, v. 16, p. 35-50.

Donovan, R. N., R. J. Foster, and T. S. Westoll, 1974, A stratigraphical revision of the Old Red Sandstone of North-eastern Caithness: Transactions of the Royal Society of Edinborough, v. 69, p. 167-201.

Froelich, A. J. and P. E. Olsen, 1984, Newark Supergroup, a revision of the Newark Group in eastern North America: USGS Bulletin 1537-A, p. A55-A58.

Gibbs, A. D., 1983, Balanced cross-section construction from seismic sections in areas of extensional tectonics: Journal of Structural Geology, v. 5, p. 153-160.

Gilbert, G. K., 1890, Lake Bonneville: USGS Monograph 1, 438 p.

Gore, P. J. W., 1989, Toward a model for open- and closed-basin deposition in ancient lacustrine sequences: the Newark Supergroup (Triassic-Jurassic), eastern North America: Palaeogeography, Palaeoclimatology, Palaeoecology, v. 70, p. 29-51.

Gorham, E. and F. M. Boyce, 1986, [Title], in B. Henderson-Sellers, and H. R. Markland, Decaying lakes—The origins and control of cultural eutrophication: New York, John Wiley and Sons, Fig. 2.20, p. 35.

Gray, J., 1988, Evolution of the freshwater ecosystem—The fossil record: Palaeogeography, Palaeoclimatology, Palaeoecology, v. 62, p. 1-214.

Groshong, R. H., 1989, Half-graben structures—Balanced models of extensional fault-bed folds. GSA Bulletin, v. 101, p. 96-105.

Hays, J. D., J. Imbrie, and N. J. Shackleton, 1976, Variations in the earth's orbit—Pacemaker of the Ice Ages. Science, v. 194, p. 1121-1132.

Hecky, R. E. and E. J. Fee, 1981, Primary production and rates of algal growth in Lake Tanganyika: Limnology and Oceanography, v. 26, p. 532-547.

Hubert, J. H., and K. A. Mertz, 1984, Eolian sandstones in Upper Triassic-Lower Jurassic red beds of the Fundy basin, Nova Scotia: Journal of Sedimentary Petrology, v. 54, p. 798-810.

Janaway, T. M., and J. Parnell, 1989, Carbonate production within the Orcadian basin, northern Scotland—A petrographic and geochemical study: Palaeogeography, Palaeoclimatology, Palaeoecology, v. 70, p. 89-105.

Johnson, T. C., J. D. Halfman, B. R. Rosendahl, and G. S. Lister, 1987, Climatic and tectonic effects on sedimentation in a rift-valley lake—Evidence from high-resolution seismic profiles, Lake Turkana, Kenya: GSA Bulletin, v. 98, p. 439-447.

Kalugina, N. S., 1980, Chaoboridae and Chironomidae from Lower Cretaceous deposits of Manlay [in Russian], in Early Cretaceous of Lake Manlay: USSR Academy of Science, Paleontological Institute Transactions, v. 13, p. 61-64.

Kutzbach, J. E., 1987, The changing pulse of the Monsoon, in J. S. Fein, and P. L., Stephens, eds., Monsoons: New York, John Wiley & Sons, p. 247-268.

Kutzbach, J. E., and R. G. Gallimore, 1989, Pangean climates—Megamonsoons of the megacontinent: Journal of Geophysical Research, v. 94, p. 3341-3358.

Lambiase, J. J., A model for tectonic control of lacustrine stratigraphic sequences in continental rift basins, this volume.

Likens, G. E., 1975, Primary production of inland aquatic ecosystems, in H. Leith, and R. H. Whittaker, eds., Primary productivity of the biosphere: New York, Springer-Verlag, p. 185-202.

Lopez, G. R., 1988, Comparative ecology of the macrofauna of freshwater and marine muds: Limnology and Oceanography, v. 33, p. 946-962.

Manspeizer, W., 1982, Triassic-Liassic basins and climate of the Atlantic passive margins: Geologische Rundshau, v. 73, p. 895-917.

Manspeizer, W., and P. E. Olsen, 1981, Rift basins of the passive margin: tectonics, organic-rich lacustrine sediments, basin analysis, in W. Hobbs, III, ed., Field guide to the geology of the Paleozoic, Mesozoic, and Tertiary rocks of New Jersey and the central Hudson Valley, New York: Petroleum Exploration Society of New York, p. 25-105.

McCune, A. R., K. S. Thomson, and P. E. Olsen, 1984, Semionotid fishes from the Mesozoic Great Lakes of North America, in A. A. Echelle, and I. Kornfield, eds., Evolution of species flocks: University of Orono Press, p. 27-44.

Olsen, P. E., 1978, On the use of the term Newark for Triassic and Early Jurassic rocks of eastern North America: Newsletters on Stratigraphy, v. 7, p. 90-95.

Olsen, P. E., 1984a, Comparative paleolimnology of the Newark Supergroup—A study of ecosystem evolution: Ph.D. Thesis, Yale University, 724 p.

Olsen, P. E., 1984b, Periodicity of lake-level cycles in the Late Triassic Lockatong Formation of the Newark Basin (Newark Supergroup, New Jersey and Pennsylvania), in A. Berger, J. Imbrie, J. Hays, G. Kukla, B. Saltzman, eds., Milankovitch and climate: NATO Symposium, D. Reidel Publishing Co., pt. 1, p. 129-146.

Olsen, P. E., 1985, Constraints on the formation of lacustrine microlaminated sediments, in G. R., Robinson, Jr., and A. J. Froelich, eds., Proceedings of the second U.S. Geological Survey Workshop on the Early Mesozoic basins of the eastern United States: USGS Circular 946, p. 34-35.

Olsen, P. E., 1986, A 40-million-year lake record of Early Mesozoic climatic forcing. Science, v. 234, p. 842-848.

Olsen, P. E., 1988, Continuity of strata in the Newark and Hartford Basins of the Newark Supergroup: USGS Bulletin 1776, p. 6-18.

Olsen, P. E., A. J. Froelich, D. L. Daniels, J. P. Smoot, and P. J. W. Gore, (in press), Chapter 9—Rift basins of Early Mesozoic age, in W. Horton, ed., Geology of the Carolinas: Carolina Geological Society, Rahleigh.

Olsen, P. E., C. L. Remington, B. Cornet, and K. S. Thomson, 1978, Cyclic change in Late Triassic lacustrine communities: Science, v. 201, p. 729-733.

Olsen, P. E., R. W. Schlische, and P. J. W Gore, eds., 1989, Tectonic, depositional, and paleoecological history of Early Mesozoic rift basins, eastern North America: Washington, American Geophysical Union, International Geological Congress Field Trip T351, 174 p.

Olsen, P. E., and R. W. Schlische, 1990. Transtensional arm of the early Mesozoic Fundy rift basin: Pene-contemporaneous Faulting and sedimentation: Geology, v. 18, p. 695-698.

Ratcliffe, N. M., and W. C. Burton, 1985, Fault reactivation models for the origin of the Newark Basin and studies related to U.S. eastern seismicity, *in* G. R. Robinson, Jr., and A. J. Froelich, eds., Proceedings of the second U.S. Geological Survey Workshop on the Early Mesozoic basins of the eastern United States: USGS Circular 946, p. 36-45.

Ratcliffe, N. M., W. C. Burton, R. M. D'Angelo, and J. K. Costain, 1986, Low-angle extension faulting, reactivated mylonites, and seismic reflection geometry of the Newark basin margin in eastern Pennsylvania: Geology, v. 14, p. 766-770.

Reynolds, D. J., and R. W. Schlische, 1989, Comparative studies of continental rift systems (abs.): GSA Abstracts with Program, v. 21, p. 61.

Robinson, P. L., 1973, Palaeoclimatology and continental drift, *in* D. H. Tarling, and S. K. Runcom, eds., Implications of continental drift to the earth sciences: London, Academic Press, pt. 1, p. 449-476.

Rossignol-Strick, M., 1983, African monsoons, an immediate climatic response to orbital insolation: Nature, v. 304, p. 46-49.

Schlische, R. W., and P. E. Olsen, 1988, Structural evolution of the Newark basin, *in* J. M. Husch, and M. J. Hozic, eds., Geology of the central Newark basin—Field guide and proceedings: Geological Association of New Jersey Fifth Annual Meeting, p. 44-65.

Schlische, R. W., and P. E. Olsen, 1990, Quantitative filling model for continental extensional basins with applications to early Mesozoic rifts of eastern North America: Journal of Geology, v. 98, p. 135-155.

Scholz, C. A., and B. R. Rosendahl, 1988, Low lake stands in Lakes Malawi and Tanganyika, east Africa, delineated with multi-fold seismic data: Science, v. 240, p. 1645-1648.

Smith, I. R., and I. J. Sinclair, 1972, Deep water waves in lakes: Freshwater Biology, v. 2, p. 387-399.

Smoot, J. P., 1985, The closed basin hypothesis and the use of working models in facies analysis of the Newark Supergroup, *in* G. R. Robinson, Jr., and A. J. Froelich, eds., Proceedings of the second U.S. Geological Survey Workshop on the Early Mesozoic basins of the eastern United States: USGS Circular 946, p. 4-10.

Smoot, J. P., and B. Castens-Seidel, 1982, Sedimentary fabrics produced in playa sediments by efflorescent salt crusts—An explanation for 'adhesion ripples' (abs.): International Association of Sedimentologists 11th International Congress, Abstracts of Papers, p. 10.

Smoot, J. P., and P. E. Olsen, 1985, Massive mudstones in basin analysis and paleoclimatic interpretation of the Newark Supergroup, *in* G. R., Robinson, Jr., and A. J. Froelich, eds., Proceedings of the second U.S. Geological Survey Workshop on the Early Mesozoic basins of the eastern United States: USGS Circular, 946, p. 29-33.

Smoot, J., and P. E. Olsen, 1988, Massive mudstones in basin analysis and paleoclimatic interpretation of the Newark Supergroup, *in* W. Manspeizer, ed., Triassic-Jurassic rifting, continental breakup and the origin of the Atlantic Ocean and passive margins: New York, Elsevier, p. 249-274.

Street-Perrott, F. A., 1986, The response of lake levels to climate change—Implications for the future, *in* C. Rosenzweig, and R. Dickenson, eds., Climate-vegetation interaction: Boulder, Colorado, University Corporation for Atmospheric Research, Office for Interdisciplinary Earth Studies, p. 77-80.

Thayer, C. W., 1979, Biological bulldozers and the evolution of marine benthic communities: Science, v. 203, p. 458-461.

Weiss, R. F., V. M. Koropalov, E. C. Carmack, and H. Craig, 1989, Geochemical studies of ventilation and deep metabolic processes in Lake Baikal: EOS, v. 70, p. 1136-1137.

Wootton, Robin, J., 1988, The historical ecology of aquatic insects—An overview, *in* J. Gray, ed., Aspects of freshwater paleoecology and biogeography: Palaeogeography, Palaeoclimatology, Palaeoecology, v. 62, p. 477-492.

Seismic Facies Analysis of Fluvial-Deltaic Lacustrine Systems—Upper Fort Union Formation (Paleocene), Wind River Basin, Wyoming

Louis M. Liro
Yvonna C. Pardus
Texaco E&P Technology Division
Houston, Texas, U.S.A.

Progradation of deltaic systems into Paleocene Lake Waltman (Wind River basin, Wyoming) is interpreted from seismic reflection data from the upper Fort Union Formation. Subdivision of the interval into seismic sequences and interpretation of variations in seismic attributes (i.e., amplitude, continuity, and frequency) allow detailed reconstruction of the depositional history of this interval. Sequences that show distinct clinoform morphology from seismic data are interpreted as lobate deltas. Higher amplitude, continuous events within the clinoforms are interpreted as prograding delta-front facies. Seismic data clearly demonstrate the time-transgressive nature of this facies. Downdip of the clinoforms, low-amplitude, generally continuous seismic events correspond to homogeneous shales that represent deposition in an areally extensive, low-energy lacustrine environment. The lacustrine facies is traceable laterally in the seismic data, which allows interpretation of changes in lake morphology through late Fort Union time. Overlying the upper Fort Union lacustrine strata are fluvial deposits interpreted from discontinuous, variable-amplitude seismic facies.

Seismic facies interpretations are compared to lithologic interpretations from a series of closely spaced wells adjacent to seismic lines across the basin to test the accuracy of depositional environment prediction from seismic data.

INTRODUCTION

The Fort Union Formation of Paleocene age in the Wind River basin of Wyoming consists of up to 7000 ft (2100 m) of fluvial and fluvial-deltaic lacustrine sediment. Various workers (Keefer, 1961, 1965a; Fox and Priestly, 1983a-e; Phillips, 1983; Robertson, 1984) have used log data and field observations to develop the stratigraphic framework for the Fort Union Formation. In a study of the Fort Union in the western Wind River basin, Ray (1982) described relevant seismic facies patterns and interpreted paleogeography and depositional environments associated with the fluvial-deltaic and lacustrine systems present.

This paper presents results of an integrated seismic facies and well log study of the upper Fort Union Formation in the eastern and central Wind River basin. Seismic facies patterns are interpreted within eight seismic sequences, which define depositional elements within this interval. Variations in seismic facies distribution within the sequences allow a reconstruction of the depositional history. To relate this regional interpretation to lithostratigraphy, a comparison of seismic facies to well log data is presented.

STUDY AREA

The Wind River basin (Figure 1) is one of a series of intermontane basins in the Rocky Mountains of west-central United States. Maximum thickness of

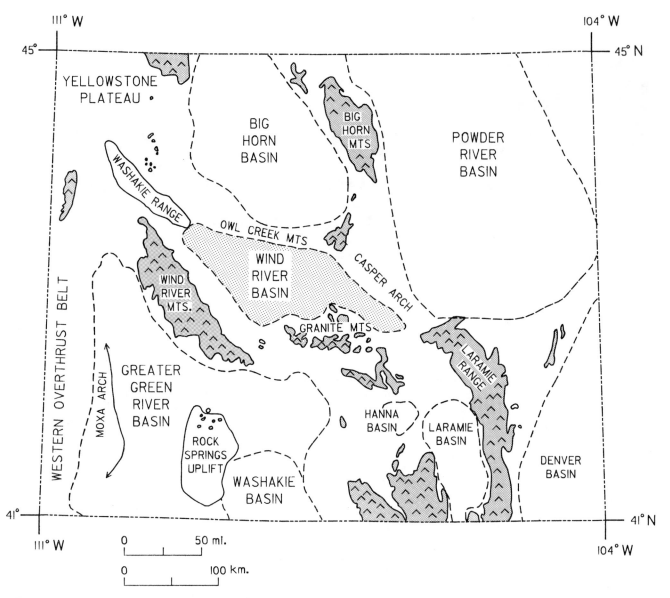

Figure 1. Location of Wind River structural basin with respect to gross tectonic elements in Wyoming (modified from Landes, 1970).

the sedimentary section (Cambrian to Tertiary) is 33000 ft (10000 m) (Keefer, 1965b).

The present outline of the structural basin developed during the Laramide orogeny of latest Cretaceous and early Tertiary time (Keefer, 1965b). Laramide deformation produced rimming thrust-belt mountain ranges and a strongly asymmetrical basin (Figure 2). The Tertiary depocenter, containing up to 17000 ft (5200 m) of strata, is located immediately south of the Owl Creek Mountains and west of the Casper Arch.

Laramide-related deposition in the Wind River basin began in Late Cretaceous time and is represented by fluvial to shallow-marine deposits of the Lance Formation, which thickens north-northeastward (Keefer, 1965b). Uplift of the Granite Mountains along the southern basin margin and the Washakie Range along the northwestern margin also began at this time and continued into the Tertiary. Continued deformation in the area uplifted the Wind River Mountains to the south and southwest, the Owl Creek Mountains to the north, and the Casper Arch to the east.

By Paleocene time the Wind River basin had become an intermontane basin with little or no open marine communication. By middle Paleocene, a large lake called Lake Waltman (Keefer, 1965b) had formed (Figure 3), which, based on interpretations of the

Liro and Pardus

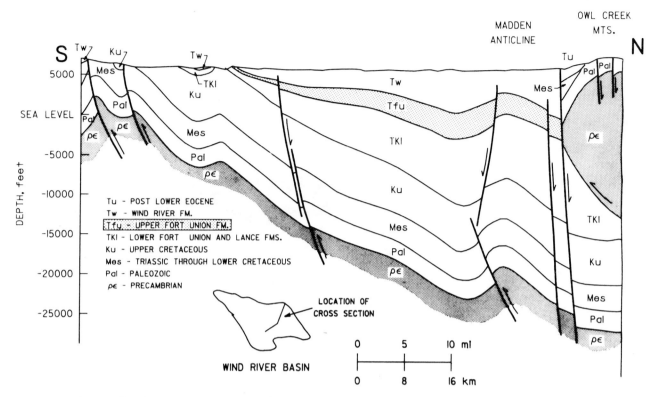

Figure 2. Generalized cross section through eastern Wind River basin (modified from Ray and Keefer, 1985).

maximum bounds of lacustrine seismic facies (Ray, 1982, and our work), may have covered as much as 1600 mi² (4100 km²). By late Paleocene time Lake Waltman had nearly filled from its southern and western margins with deltaic sediment fed by rivers from the adjacent highlands.

The upper Fort Union Formation is divided into the (1) lacustrine Waltman Shale Member, and (2) Shotgun Member (Figure 4), fluvial and deltaic strata deposited along the fringes of Lake Waltman. Deposition of a maximum 7000 ft (2100 m) of Fort Union strata accompanied extensive erosion of the uplifted regions, and by early Eocene Lake Waltman had been completely infilled, and drainage out of the basin became reestablished.

PREVIOUS STUDY

Keefer (1961) first detailed the stratigraphy and depositional environments of the Fort Union Formation in the Wind River basin, including definition of the Shotgun and Waltman Shale Members. Later studies (Keefer, 1965a, 1965b, 1969)

refined the depositional model and demonstrated the potential for both hydrocarbon source and reservoir intervals within the upper Fort Union Formation. Hyne et al. (1979) drew analogies between the fluvial-deltaic lacustrine setting of the upper Fort Union Formation and a possible modern analog, the Catatumbo River and Lake Maracaibo, Venezuela.

Interest in the hydrocarbon exploration potential of the upper Fort Union Formation resulted in several subsequent studies. Ray (1982) used seismic stratigraphic techniques to infer depositional environments of the upper Fort Union Formation in the western Wind River basin. Phillips (1983) correlated outcrop stratigraphy to subsurface control in the eastern Wind River basin and described an upper Fort Union depositional model influenced by regional tectonism. Robertson (1984) described the depositional environments and reservoir quality of Shotgun Member sandstones in the Haybarn field. All these studies agree on a general framework for the upper Fort Union Formation—a large lake with fringing delta systems along its southern and western margins. In this study we further detail this fluvial-deltaic setting in the eastern and central Wind River basin and present seismic facies interpretations for the lake's filling by deltaic sedimentation.

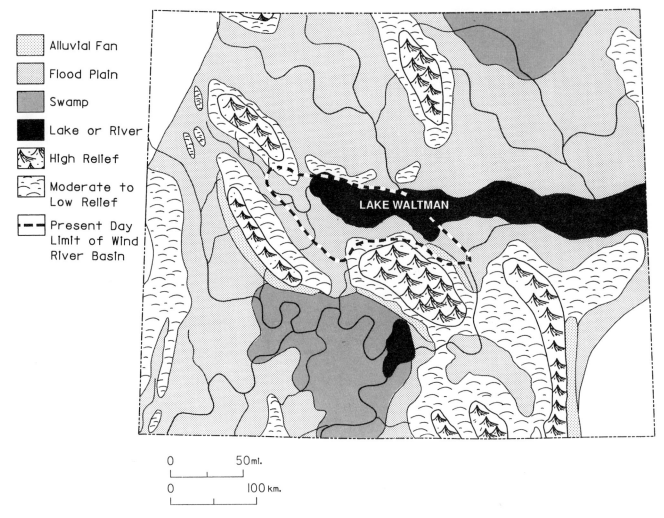

Figure 3. Paleogeography of Wyoming in late Paleocene time. Dashed line indicates present limits of the Wind River basin. Modified from McDonald (1972).

Legend:
- Alluvial Fan
- Flood Plain
- Swamp
- Lake or River
- High Relief
- Moderate to Low Relief
- Present Day Limit of Wind River Basin

LAKE WALTMAN

0   50 mi.

0   100 km.

TECHNIQUES OF ANALYSIS

Seismic facies and geophysical log analysis were used to interpret depositional environments in the upper Fort Union Formation. After nine seismic sequence boundaries had been identified and correlated, variations in seismic amplitude, continuity, and frequency content within each sequence were mapped and interpreted in terms of prevailing depositional environment. We then compared these seismic facies units with depositional environments and lithologic variations interpreted from logs adjacent to the seismic lines. This also allowed well-to-well correlations that otherwise would have been difficult because of discontinuous strata.

Approximately 600 mi (970 km) of seismic line were run across the north-central part of the basin. The 29 lines, which were spaced 3–5 mi (5–7 km) apart, generally paralleled structural dip of the thickest

Tertiary section (Figure 5). For interpretation of regional seismic stratigraphy, the data were displayed with 500-ms timing lines at a scale of 3.75 in./sec and 32 traces/in. In general, the quality of the seismic data is excellent. Additional data would be needed west of this grid for a basinwide study of the Tertiary fill.

From a synthetic seismogram package (Geo-Seismic Services, unpublished data) of sonic logs from approximately 200 wells throughout the basin; of these, more than 100 were used in the study (Figure 5). Calculated reflection coefficients were convolved with a seismic wavelet to create a synthetic seismic trace, our main correlation tool for tying well control to seismic data. This trace and the availability of numerous sonic logs greatly enhanced the results. In addition, other available logs (SP, gamma ray, resistivity) of the upper Fort Union Formation were used to further evaluate lithologic interpretations from the sonic logs and seismic data.

Liro and Pardus

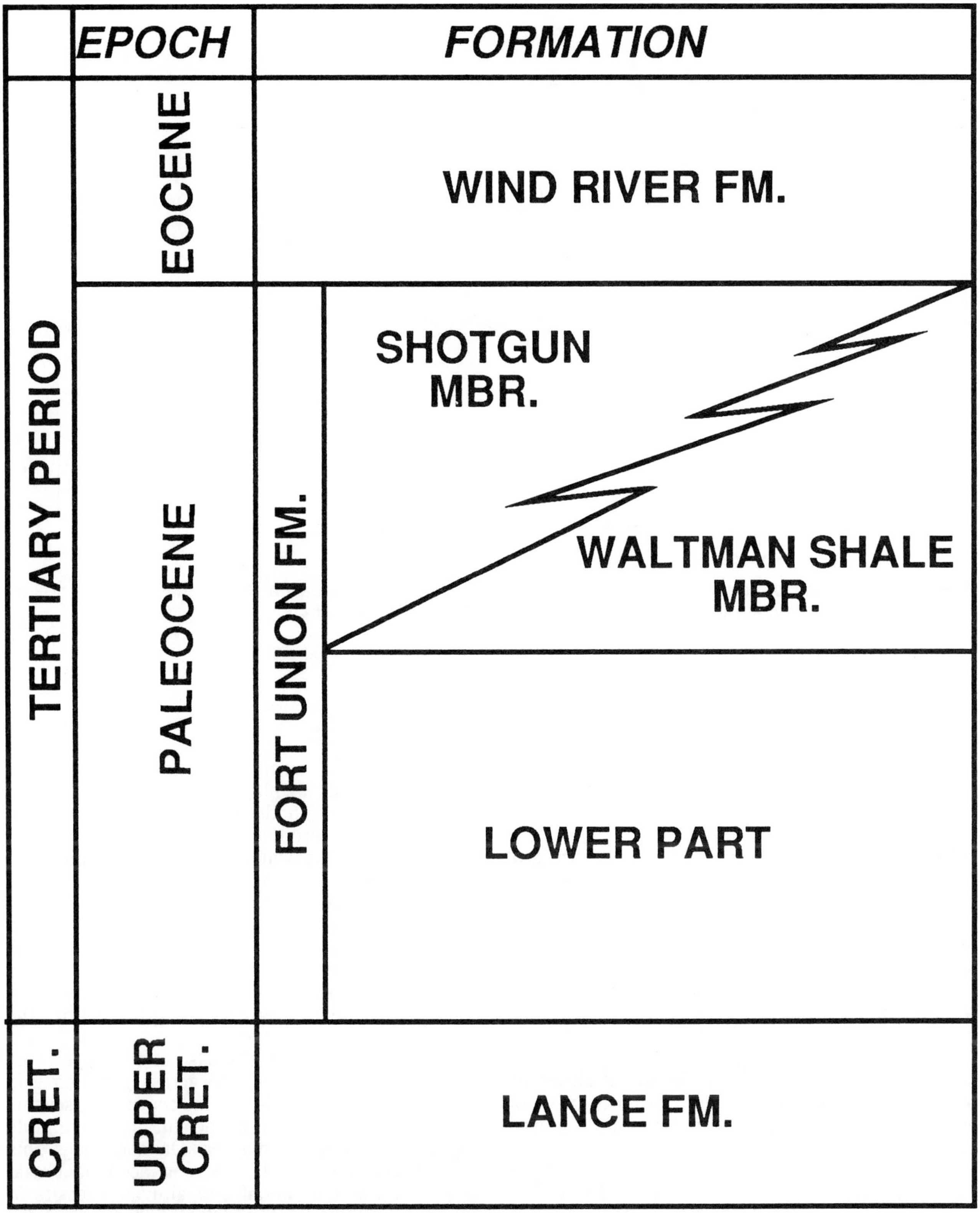

EPOCH		FORMATION
TERTIARY PERIOD	EOCENE	WIND RIVER FM.
	PALEOCENE / FORT UNION FM.	SHOTGUN MBR. / WALTMAN SHALE MBR.
		LOWER PART
CRET.	UPPER CRET.	LANCE FM.

Figure 4. Stratigraphic column for the lower Tertiary in the Wind River basin. Modified from Keefer (1961) and Ray (1982).

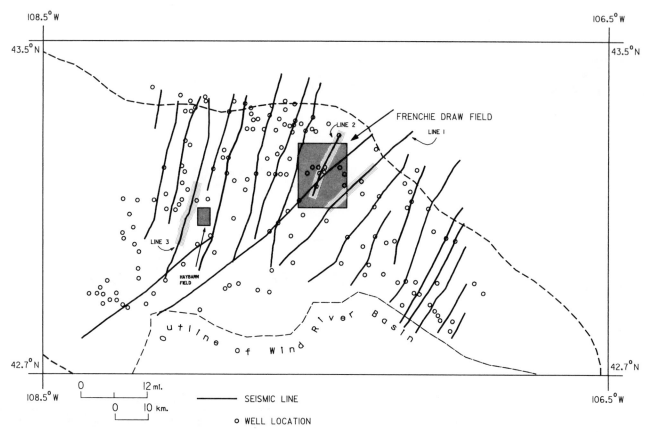

Figure 5. Locations of seismic lines and wells and Frenchie Draw and Haybarn fields in north-central Wind River basin, and location. Seismic data courtesy of Geo-Seismic Services Inc.

SEISMIC SEQUENCE ANALYSIS

Primary seismic reflections follow chronostratigraphic surfaces (Vail et al., 1977). This relationship allows for definition and study of depositional sequences on seismic data. Such seismic sequences were defined in the Wind River basin by methods originally outlined in Mitchum and Vail (1977). Identification of systematic patterns of seismic event terminations led to the delineation of bounding events for each sequence, which then were correlated throughout the seismic grid. In this way, seismic sequence boundaries could be identified either on other lines where termination patterns are not evident or where sequence boundaries are conformable with surrounding reflectors.

After a review of the entire data set, lines in the central seismic grid were studied in detail first because the clinoform stratal geometry there typically exhibits numerous reflection terminations (Figure 6), allowing for reliable identification of seismic sequence boundaries. On these and adjacent lines, the following termination styles were observed:

1. Toplap and downlap terminations within the clinoforms, used to define most of the seismic sequences in this study;

2. Updip, low-angle onlap terminations against the southern basin margin, which served to confirm sequence boundaries identified in (1) above;

3. Less numerous and more irregularly distributed, updip, low-angle onlap terminations against the structural northern basin margin.

Nine seismic sequence boundaries, encompassing eight seismic sequences were identified within the upper Fort Union Formation (Figure 7). The lowest sequence boundary (1), the base of the upper Fort Union, was easily correlated across the data grid because of its high amplitude and continuous seismic character and proved to be the most reliable regionally correlative marker. On sonic logs this marker correlates with a strong acoustic break, where interpreted fluvial and alluvial deposits of velocity 10000–12000 ft/sec (3000–3700 m/sec) in the lower Fort Union are overlain by homogeneous shale of the Waltman Shale Member, velocity 8000–10000 ft/sec (2400–3000 m/sec).

Liro and Pardus

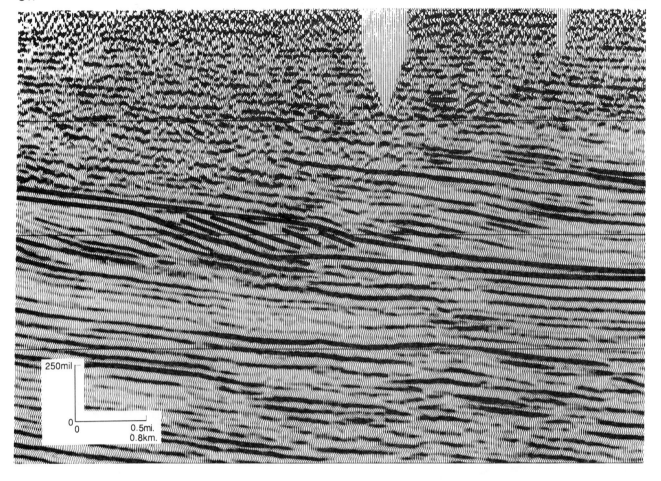

Figure 6. Clinoform stratal geometry in the upper Fort Union Formation, illustrating higher angle reflection terminations characteristic of toplap, downlap, and onlap relationships. Seismic line 1 was run approximately parallel to structural dip. See Figure 5 for line location.

Above the base of the lower Fort Union, clinoform geometries are apparent. In the central seismic data grid, seismic facies within the clinoforms pass through the seismic sequence boundaries of the upper Fort Union, defining a time-transgressive facies pattern (Figure 8). Low-angle, high-amplitude reflectors comprise the initial clinoforms; in successive sequences well defined foreset patterns are observed within the clinoforms.

Clinoforms from the western part of the study area built eastward. Several sequences higher and farther east in the vicinity of the Frenchie Draw field (Figure 5), a separate series of clinoforms prograded generally from south to north. This pattern suggests that clinoform development ceased first in the west and somewhat later in the east. The inferred top of the Fort Union Formation is recorded by cessation of clinoform development (particularly high-angle foreset patterns) throughout the area, which is consistent with progressive infilling of Lake Waltman and decreasing subsidence in the basinal trough (Paape, 1968). Concordant events are associated with fluvial deposition in the overlying Wind River Formation (Eocene).

SEISMIC FACIES ANALYSIS

After delineating and correlating sequence boundaries through the seismic grid, we analyzed the seismic facies to infer lithologies and depositional environments. Four major facies patterns (fluvial, prograding delta front, distal delta, lacustrine) were defined and compared to well log responses in the same intervals.

Fluvial Facies

Most of the lower Fort Union Formation, updip sections (part of the Shotgun Member) of the upper Fort Union, and overlying Wind River Formation are characterized seismically by discontinuous to

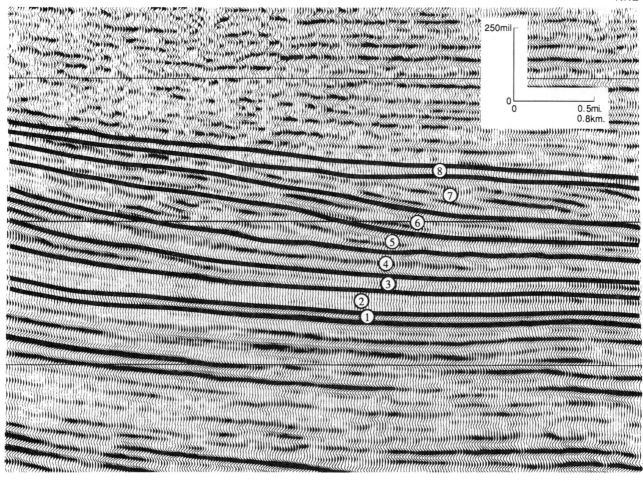

Figure 7. Eight seismic sequences (nine boundaries) identified in the upper Fort Union Formation along line 2 across Frenchie Draw field. See Figure 5 for line location.

moderately continuous events with variable- to high-amplitude response and variable to low frequency (Figure 9). Seismic geometries for the parallel reflectors of this facies are not well developed. This seismic response is interpreted to represent discontinuous, intercalated sandstone, shale, and coal of fluvial and nearshore environments, similar to seismic facies described by Sangree and Widmier (1977) for nonmarine and marginal-marine environments. This interpretation is consistent with Ray's (1982) fluvial seismic facies and with the fan-plain facies (perhaps including the braided stream channel facies) that Phillips (1983) observed in outcrop.

In well logs, the fluvial facies is represented by intervals of thin-bedded, "ratty" log response, often with thin, highly resistive layers suggestive of coal or carbonaceous shale. In his study of lower Fort Union reservoirs in the Fuller Reservoir field, Pirner (1978) found that individual sandstone beds are thin and discontinuous, difficult to correlate, and usually lacking distinctive log signature.

Prograding Delta-Front Facies

The Shotgun Member of the upper Fort Union contains numerous basinward-prograding clinoforms. Where a seismic line crosses a clinoform along depositional dip and through its thickest sections (Figure 6), the clinoform resembles the distinctive geometry of lateral outbuilding of successive basinward-sloping depositional surfaces (Mitchum et al., 1977) associated with an isochronally thickened section of the sequence. Where the seismic line crosses the clinoform in a less optimum position, such as along its edges or at some angle to depositional dip, evidence for its existence may be only a subtle isochronal thick (Figure 10). Continuous, moderate- to high-amplitude seismic events, terminating in toplaps and downlaps within the clinoform, are interpreted to represent delta-front sandstones and siltstones interbedded with lacustrine shale. This seismic facies is equivalent to Phillips's (1983) distal fan facies. In Figure 8 the prograding (lateral

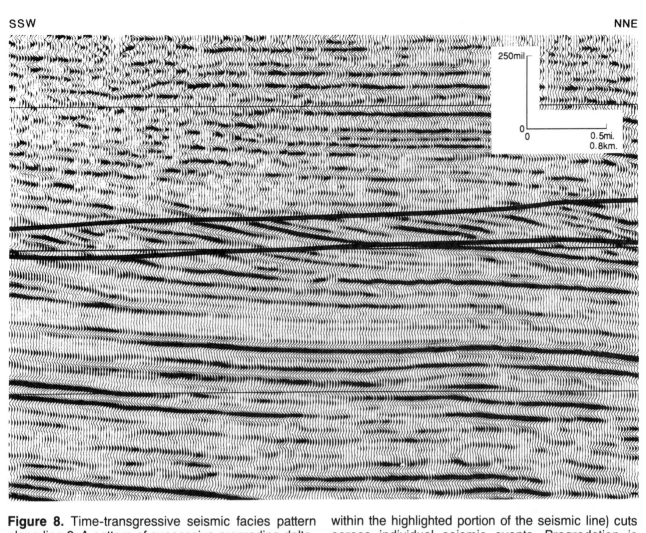

250mil

0
0   0.5mi.
0.8km.

Figure 8. Time-transgressive seismic facies pattern along line 2. A pattern of successive prograding delta-front facies (inclined high-amplitude event segments within the highlighted portion of the seismic line) cuts across individual seismic events. Progradation is basinward, or to the right. See Figure 5 for line location.

component) and aggrading (vertical component) nature of the delta-front facies, as it crosses individual seismic events and sequences, is strong evidence that this facies is indeed time transgressive.

On electric logs individual prograding delta-front sandstones exhibit distinctive "funnel" shapes (Robertson, 1984) attributed to coarsening-upward grain size. Delta-front sandstones and siltstones also are interpreted where sonic velocity increases far above that of the enclosing lacustrine shale. High-amplitude events within clinoform packages are due to delta-front intercalations of sandstone and siltstone with lacustrine shale. Juxtaposition of coarser grained clastics of the Shotgun Member with potential organic-rich source facies of the Waltman Shale Member (Keefer, 1969) may allow migration of hydrocarbons into stratigraphically isolated reservoir sandstones, as Robertson (1984) suggested for the Haybarn field.

Distal Delta Facies

The distal delta facies commonly occurs where clinoforms thin downdip or between the fluvial and lacustrine facies (described below). It is characterized by low to moderate amplitude, and moderate to good continuity (Figure 9). Its spatial position and areal distribution suggest that it represents distal toes of associated deltaic systems or possibly wave- or current-reworked sediment parallel to the shoreline. In some cases, internal terminations suggest localized deltaic deposition.

In sonic and electric logs the distal facies typically consists of isolated lacustrine shale-enclosed sandstones, whose velocities are similar to those of delta-front sandstones. This facies likely coincides with productive upper Fort Union sandstones in the Haybarn field (Robertson, 1984).

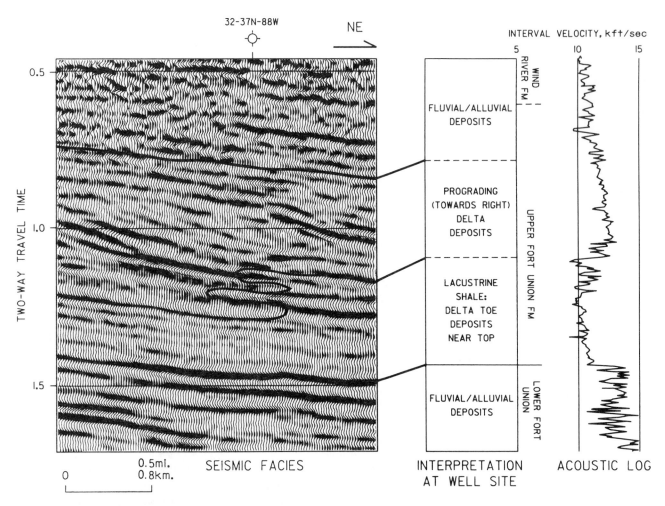

Figure 9. Interpreted facies patterns on seismic and correlative well log data, seismic line 1, sec. 32, T37N, R88W. See Figure 5 for line location.

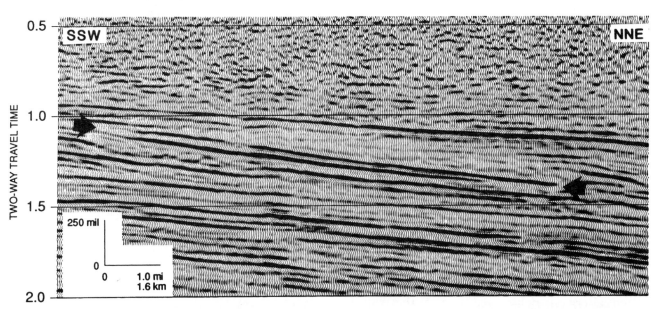

Figure 10. Interpreted clinoform geometry (between arrows) on seismic line 3. Although the line approximately follows general structural dip, it appears to cross the interpreted clinoform along depositional strike. High amplitude associated with highlighted seismic events is interpreted as prograding delta-front facies. Compare to Figure 8, which shows the same facies parallel to dip. See Figure 5 for line location.

Liro and Pardus

Lacustrine Facies

The lacustrine Waltman Shale Member interval seismically is represented by continuous, low-amplitude events. Because the facies is stratigraphically bounded by higher amplitude delta-front and fluvial facies, it is easily identified on seismic sections when sufficiently thick. Near the southern basin margin, however, the shale thins below one seismic cycle—approximately 250–350 ft (80–100 m) at 30-Hz dominant frequency—and becomes difficult to distinguish from fluvial deposits.

Sonic velocity in the Waltman Shale Member is significantly lower than in adjacent rocks (Figure 9). In other logs, the facies has a distinctively homogeneous appearance, with low SP response and uniform, typically low resistivity. Gamma response is high, but not as high as gamma anomalies observed in organic-rich marine shales (Meyer and Nederlof, 1984). That these log characteristics are best developed downdip in thicker sections of the shale suggests that the Waltman Shale Member is more uniform there, devoid of significant siltstone or sandstone. This inference is supported by field observations (Phillips, 1983) and well cuttings examinations (Robertson, 1984).

FACIES TRANSITIONS

The transition from lower Fort Union fluvial deposits into upper Fort Union deltaic and lacustrine deposits occurs along one seismic event, which implies that Lake Waltman formed rapidly or that initial transgression of the lake is below seismic resolution. Sonic logs indicate that the acoustic break between the low-velocity Waltman Shale Member and the lower Fort Union Formation is generally very sharp, typically occurring over several meters (Figure 9). We conclude that Lake Waltman formed very rapidly, with little or no associated transgressive facies, similar to the formation of Paleocene Lake Lebo in the Powder River basin (Ayers, 1986).

Facies transitions within the upper Fort Union Formation are gradual, as seen at the Humble 2 Highland Unit well (sec. 32, T37N, R88W) in Figure 9. Seismic facies patterns correlate strongly with sonic log response. The boundary between lacustrine shale and the fluvial facies of the lower Fort Union is marked by a sharp acoustic break. Higher velocity, interpreted coarser grained clastics increase upward in the Waltman Shale Member. In seismic section this is evidenced by the toes of approaching clinoforms and a general increase in amplitude content of the seismic events. Deltaic facies coinciding with seismic clinoforms show markedly increased velocity and thicker individual sandstone and siltstone units. Above the delta lobe, rapid velocity changes and thin strata characterize fluvial intervals of the uppermost Fort Union Formation and Wind River Formation.

FACIES MAPPING

The seismic facies patterns noted above were delineated for each seismic line in the study grid, and a depositional environment map interpreted for each seismic sequence. Of the eight sequences identified in the upper Fort Union Formation, we present results for three representative sequences, approximately at the base, middle, and top of this interval.

Because the sequence at the base of the upper Fort Union is seismically thin (typically only one or two cycles thick), conventional seismic facies analysis is impractical. Based on interpretations from well control that the underlying lower Fort Union is composed of dominantly high-velocity fluvial deposits, amplitude variations along the base of sequence 1 (Figure 7) were mapped and compared to modeled responses. Strong amplitude at the base of sequence 1 is interpreted to represent low-velocity lacustrine shale (Figure 11). Fluvial facies are interpreted where the amplitude at the base of sequence 1 is weak, that is, where the acoustic contrast is small. Amplitude response intermediate to the fluvial and lacustrine facies is interpreted to represent a possible shallow-lacustrine facies.

Sequence 1 is interpreted to represent the transition from the fluvial plain setting to a lacustrine setting. Sonic log control in the northern half of the study area (where the upper Fort Union generally is thicker) supports the interpretation that Lake Waltman formed rapidly, as the acoustically slow Waltman Shale Member immediately overlies the lower Fort Union over a wide area. Sonic log control in the south (generally corresponding to the shallow-lacustrine facies in Figure 11) suggests, however, a more gradual transition—the transition from acoustically fast lower Fort Union to the slower Waltman Shale Member occurs over an interval as thick as 70 m. In this sequence and the others above, poor seismic imaging around the Madden anticline (Figures 2, 11) precluded effective seismic facies study of the northern basin margin .

Clinoform development in the middle upper Fort Union Formation becomes increasingly evident in stratigraphically higher sequences. Sequence 4 (Figure 7) shows two interpreted clinoform intervals—one interpreted from depositional dip-oriented lines and another from strike-oriented lines. Dip-oriented lines show pronounced thickening of the sequence in the clinoforms, which gives them their geometry. Toplap and downlap terminations are numerous, and reflectors within the clinoforms generally are high amplitude. Strike-oriented lines demonstrate elongate, mildly lobate forms defined by high-amplitude events. In strike view, no obvious clinoform geometry is apparent. In map view (Figure 12) distribution of the high-amplitude facies suggests the outline of a delta form.

Deltaic fill of Lake Waltman is interpreted from prograding clinoform development and time-transgressive delta-front seismic facies. The presence

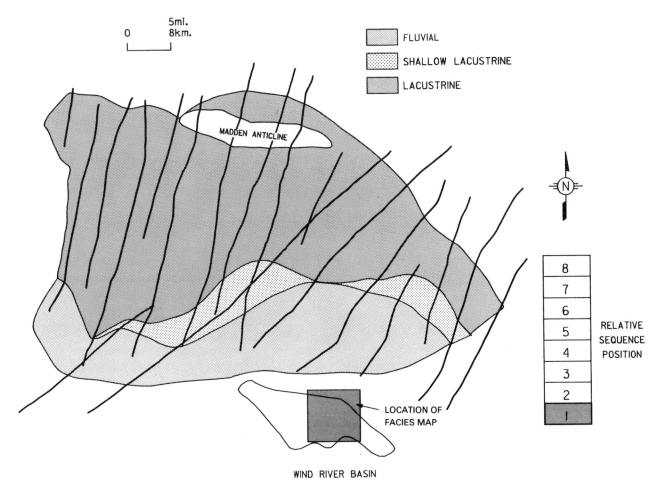

Figure 11. Generalized depositional environments interpreted from seismic facies distribution near the base of the upper Fort Union Formation (sequence 1, Figure 7).

of areally extensive lacustrine facies basinward of this facies suggests a low-energy lake environment. In this sequence the interpreted delta system in the eastern study area is lobate, perhaps similar in depositional style to the Catatumbo River delta of Lake Maracaibo, Venezuela (Hyne et al., 1979). The map view of the delta system in the western study area (Figure 12) is less symmetrical, perhaps indicating the presence of a contemporaneous structural nose or sediment-reworking currents. Productive upper Fort Union sandstones in the Haybarn field (Robertson, 1984) are located in the distal delta facies of the western delta system.

Sequence 7 (Figure 7) near the top of the upper Fort Union displays seismic facies and geometry similar to that of sequence 4, suggesting a continuation of the same or similar depositional environment. Deltaic encroachment, interpreted from clinoform patterns, reduced Lake Waltman to an east-west-elongated lake or embayment (Figure 13).

The subsequent (and stratigraphically highest) sequence interpreted in the upper Fort Union Formation shows little indication of lacustrine facies,

which suggests minimal lacustrine influence on deposition and a return to a dominantly fluvial facies.

FIELD STUDY

To relate seismic sequences and facies to lithostratigraphy in the upper Fort Union Formation, logs from five wells in and near the Frenchie Draw field (Figure 5) were correlated to seismically defined sequence boundaries. Four wells near the seismic lines within the field (Figure 14) and one well outside the field (Humble Hiland Unit 3) were selected to study dip-oriented facies development and transition in the upper Fort Union. The Unit 11 and Unit 17 wells were not, however, part of the original seismic facies study. (See Normark, 1978, for a more complete description of this field.)

Synthetic seismograms were calculated from sonic logs for correlating log data with seismic data—sequence boundaries, in particular. The well synthetics were tied to the nearest seismic line; all

Liro and Pardus

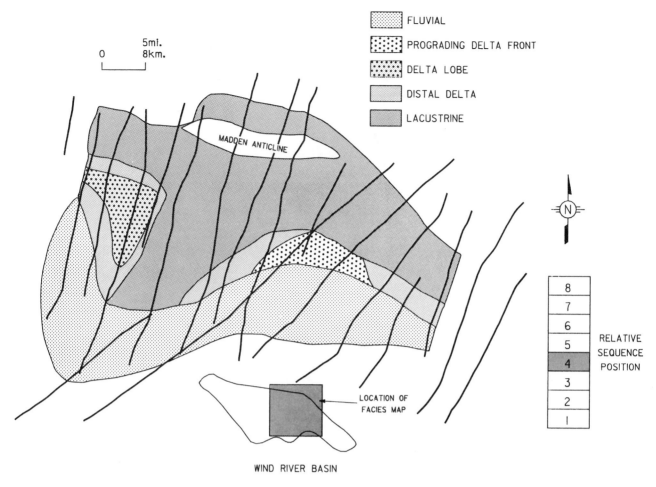

FLUVIAL
PROGRADING DELTA FRONT
DELTA LOBE
DISTAL DELTA
LACUSTRINE

MADDEN ANTICLINE

5mi.
8km.
0

N

8
7
6
5
4
3
2
1

RELATIVE
SEQUENCE
POSITION

LOCATION OF
FACIES MAP

WIND RIVER BASIN

Figure 12. Generalized depositional environments interpreted from seismic facies distribution approximately in the middle upper Fort Union Formation sequence 4, Figure 7).

are located within about 0.25 mi (400 m) of the seismic line to which they are correlated.

For synthetic computation, a zero-phase Ricker wavelet with a 30-Hz center frequency was chosen to closely approximate the shape and frequency bandwidth of the seismic wavelet present in the seismic data (W. W. Krauter, pers. comm.). A Gardner density relationship was assumed. Sonic logs were available through the entire upper Fort Union in three of the wells; the other two wells (Unit 11 and Unit 17) had sonic control only for the lower part of the upper Fort Union. For the rest of the section in these wells, time-to-thickness conversions were approximated based on interpreted lithology and average velocities, which allowed sequence boundary identification by other logs (i.e., resistivity) that penetrated the entire upper Fort Union.

The synthetic seismogram and log data for the Unit 3 well are shown in Figure 15, in which the upper Fort Union Formation is located approximately between 0.58 and 1.13 sec. In general, interpreted sandstones are acoustically faster (i.e., slower interval transit time) than shale of the Waltman Shale Member. Sandstones have interval transit

times of approximately 80–90 μs/ft or 12000 ft/sec (~3700 m/sec), compared to shales with velocities of 90–100 μs/ft or 10000 ft/sec (~3000 m/sec). In addition, sandy intervals interpreted from sonic logs generally are more resistive than adjacent shaly intervals. Apparent on the sonic log of well Unit 3 is a vertical transition from shaly to sandy intervals in the upper Fort Union, interpreted as recording the gradual regression of the lake margin.

A reflector at the base of the Waltman was selected as the primary correlation datum between synthetics and seismic data because of its pronounced seismic amplitude corresponding to an easily identified lithologic change inferred on the logs.

Figure 16 shows resistivity log correlation of seismically defined sequence boundaries in the upper Fort Union Formation. A general basinward (left to right) progradation is interpreted, based on clinoform progradation recognized on adjacent seismic line 2 (Figure 17). The clinoform near the top (sequence 7) appears to be divided into sandy and shaly subunits.

From a comparison of seismic facies interpretations with log data (Figures 18a, 18b), we see that the distribution of seismic facies patterns closely

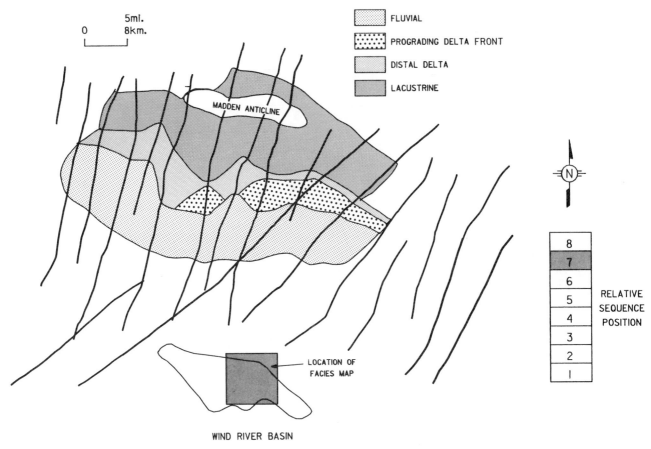

FLUVIAL

PROGRADING DELTA FRONT

DISTAL DELTA

LACUSTRINE

MADDEN ANTICLINE

N

8
7
6
5
4
3
2
1

RELATIVE
SEQUENCE
POSITION

LOCATION OF
FACIES MAP

0   5mi.
    8km.

WIND RIVER BASIN

Figure 13. Generalized depositional environments interpreted from seismic facies distribution near the top of the upper Fort Union Formation (sequence 7, Figure 7).

matches the types of log character (described earlier) in the upper Fort Union. In particular is a strong correlation between the occurrences of interpreted massive shale intervals and lacustrine seismic facies (see also Figure 17). Interpreted thick sandstone and siltstone intervals in the log data correlate with the regressive clinoforms and fluvial facies in the uppermost Fort Union. Seismic facies interpretations on a sequence scale create the framework in which finer scale variations (i.e., within the seismic sequence) in lithology can be interpreted and correlated with logs.

CONCLUSIONS

Discontinuous sandstone, shale, and coal in the lower Fort Union Formation (lower to middle Paleocene) are overlain by lacustrine shale of the upper Fort Union (middle upper Paleocene). Lacustrine shale grades upward and interfingers laterally with fluvial-deltaic sandstone and siltstone deposited by rivers debouching from the west and south. Deposition by these delta systems successively reduced the area of the lake. Seismic and well data

in the Frenchie Draw field area indicate that relative progradation and aggradation varied, and that finer scale subsequence transgressions and regressions of the lake margin may be evident and mappable.

Downdip of the upper Fort Union fluvial and deltaic deposits, homogeneous muds accumulated in a low-energy, areally extensive lake. The Waltman Shale Member, a potential hydrocarbon source rock (Keefer, 1969), shows distinctive and easily recognizable seismic character that allows mapping of the unit.

An integrated study of seismic facies and geophysical log data allows definition of depositional elements within seismic sequences, on both regional and smaller (field) scales. Seismic facies and geometries in lacustrine and fluvial-deltaic strata can be correlated and mapped, allowing reconstruction of facies and depositional environment variations through time. Such a seismically defined framework allows further interpretation and correlation of subsequence variations in facies patterns and lithology from log data.

The results of this study are in agreement with Keefer's (1965b) regional work, Ray's (1982) seismic stratigraphic study, and Phillips's (1983) field work. Our findings add to the body of knowledge for the upper Fort Union Formation in the Wind River basin.

Liro and Pardus

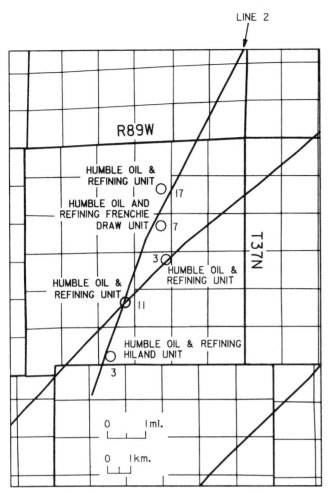

Figure 14. Locations of wells and seismic lines used for detailed field study of Frenchie Draw field. See Figure 5 for field location.

In particular, seismic data have been used to describe the development of Lake Waltman and its associated fluvial-deltaic systems by analysis of seismic sequences.

The seismic stratigraphic method, originally applied to interpretation of marine strata (Vail and Mitchum, 1977), has been shown to be effective in nonmarine strata as well. Perhaps one underutilized aspect of seismic stratigraphy is the potential to tie well data using chronostratigraphically consistent markers (sequence boundaries) initially identified and correlated on seismic data; this should be particularly useful in areas where correlation of individual lithologic units or formation tops on well logs is not apparent.

We believe that the use of seismic stratigraphic techniques in the Wind River basin will help to delineate possible stratigraphic and structural-stratigraphic hydrocarbon traps in the upper Fort Union Formation, similar to the Haybarn field (Robertson, 1984). The techniques should be equally applicable in other basins containing lacustrine and associated nonmarine strata.

ACKNOWLEDGMENTS

The authors acknowledge beneficial technical discussions with Texaco research personnel B. J. Katz, T.-C. Shih, G. S. Edwards, and G. Purnell, as well as with Texaco Denver Division explorationists R. Borcherding, C. Metzgar, D. Ephraim, and W. Hay. Figures were drafted by J. Mulvaney, N. Zeitlin, G. Griffith, and G. Novaez. M. Hill assisted in manuscript preparation. We also acknowledge reviews of the draft manuscript by James W. Castle and Walter B. Ayers, Jr.

We express sincere thanks to Geo-Seismic Services Inc. for permission to display their seismic data from a 1982 seismic survey, and to Texaco Inc. and its management in Houston and Denver for permission to present and publish these results.

REFERENCES

Ayers, W. B., Jr., 1986, Lacustrine and fluvial-deltaic depositional systems, Fort Union Formation (Paleocene), Powder River basin, Wyoming and Montana: AAPG Bulletin v. 70, p. 1651–1673.

Fox, J. E., and R. L. Priestley, 1983a, Preliminary chart A–A′ showing electric log correlation, facies, and test data of some Cretaceous and Tertiary rocks, Wind River basin, Wyoming: USGS Open-File Report 83--624-A, 1 pl.

Fox, J. E., and R. L. Priestley, 1983b, Preliminary chart B–B′ showing electric log correlation, facies, and test data of some Cretaceous and Tertiary rocks, Wind River basin, Wyoming: USGS Open-File Report 83-624-B, 1 pl.

Fox, J. E., and R. L. Priestley, 1983c, Preliminary chart C–C′ showing electric log correlation, facies, and test data of some Cretaceous and Tertiary rocks, Wind River basin, Wyoming: USGS Open-File Report 83-624-C, 1 pl.

Fox, J. E., and R. L. Priestley, 1983d, Preliminary chart D–D′ showing electric log correlation, facies, and test data of some Cretaceous and Tertiary rocks, Wind River basin, Wyoming: USGS Open-File Report 83-624-D, 1 pl.

Fox, J. E., and R. L. Priestley, 1983e, Preliminary chart E–E′ showing electric log correlation, facies, and test data of some Cretaceous and Tertiary rocks, Wind River basin, Wyoming: USGS Open-File Report 83-624-E, 1 pl.

Hyne, N. J., W. A. Cooper, and P. A. Dickey, 1979, Stratigraphy of intermontane, lacustrine delta, Catatumbo River, Lake Maracaibo, Venezuela: AAPG Bulletin v. 63, p. 2042-2057.

Keefer, W. R., 1961, Waltman Shale and Shotgun Members of Fort Union Formation (Paleocene) in Wind River Basin, Wyoming: AAPG Bulletin v. 45, p. 1310-1323.

Keefer, W. R., 1965a, Stratigraphy and geologic history of the uppermost Cretaceous, Paleocene, and lower Eocene rocks in the Wind River basin, Wyoming: USGS Professional Paper 495-A, 77 p.

Keefer, W. R., 1965b, Geologic history of the Wind River basin, central Wyoming: AAPG Bulletin v. 49, p. 1878-1892.

Keefer, W. R., 1969, Geology of petroleum in Wind River basin, central Wyoming: AAPG Bulletin v. 53, p. 1839-1865.

Landes, K. K., 1970, Petroleum geology of the United States: New York, Wiley-Interscience, 571 p.

McDonald, R. E., 1972, Eocene and Paleocene rocks of the southern and central basin, *in* W. M. Mallory, ed., Geologic atlas of the Rocky Mountain region: Rocky Mountain Association of Geologists, p. 243-256.

Meyer, B. L., and M. H. Nederlof, 1984, Identification of source rocks on wireline logs by density/resistivity and sonic transit time/resistivity crossplots: AAPG Bulletin v. 68, p. 121-129.

Mitchum, R. M., Jr., and P. R. Vail, 1977, Seismic stratigraphic interpretation procedure, *in* C. E. Payton, ed., Seismic stratigraphy—Applications to hydrocarbon exploration: AAPG Memoir 26, p. 135-143.

Lacustrine Basin Exploration: Case Studies and Modern Analogs

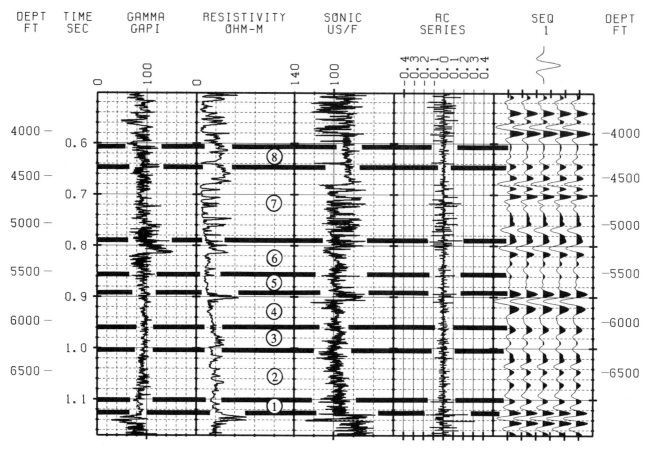

Figure 15. Seismic sequence boundaries displayed on synthetic seismogram and well log data, Humble Oil and Refining Frenchie Draw Unit 3 well, sec. 22, T37N, R89W. Sequence numbers are the same as in Figure 7. The upper Fort Union Formation is located approximately between 0.58 and 1.13 sec. The interval between 0.58 and 0.80 sec (sequences 7-8) is interpreted to be dominantly sandstone and siltstone with intercalated shale, associated with clinoforms of a deltaic system. See Figure 14 for well location.

Mitchum, R. M., Jr., P. R. Vail, and J. B. Sangree, 1977, Stratigraphic interpretation of seismic reflection patterns in depositional sequences, *in* C. E. Payton, ed., Seismic stratigraphy—Applications to hydrocarbon exploration: AAPG Memoir 26, p. 117-133.

Normark, R. M., 1978, Frenchie Draw gas field, *in* R. G. Boyd, G. M. Olson, and W. W. Boberg, eds., Resources of the Wind River basin: Wyoming Geological Association 30th Annual Field Conference Guidebook, p. 277-280.

Paape, D. W., 1968, Geology of Wind River basin of Wyoming and its relationship to natural gas accumulation, *in* B. W. Beebe, ed., Natural gases of North America: AAPG Memoir 9, p. 760-779.

Phillips, S. T., 1983, Tectonic influence on sedimentation, Waltman Member, Fort Union Formation, Wind River basin, Wyoming, *in* J. D. Lowell, and R. Gries, eds., Rocky Mountain foreland basins and uplifts: Rocky Mountain Association of Geologists, p. 149-167.

Pirner, C. F., 1978, Geology of the Fuller Reservoir Unit, Fremont County, Wyoming, *in* R. G. Boyd, G. M. Olson, and W. W. Boberg, eds., Resources of the Wind River basin: Wyoming Geological Association 30th Annual Field Conference Guidebook, p. 281-288.

Ray, R. R., 1982, Seismic stratigraphic interpretation of the Fort Union Formation, western Wind River basin—Example of subtle trap exploration in a nonmarine sequence, *in* M. T. Halbouty, ed., The deliberate search for the subtle trap: AAPG Memoir 32, p. 169-180.

Ray, R. R., and W. R. Keefer, 1985, Wind River basin, central Wyoming, *in* R. R. Gries and R. C. Dyer, eds., Seismic exploration in the Rocky Mountain region: Rocky Mountain Association of Geologists and Denver Geophysical Society, p. 201-212.

Robertson, R. D., 1984, Haybarn field, Fremont county, Wyoming—An upper Fort Union (Paleocene) stratigraphic trap: Mountain Geologist v. 21, p. 47-56.

Sangree, J. B., and J. M. Widmier, 1977, Seismic interpretation of clastic depositional facies, *in* C. E. Payton, ed., Seismic stratigraphy—Applications to hydrocarbon exploration: AAPG Memoir 26, p. 165-184.

Vail, P. R., and R. M. Mitchum, Jr., 1977, Seismic stratigraphy and global changes of sea level, part 1—Overview, *in* C. E. Payton, ed., Seismic stratigraphy—Applications to hydrocarbon exploration: AAPG Memoir 26, p. 135-143.

Vail, P. R., R. G. Todd, and J. B. Sangree, 1977, Chronostratigraphic significance of seismic reflections, *in* C. E. Payton, ed., Seismic stratigraphy—Applications to hydrocarbon exploration: AAPG Memoir 26, p. 99-116.

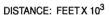

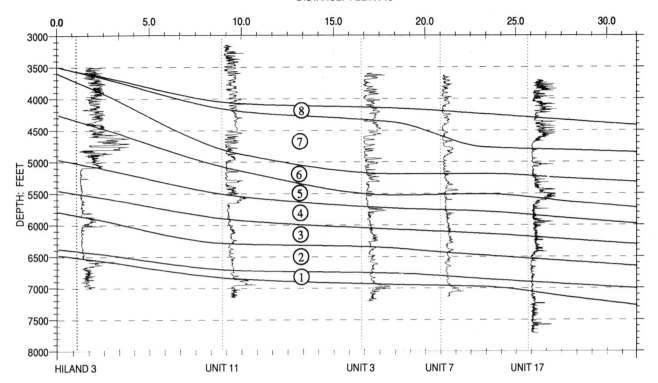

Figure 16. Resistivity log correlation of seismic sequence boundaries in the upper Fort Union Formation at Frenchie Draw field. See Figure 14 for well locations.

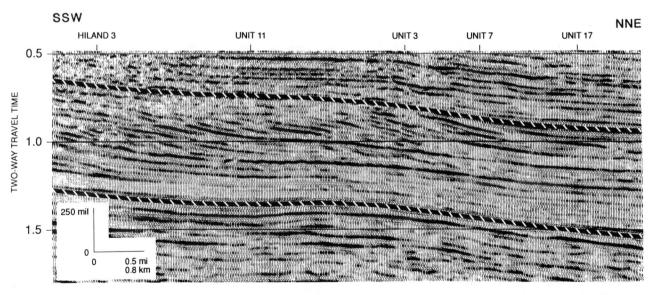

Figure 17. Seismic line 2 through the upper Fort Union Formation. Heavy lines indicate base and interpreted top of the upper Fort Union Formation, corresponding to the base of sequence 1 and the top of sequence 8. Clinoform geometry is apparent in the upper half of the interval. See Figure 14 for seismic line and well locations.

Lacustrine Basin Exploration: Case Studies and Modern Analogs

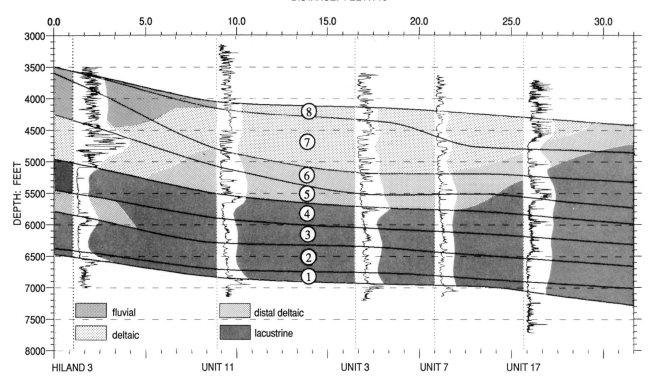

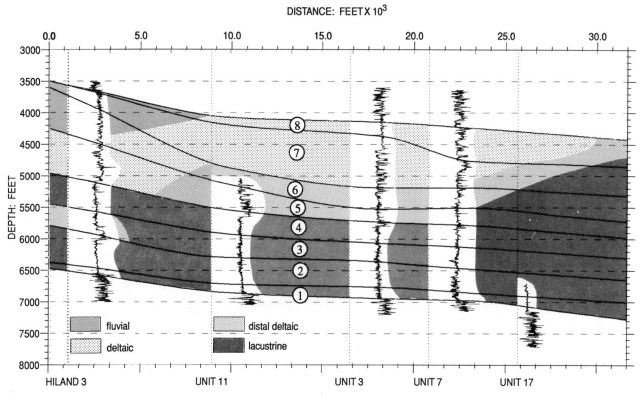

Figure 18. Depositional environments of the upper Fort Union Formation in Frenchie Draw field, inferred from seismic facies, overlain on seismic sequence- correlated logs. a, resistivity logs; b, velocity (sonic) logs. See Figure 14 for well locations.

Liro and Pardus

Sedimentation in Eocene Lake Uinta (Lower Green River Formation), Northeastern Uinta Basin, Utah

James W. Castle
Cabot Oil & Gas Corporation
Pittsburgh, Pennsylvania, U.S.A.

Detailed sedimentological analysis and subsurface paleoenvironmental mapping of Eocene sandstones in the upper Douglas Creek Member of the Green River Formation indicate that two major depositional systems, fluvial-deltaic and wave-dominated lacustrine shoreline, are represented along the northeastern margin of ancient Lake Uinta. Within correlative units, which include important oil-producing sandstones, depositional environments can be traced laterally from onshore fluvial-deltaic, to lacustrine barrier-beach, and to low-energy offshore lacustrine.

Sandstone-to-shale sequences that become finer grained upward above a sharp basal contact are interpreted to represent fluvial-channel deposition. Sandstone is quartzose in composition, commonly contains plant fragments, and is cross-bedded. Channel sandstone thickness is 1.5–3 m. Mapping of fluvial deposits indicates that elongate channel sandstone bodies trend southwest, which is consistent with a Uinta uplift source area to the north. General southwestward progradation of the fluvial-deltaic system was interrupted by lacustrine transgressions.

Sequences that gradually become coarser grained upward from shale and siltstone at the base to cross-bedded, fine- and medium-grained sandstone at the top are interpreted as prograding shoreface deposits. Sandstone is quartzose and commonly contains ooids. Thickness of shoreface sequences ranges from approximately 3 to 11 m. At the top of many coarsening-upward sequences, algal structures and probable mud cracks in thinly laminated dolomitic mudstone indicate very shallow water and possible subaerial exposure. Prograding shoreface deposits form thick, laterally continuous sandstone bodies interpreted as barrier-beach complexes. Paleoenvironmental mapping, based on data from 516 wells, shows that the barrier-beach facies occurs between offshore-lacustrine facies to the west-southwest and mud flat-lagoonal facies to the east-northeast. Maximum lateral dimensions of barrier-beach complexes are approximately 15 km parallel to paleoshoreline and 5 km across.

Comparison of these results with those with from previous studies of the Green River Formation in the Uinta basin indicates that the distribution of wave-produced sequences was restricted along the basin margin by local conditions related to shoreline configuration and lake bathymetry. Favoring development of shoreface sequences in certain areas were large wave fetch, shoreline stability, clastic sediment input, and deep water offshore (i.e., depth below wave base). During at least some of the time represented by deposition of the lower Green River Formation, those conditions persisted along the northeastern margin of Lake Uinta but not along other lake margin areas.

The description of lacustrine sandstones from this study can be used as a comparison with lacustrine deposits in other areas. As demonstrated for the lower Green River Formation, fluvial channel sandstones and shoreface sandstones form major hydrocarbon reservoirs in proximity to one another along margins of ancient lakes.

INTRODUCTION

In previous studies of Green River Formation reservoir sandstones, various depositional environments, including fluvial, deltaic, and lacustrine shoreline, have been interpreted. Questions regarding the origin of Green River sandstones and recognition of the importance of distinguishing fluvial from lacustrine, have persisted for more than thirty years (e.g., Picard, 1957b). This study describes an example of marginal-lacustrine deposits to help explain some of the diversity among previous depositional interpretations of Eocene clastics that were deposited around the flanks of the Uinta basin.

The purpose of this investigation is to interpret the detailed sedimentology of cored sequences and the lateral distribution of paleoenvironments for clastic reservoirs in the lower Green River Formation over an area of approximately 650 km² in the northeastern Uinta basin (Figure 1). Oil field development at Red Wash, Walker Hollow, and Wonsits Valley has provided abundant subsurface data for the study.

Most of Uinta basin oil production is from Paleocene and Eocene strata. In the study area nearly all production is from the Douglas Creek and Garden Gulch Members of the Green River Formation at depths of 1200 to 1850 m. The Green River Formation, which is approximately 900 m thick in northeastern Uinta basin, conformably overlies and, in places, intertongues with the fluvial Wasatch Formation. Stratigraphic nomenclature used in this study follows Bradley (1931) and most previous investigators (Figure 2).

In general, structural dip in the area is 1–2°NW (Figure 3). An anticlinal nose, which plunges gently westward, occurs in Red Wash field. A northwest-plunging syncline is present near the juncture of Red Wash and Wonsits Valley fields. Hydrocarbon accumulation generally is limited updip by stratigraphic pinch-out of individual reservoirs. Based on core studies, geophysical logs, and production data, the effect of natural fracturing on hydrocarbon accumulation and production generally appears to be small.

In the study area multiple sandstones and minor carbonate rocks are interbedded with dolomitic mudstone and shale. In the northeast, upper Douglas Creek sandstone reservoirs generally are thin and laterally discontinuous. In contrast, reservoirs are thicker and more continuous in the southwest (Figure 4). Farther southwest, sandstone grades into oolitic grainstone and shale. The source of hydrocarbons produced at the Red Wash and Wonsits Valley fields probably is offshore-lacustrine shale.

Many depositional environments have been interpreted for the Green River Formation. Bradley (1931) suggested that Green River shale represents deposition in a deep, permanently stratified lake, with laminated oil shale deposited near the lake's center. Dolomitic mudstone in the Green River Formation of Wyoming was interpreted by Eugster and Surdam (1973), Eugster and Hardie (1975), and Smoot (1978) to represent playa lake deposition. For sandstones, various lacustrine shoreline environments have been proposed (Sanborn and Goodwin, 1965; Picard 1967; Picard and High, 1970). The Green River Formation at Red Wash was interpreted as deltaic by Koeso-emadinata (1970) and as barrier-bar by Webb (1978). A fluvial origin of Green River reservoirs has been interpreted more recently (Pitman et al., 1982; Fouch, 1985; Franczyk et al., 1989). Picard (1957b), Picard and High (1970, 1972), and High and Picard (1971) cited several criteria for differentiating lacustrine and fluvial deposits in the Green River Formation.

In general, clastic sediment deposited along the northeastern flank of the Uinta basin was derived from erosion of the Uinta uplift and transported south through fluvial systems to Lake Uinta. Sedimentation patterns and sequences in the study area formed in response to transgressions and regressions of the lake (Bradley, 1931; Picard, 1955). Lacustrine transgressions generally proceeded from the lake's center northeastward toward the alluvial plains.

METHODS OF STUDY

The upper 30 m of the Douglas Creek Member was selected for detailed stratigraphic study (Figures 2, 4). The interval includes the "K" interval and adjacent parts of the overlying "J" and underlying "KA" intervals defined in the Red Wash field (Chevron unpublished reports) and occurs approximately midway between the middle marker and carbonate marker of Fouch (1975) and Ryder et al. (1976). The base of the zone lies about 100 m above the top of the Wasatch Formation. This interval is a major producing zone in both the Red Wash and Wonsits Valley units. Log character and sandstone distribution indicate that the interval generally is representative of reservoirs within the study area.

All available cores through the interval were examined (Figure 3). The approximately 450 m of cored rock described and interpreted includes 165 m from eight Wonsits Valley wells and 180 m from nine Red Wash wells. Selected cores from other lower Green River producing zones were examined briefly for comparison.

Core description emphasized identifying features for sedimentological analysis and relating lithologic characteristics to reservoir quality and logs (gamma-ray, SP, resistivity, sonic, density, and neutron). Lithofacies were assigned to facilitate description and interpretation of various rock types and sequences. Seventy samples were analyzed in thin section and 128 samples by x-ray diffraction (XRD). Organic geochemical analyses, including total organic carbon (TOC) and microscopic organic analysis, were performed on 96 samples. Porosity and permeability had been measured previously at 30-cm intervals in sandstone beds. Depositional environments were interpreted from the vertical succession of grain size, sand percentage, and sedimentary structures, and from the analytical data.

Following comparison of logs to results of core analysis, depositional environments were interpreted from log patterns of 516 uncored wells (Figure 5). Paleoenvironments were mapped for each of three stratigraphic units (Figure 4) by using interpretations from log patterns to tract depositional environments identified in and calibrated to described cores. Lateral facies relationships and a depositional model then were developed from the

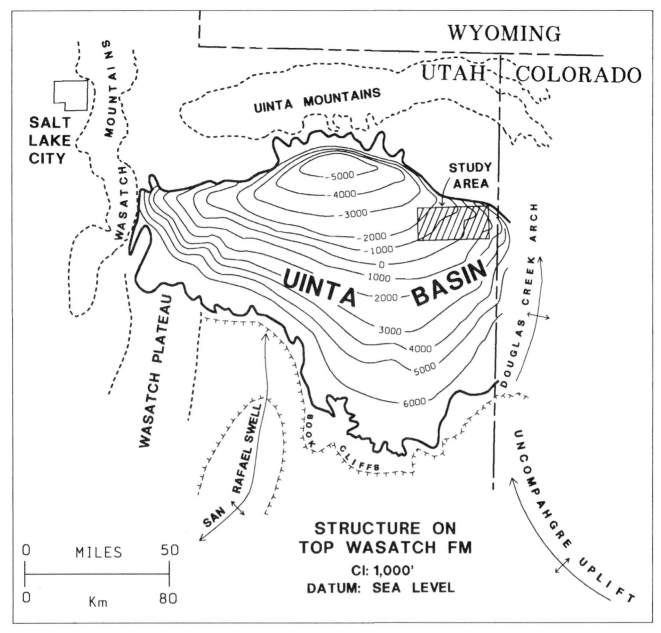

Figure 1. Location of Uinta basin and vicinity, northeastern Utah, showing regional tectonic elements and structure on top of the Wasatch Formation (after Osmond, 1964).

paleoenvironmental maps. Seismic data could not be used because individual reservoir units are too thin to be resolved with available seismic data.

GENERAL LITHOLOGIC DESCRIPTION

Sandstone in the cores examined generally is very fine- to medium-grained and moderately to well sorted. Minor, poorly sorted, coarse- to very coarse-grained sandstone contains occasional granule- and pebble-size rock fragments. Cross-bedding, usually in sets 5–15 cm thick, is common in fine- to medium-grained sandstone; low-angle, inclined cross-bedding and flat bedding occur less commonly. In very fine-grained sandstone, wave-ripple bedding, current-ripple bedding, and burrows occur commonly.

Mineralogically, the sandstone is 70–90% detrital quartz with minor feldspar and lithic fragments. Most rock fragments are composed of dolomitic mudstone of probable intraformational origin. Volcanic and metamorphic rock fragments are rare. Carbonate ooids and oolitically coated quartz grains present in some sandstones form as much as 25% of total rock volume. The amount and distribution of detrital clay, which occurs as intergranular matrix and thin interlaminae, are highly variable. Detrital

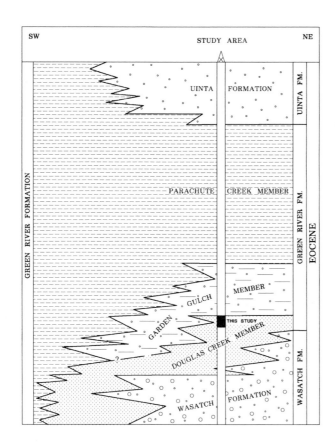

clay minerals include illite, mixed-layer illite/smectite, and kaolinite.

In most sandstone beds, authigenic mineral growth has caused minor reduction in porosity and permeability. Quartz and iron-rich calcite, the most commonly occurring mineral cements, each usually form 5–15% of total rock volume. Minor sandstones are extensively cemented with ferroan dolomite. Authigenic clay minerals include illite and illite/smectite, but Ray (1985) reported the occurrence of minor authigenic kaolinite in Red Wash sandstones. In contrast to Green River sandstones studied elsewhere in the Uinta basin (e.g., Pariette Bench field, Pitman et al., 1982), authigenic clay in most Red Wash and Wonsits Valley sandstones occurs in only small amounts and generally has not greatly reduced porosity or permeability.

Although most pore spaces in Red Wash and Wonsits Valley sandstones probably are primary in

Figure 2. Generalized stratigraphy of lower Tertiary rocks in northeastern Uinta basin. Stratigraphic nomenclature follows that used traditionally for the Red Wash area (Bradley, 1931; Chatfield, 1972). Subdivision of the Green River Formation into four members is consistent with most previous studies (e.g., Picard, 1957a; Cashion, 1967; Hintze, 1980).

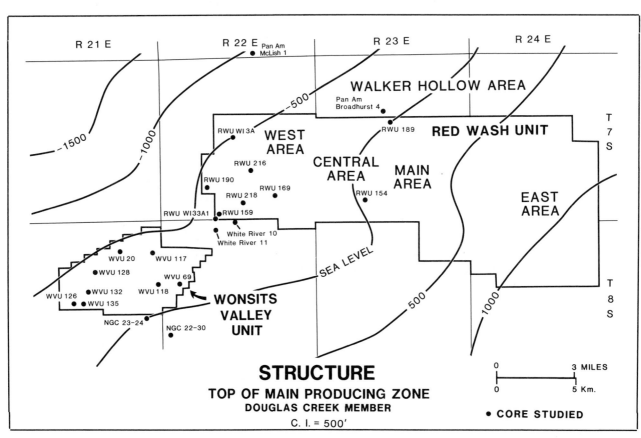

Figure 3. Simplified structure map of study area showing locations of cores studied and four principal areas of the Red Wash unit.

Castle

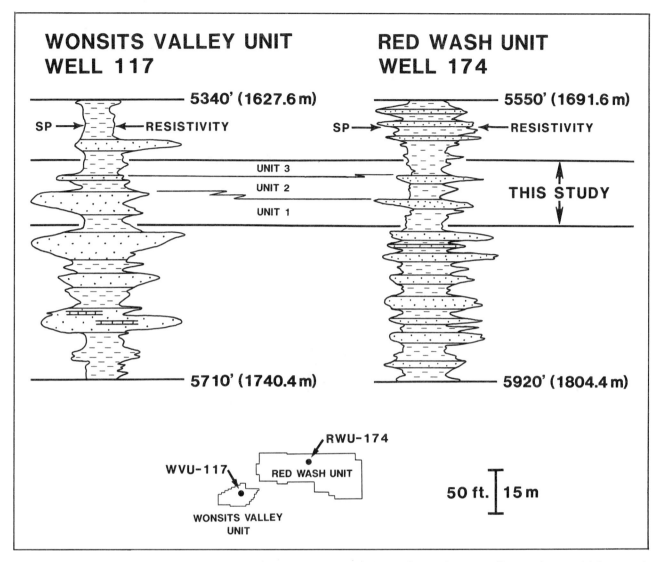

WONSITS VALLEY UNIT WELL 117

5340' (1627.6 m)

SP → ← RESISTIVITY

UNIT 3
UNIT 2
UNIT 1

5710' (1740.4 m)

RED WASH UNIT WELL 174

5550' (1691.6 m)

SP → ← RESISTIVITY

THIS STUDY

5920' (1804.4 m)

RWU-174
RED WASH UNIT
WVU-117
WONSITS VALLEY UNIT

50 ft. 15 m

Figure 4. Typical logs of the uppermost Douglas Creek Member in Wonsits Valley and Red Wash units. Compared to Red Wash unit, reservoir sandstones in Wonsits Valley unit generally are fewer, thicker, and more laterally continuous. Stratigraphic units 1, 2, and 3 are correlatable intervals mapped in this study.

origin, large elongated pores indicate some secondary origin. From observations, it is estimated that 25% of total porosity may be the result of dissolution of mineral grains and cements. It is possible, but unproven, that carbonate dissolution may have been induced by waterflooding. However, porosity and permeability values measured in cores are consistent with production volumes, which suggests that core values have not been significantly altered.

Shale beds are highly diverse in composition, internal structures, and color. Five shale subfacies have been identified (Figure 6D). Shale beds contain variable amounts of quartz silt and very fine-grained sand, which occur in lenses, interlaminae, and thin interbeds. In shales analyzed by XRD, common illite/smectite and dolomite were detected together with variable, but usually minor, illite, calcite, kaolinite, and analcime. Minor plant fragments, ostracodes,

fish bones, and fish scales also occur in some beds, as well as mud cracks and small, vertical root structures in intervals of interlaminated shale and siltstone.

Microscopic organic analysis indicates that the organic content of most shale samples consists of a large proportion of alginite that probably was produced by freshwater algae in a lacustrine environment and preserved in that subaqueous environment. In several samples, most of the organic matter consists of vitrinite, probably derived from land plants and transported to the site of deposition. The proportion of alginite to vitrinite within correlative shales generally increases southwestward or basinward.

Small, probably incipient fractures present in some shale intervals may have formed during regional uplift or possibly during coring and core recovery.

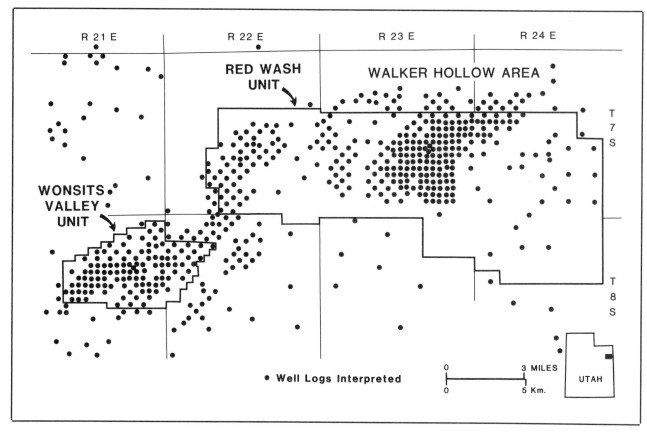

Figure 5. Locations of wells for which depositional environments were interpreted in Red Wash and Wonsits Valley units and vicinity.

SEQUENCE DESCRIPTION AND INTERPRETATION

Definition of Terms

Interpretive terminology used in this study includes not only terms commonly applied to alluvial and lacustrine deposits but also those applied to nearshore-marine deposits, including lower shore-face, upper shoreface, swash zone, and backshore (back-barrier). Other environmental terms include offshore lacustrine, nearshore lacustrine, mud flat-lagoon, and low-energy bar.

Offshore-lacustrine environment refers to an area of normally quiet-water deposition wherein the sediment surface lies below fair-weather wave base. The *nearshore-lacustrine* environment is defined as a low-energy, normally shallow-water regime with some wave reworking and possible minor subaerial exposure. A *lagoon* environment refers to an area of shallow water parallel to the shoreline and connected via an outlet to a larger water body. A *mud flat* is defined as a flat, very shallow to exposed area of low-energy sedimentation. Because of the difficulty of distinguishing lagoonal from mud flat deposits,

and their probable interbedding, the environments are usually grouped as *mud flat-lagoon*. Finally, a *low-energy bar* forms where wave energy is sufficient for minor reworking of sand, usually resulting in a coarsening-upward sequence.

Sandstone bodies mapped in this study are considered to be barrier-beach deposits if the shoreface sequences are positioned laterally between offshore-lacustrine and lagoonal or onshore deposits. A *barrier-beach complex* is defined as a long, narrow sand accumulation aligned parallel to shoreline (after Elliott, 1978). If connected to land, either subaerially or subaqueously, the accumulation is considered to be a beach. On the other hand, if the accumulation is separated from land by a shallow lagoon, then it is classified as a barrier island. Following Elliott's terminology, barrier-beach sands may occur within deltas, along depositional strike from deltas, or in an oceanic or lacustrine context completely devoid of deltas. Although high-energy, stable-shoreline conditions favorable for barrier-beach deposition occur more commonly in shallow-marine settings than in lakes, conditions favorable for barrier-beach complexes apparently were present, at times, along the northeastern shore of Lake Uinta. In this paper, *fluvial-deltaic* refers to a regime in which multiple

248   Castle

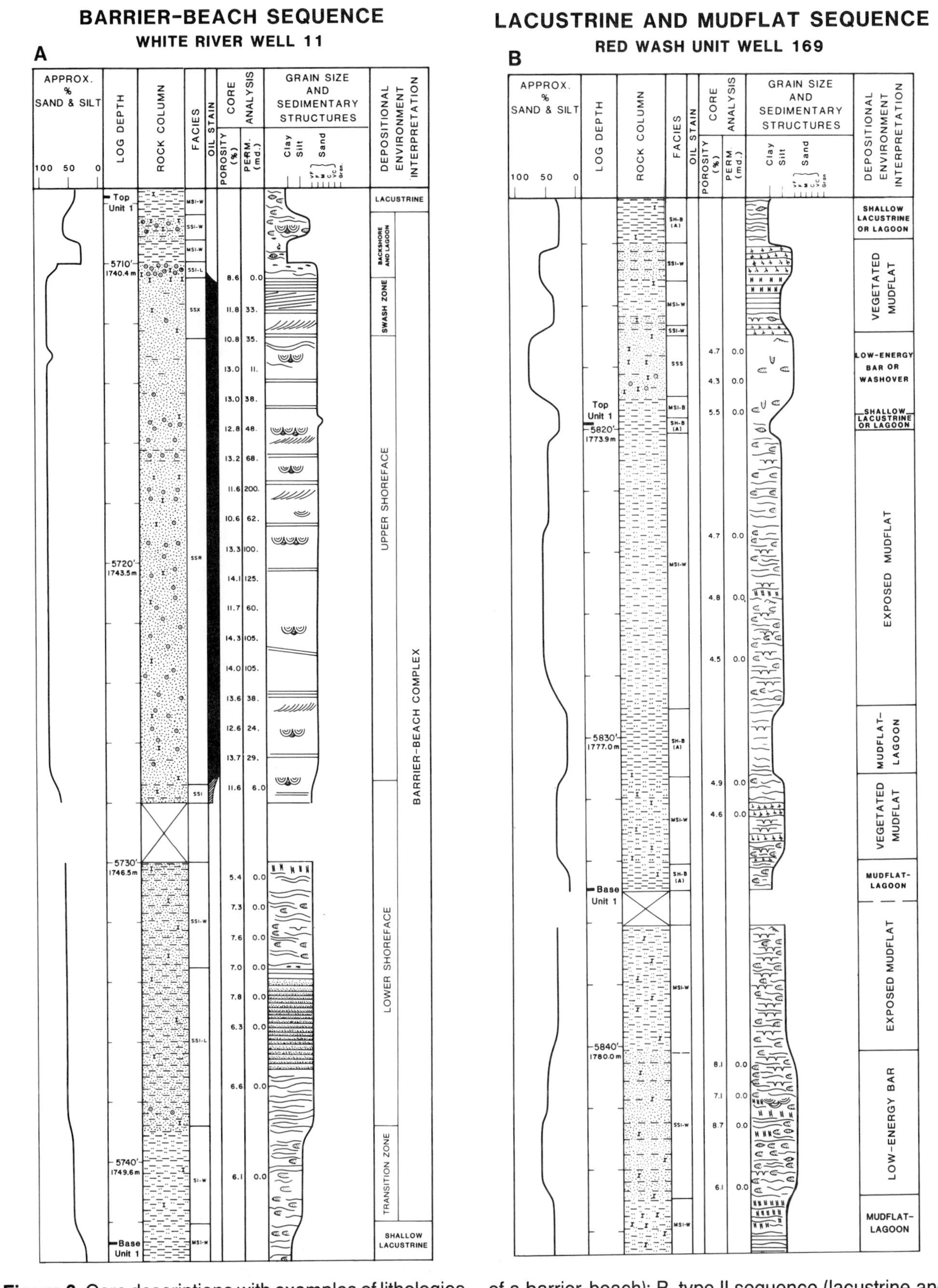

Figure 6. Core descriptions with examples of lithologies and sequences for various depositional environments interpreted. A, type I sequence (prograding shoreface of a barrier-beach); B, type II sequence (lacustrine and mud flat).

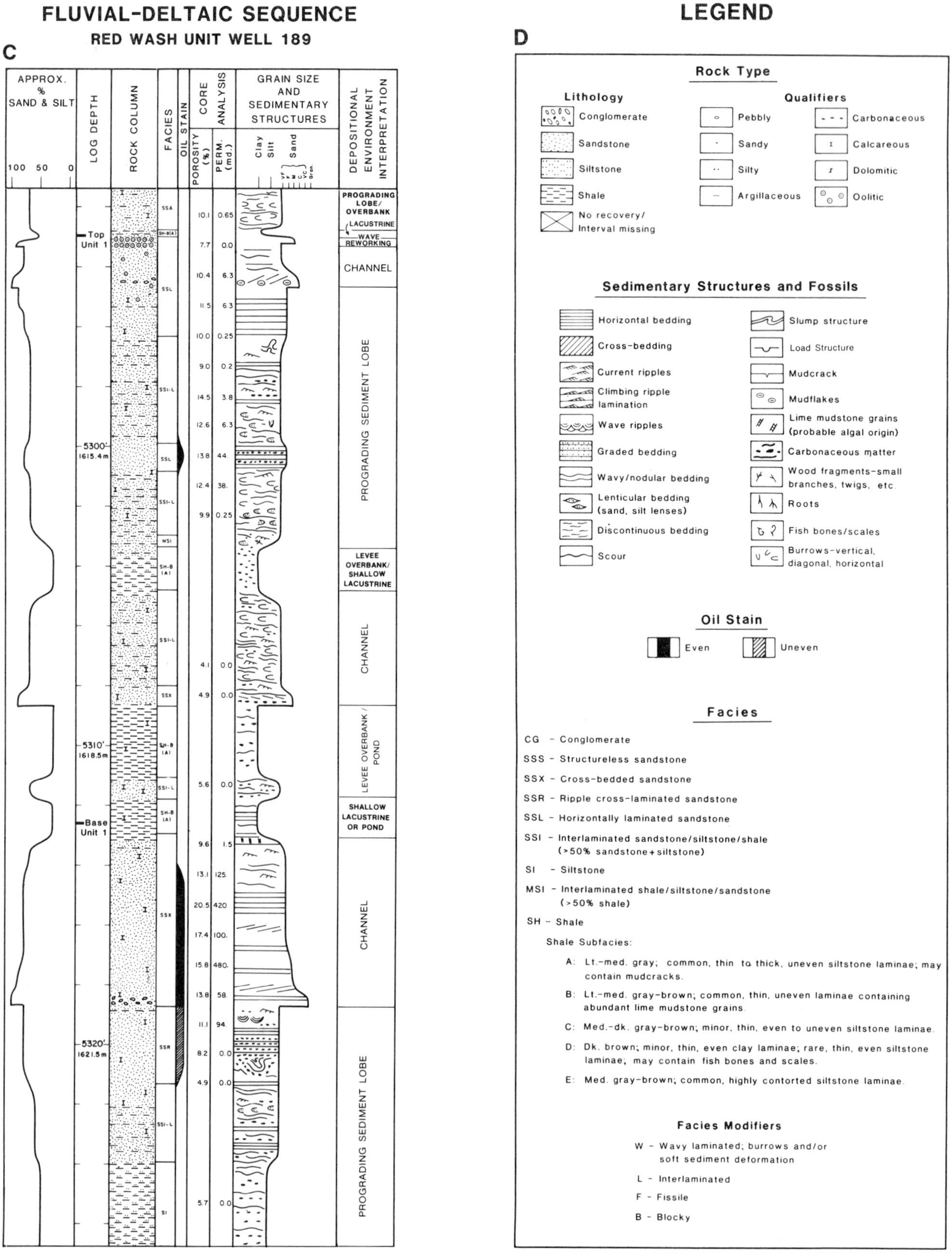
FLUVIAL-DELTAIC SEQUENCE
RED WASH UNIT WELL 189
C
APPROX. % SAND & SILT
100 50 0
LOG DEPTH
ROCK COLUMN
FACIES
OIL STAIN
CORE ANALYSIS
POROSITY (%)
PERM. (md.)
GRAIN SIZE AND SEDIMENTARY STRUCTURES
Clay Silt Sand
DEPOSITIONAL ENVIRONMENT INTERPRETATION
Top Unit 1
5300' 1615.4m
5310' 1618.5m
Base Unit 1
5320' 1621.5m
PROGRADING LOBE/ OVERBANK
LACUSTRINE
WAVE REWORKING
CHANNEL
PROGRADING SEDIMENT LOBE
LEVEE OVERBANK/ SHALLOW LACUSTRINE
CHANNEL
LEVEE OVERBANK / POND
SHALLOW LACUSTRINE OR POND
CHANNEL
PROGRADING SEDIMENT LOBE
LEGEND
D
Rock Type
Lithology
Conglomerate
Sandstone
Siltstone
Shale
No recovery/ Interval missing
Qualifiers
Pebbly
Sandy
Silty
Argillaceous
Carbonaceous
Calcareous
Dolomitic
Oolitic
Sedimentary Structures and Fossils
Horizontal bedding
Cross-bedding
Current ripples
Climbing ripple lamination
Wave ripples
Graded bedding
Wavy/nodular bedding
Lenticular bedding (sand, silt lenses)
Discontinuous bedding
Scour
Slump structure
Load Structure
Mudcrack
Mudflakes
Lime mudstone grains (probable algal origin)
Carbonaceous matter
Wood fragments-small branches, twigs, etc.
Roots
Fish bones/scales
Burrows-vertical, diagonal, horizontal
Oil Stain
Even
Uneven
Facies
CG - Conglomerate
SSS - Structureless sandstone
SSX - Cross-bedded sandstone
SSR - Ripple cross-laminated sandstone
SSL - Horizontally laminated sandstone
SSI - Interlaminated sandstone/siltstone/shale (>50% sandstone+siltstone)
SI - Siltstone
MSI - Interlaminated shale/siltstone/sandstone (>50% shale)
SH - Shale
Shale Subfacies:
A: Lt.-med. gray; common, thin to thick, uneven siltstone laminae; may contain mudcracks.
B: Lt.-med. gray-brown; common, thin, uneven laminae containing abundant lime mudstone grains.
C: Med.-dk. gray-brown; minor, thin, even to uneven siltstone laminae.
D: Dk. brown; minor, thin, even clay laminae; rare, thin, even siltstone laminae; may contain fish bones and scales.
E: Med. gray-brown; common, highly contorted siltstone laminae.
Facies Modifiers
W - Wavy laminated; burrows and/or soft sediment deformation
L - Interlaminated
F - Fissile
B - Blocky

Figure 6. C, type III sequence (fluvial-deltaic). Type IV sequence (not illustrated) consists entirely of shale, which represents deposition in offshore- and nearshore-lacustrine environments. Log depths shown are adjusted from core depths by detailed comparison between log curves and core lithologies. Lithology and oil stain were determined visually. Helium porosity and air permeability were measured by mechanical core-analysis techniques.

streams flowed across an alluvial plain and then into a lake (after Franczyk et al., 1989). The term applies to the general depositional system, which includes multiple deposits that formed at or near the lake's edge. *Prograding lobe* refers to a preserved progradational deposit of a stream-mouth bar that formed where a channel entered the lake.

Table 1 compares the terminology of depositional environments interpreted in this study to that used by Ryder et al. (1976) for the Green River Formation in the western Uinta basin.

Sequence Type I

Description

The most common type of sequence observed in cores begins with shale and siltstone at the base and coarsens upward into fine- and medium-grained sandstone (Figure 6A). The sequences range in thickness from approximately 3 to 11 m. Near the base of the type I sequence, shale is medium to dark gray in color, silty, and laminated. The shale contains a large proportion of alginite relative to vitrinite. Light-gray siltstone is interbedded with the shale. In the lower part of the coarsening-upward sequences, bedding usually is wavy and disrupted by small (<5 mm across) burrows typically oriented parallel to bedding.

The sequence gradually coarsens upward into light-gray, very fine- to fine-grained sandstone (Figure 7A). Wave-ripple and current-ripple bedding occur commonly. Also present are minor, horizontal and vertical burrows and rare, thin graded beds, which fine upward from fine-grained sandstone at the base to siltstone and shale at the top.

Grain size and sand percentage continue to increase upward into light-gray, fine- and medium-grained sandstone (Figure 7B). Primary structures include cross-bedding in 5- to 15-cm-thick sets, asymmetrical wave-ripple bedding, and current-ripple bedding. Flat bedding, which is probably upper flow-regime bedding, and low-angle cross-bedding are common in the upper part of the sandstone (Figure 7C). White, fine- to medium-grained carbonate ooids and oolitically coated quartz grains occur commonly in the sandstone. The percentage of ooids generally increases upward to a maximum 25% of total rock volume.

Minor scoured surfaces overlain by fine- to coarse-grained sandstone occur within the upper type I sequences in cores RWU-190, RWU-216, RWU-218, and RWU-WI33A. Grain size decreases sharply within several centimeters above the scoured contact to very fine and fine. Cross-bedding and sometimes flat bedding occur immediately above some scoured surfaces. Small mud chips occur rarely in the sandstone. Based on petrographic study, the mineralogy of sandstone above the scoured surface appears to be identical to that below. Ooids and oolitically coated quartz grains occur in both the overlying and underlying sandstone.

Table 1. Comparison of depositional environment terminology between northeastern Uinta basin (this study) and western Uinta basin (Ryder et al., 1976)

Northeastern Uinta Basin	Western Uinta Basin
Offshore Lacustrine	Open Lacustrine
Nearshore Lacustrine	Open Nearshore Lacustrine
Barrier-Beach	Not Observed*
Clastic Mud Flat-Lagoon**	Lake-Margin Carbonate Flat**
Low-Energy Bar	Lake-Margin Carbonate Flat
Fluvial-Deltaic	Deltaic and Interdeltaic
Channel	Channel
Overbank	Overbank
Prograding Sediment Lobe	Delta Front
Not Observed	Alluvial Facies

*Major, clastic barrier-beach deposits are not reported in western Uinta basin. However, carbonate grainstone units are interpreted as beach and shoal deposits of the lake-margin carbonate-flat environment (Ryder et al., 1976).

**Differences between clastic mud flat-lagoon deposits and lake-margin carbonate-flat deposits are caused primarily by the amount of clastic input and volume of freshwater input to the depositional environment. Clastic sediment deposition predominates in the clastic mud flat-lagoon environment, and carbonate deposition predominates in the lake-margin carbonate-flat environment.

In many cores a 0.3- to 1.5-m-thick interval of thinly interbedded and interlaminated, gray, fine-grained sandstone and light-brown dolomitic mudstone overlies the coarsening-upward sequence (Figure 7D). Thinly laminated dolomitic mudstone contains burrows and probable desiccation structures. In some cores a single 0.3- to 1-m thick, white to light-gray, fine- to medium-grained sandstone lies just above the coarsening-upward sequence but is separated from it by a thin shale. This well sorted sandstone shows both flat bedding and cross-bedding and contains common to abundant medium-grained ooids and oolitically coated quartz grains. The sandstone usually is tightly cemented by calcite and dolomite.

Interpretation

This vertical sequence of grain size and sedimentary structures indicates an upward increase in energy level, which is interpreted as a gradation from offshore-lacustrine to wave-dominated shoreline conditions (Figure 8). In the lower type I sequence, interbedded shale, siltstone, and very fine-grained sandstone represent a transition zone between a sub-wave base, offshore-lacustrine environment and a lower shoreface environment. The transition beds have a maximum thickness of 3 m and average approximately 0.6 m. In the shale interbeds, high alginite content, which averages 80% of the organic material, is more consistent with lacustrine deposition than with fluvial or deltaic deposition. Above the transition interval, very fine- to fine-grained

sandstone represents lower shoreface deposition because of grain size and primary structures, which include wave-ripple and current-ripple bedding. Thickness of the lower shoreface interval ranges from approximately 1 to 6 m. Thin graded beds that occur in this part of the sequence may have originated by redeposition of sediment that was eroded and transported by storm waves.

The increase in energy level from lower shoreface to upper shoreface environments is indicated by upward increasing grain size and sand content and by sedimentary structures. High-energy flat bedding and low-angle cross-bedding near the top of some upper shoreface deposits suggests deposition in the swash-backwash zone of a beach. Thickness of the upper shoreface-swash zone interval ranges from 1 to 5 m. Multiple coarsening-upward sequences observed in some cores may represent bars developed on the shoreface. The upward increase in percentage of ooids and oolitically coated quartz grains is consistent with shoaling conditions. Preservation of thin, apparently fragile, concentric coatings on many ooids indicates that the ooids are not reworked from older deposits. A foreshore zone, defined as the interval between high mean tide and low mean tide, was not recognized in any cores probably because of the absence of tidal processes along the lacustrine shoreline.

Minor scoured surfaces overlain by cross-bedded, fine- to coarse-grained sandstone probably represent erosion and deposition in small, locally developed runoff channels. Bed thinness and similarity in composition, including ooid content above and below the scoured surface, suggest that the channels were not major conduits for sediment transport.

Well-sorted, fine-grained sandstone that is thinly interbedded and interlaminated with dolomitic mudstone and that overlies the coarsening-upward sequences represents backshore deposition. Sand grains probably were reworked from upper shoreface and swash-zone deposits by wind and water, including storm waves and washover. The dolomitic mudstone probably represents deposition in a backshore mud flat or lagoon. Growth of algal mats may have contributed the fine-grained carbonate and

produced some of the thin lamination seen in the mudstone.

The 0.3- to 1-m-thick, fine- to medium-grained sandstone beds that overlie the shoreface sequences are interpreted as deposits reworked during a rise in lake level. Evidence for reworking includes grain size, good sorting, abundance of ooids and oolitically coated quartz grains, and position in the sequence. The sandstone usually is underlain by backshore or nearshore-lacustrine deposits and overlain by offshore-lacustrine shale.

Sequence Type II

Description

Interlaminated and interbedded shale, siltstone, and very fine- to fine-grained sandstone represent a second major sequence type (Figure 6B). The thicknesses are commonly 5 to 10 m thick. Medium- to dark-gray shale and light- to medium-gray siltstone are the most common lithologies. Minor light-gray sandstone and rare light-brown dolomitic mudstone occur.

Sandstone and siltstone beds, commonly less than 1 m but up to 4 m thick, contain medium- to dark-gray shale laminae. Some beds gradually coarsen upward from siltstone at the base to fine-grained sandstone at the top. Upper contacts of the beds are sharp to gradational. Rarely, sand-sized dolomitic mudstone grains are concentrated in layers within the sandstone. Ooids occur rarely. Minor wave-ripple bedding and cross-bedding in sets up to 7 cm thick are present. Small burrows commonly disrupt the bedding surfaces. Probable mud cracks and small, black, vertical structures (1 cm long and 2 mm wide) are common within discrete intervals (Figures 7E, 7F). Composition of microscopic organic material in the shale is highly variable in total quantity and in proportion of alginite to vitrinite.

Interpretation

In the type II sequence, the large proportion of shale and shaly siltstone to sandstone indicates a predominantly low-energy environment, with deposition

Figure 7. Representative cores of major depositional environments. A, ripple-laminated, very fine-grained sandstone, which grades upward into siltstone and shale; shale laminae burrowed; interpreted as lower shoreface (WVU–117, 1663 m core depth). B, ripple-laminated, fine- to medium-grained sandstone of upper shoreface (WVU–117, 1670.5 m core depth). C, well-sorted, fine-grained sandstone; high-energy flat bedding overlain by low-angle cross-bedding interpreted as upper shoreface/swash zone (White River–11, 1741 m core depth). D, interlaminated oolitic fine-grained sandstone and dolomitic mudstone; some ooids are black, others gray; interpreted as backshore/mud flat-lagoon (RWU–216, 1760.4 m core depth). E, interlaminated siltstone and dolomitic mudstone; vertical, black root structures in siltstone; algal lamination in mudstone; stromatolitic algal structure approximately 3 cm below top; interpreted as mud flat-lagoon (RWU–218, 1753 m core depth). F, interlaminated siltstone and shale; burrows and probable deformed mud cracks; interpreted as exposed mud flat (RWU–169, 1779.6 m core depth). G, laminated shale; thin, silty interlaminae; interpreted as nearshore-lacustrine (WVU–117, 1675 m core depth). H, pebbly, coarse-grained sandstone above sharp basal contact; interpreted as fluvial channel (RWU–189, 1618 m core depth). I, interlaminated very fine-grained sandstone and shale; soft-sediment deformation; common, black plant fragments; interpreted as levee-overbank (RWU–189, 1618.9 m core depth). Scale bar is 5 cm long.

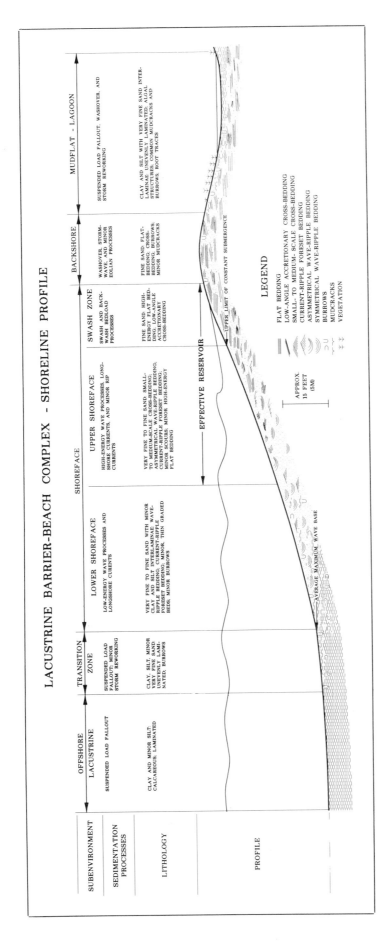

Figure 8. Schematic profile of lacustrine barrier-beach shoreline, showing processes and lithologies of major depositional environments and subenvironments. Prograding shoreface deposition is interpreted from detailed core study. Barrier-beach interpretation is based on overall vertical sequence and lateral distribution of lithologies. Basinward shoreline progradation produced clastic coarsening-upward (type I) sequences, which are common in producing zones of Wonsits Valley-west Red Wash area. Hydrocarbon production is from upper shoreface-swash zone, which is indicated as effective reservoir. Shales shown contain abundant algal organic material. Sediment-surface irregularities caused by local variations in transportational and depositional processes probably occurred but are not shown. Boundaries between subenvironments are gradational. Vertical scale approximate; no horizontal scale.

Castle

mainly from suspended-load fallout. However, periodic wave and current reworking is suggested by wave-ripple bedding and cross-bedding in the sandstone. Sedimentation rate and organic input were favorable for survival of burrowing organisms during deposition of the fine-grained sediment. Dolomitic mudstone laminae are interpreted to have formed by carbonate production associated with algal growth or with evaporative conditions. Rare mounded structures, 2 cm in diameter, probably represent algal buildups (Figure 7E). The dolomitic mudstone grains in the sandstone may be analogous to dolomite peloidal intraclasts described by Smoot (1978) from the Wilkins Peak Member of the Green River Formation, Wyoming. He suggested formation of the intraclasts by disintegration of dolomitic surface crusts on an evaporative mud flat.

Based on lithologies, organic content, and primary structures, the type II sequence is interpreted to represent deposition mainly on a clastic mud flat. The occurrence of dolomitic intraclasts, mud cracks, and small, black vertical structures, interpreted as root traces, indicates that the mud flat was periodically exposed and vegetated. Shale intervals with small percentages of silt and sand grains and that lack evidence of exposure are interpreted as shallow (nearshore) lacustrine or bay/lagoon deposits (e.g., Figure 6B, 1771.5–1772 m depth; Figure 7G). Coarsening-upward intervals are interpreted as prograding low-energy bars. Occasional wave reworking may have occurred during storms.

Sequence Type III

Description

Fining-upward sequences, the third sequence type, range from 1.5 to 3 m thick and are characterized by a sharp, usually scoured base and an upward decrease in grain size and sand percentage (Figures 6C, 7H). Sandstone is light gray and brown and very fine to medium grained. Modal grain size at the base of the sequences is as large as very coarse to granular. Pebble-sized shale rip-up clasts occur as basal lag deposits. Cross-bedding in sets up to 50 cm thick and high-energy flat bedding occur in the lower parts of sandstone beds. In the upper part, very fine-grained sandstone with current-ripple cross-lamination is interbedded with siltstone and shale. Abundant dish-and-pillar fluid-escape structures were seen in the Pan Am Broadhurst No. 4 core.

Separating the fining-upward sequences are light-gray and brown shale and siltstone. The shale typically is sandy and contains siltstone and sandstone lenses and laminae (Figure 7I). Minor ripple cross-lamination occurs in thin interbeds of siltstone and very fine-grained sandstone. Burrows are typically absent to minor. Coarsening-upward shale-to-sandstone intervals ranging from 1.5 to 4.5 m thick lie immediately below some fining-upward sequences (e.g., Figure 6C, 1621.2–1623.6 m depth).

The microscopic organic material in most shale samples consists almost entirely of vitrinite. TOC content of vitrinitic shale averages less than 0.5%. Common macroscopic plant fragments occur in both the shale and sandstone. Small leaf and stem fragments were observed in the sandstone.

Interpretation

Type III sequences are interpreted as fluvial-channel and overbank deposits because of a sharp and usually scoured base, basal lag with mudstone chips, common land-plant debris, and upward decrease in grain size. Interbedded vitrinitic shale represents flood-plain deposition. Minor alginite-rich shale probably formed from sediment that accumulated in a lake or pond, which may have developed locally on the flood plain or may have been connected to Lake Uinta. The alginitic shales generally are more laterally extensive than the vitrinitic shales and may have formed during lacustrine transgressions.

The minor coarsening-upward intervals associated with fluvial-channel sequences are interpreted as lobe deposits that prograded where a channel entered a standing body of water. Common macroscopic plant fragments (vitrinite), highly deformed bedding, and association with fining-upward sequences differentiate the prograding-lobe sequences from type I coarsening-upward sequences (shoreface deposits).

Sequence Type IV

Description

Thick shale sequences that show cyclic variation in color, carbonate content, and internal bedding occur particularly in cores from the Wonsits Valley unit. Although these sequences are approximately 3–15 m thick, thicker shale sequences occur in some uncored wells. Two principal types of shale are recognized—medium- to dark-gray and brown shale with less than 20% silt and sand grains; and light- to medium-gray and brown shale containing wavy laminae and lenses of siltstone and very fine-grained sandstone.

The medium- to dark-gray and brown shale typically shows very thin, horizontal internal lamination defined by variations in clay, carbonate, and siltstone content. Burrows are generally absent, and disarticulated fish bones and scales occur rarely. Two shale subfacies (SH–C and SH–D) of the medium- to dark-gray and brown shale were identified based on differences in lamination (Figure 6D).

In contrast to the darker shale, light- to medium-gray and brown shale (subfacies SH–A and SH–B) contains more common silt to very fine-grained sand with wavy to lenticular bedding. Very fine-grained dolomite crystals are common. Laminae in the light- to medium-gray and brown shale have been deformed by desiccation, burrowing, and minor loading. In some thin intervals, laminae are highly contorted probably because of loading (subfacies SH–E). Some beds contain abundant, light-brown, dolomitic mudstone laminae, which probably formed by algal activity or by evaporative conditions. Rare occurrences of sand-sized intraclasts of dolomitic mudstone concentrated in thin laminae were observed within intervals of light- to medium-gray and brown shale. Rare stromatolites up to 5 cm high and 5 cm across occur in Wonsits Valley cores.

Microscopic organic analysis indicates that organic material in the shale consists mostly (60–90%) of alginite (type I kerogen) with minor but variable amounts of vitrinite and inertinite. TOC content averages approximately 3% in the medium- to dark-colored shale and approximately 1.5% in the lighter colored shale.

Interpretation

Medium- to dark-gray and brown shale is interpreted to have formed below wave base in an open, offshore-lacustrine environment. Evidence for this interpretation includes thinness and evenness of laminations, very low percentage of silt and sand grains, and high alginite content. Absence of bioturbation in the dark shale probably was the result of anaerobic conditions.

In contrast, the light- to medium-gray and brown shale, which contains a larger percentage of silt and sand grains, a larger percentage of dolomite, and a smaller percentage of organic material, is interpreted as a nearshore-lacustrine deposit. Periodic clastic input and storm-wave or current reworking produced minor, thin sandstone and siltstone beds. The presence of burrowing and occurrence of mollusks and ostracodes indicate that the water was oxygenated. The smaller content of organic material may be the result of oxidation in the depositional environment.

WELL LOG PATTERNS

In order to identify a characteristic log pattern for each type of depositional sequence, well log curves were compared to the corresponding core descriptions. Log shapes are more characteristic for recognizing the sandstone depositional sequences than for differentiating various types of shale because some shale types show similar log character.

Sequence Type I

Patterns on the gamma-ray, SP, resistivity, and porosity curves reflect several characteristics of type I sequences—gradational base, upward-increasing sand content and porosity, and sharp upper contact (Figures 9A, 9B). For type I sequences, which are interpreted as prograding-shoreface deposits, resistivity values gradually increase upward with a

maximum of 50–100 ohm-m in oil zones. Based on their well defined lithologic characteristics and corresponding log shapes, these sequences usually can be recognized even when SP and resistivity curves only are available, which is often the case for older wells in the area.

Carbonate-cemented backshore and transgressive-reworked sandstone beds, which occur above some shoreface sequences, can be differentiated from upper shoreface sandstone by smaller separation between resistivity curves and by smaller log-porosity values.

Sequence Type II

Type II sequences, interpreted as mud flat and lagoonal deposits, show log patterns corresponding to shale interbedded with siltstone and sandstone (Figures 9B, 9C). Logs show high gamma radiation with common low-gamma spikes, and low resistivity with common higher resistivity peaks. Compared to type IV lacustrine shale sequences, type II sequences can be recognized in most cases by log patterns indicative of greater sandstone content and less carbonate content.

In many low-energy bar sequences, log curves reflect the slight, gradational increase upward in sand-to-clay ratio. Compared to shoreface sandstones, low-energy bars generally show higher gamma-ray values and are typically thinner.

Sequence Type III

Log patterns of type III (fluvial) sequences indicate a sharp base, gradual upward increase in clay content, and upward decrease in both porosity and permeability (Figures 9B, 9D, 9E). Because permeability values commonly are smaller in channel sandstones than in upper shoreface sandstones, maximum resistivities in oil-saturated fluvial zones also tend to be lower than in upper shoreface reservoirs. Flood-plain shale exhibits a log character similar to that of mud flat shale. However, unlike mud flat sequences, flood-plain deposits sometimes exhibit a slight upward decrease in gamma radiation and a slight upward increase in resistivity. The vertical gradation in log character corresponds to upward increase in silt and sand percentage, probably the result of progradation of levee-overbank sequences associated with nearby channels.

Log patterns that correspond to prograding-lobe deposits reflect upward-increasing sand content (Figure 9E). Because of shale interbeds and clay matrix in the sandstone, these patterns indicate less clean sandstone (e.g., higher gamma values) than those corresponding to prograding-shoreface sequences. Log patterns of prograding lobes may resemble patterns of low-energy bars because of general shaliness and upward increase in sand percentage.

WELL LOG PATTERNS WITH DEPOSITIONAL ENVIRONMENTS INTERPRETED FROM CORES

WELL LOG PATTERNS WITH DEPOSITIONAL ENVIRONMENTS INTERPRETED FROM LOGS

A. WONSITS VALLEY UNIT WELL 117

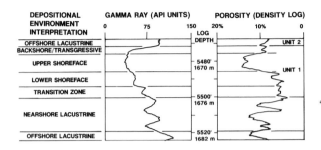

B. RED WASH UNIT WELL 118

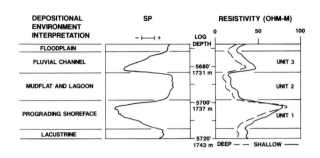

C. RED WASH UNIT WELL 169

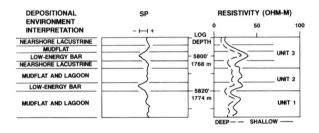

D. RED WASH UNIT WELL 285

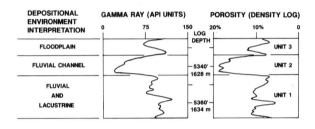

E. RED WASH UNIT WELL 189

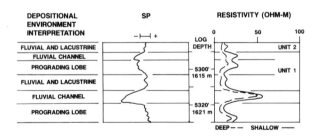

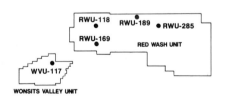

Figure 9. Examples of well log curves showing characteristic shapes corresponding to depositional environments of the lower Green River Formation. A, gamma-ray and porosity for prograding-shoreface deposits (sequence type I) and lacustrine shale (sequence type IV). Low-porosity sandstone overlying upper shoreface interval is calcite and dolomite cemented and represents backshore deposition with possible transgressive reworking. B, SP and resistivity for prograding shoreface deposits (type I), mud flat and lagoonal shale (type II) and fluvial-channel and flood-plain deposits (type III); resistivity values in shoreface sandstones usually are greater than in fluvial-channel sandstones. C, SP and resistivity for interbedded mud flat-lagoon, nearshore-lacustrine, and low-energy bar deposits (type II); log characteristics of mud flat and lagoonal shale are similar to those of nearshore-lacustrine shale. D, gamma-ray and porosity for fluvial-channel and flood-plain deposits (type III); lacustrine shale may occur in the lower part of the interval illustrated. E, SP and resistivity patterns corresponding to prograding-lobe, fluvial-channel, and flood-plain environments (type III).

Lacustrine Basin Exploration: Case Studies and Modern Analogs

Sequence Type IV

Nearshore- and offshore-lacustrine shales commonly show log pattern cyclicity caused by variation in clay mineral and dolomite content (Figure 9A). Those variations are most apparent on gamma-ray and density logs through thick shale sequences in southwestern Wonsits Valley unit. Zones of lower gamma radiation, lower porosity values, and higher resistivity values correspond to intervals of greater dolomite content, which occurs in nearshore-lacustrine shale.

DEPOSITIONAL SYSTEMS

Interpretation of Depositional Systems

Core and log interpretations were used to map paleoenvironments for each of three correlatable units within the interval studied. These maps define two major depositional systems—wave-dominated lacustrine shoreline and fluvial-deltaic. In the lacustrine shoreline system, silt, clay, and minor sand were deposited in nearshore-lacustrine and mud flat-lagoonal environments. Because mapping of units 1 and 2 shows that shoreface deposits are separated from the onshore area by mud flat-lagoonal facies, the shoreface deposits are interpreted as barrier islands (Figures 10A, 10B). Beaches, which are shoreline attached and smaller than barrier complexes, are mapped in units 2 and 3 (Figures 10B, 10C).

An alternative interpretation of the barrier-beach sandstones is that they represent sheet deposits formed by coalescing mouth bars of distributary channels. Lateral continuity and upward-coarsening grain size are consistent with mouth-bar deposition as well as with barrier-beach deposition. However, barrier-beach deposition is more likely because of (1) evidence of wave action; (2) ooids and algal-produced sediment; (3) high alginite content relative to vitrinite content in interbedded shale; (4) general absence of associated distributary-channel deposits; (5) shale that encloses the coarsening-upward sequences and that shows sedimentological evidence for offshore-lacustrine deposition rather than flood-plain or delta-plain deposition; and (6) a broad area of shale and siltstone interpreted as mud flat-lagoonal facies that occurs laterally between the sandstone bodies and onshore deposits.

The fluvial-deltaic depositional system identified in the study area includes channel, prograding-lobe, and levee-overbank environments. Individual channels were too small (i.e., narrower than well spacing) to be mapped. Based on changes in distribution of paleoenvironments from unit 1 to unit 3 (Figure 10), the fluvial-deltaic system prograded generally westward and southwestward over mud flat-lagoon and lacustrine deposits through time. Relief on depositional and erosional surfaces in the fluvial-deltaic environment probably was low, which would have led to rapid flooding of large areas during lacustrine transgressions.

Water Depth

Water depth probably was important in controlling wave energy, and therefore sedimentation, along the margin of ancient Lake Uinta. Based on thicknesses of the described shoreface sequences (approximately 12 m measured in cores), maximum depth of the lower shoreface sediment bottom below lake level probably was 15–20 m. The maximum depth of the rippled lower shoreface sediment bottom can be inferred from expected wave periodicity. According to Clifton (1976, p. 141), wave period in lakes usually is short (2 to 4 sec), so that wave ripples form in water up to about 25 m deep. Water depth in the offshore-lacustrine environment interpreted for shales in the southwestern part of the study area probably was 20–25 m or perhaps greater. During deposition of the lower Green River Formation, wave fetch across northern Lake Uinta was large, probably at least 80 km from west to east (Picard, 1957a, p. 126).

These interpretations of water depth are consistent with others for the Green River Formation. Ryder et al. (1976), for example, interpreted a water depth of 5–30 m for organic-rich shale representing quiet-water, open-lacustrine deposition in Lake Uinta. A maximum depth of 20-30 m was suggested for varved lacustrine shale by Bradley (1929). Surdam and Stanley (1979) interpreted a water depth of possibly 25 m or greater in the center of Lake Gosiute in Wyoming.

Sandstone Distribution and External Geometry

Important differences in sandstone distribution, trend, and continuity occur between the lacustrine-shoreline and fluvial-deltaic systems. Paleoenvironmental maps of units 1 and 2 (Figures 10A, 10B) show that the lacustrine barrier islands trend northeast to southwest in the Wonsits Valley-west Red Wash area. Depositional strike of the wave-dominated shoreline was generally southwestward. Sand deposited on the barrier complexes probably was transported from the north-northwest by longshore currents. Progradation direction was westward and southwestward. Shoreline orientation, as interpreted in this study, is consistent with trends interpreted by Campbell (1966) and Picard (1967) for Douglas Creek lake-margin sandstones. Correlation and mapping of subsurface data indicate that the barrier-beach sandstones are laterally continuous reservoirs across the Wonsits Valley-west Red Wash area.

Sediment was deposited by the fluvial-deltaic system in the northern Red Wash field, particularly

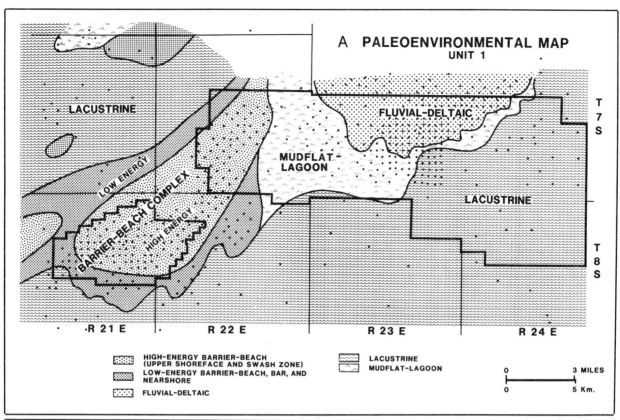

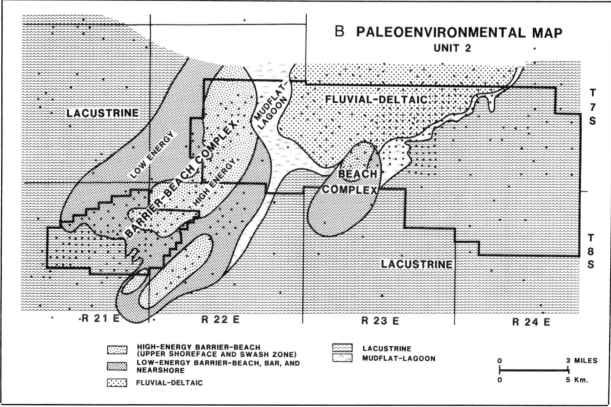

Figure 10. Paleoenvironmental maps of three stratigraphic units in the upper Douglas Creek Member of
Green River Formation—A, Unit 1; B, Unit 2; and C,
Unit 3 (shown on next page). Interpretations are based
on core and log control (Figures 3, 5). Areas of
sandstone representing lacustrine-shoreline (shoreface) deposition are labeled as beach if physically
connected to onshore areas or as barrier-beach if
physical connection is uncertain.

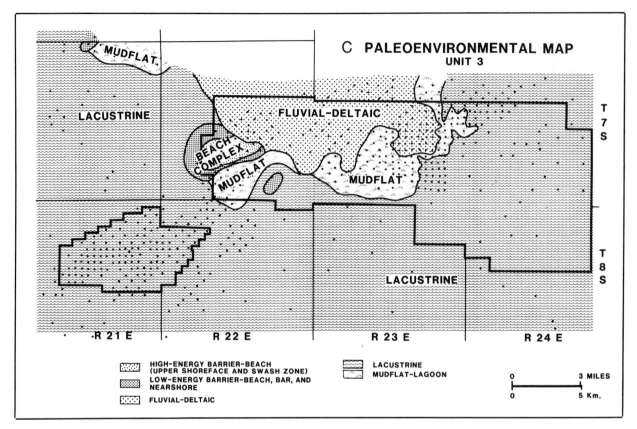

MUDFLAT

LACUSTRINE

FLUVIAL-DELTAIC

BEACH COMPLEX

MUDFLAT

MUDFLAT

T 7 S

LACUSTRINE

T 8 S

R 21 E   R 22 E   R 23 E   R 24 E

	HIGH-ENERGY BARRIER-BEACH (UPPER SHOREFACE AND SWASH ZONE)
	LOW-ENERGY BARRIER-BEACH, BAR, AND NEARSHORE
	FLUVIAL-DELTAIC
	LACUSTRINE
	MUDFLAT-LAGOON

0 ——— 3 MILES
0 ——— 5 Km.

Figure 10. C, Unit 3. Each beach/barrier-beach complex is a separate sandstone body built by composite, prograding-shoreface deposits. Barrier-beach complexes are separated vertically by transgressive lacustrine shale. Channels, flood plains, and prograding lobes in basinward-prograding (west-southwest) fluvial-deltaic areas are undifferentiated because of close genetic relationship and mutual occurrence in many wells.

in the main and central areas, and in the Walker Hollow area (Figures 3, 10). Cores and logs indicate that the fluvial channels probably were shallow, narrow, more or less straight, and possibly ephemeral. Because of their small size, it is unlikely that they served as major conduits for sediment transport. Subsurface correlations indicate that the channel sandstones are laterally discontinuous, with downstream orientation toward the south-southwest, which is consistent with previous interpretations of regional transport directions (Picard, 1967; Koeso-emadinata, 1970). The sands were derived from the Uinta uplift to the north.

It appears that the contrasts in depositional systems between the western and eastern parts of the study area may recur in other producing intervals within the Red Wash and Wonsits Valley fields. Four additional reservoirs, which are coarsening-upward sequences probably representing shoreface deposition, occur in Wonsits Valley field. In the main area of the Red Wash field (Figure 3), minor shoal and shoreline sandstones and carbonates, as well as fluvial-deltaic deposits, also have been recognized (Thompson, 1988; Chevron unpublished reports). There, lithologies and depositional environments are more variable both vertically and laterally compared

to the Wonsits Valley-west Red Wash area. In the Red Wash main area, sedimentation occurred at or near the lake margin, where processes associated with fluvial-deltaic and nearshore-lacustrine environments changed over small distances.

REGIONAL GEOLOGY OF UINTA BASIN

Outcrop Comparisons

Shoreface and beach origins of siliciclastic deposits in the Green River Formation have been recognized in previous studies of outcrops along the northern and southeastern flanks of the Uinta basin. However, no siliciclastic shoreface and beach deposits have been reported from the western or southwestern areas of the basin.

The Green River Formation crops out nearest the Red Wash field at Raven Ridge, 5–8 km to the east. Rocks exposed there include the Douglas Creek through the Evacuation Creek members (Picard, 1967). The section includes rocks considered stratigraphically equivalent to the producing interval at Red Wash (Sanborn and Goodwin, 1965). However, precise correlations from outcrop to subsurface have not been established. Approximately 750 m of section are exposed along the ridge, which extends approximately 5 km. Sanborn and Goodwin (1965) interpreted the depositional environment of the thickest sandstone beds as beach, with fluvial-deltaic and low-energy nearshore-lacustrine environments for thinner beds. Based on upward-increasing grain size and on the sequence of primary structures, which include ripple bedding, cross-bedding, and high-energy flat bedding, some exposed sandstones represent upward shallowing and increasing energy (brief examination, this study). Such coarsening-upward sequences at Raven Ridge appear to form composite prograding-shoreface sandstone units that are traceable for several kilometers along strike. Picard and High (1981) reported that individual transgressive/regressive cycles at Raven Ridge can be traced from offshore to nearshore and shoal, to backshore and lagoonal, and to fluvial-deltaic. In a detailed analysis of paleocurrent indicators at Raven Ridge, Picard (1967) interpreted lacustrine-shoreline sandstone with longshore trends from 31° to 120°, which are consistent with subsurface trends of barrier-beach deposits mapped in this study. Sandstones interpreted as fluvial are characterized by southwestward paleocurrent directions parallel to depositional slope (Picard, 1967). Southwestward orientation of channels at Raven Ridge matches that of subsurface channels at Red Wash field.

Another example of fluvial and lacustrine-shoreline deposits, including beach sandstones, is in the Douglas Creek and Garden Gulch members of the P. R. Springs area (Picard and High, 1970) in southeastern Uinta basin approximately 65 km south of Red Wash field. Sandstone bedding characteristics, lateral continuity, and composition there are similar to what was observed in west Red Wash and Wonsits Valley barrier-beach sequences. At P. R. Springs, lacustrine sandstone is described as cross-bedded and flat bedded, algal-laminated, oolitic, burrowed, and laterally continuous for several kilometers (Picard and High, 1970). Fluvial sandstones show channeling and contain basal lag deposits.

Along the western and southwestern flanks of the Uinta basin, the influence of wave processes on carbonate deposition has been interpreted from outcrops. Ryder et al. (1976) described oolitic and ostracodal grainstones that occur in 2- to 5-m-thick upward-coarsening cycles. Algal structures and mud cracks near the tops of some cycles are analogous to those at the tops of clastic shoreface sequences observed in Wonsits Valley-west Red Wash cores. The carbonate sequences are interpreted as shoal and beach deposits that developed during local and regional shoreline transgressions (Ryder et al., 1976). From outcrops along the southwestern basin margin, Franczyk et al. (1989) interpreted a laterally continuous oolitic grainstone as a shoaling-upward sequence that formed nearshore or at the lake margin. In the Wonsits Valley field farther northeast is an oolitic, ostracodal grainstone (G1 limestone) that is similar in composition, texture, and sequence but is thicker (~10 m). The G1 limestone is a laterally continuous carbonate-shoal reservoir about 30 m below the base of unit 1. Toward the northeast, the direction of the clastic detritus source, carbonate grainstone in the G1 grades into quartzose sandstone of a coarsening-upward sequence. The G1 limestone and laterally equivalent sandstone represent wave-dominated lake-margin deposition but of higher wave energy than the carbonate-shoal deposits to the west and southwest.

Subsurface Comparisons

In the northern Uinta basin a large volume of hydrocarbons has been produced from the Green River Formation in the Bluebell-Altamont field, located about 65 km west of the Red Wash field. Shallow lacustrine-shoreline and fluvial-deltaic sandstones form important producing intervals (Lucas and Drexler, 1976). However, production comes primarily from fractured reservoirs with low matrix permeability, in contrast to the Red Wash and Wonsits Valley fields, where production follows stratigraphically controlled porosity distribution.

Other subsurface comparisons include the Douglas Creek Member at Pariette Bench field in southeastern Uinta basin and a tongue of the Wasatch Formation at the Natural Buttes field in eastern Uinta basin (Pitman et al., 1982, 1986). In both areas sandstone was interpreted as alluvial and lower delta plain in origin. Bedding contacts, internal sedimentary structures, and grain-size profiles of the fluvial-channel deposits in those areas are similar to characteristics of channel sandstones in the Red Wash area. Furthermore, individual reservoirs were described as thin or lenticular, which also is analogous to Red Wash fluvial reservoirs; however, no siliciclastic lacustrine bars have been reported.

The degree of diagenesis, including secondary porosity development and authigenic mineral growth, in fluvial sandstones of the Pariette Bench and Natural Buttes fields (Pitman et al., 1982, 1986) apparently is greater than in most sandstones of the Wonsits Valley-Red Wash area. Green River barrier-beach sandstones may in general be less diagenetically altered than fluvial sandstones because of greater mineralogical and textural maturity related to wave reworking in the shoreface environment, particularly in the upper shoreface and swash zone. In addition, the higher content of physically and chemically less stable constituents in fluvial deposits would favor diagenetic reactions such as grain dissolution and cementation.

Regional Controls on Clastic Sedimentation

The extent of siliciclastic shoreface deposits and major bar complexes along the margin of Lake Uinta apparently is restricted because of the erratic distribution of conditions that controlled wave energy and rate of sand accumulation at the shoreline. This investigation indicates that wave energy along sections of the northeastern paleoshoreline of Lake Uinta was great enough to form barrier-beach complexes. Compared to other areas, the northeastern margin apparently experienced greater wave fetch, deeper water offshore, possibly greater clastic input, and greater shoreline stability. Those characteristics not only are indicated by sedimentology of basin-margin deposits but also are consistent with the basin's tectonic history. That subsidence was greater along the northern flank than along the southern flank may have resulted in overall greater water depth in the northern part of the lake. Present structural depth of the basin certainly is greater in the north than in the south (Figure 1). The occurrence of thick, dark, organic-rich offshore-lacustrine shale in southern and southwestern Wonsits Valley field and adjoining areas is consistent with deeper water in that offshore area.

The southern lake floor's slight regional depositional slope resulted in unstable shorelines that could have migrated rapidly with only minor changes in lake level (Picard and High, 1981; Franczyk et al., 1989). Shallow water in that area would have inhibited normal formation of large waves, and shoreline instability would have minimized the time that sediments were subjected to wave reworking. Consequently, wave-produced deposits are less common along the southern margin.

CONCLUSIONS

In previous studies of the Uinta basin (Pitman et al., 1982; Fouch, 1985; Franczyk et al., 1989), Upper Cretaceous and Tertiary reservoirs were described as low-permeability, fractured, fluvial sandstones with secondary porosity. Although those characteristics accurately describe reservoirs in some areas, they generally do not apply to producing Green River sandstones in the Red Wash and Wonsits Valley fields. In those areas the primary control on reservoir geometry and quality is depositional; diagenesis is of secondary importance. Because of possible differences in primary mineralogy and texture, diagenesis may, however, have affected porosity and permeability to a greater extent in fluvial sandstones than in barrier-beach sandstones. Additional detailed study would help to document relationships between the depositional facies and diagenesis.

This study incorporates a large volume of subsurface data collected throughout northeastern Uinta basin, which helps to explain differences among other sedimentological interpretations for clastic deposits in the Green River Formation. The paleoshoreline of northeastern Lake Uinta exhibits a wide variety of facies. Lithologies in Wonsits Valley and Red Wash cores are analogous to deposits previously interpreted as lacustrine shoreline and beach, carbonate shoal,

fluvial, deltaic, playa lake, and deep lacustrine. The variable distribution of processes resulted in complex facies associations in an area where channelized flow, wave processes, and lake currents all were present. Local conditions were controlled by water depth, sediment input, wave fetch, and shoreline configuration.

In other siliciclastic lake systems the occurrence of beach deposits may be restricted to certain stretches of shoreline by the same conditions that controlled deposition along the paleoshoreline of Lake Uinta. Barrier-beach complexes can be expected to have formed along stable paleoshorelines subject to long wave fetch across an area where the lake bottom was deeper than wave base. Clastic-sediment input from streams or by longshore currents is needed to form large clastic-lacustrine sand bodies along the lake margin.

The processes active along many lake margins, such as along northeastern Lake Uinta, can effect a combination of reservoir, source, and seal rocks that is favorable for hydrocarbon occurrence. In the example described in this study, barrier-beach and fluvial sandstone reservoirs occur within the same correlatable interval. Offshore-lacustrine shale represents a potential hydrocarbon source rock. Reservoir seals were formed by lacustrine transgressions over both lacustrine-shoreline and fluvial-deltaic facies. Because of depositional controls, the occurrence of reservoir, source, and seal rocks are predictable based on detailed sedimentological analysis and paleoenvironmental mapping.

The lacustrine depositional model presented in this study may be common in ancient clastic lacustrine sequences, where sufficiently high wave energy was expended along the lake shoreline. The association of lacustrine barrier-beach facies with fluvial-deltaic facies in northeastern Uinta basin provides a model that can be applied to exploration of lacustrine deposits in other basins.

ACKNOWLEDGMENTS

The author conducted this investigation at Chevron U.S.A., Denver, Colorado. Appreciation is extended to Chevron for permission to publish and for providing drafting support. Numerous people at Chevron contributed through their thoughtful discussions. Helpful comments and suggestions were made by Norbert Cygan, John Kelly, and Diana Thompson. Steve Jacobson, Sheldon Nelson, and Rick Haack contributed to the study by providing and discussing organic geochemical data. Invaluable assistance with laboratory work, including thin sections and core handling, was provided by Pat Flynn, Skip Nielsen, and Steve Petersen. The author also thanks Tom Fouch and Ken Stanley, who contributed significantly to the quality and content of this paper by reviewing an earlier version of the manuscript. The paper was improved as the result of reviews by William C. Dawson and Louis Liro. Word processing by Cathy Collins, Cabot Oil and Gas Corporation, is gratefully acknowledged.

REFERENCES

Bradley, W. H., 1929, The varves and climate of the Green River epoch: USGS Professional Paper 158-E, p. 87-110.

Bradley, W. H., 1931, Origin and microfossils of the oil shale of the Green River Formation of Colorado and Utah: USGS Professional Paper 168, 58 p.

Campbell, G. S., 1966, Douglas Creek trend, case history, Uinta basin, Utah (abs.): AAPG Bulletin, v. 50, p. 2038.

Cashion, W. B., 1967, Geology and fuel resources of the Green River Formation, southeastern Uinta basin, Utah and Colorado: USGS Professional Paper 548, 48 p.

Chatfield, J., 1972, Case history of Red Wash field, Uintah County, Utah, in R. E. King, ed., Stratigraphic oil and gas fields—Classification, exploration methods, and case histories: AAPG Memoir 16, p. 342-353.

Clifton, H. E., 1976, Wave-formed sedimentary structures—A conceptual model, in R. A. Davis, and R. L. Ethington, eds., Beach and nearshore sedimentation: SEPM Special Publication 24, p. 126-148.

Elliott, T., 1978, Clastic shorelines, in H. G. Reading, ed., Sedimentary environments and facies: New York, Elsevier, p. 143-177.

Eugster, H. P., and L.A. Hardie, 1975, Sedimentation in an ancient playa-lake complex—The Wilkins Peak Member of the Green River Formation of Wyoming: GSA Bulletin, v. 86, p. 319-334.

Eugster, H. P., and Surdam, R.C., 1973, Depositional environment of the Green River Formation of Wyoming—A preliminary report: GSA Bulletin, v. 84, p. 1115-1120.

Fouch, T. D., 1975, Lithofacies and related hydrocarbon accumulations in Tertiary strata of the western and central Uinta basin, Utah, in D. W. Bolyard, ed., Symposium on deep drilling frontiers in the central Rocky Mountains: Rocky Mountain Association of Geologists Special Publication, p. 163-173.

Fouch, T. D., 1985, Oil and gas-bearing upper Cretaceous and Paleogene fluvial rocks in central and northeast Utah: SEPM Short Course Notes 19, p. 241-271.

Franczyk, K. J., J. K. Pitman, W. B. Cashion, J. R. Dyni, T. D. Fouch, R. C. Johnson, M. A. Chan, J. R. Donnell, T. F. Lawton, and R. R. Remy, 1989, Evolution of resource-rich foreland and intermontane basins in eastern Utah and western Colorado in the collection P. M. Hanshaw, ed., 28th International Geological Congress Field Trip Program: American Geophysical Union, 53 p.

High, L. R., Jr. and M. D. Picard, 1971, Nearshore facies relations, Eocene Lake Uinta, Utah (abs): AAPG Bulletin, v. 55, p. 343.

Hintze, L. F. (compiler), 1980, Geologic map of Utah: Utah Geological and Mineral Survey, 1:500,000.

Koesoemadinata, R. P., 1970, Stratigraphy and petroleum occurrence, Green River Formation, Red Wash Field, Utah: Colorado School of Mines Quarterly, v. 65, no. 1, p. 1-77.

Lucas, P. T., and J. M. Drexler, 1976, Altamont-Bluebell—A major, naturally fractured stratigraphic trap, in J. Braunstein, ed., North American oil and gas fields: AAPG Memoir 24, p. 121-135.

Osmond, J. C., 1964, Tectonic history of the Uinta Basin, Utah, in E. F. Sabatka, ed., Guidebook to the geology and mineral resources of the Uinta basin: Intermountain Association of Petroleum Geologists 13th Annual Field Conference, p. 47-58.

Picard, M. D., 1955, Subsurface stratigraphy and lithology of Green River Formation in Uinta basin, Utah: AAPG Bulletin, v. 39, p. 75-102.

Picard, M. D., 1957a, Green River and lower Uinta Formations—Subsurface stratigraphic changes in central and eastern Uinta basin, Utah, in O. G. Seal, ed., Guidebook to the geology of the Uinta basin: Intermountain Association of Petroleum Geologists 8th Annual Field Conference, p. 116-130.

Picard, M. D., 1957b, Criteria used for distinguishing lacustrine and fluvial sediments in Tertiary beds of Uinta basin, Utah: Journal of Sedimentary Petrology, v. 27, p. 373-377.

Picard, M. D., 1967, Paleocurrents and shoreline orientations in Green River Formation (Eocene), Raven Ridge and Red Wash areas, northeastern Uinta basin, Utah: AAPG Bulletin, v. 51, p. 383-392.

Picard, M. D., and L. R. High, Jr., 1970, Sedimentology of oil-impregnated lacustrine and fluvial sandstone, P.R. Spring area, southeast Uinta basin, Utah: Utah Geological and Mineralogical Survey Special Studies 33, 32 p.

Picard, M. D., and L. R. High, Jr., 1972, Criteria for recognizing lacustrine rocks, in J. K. Rigby, and W. K. Hamblin, eds., Recognition of ancient sedimentary environments: SEPM Special Publication 16, p. 108-145.

Picard, M. D., and L. R. High, Jr., 1981, Physical stratigraphy of ancient lacustrine deposits, in F. G. Ethridge, and R. M. Flores, eds., Recent and ancient non-marine depositional environments—Models for exploration: SEPM Special Publication 31, p. 233-259.

Pitman, J. K., D. E. Anders, T. D. Fouch, and D. J. Nichols, 1986, Hydrocarbon potential of nonmarine Upper Cretaceous and Lower Tertiary rocks, eastern Uinta basin, Utah, in C. W. Spencer, and R. F. Mast, eds., Geology of tight gas reservoirs: AAPG Studies in Geology 24, p. 235-252.

Pitman, J. K., T. D. Fouch, and M. B. Goldhaber, 1982, Depositional setting and diagenetic evolution of some Tertiary unconventional reservoir rocks, Uinta basin, Utah: AAPG Bulletin, v. 66, p. 1581-1596.

Ray, E. S., 1985, Diagenesis of sandstones from the Douglas Creek Member of the Green River Formation (Eocene) at Red Wash field, Uintah County, Utah: M.S. thesis, Texas A & M University.

Ryder, R. T., T. D. Fouch, and J. H. Elison, 1976, Early Tertiary sedimentation in the western Uinta basin, Utah: GSA Bulletin, v. 87, p. 496-512.

Sanborn, A. F., and J. C. Goodwin, 1965, Green River Formation at Raven Ridge, Uintah County, Utah: Mountain Geologist, v. 2, p. 109-114.

Smoot, J. P., 1978, Origin of the carbonate sediments in the Wilkins Peak Member of the Green River Formation (Eocene), Wyoming, U.S.A., in A. Matter, and M. F. Tucker, Modern and ancient lake sediments: International Association of Sedimentologists Special Publication 2, p. 109-127.

Surdam, R. C., and K. O. Stanley, 1979, Lacustrine sedimentation during the culminating phase of Eocene Lake Gosiute, Wyoming (Green River Formation): GSA Bulletin, v. 90, p. 93-110.

Thompson, D. M., 1988, Determining reservoir quality, distribution, and continuity in complex lacustrine margin sandstones, Red Wash (main area), Uintah County, Utah (abs): AAPG Bulletin, v. 72, p. 253.

Webb, M. G., 1978, Reservoir description, K_f Sandstone, Red Wash Field, Utah, in Proceedings, SPE Improved Oil Recovery Symposium: Society of Petroleum Engineers Paper 7046, p. 97-101.

A Model for Tectonic Control of Lacustrine Stratigraphic Sequences in Continental Rift Basins

Joseph J. Lambiase
Marathon Oil Company
Houston, Texas, U.S.A.

Lake deposits have similar sedimentary characteristics and occupy similar positions within stratigraphic sequences of numerous continental rift basins that represent a wide range of geologic age, climate, and geographic location. Each lacustrine unit deposited during a phase of large lake development corresponds to a distinct rifting episode. New models for the structural evolution of continental rifts, together with changing depositional patterns through time, suggest that the topography required for large lake development only occurs early in rift history. Temporal changes in relative rates of subsidence and deposition can terminate lake occurrence by filling topographic lows, and preclude further large lake development. Thus, structural evolution, and resulting depositional patterns, limit large lake development to a specific interval in a rift's history, and are a primary control on large lake occurrence.

INTRODUCTION

Lacustrine sediments are widely recognized as important components of continental rift-basin stratigraphic sequences. These deposits, as described in numerous basins, represent a wide range of lake systems that varied greatly in areal extent and water depth. Models have been proposed that predict the lateral distribution of lacustrine and associated facies (e.g., Leeder and Gawthorpe, 1987), but less attention has been given to the vertical sequences generated during stratigraphic evolution of continental rifts. The entire rift-fill sequence can be several thousands of meters thick and include sediment deposited throughout the basin's history—from the initial synrift phase through postrift regional subsidence. Lacustrine facies can occur throughout the section.

Variations in occurrence, distribution, and type of lacustrine deposits are attributed to many factors, but climate generally is regarded as the most important influence. Although climate is unquestionably important, the present study indicates that tectonically induced structural geometry also plays a major role. Furthermore, large, deep lakes can originate only during a specific phase of a rift's tectonic evolution, and climate and other factors determine whether they actually do occur during that phase.

Of the various types of lacustrine facies, those that represent lakes of sufficient size and persistence to have accumulated large volumes of organic-rich shale are of special interest for hydrocarbon exploration. Major lake sequences certainly do not occur in all basins, but where they do, they occupy equivalent stratigraphic positions. These sequences, and the controls on their distribution, constitute the primary focus of this study. Based on the stratigraphy of numerous rifts ranging in age from Precambrian to Holocene and spanning six continents, a model is proposed that recognizes tectonically induced structural geometry as a control that can both allow and restrict the occurrence of large lakes and their deposits. The examples discussed below were drawn from the literature and supplemented by field studies (Newark basin, Mombasa basin, Mid-Continent rift, Morondava basin) and borehole and seismic studies (Sumatra, west Africa, Sudan).

STRUCTURAL GEOMETRY AND DEPOSITIONAL PATTERNS

An important element in the model presented here is the structural geometry in continental rift basins. From recent analysis of modern and ancient rifts, many workers conclude that the basic structural unit of continental rifts is the half-graben (e.g., Bosworth,

1985; Rosendahl, 1987). In these models, each half-graben is a distinct structural and depositional basin whose features affect depositional patterns. Half-grabens are generally about 100 km long and 50 km wide. Both sides usually are fault-bounded, but the fault on the downdip side has much greater throw than that on the updip or flexural side, giving the basin its characteristic asymmetry (Figure 1A). Other features that bear on sedimentation patterns are the rift shoulders and the fault-block geometry of the basin floor. Rift shoulders form the highest topography and tend to be higher on the downdip side than on the flexural side (Figure 1A). They also slope away from the rift. The basin floor is faulted into blocks 5 to 10 km wide and broken along strike by cross-faults at variable spacing.

Many rift systems consist of a series of half-grabens, often arranged such that adjacent graben floors slope in opposite directions. The linkage between two half-grabens is a structural high known as an accommodation zone or transfer zone (Rosendahl et al., 1986) (Figure 1B). Several workers (e.g., Bosworth, 1985) recently have discussed the exact structural origin and geometry of these features. For this discussion the important point is that accommodation zones generally are structurally higher than the basin floor but not as high as rift shoulders. The relative heights of these features vary from basin to basin, but in Lake Tanganyika, for example, accommodation zones are as high as 700 m above the adjacent basin floor but 1500 m lower than the rift shoulders. Another important aspect of the structural geometry is that each half-graben has an axial plunge, and adjacent half-grabens often plunge in opposite directions. In addition, the entire rift system may plunge along its strike.

Each aspect of the structural geometry described above exerts a strong influence on depositional patterns. Rift shoulders deflect regional drainage away from the basin, thereby limiting sediment influx (Frostick and Reid, 1989). This effect is more pronounced on the downdip sides of half-grabens where rift shoulders are highest (Figure 1A); thus, much more sediment enters the rift from its updip margins (Frostick and Reid, 1987). Generally, because no structural barriers mark the ends of a rift system, sediment has relatively easy access into the system-terminating half-graben. In many systems this axial transport contributes much of the sediment. Another site of significant sediment input into a rift is near accommodation zones (Figure 1B). Rift shoulders are not excessively high here, and cross-faults associated with accommodation features can extend beyond the rift margin and serve as conduits for sediment transport across or through the rift shoulders. This is especially true where accommodation zones are reactivated antecedent structures, such as in northern Lake Malawi where the Permo-Triassic fault that bounds the northern side of the Ruhuhu trough serves as an accommodation zone within the modern lake and also controls the position of the

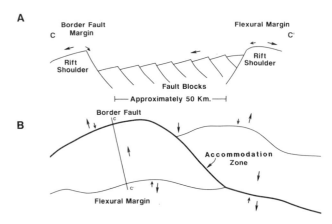

Figure 1. Diagrammatic section (A) and map view (B) of generalized structural geometry of a continental rift. Arrows indicate sediment transport directions and relative transport rates.

Ruhuhu River, which transports sediment into the lake.

Because of the asymmetrical dip and plunge of the half-graben, a topographic low forms that becomes a depocenter as drainage is focused into it. However, fault-block topography on the basin floor impedes downdip sediment transport by creating smaller depocenters in the hanging wall against a rift shoulder or an adjacent block (Figure 2). Whenever the rate of topographic rejuvenation exceeds the rate of deposition, most sediment moving downdip becomes trapped in this manner, although some bypasses these depocenters along cross-faults that separate the blocks. The updip ends of topographically prominent fault blocks are vulnerable to erosion and thus contribute sediment to the system.

As both structural and topographic highs, accommodation zones inhibit rift-axial transport of sediment between adjacent half-grabens. Therefore, each half-graben becomes, in essence, an isolated depositional basin. Furthermore, because half-grabens bounded by rift shoulders and accommodation zones have limited sediment access on all sides, overall sedimentation rate is relatively low.

An important assumption in the structural control of depositional patterns, as described above, is that the rate of tectonically controlled topographic rejuvenation exceeds the rate of deposition. Obviously, this condition must exist as the basin is initiated, otherwise a topographic low could not form, but it cannot persist indefinitely because the basin then could not fill.

Depositional patterns and facies distribution change when deposition rate exceeds the rate of topographic rejuvenation. As a basin fills, fault-block topography quickly becomes buried, and basinwide depositional units form. Eventually, rift-axial transport becomes more dominant after the accommodation zones have been buried, and, finally, rift shoulders are eroded and no longer affect depositional patterns. Each of these phases, plus the initial

Lambiase

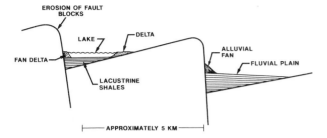

Figure 2. Idealized section showing sedimentation patterns on downdip edges of fault blocks.

condition that topographic rejuvenation exceeds sedimentation rate, results in distinct facies assemblages, which succeed each other in the same order during the evolution of most rifts.

STRATIGRAPHIC SEQUENCES

Observed stratigraphic sequences in continental rifts consist of materials deposited in a variety of nonmarine environments, including large lakes. Major lacustrine sequences contained therein consist of a suite of time-equivalent subaqueous deposits that covers nearly all of the basin, is at least several hundred meters thick, and includes sediment deposited below wave base. These sequences do not occur in all continental rifts; the percentage of basins that contain major lake deposits varies from rift system to rift system but generally is less than 50%. As a modern analog, major lakes of the east African rift system—Tanganyika, Turkana, and Malawi—occupy less than 25% of its total area. This discussion of stratigraphic sequences will be limited to basins that contain major lacustrine sequences.

Another aspect of this discussion is that it deals with gross stratigraphic sequences only. For example, a major lacustrine sequence is considered to include all strata deposited after the onset of lacustrine conditions and before subaerially deposited sediment covered the entire basin. As defined, nonlacustrine sediments may be included in a lacustrine sequence because fluctuating lake levels cause interbedding of subaerially deposited and lacustrine sediment along the basin margins. Although these effects can be large, as evidenced in Lake Tanganyika (Scholz and Rosendahl, 1988), fluctuating lake levels are considered relatively minor perturbations during a lacustrine depositional interval rather than boundaries of successive cycles. In addition, volcanism can be important, but it is not discussed in this report. The sequence descriptions and environmental interpretations presented here were based primarily on field observations and on borehole and seismic studies supplemented by published reports.

Newark Basin

Complete stratigraphic sequences are not known from many rift basins because relatively few are exposed, and drilling has not penetrated many complete sequences. Among the better known rift systems is that along the United States' East Coast. Of the Early Mesozoic basins that comprise the East Coast system, about 50% contain major lake sequences, and Newark basin is the most well known. Numerous authors have discussed its stratigraphy (e.g., Van Houten, 1969; Olsen, 1980a), and seismic data have helped define its structural geometry (Ratcliffe et al., 1986; Unger, 1988).

Each of the two stratigraphic sequences that occur in Newark basin is related to a distinct tectonic event. The older, Upper Triassic sequence contains a basal conglomerate and sandstone, the Stockton Formation, which is 1800 m thick in outcrop and probably considerably thicker in the subsurface (Manspeizer and Olsen, 1981). The Stockton is a complex of braided- and meandering-stream deposits with braided-stream deposits dominating the lower part. Meandering-river sandstones and associated overbank shale are more prevalent near the top . Laterally equivalent alluvial fan deposits occur along the downdip western border (Arguden and Rodolfo, 1986).

Overlying the Stockton is the Lockatong Formation, 1100 m of lacustrine shale deposited in a wide range of water depths as a result of cyclical climate-driven lake-level fluctuations (Olsen, 1984). Superimposed on the cyclicity is a gradual shallowing from near the base of the Lockatong into the overlying Passaic Formation. The Passaic, the uppermost unit of the Upper Triassic sequence, is more than 2000 m thick. Mudstone and siltstone that dominate the section are mainly of shallow lacustrine origin but include some alluvial and fluvial deposits.

The younger stratigraphic sequence in the Newark basin is of Early Jurassic age and consists of three sedimentary formations interbedded with basalt. All three have lacustrine, deltaic, and fluvial facies, but the basal Feltville and overlying Towaco Formations are more indicative of deep-water environments, including turbidites (Olsen, 1980b), than the uppermost Boonton Formation. Generally, the sequence shallows upward with the Boonton having proportionately fewer deep-water lacustrine strata than the older Jurassic formations.

Notable similarities and differences exist between Triassic and Jurassic sequences in the Newark basin. The deepest water lacustrine facies in each sequence occurs at or near the base of lacustrine deposits, where the transition from underlying nonlacustrine rocks into deep-water conditions is relatively rapid. In contrast, the upper boundary of lacustrine strata in each sequence is a gradual transition into shallower water deposits. A fluvial unit marks the basal Triassic sequence but not the Jurassic sequence.

Mombasa Basin

Many of the Permo-Triassic Karroo sediments of southern and eastern Africa were deposited in continental rifts. Major lacustrine sequences have been documented in few basins within that system except for the Mombasa basin in Kenya. Of the four formations that comprise the rift-fill sequence, the lower two are lacustrine in origin. Deposition began with siltstone, sandstone, and shale of the Taru Formation. From field observations, a variety of lacustrine environments are represented, including turbidites. Much of the 2700-m-thick unit was deposited in deep water (Walters and Linton, 1973). Similar facies were observed in the overlying 1500-m-thick Maji Ya Chumvi Formation, but the proportion of deep-water facies decreases upward, and most of the upper formation represents shallow-water environments.

The upper two units in the sequence, the Mariakani and Mazeras Formations, are primarily sandstones of approximate thicknesses of 3000 m and 1000 m, respectively (Karanja, 1984). Shallow lacustrine beds dominate the lower Mariakani Formation but quickly give way to a fluviodeltaic complex that becomes entirely fluvial near the top of the unit. Mazeras sandstones are all of fluvial origin.

A petrographic change within the upper Mazeras Formation is accompanied by a relative increase in overbank shale. This suggests a change in sediment source area, possibly coincident with a transition from synrift to postrift deposition. Overall, the Karroo sequence coarsens upward through the gradual transition from deep to shallow lacustrine facies, and finally into fluvial strata.

Mid-Continent Rift

Strata of the Precambrian Mid-Continent or Keweenawan rift that are exposed in Minnesota and Wisconsin include the synrift Oronto Group and the postrift Bayfield Group. The Oronto Group consists of the basal Copper Harbor Conglomerate, a fluvial/alluvial conglomerate and sandstone sequence (Dickas, 1986). Lacustrine shale forms the middle unit, the Nonesuch Shale, which is overlain by the fluvial Freda Sandstone, the uppermost synrift unit. All three postrift units, the Orienta Sandstone, Devil's Island Sandstone, and Chequamegan Sandstone, are fluvial sandstones that unconformably overlie the synrift sequence (Dickas, 1986).

Central Sumatra

Several other continental rift-basin stratigraphic sequences bear marked similarities to the sequences described above. These encompass a wide range of ages and climatic settings and are widely distributed. Paleogene rifts in central Sumatra have a three-part stratigraphic sequence similar to those in the Mid-Continent rift and Newark basin. Mudstone,

siltstone, and sandstone of the Lower Red Beds Formation, which forms the base of the sequence, were deposited in marsh/shallow-water and deltaic environments (Williams et al., 1985). Conformably overlying the Lower Red Beds are at least 580 m of deep-water lacustrine shale and possible interbedded turbidites known as the Brown Shale Formation. More than 600 m of Lake Fill Formation fluvial and deltaic sandstone and conglomerate with lacustrine shale conformably overlie the Brown Shale Formation (Williams et al., 1985). The Fanglomerate Formation along the downdip edge of the basin consists of sandstone and conglomerate that are laterally equivalent to the other synrift formations.

Reconcavo Basin

Reconcavo basin, Brazil, shares stratigraphic attributes with the preceding continental rift basins. Candeias Formation shale and sandstone overlie an extensive prerift sequence. A deep lake developed rapidly, and many sandstones interbedded with the shale are turbidites (Milani and Davison, 1988). Thick, laterally equivalent conglomerates also were deposited along the downdip side of the basin. Prograding deltaic sediments of the Ilhas Group filled the lake, and synrift deposition concluded with Sao Sebastiao Formation fluvial sediments (Milani et al., 1988; Milani and Davison, 1988). Postrift fluvial conglomerate of the Marizal Formation unconformably overlies the synrift sequence. The early onset of deep-water conditions, including turbidite sandstones, and the overall coarsening-upward sequence are similar to other basins, especially Mombasa.

Morondava Basin

A three-part lithologic sequence exists in the Permo-Triassic section of Morondava basin in southwestern Madagascar. The basal unit is the Lower Sakamena Formation (Upper Permian), which consists of sandstone, conglomerate, and siltstone facies that exceed 2000 m in thickness. Deposition occurred in meandering and braided fluvial systems, alluvial fans, deltas, and lakes (Besairie and Collignon, 1972). Lacustrine shale dominates the overlying Middle Sakamena Formation (Lower Triassic). These vary in thickness and are transitional into the fluviodeltaic sandstone and shale of the Upper Sakamena Formation. The Upper Sakamena coarsens upward, a trend that continues into the overlying Isalo I fluvial sandstones, which are more than 1000 m thick (Besairie and Collignon, 1972). A postrift fluvial sandstone, Isalo II, caps the sequence. Thus, the Permo-Triassic sequence in Morondava basin consists of a basal sand followed by lacustrine shale that coarsens and shallows upward into deltaic and finally fluvial sands.

West Africa

Similar sequences occur in rifts that have experienced more than one extensional tectonic episode. One example is the Cretaceous continental rift basins of west Africa. This system includes a number of basins that extend along the continental margin from Gabon to Angola. Although formation names vary throughout the system, the synrift stratigraphic section consists of two distinct sequences separated by an unconformity (Brice et al., 1982). Each sequence consists of a relatively thin, basal fluvial sandstone overlain by lacustrine shale, which coarsens upward into a deltaic sandstone complex. The lower sequence was deposited in response to initial rifting in the Late Jurassic and Early cretaceous time (Brice et al., 1982). Renewed tectonism induced the unconformity between the sequences and subsequent deposition of the upper sequence.

Sudan

Continental rifts in southern Sudan also underwent two periods of tectonism during the Cretaceous (Schull, 1988). Fluvial and lacustrine claystone, siltstone, and sandstone of the Sharaf Formation form the base of the lower sequence. Thick (1800 m) Abu Gabra Formation lacustrine shale with interbedded silt and sand comprise the middle of the sequence and are capped by as much as 1500 m of Bentiu Formation fluvial sandstone (Schull, 1988).

A second sequence was initiated in response to renewed tectonism in the Turonian. It consists of up to 1800 m of coarsening-upward lacustrine, deltaic, and fluvial deposits. The Aradeiba and Zarqa Formations are mainly claystone, shale, and siltstone overlain by the Ghazal and Baraka Formations, which are predominantly sand (Schull, 1988).

Synthesis

It is clear from the preceding descriptions of stratigraphic sequences in continental rifts that include major lake sequences that these basins share several attributes. Many sequences have a basal sandstone, generally of fluvial origin, with subordinate, generally thin lacustrine deposits (Figure 3). Lacustrine shale directly overlies either the basal sand in sharp contact or prerift rocks if the basal sand is missing. The deepest water lacustrine sediments tend to occur near the base of the unit, with water depths decreasing upward. This shallowing is accompanied by an increase in grain size. Thickness of the lacustrine shale unit varies from basin to basin.

The upper boundary of the lacustrine shale unit is transitional into an overlying fluviodeltaic complex of silt and sand. Fluvial sand and shale dominate the entire upper rift sequence, although significant shallow-water lacustrine, paludal, evaporitic, and/or eolian deposits can occur. These strata actually are composed of two distinct sequences. Synrift fluvial sand directly overlies lacustrine and deltaic units, and its distribution is confined within the rift basin's bounding faults. The second fluvial-dominated unit consists of postrift sediments deposited in a broad downwarp following regional subsidence but not confined by basin-bounding faults. Generally, the postrift unit tends to include more fine-grained sediment than the synrift unit because of lower postrift stream gradients.

Thus, five general units comprise stratigraphic sequences in continental rifts containing major lacustrine deposits. In order of deposition, these include basal sand, lacustrine shale, fluviodeltaic complex, synrift fluvial sand, and postrift fluvial sand (Figure 4). The stratigraphic sections in Figure 3 show this general pattern for each sequence except where deposition has been interrupted by renewed tectonism, as demonstrated by the Newark basin, Sudan, and west Africa. Renewed tectonism can abruptly terminate deposition of one sequence and initiate deposition of a second sequence.

Each basin varies from the ideal sequence to some degree. This is expected because each basin has a unique history with regard to amount and rate of extension and subsidence, climate and sediment supply, and provenance. Considering the variations in these parameters during the different ages represented in Figure 3, the stratigraphic sections are remarkably similar. Of particular importance is that the only stratigraphic sections with more than one major lacustrine unit are from basins that endured more than one tectonic episode; there is an exact correlation between major lacustrine units and the number of tectonic events. In addition, not only do the lacustrine units all occur in the same relative stratigraphic position near the base of the sequence, but they all exhibit deep-water conditions near their bases and shallow upward. This consistency in form and occurrence suggests that structural geometry and climate are equally important as controls on lake distribution.

If climate were the most important control, observed sequences should contain multiple major lake units deposited during a single tectonic episode and less consistent, deep to shallow progressions within those units. Obviously, climate is very important; enough water must be available to form and maintain a large lake, but observed stratigraphic sequences suggest that when climatic conditions permit, structural geometry and the evolution of that geometry control the occurrence of major lakes.

TECTONO-STRATIGRAPHIC MODEL

Any successful model for lake deposition in continental rift basins must address the evolutionary

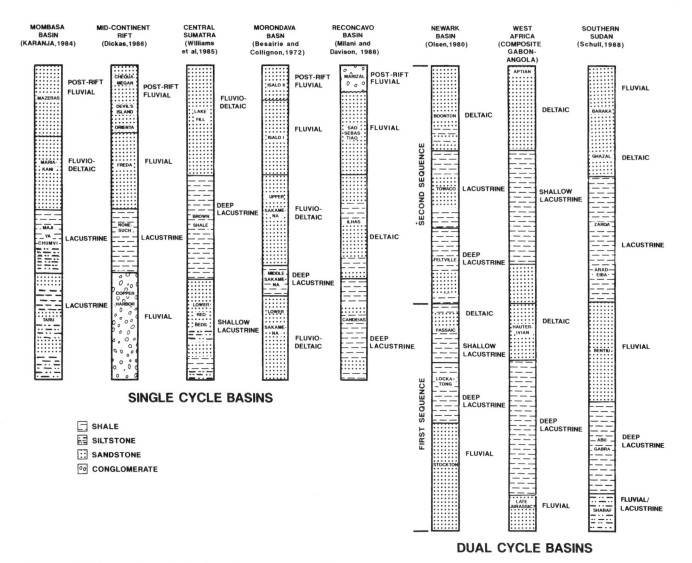

Figure 3. Comparison of stratigraphic sequences in eight rift basins, showing dominant lithologies and facies. Formation thicknesses not to scale.

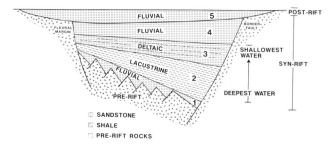

Figure 4. Idealized section of a half-graben containing a nonmarine rift stratigraphic sequence of five dominant lithologies and facies deposited from prerift through postrift phases.

nature of both the deposits and the sequences within which they occur. In Figure 3 note that each lacustrine unit is not uniform through time, nor is the overall basin-fill sequence. This implies a dynamic component of basin evolution that is responsible for the observed character of the stratigraphy. Thus, it is not simply a matter of modeling how structural geometry controls stratigraphic development but of modeling the evolution of that geometry through time and recognizing the effects of that evolution on stratigraphy. In this respect, consider the following tectono-stratigraphic model.

A rift basin passes through several tectonic and structural phases during its evolution from initial

Lambiase

rifting to postrift subsidence, and each phase is accompanied by a particular depositional style. Possibly the least well understood aspect of rifting tectonics is the initial rifting process. Rift-shoulder uplift is attributed to thermal activity associated with crustal thinning (Steckler, 1985; Weissel and Karner, 1989), but the exact timing of this uplift relative to the onset of sedimentation is important because sediment-supply rate is altered significantly by development of rift shoulders. Before they form, regional drainage freely enters the basin, and sediment-supply rate can be quite high; after uplift, sediment supply is restricted to internal sources. Another consequence of the timing of uplift is that the onset of rift shoulder-induced climate modification, as described by Hay et al. (1982), and the onset of its impact on sedimentation, also will be affected. Variations in the timing of the uplift and its effects on local climate and sediment supply probably account for some of the variations observed in basal synrift units, as illustrated in Figure 3.

Regardless of when rift shoulders evolve, block faulting probably occurs prior to significant synrift sedimentation because that faulting most likely establishes the rift as a basin capable of accumulating sediment. Initially the basin floor is a complex of tilted fault blocks at slightly lower elevation than the rift margins because basin asymmetry is just beginning to evolve. Sediment is deposited on the downdip edges of fault blocks. Although shallow lakes can form in topographic lows bounded by fault blocks, the lacustrine deposits are laterally restricted and relatively thin. Most deposits are fluvial and alluvial sand and gravel, which form the basal unit of the generalized stratigraphic sequence (Figure 4; Table 1). As subsidence along the basin-bounding fault continues, basin asymmetry develops rapidly relative to sedimentation rate. During this transition sedimentation patterns remain as at the outset except that some sediment bypasses the fault-block lows and moves downdip. Lakes remain restricted to topographic lows between fault blocks. Lake Bogoria in the Gregory rift, Kenya, is a modern example of a lake that occupies the downdip edge of a fault block (Tiercelin et al., 1987).

Accommodation zones become topographic highs as basin-bounding fault subsidence creates half-graben morphology, and their emergence causes a major change in topography that sets the stage for large lake development. For the first time in the rift's brief history, topographic barriers emerge along strike to form basins of half-graben size that are closed on all sides—laterally by rift shoulders and by accommodation zones at the ends, which can be up to several thousand meters deep. A large lake will form in any basin with an adequate supply of water, and lakes can occupy one or more adjacent half-grabens depending on the amount of water available and the relative heights of intervening accommodation zones. Lake Mobutu Sese-Seko (Albert) is a modern example of a lake that occupies a single half-

graben; Lakes Tanganyika and Malawi each extend over several. Maximum lake depth is limited by the height of the lowest lake-bounding accommodation zone.

An important element in the evolution of this geometry is that the entire process from the onset of faulting to large lake development takes place geologically very quickly. This usually limits the thickness of basal units and explains the observed abrupt transition from subaerial to deep-water conditions (Figure 3). Subsidence proceeds so rapidly that lakes become deep almost immediately.

Another consequence of structural geometry during lake formation is that neither of the geomorphic features that confine the lake serves as a significant source for sediment because most drainage on rift shoulders is away from the basin, and accommodation zones are narrow (minimal drainage area). Together with continuing basin subsidence, low sediment-supply rates create a sediment-starved basin (Figure 5) that persists long enough to accumulate thick sequences of fine-grained lacustrine sediment that typify this period of basin evolution (Figure 4; Table 1).

Sediment-starved half-grabens comprise most of a rift system during this stage and include those basins that may or may not contain lakes. The exceptions are those half-grabens that mark the ends of the rift system. Because no accommodation zones are found at the ends of the rift, the end half-graben generally forms a ramp into the surrounding region. This readily allows sediment to enter the rift along its axis from either or both ends depending on regional drainage. However, as long as subsidence rate equals or exceeds sediment-supply rate, the rift remains sediment starved.

At some point sediment-supply rate will surpass subsidence rate. Either regional drainage systems become better developed and deliver more sediment into the rift, or subsidence rate decreases, or both. Generally this happens in one or both end half-grabens first. Sediment supply will be greatest there if one assumes that nearly equal amounts of sediment enter the rift from its flanks along its entire length; the added component from axial sources boosts sedimentation rate in the end half-graben above that of the other basins (Figure 5). When the critical condition of sedimentation exceeding subsidence rate is achieved, the basin begins to fill and continues to do so unless renewed tectonism accelerates subsidence rate above the sediment-supply rate.

Basin-filling occurs as progradation of primarily sandy sediment into the basin. Fluvial and alluvial deposition dominate in half-grabens without lakes, whereas deltaic and prodelta deposits fill lake-bearing basins (Figure 4; Table 1). Sediment entering any given half-graben is confined to that basin until the sediment surface reaches the height of the lower of either adjacent accommodation zone. A modern example of a half-graben in this phase of filling appears to be the Ruzizi basin, the northernmost

Table 1. Summary of relationships among tectonic phases, topography, and stratigraphy in continental rifts (See text and Figure 4 for descriptions of stratigraphic units.)

Tectonic Phase	Topography	Unit Number and Depositional Style
Regional subsidence	Gentle regional downwarp	5—Mostly fluvial/alluvial; some small shallow lakes
Less active faulting, slower subsidence	Prominent rift shoulders, nearly flat basin floor	4—Mostly fluvial/alluvial; some small shallow lakes
Very active faulting, rapid subsidence	Prominent rift shoulders and accommodation zones, asymmetrical basins	3—Mostly deltaic and prodeltaic; some fluvial and lacustrine
		2—Lacustrine in large, deep lakes
	Fault-block topography, no rift shoulders or accommodation zones	1—Mostly fluvial/alluvial; some shallow lakes

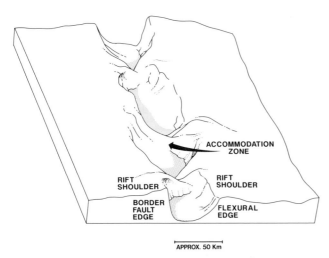

APPROX. 50 Km

Figure 5. Block diagram of a continental rift following rapid initial subsidence. Stippled areas are downdip sides of half-grabens. Basins are sediment starved, and large lakes form if enough water is available. Dimensions are scaled to the east African rift with a vertical exaggeration of 10.

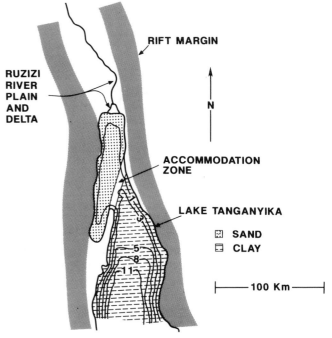

Figure 6. Northern end of Lake Tanganyika, east Africa, showing shallow sand-filled northern half-graben and deep clayey subbasin immediately south of the intervening accommodation zone. Bathymetric contours are in 100 m.

subbasin in Lake Tanganyika (Figure 6). Sediment enters the basin along its axis via the Ruzizi River and along its flexural margin via several smaller rivers. Water depths are shallower than in the rest of the lake, and seismic data suggest that a thick sandy sequence has partly filled the basin (Rosendahl et al., 1986). The Ruzizi River has built into the basin a prograding delta that may have supplied much of

Lambiase

the sandy prodelta sediment in the basin, although most of its present sediment load is fine grained (A. S. Cohen, 1989, pers. comm.). While the Ruzizi basin has been filling, the half-graben immediately to the south has remained sediment starved and is accumulating primarily fine-grained sediment (Figure 6: Huc and Vandenbroucke, 1988).

As a basin fills with sediment, the displaced volume causes the water surface level to rise until it reaches and stabilizes at the height of the lowest confining structural barrier (generally an accommodation zone), thereby limiting water depth (Olsen and Schlische, 1988). Continued prodeltaic sedimentation requires a decreasing water depth with time until the basin is full, and a lake can no longer exist. This accounts for the gradual shallowing and transition to deltaic and, finally, fluvial sedimentation observed in lacustrine sequences. The onset of fluvial sedimentation corresponds to the boundary between units 3 and 4 (Figure 4; Table 1). Burial of the topographic barrier (accommodation zone) precludes formation of another large, deep lake, but small, shallow lakes often occur during the subsequent fluvial-dominated depositional phases.

When a half-graben has filled to the height of the adjacent accommodation zone, a major change occurs in depositional style. Sediment is no longer confined but is carried into the adjacent basin (Figure 7). The floor of the filled basin upstream becomes a fluvial plain, and deposition shifts to the next downstream basin. Thus, the filled half-graben has become a sediment bypass zone despite the presence of high rift shoulders (Figure 7). Modern examples of this phase of basin filling are the Semliki River plain south of Lake Mobutu and the Ruzizi River plain north of Lake Tanganyika (Figures 6, 8). In both cases the rivers transport sediment across fluvial/alluvial plains overlying buried accommodation zones and deposit it in lake-filled half-grabens.

Basin-filling continues successively along the rift system until all accommodation zones have been completely or nearly buried, and subaerial deposition prevails. At that time the basin floor has become a fluvial/alluvial plain confined by the rift shoulders and often has a through-flowing river system; the entire basin looks like the upstream part of the block diagram in Figure 7. The Rio Grande rift is a modern example of a system that has reached this phase of filling (Cavazza, 1989), although ongoing regional uplift is causing erosion of some previously deposited basin fill.

Erosion of the rift shoulders and influx of sediment from outside the basin, together with continued local and/or regional subsidence, account for deposition of observed thick fluvial/alluvial sequences. These processes continue until rift filling and erosion have subdued the topography of the rift shoulders. Further regional subsidence results in a broad depression that fills primarily with fluvial/alluvial strata corresponding to unit 5 (Figure 4; Table 1). These units often

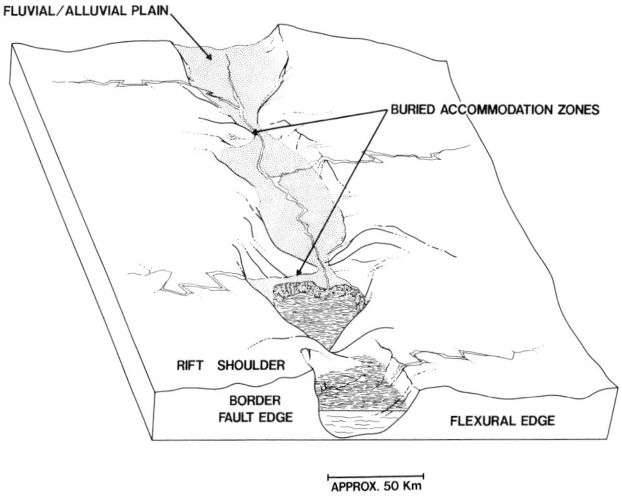

Figure 7. Block diagram from Figure 5 during basin-filling phase. Half-graben basin fills to height of the adjacent accommodation zone, at which point lacustrine deposition ceases, and sediment is carried into the adjacent downstream basin.

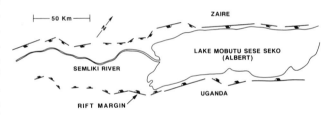

Figure 8. Semliki River plain southwest of Lake Mobutu, east Africa. The plain is a bypass zone for sediment entering the lake.

contain significant shallow-lacustrine deposits, e.g., the Barat Shale of the Sunda Shelf region (Pollock et al., 1984). Ongoing regional postrift subsidence in southern Sudan accounts for the topographic low that contains the Sudd swamp and its widespread paludal and lacustrine deposits.

The generalized five-part stratigraphic sequence depicted in Figure 4 and summarized in Table 1 relates directly to tectonic evolution of a rift but not as a one-to-one correlation between tectonic phases and stratigraphic units. Indeed, the tectonic phase boundaries occur within the stratigraphic units, and unit boundaries are more closely tied to tectonically induced topography. Thus, tectonism controls stratigraphy, but the actual relationship is tectonism controls topography, which controls stratigraphy.

IMPLICATIONS OF
THE MODEL

As described above, the basin-filling model adequately accounts for the general stratigraphic sequences discussed earlier. It identifies each stratigraphic component and the transitions between components as a product of a distinct phase in tectonic and topographic evolution of the rift system. The model characterizes the entire sequence as a necessary consequence of that basin's evolution. One should note several important implications of a model that asserts tectonically generated topography as a primary control on large lake occurrence. Large lakes can occur only after an accommodation zone and rift shoulder-bounded topographic depression form, and they can persist only as long as that depression does not fill with sediment. These conditions occur only during one phase of a basin's tectono-stratigraphic evolution and therefore restrict large lake development to that phase. As a necessary consequence of this, barring major climatic changes, only one large lake will occupy a given basin per tectonic event, a condition satisfied by the observed stratigraphic sequences described above.

An obvious consequence of the proposed model is that each subbasin fills sequentially. Assuming that the onset of lacustrine deposition is nearly coeval throughout the rift, lacustrine deposition should persist longer, and thick lacustrine sequences should accumulate in the "downstream" or downrift subbasins relative to those that fill first. Consequently, despite their similarity, lacustrine sequences in various subbasins are not time equivalent and cannot be correlated as such because equivalent basin-filling facies were not deposited simultaneously in all subbasins.

Isostatic rebound of the basin floor following extension (Spencer, 1984) and/or climatically controlled lake-level changes render subbasin correlation even less tenable, especially in subsurface studies. In Figure 7, lowered base level, of any origin, can induce erosion of the upstream half-graben basin fill and redeposition into the basin that is actively filling. This results in time-equivalent erosion and deposition, characterized by fossil assemblages that may be composed largely of reworked material.

The process of basin-filling, according to this model, requires an axial component of sediment supply, although not necessarily a large one. Assuming equal sediment-supply rates across the rift shoulders for each half-graben within the system, the addition of relatively small amounts of sediment into a terminal subbasin would increase the local sedimentation rate enough to ensure that the end basin fills before the others. After it fills, sediment is carried into the next half-graben and again increases local sedimentation rate by the introduction of the rift-axial component. This new influx causes that subbasin to fill more rapidly than adjacent subbasins supplied only from the rift shoulders. The axial component is more important in this second subbasin than in the first because its supply rate equals the sum of the basin-end and rift-shoulder supply rates for the first half-graben (Figure 9). This process is repeated along the length of the rift system with axial sediment supply becoming cumulatively more important.

The importance of axial versus lateral sediment supply has been debated for Lake Tanganyika and the Newark basin. The Ruzizi basin is, as the Newark appears to be, the "upstream" end of a series of half-grabens. In neither case is axial sediment supply believed to be large relative to lateral supply (A. S. Cohen and P. E. Olsen, 1989, pers. comm.). However, the occurrence of rift-axial transport is confirmed at (1) the northern end of the Newark basin by paleocurrent data from the Passaic Formation (W. Manspeizer, 1989, pers. comm.); and (2) by the very presence of the Ruzizi River delta in Lake Tanganyika. The proposed model suggests that axial transport probably contributed significantly to filling of the Newark basin and that sediment supply from the Ruzizi River may be instrumental when Lake Tanganyika eventually fills.

Sampling bias also can lead to an underestimation of the importance of axial transport. Because rift-axial sediments are preferentially deposited toward the downdip edge of a half-graben, they have a wedge-shaped cross-sectional geometry. Cavazza (1989) observed similar geometry in the Rio Grande rift, where axially derived sediment was deposited near the flexural margin only in later stages of deposition, even though it accounts for a significant volume of the basin fill. Because outcrops generally expose a greater proportion of the total rift-fill sequence near the flexural margin than elsewhere in a basin, the importance of axial transport can be underestimated.

Variations in sedimentation rate are expected during the phases of rift evolution. Basal fluvial sands are deposited relatively quickly because no topographic barriers have yet formed to limit sediment supply, but lacustrine sequences accumulate more slowly under starved-basin conditions. Coarse-grained clastic basin filling via delta progradation accounts for the fastest sedimentation rates, but those rates drop rapidly to nearly zero when a subbasin fills and becomes a sediment bypass zone. Rift-shoulder topography will be subdued or buried at moderate to high rates whose magnitudes depend largely on erosion rates. Regional subsidence rates are the most important control on sedimentation rate during the final, sag-basin phase of deposition. Very few data are available that allow accurate calculation of sedimentation rates in nonmarine rifts because of poor paleontological control, especially in subaerially deposited sequences. However, drill-hole data from Sudan support relatively low rates for lacustrine intervals, high rates for deltaic sequences, and low rates for overlying fluvial sequences, which include times of sediment bypass (M. R. Rodgers, 1988, pers. comm.).

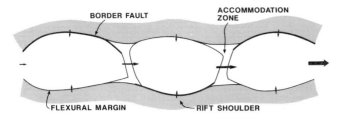

Figure 9. Schematic illustration depicting relative importance of rift-axial sediment transport (left to right) in successive half-grabens along a rift system. Arrow size is proportional to sediment-supply rate.

Discussion

The proposed model for tectonic control of major lacustrine sequences in continental rift basins adequately explains the general sequences observed in several basins. However, its assumptions about certain factors that control lacustrine deposition may not be valid in all cases. Indeed, all the assumptions implicit in the model may be valid in only very few basins. Thus, it is expected that substantial variations will be in observed lacustrine sequences.

One underlying assumption is that enough water always is available to form a large lake if the topographic configuration is favorable. Clearly, this is invalid, as climate can severely limit water availability. Changes in water availability during a basin's history controls the thickness, facies, and lithology of any lacustrine sequence that accumulates in it and may be responsible for some of the variations depicted in Figure 3.

Another important assumption is that sediment-supply rate across the rift shoulders is equivalent for all subbasins within a rift system and that supply rate is low relative to subsidence rate during most of the lacustrine phase. Sediment-supply rate data are lacking for modern rifts, but variations in stream size suggest corresponding variations in sediment-supply rate. Consider also examples such as the Ruhuhu River at Lake Malawi, where a disproportionately large antecedent drainage has been captured by later rifting. A relatively high sediment-supply rate into one half-graben could cause that subbasin to fill with no axial input, or, in the extreme, could prevent a lake from forming. The model should remain valid for adjacent half-grabens because any sediment introduced into them from a rapidly filled subbasin would constitute an axial supply.

It also was assumed that tectonically induced topography evolves approximately the same basin shapes and rift shoulder and accommodation zone heights in the same timing sequence and at approximately the same rates in all basins. Variations in tectonic history and structural style in continental rifts are expected to cause comparable differences in lacustrine stratigraphy. Other factors that affect the character of lacustrine sequences, such as sediment provenance, were not considered. Thus, by incorporating the assumptions noted above, the model demonstrates that, given an adequate supply of water, the occurrence and general character of lacustrine sequences in continental rift basins is a product of structural evolution and subsidence history.

CONCLUSIONS

The general conclusion of this study is that the occurrence and character of major lacustrine sequences in continental rift basins is strongly controlled by changing topography generated during tectono-stratigraphic evolution of these basins. Lakes can occur only during that phase of a basin's history when rift shoulders and accommodation zones form a closed, sediment-starved basin. Later burial of the accommodation zones, in response to increased sedimentation rate relative to subsidence rate, terminates any ongoing lacustrine deposition and precludes formation of another large lake.

Major lacustrine sequences generally have sharp or unconformable contacts with underlying units and reflect rapid development of maximum water depths followed by gradual shallowing upward into subaerial deposits. This progression also is a consequence of the interaction among topography, subsidence rate, and sedimentation rate. Tectono-stratigraphically generated topography is equal in importance to water availability as a control on the occurrence of major rift lakes.

ACKNOWLEDGMENTS

Discussions with a number of people contributed significantly to this study. Especially helpful were W. P. Bosworth, M. R. Rodgers, A. S. Cohen, and P. E. Olsen. W. P. Bosworth, C. J. Ebinger, and L. G. Kessler read a preliminary version of the manuscript and made valuable suggestions, as did reviewers L. E. Frostick and P. E. Olsen. However, the author is responsible for all interpretations as well as any errors or oversights. D. J. Klages typed the manuscript, and J. M. Delanoix drafted the figures. The author thanks Marathon Oil Company for permission to publish this paper.

REFERENCES

Arguden, A. T., and K. S. Rodolfo, 1986, Sedimentary facies and tectonic implications of lower Mesozoic alluvial-fan conglomerates of the Newark Basin, northeastern United States: Sedimentary Geology, v. 51, p. 97–118.

Besairie, H., and M. Collignon, 1972, Geologie de Madagascar I. les terrains sedimentaires: Annal. Geologie Madagascar, v. 35, p. 552.

Bosworth, W. P., 1985, Geometry of propagating continental rifts: Nature, v. 316, p. 625–627.

Brice, S. E., M. D. Cochran, G. Pardo, and A. D. Edwards, 1982,

Tectonics and sedimentation of the South Atlantic rift sequence—Cabinda, Angola, *in* J. S. Watkins, and C. L. Drake, eds., Studies in continental margin geology: AAPG Memoir 34, p. 5–18.

Cavazza, W., 1989, Sedimentation pattern of a rift-filling unit, Tesuque Formation (Miocene), Española Basin, Rio Grande Rift, New Mexico: Journal of Sedimentary Petrology, v. 59, p. 287–296.

Dickas, A. B., 1986, Comparative Precambrian stratigraphy and structure along the Mid-Continent Rift: AAPG Bulletin, v. 70, p. 225–238.

Frostick, L. E., and I. Reid, 1987, Tectonic control of desert sediment in rift basins ancient and modern, *in* L. E. Frostick, and I. Reid, eds., Desert sediments—Ancient and modern: Geological Society Special Publication 35, p. 53–68.

Frostick, L. E., and I. Reid, 1989, Is structure the main control of river drainage and sedimentation in rifts?: Journal of African Earth Sciences (and the Middle East), v. 8, p. 165–182.

Hay, W. W., J. F. Behensky, Jr., E. J. Barron, and J. L. Sloan, Jr., 1982, Late Triassic-Liassic paleoclimatology of the protocentral North Atlantic rift system: Palaeogeography, Palaeoclimatology, Palaeoecology, v. 40, p. 13–30.

Huc, A. Y., and M. Vandenbroucke, 1988, Northern Lake Tanganyika—A conceptual model of organic sedimentation in a rift lake (abs.), *in* Lacustrine exploration—Case studies and modern analogues: AAPG Research Conference Abstracts with Program.

Karanja, F. M., 1984, Excursion guide to the geology of Mombasa-Lamu Basin west of Mombasa Island: Kenya Ministry of Energy and Regional Development, 37 p.

Leeder, M., and R. Gawthorpe, 1987, Sedimentary models for extensional tilt-block/half-graben basins, *in* M. P. Coward, J. F. Dewey, and P. L. Hancock, eds., Continental extensional tectonics: Geological Society Special Publication 28, p. 139–152.

Manspeizer, W., and P. E. Olsen, 1981, Rift basins of the passive margin—Tectonics, organic-rich lacustrine sediments, basin analysis, *in* W. Hobbs, III, ed., Field guide to the geology of the Paleozoic, Mesozoic, and Tertiary rocks of New Jersey and the central Hudson Valley, New York: Petroleum Exploration Society of New York, p. 25–105.

Milani, E. J., and I. Davison, 1988, Basement control and transfer tectonics in the Reconcavo-Tucano-Jabota rift, Northeast central Brazil: Tectonophysics, v. 154, p. 41–70.

Milani, E. J., M. C. Lane, and P. Szatmari, 1988, Mesozoic rift basins around the northeast Brazilian microplate (Reconcavo-Tucano-Jabota, Sergipe-Alagoas), *in* W. Manspeizer, ed., Triassic-Jurassic rifting, continental breakup and the origin of the Atlantic Ocean and passive margins: New York, Elsevier, pt. B, p. 833–858.

Olsen, P. E., 1980a, Triassic and Jurassic formations of the Newark Basin, *in* W. Manspeizer, ed., Field studies in New Jersey geology and guide to field trips, 52d Annual Meeting: New York State Geological Association, p. 2–39.

Olsen, P. E., 1980b, Fossil great lakes of the Newark Supergroup in New Jersey, *in* W. Manspeizer, ed., Field studies in New Jersey geology and guide to field trips, 52d Annual Meeting: New York State Geological Association, p. 352–398.

Olsen, P. E., 1984, Periodicity of lake-level cycles in the Late Triassic Lockatong Formation of the Newark Basin (Newark Supergroup, New Jersey and Pennsylvania), *in* A. Berger, J.

Imbrie, J. Hays, G. Kukla, and B. Saltzman, eds., Milankovitch and climate—Understanding the response to astronomical forcing: Boston, D. Reidel Publishing Co., pt. 1, p. 129–146.

Olsen, P. E., and R. W. Schlische, 1988, Unraveling the rules of rifts, *in* Yearbook, Lamont-Doherty Geological Observatory: Columbia University, p. 26–31.

Pollock, R. E., J. B. Hayes, K. P. Williams, and R. A. Young, 1984, The petroleum geology of the KH Field, Kakap, Indonesia: Indonesian Petroleum Association 13th Annual Convention, Proceedings, p. 407–423.

Ratcliffe, N. M., W. C. Burton, R. M. D'Angelo, and J. K. Costain, 1986, Low-angle extensional faulting, reactivated mylonites, and seismic reflection geometry of the Newark basin margin in eastern Pennsylvania: Geology, v. 14, p. 766–770.

Rosendahl, B. R., 1987, Architecture of continental rifts with special reference to East Africa: Annual Review of Earth and Planetary Sciences, v. 15, p. 445–503.

Rosendahl, B. R., D. J. Reynolds, P. M. Lorber, C. F. Burgess, J. McGill, D. Scott, J. J. Lambiase, and S. J. Derksen, 1986, Structural expressions of rifting—Lessons from Lake Tanganyika, Africa, *in* L. E. Frostick, R. W. Renaut, I. Reid, and J. J. Tiercelin, eds., Sedimentation in African rifts: Geological Society Special Publication 25, p. 29–43.

Scholz, C. A., and B. R. Rosendahl, 1988, Low lake stands in Lakes Malawi and Tanganyika, East Africa, delineated with multi-fold seismic data: Science, v. 240, p. 1645–1648.

Schull, T. J., 1988, Rift basins of interior Sudan—Petroleum exploration and discovery: AAPG Bulletin, v. 72, p. 1128–1142.

Spencer, J. E., 1984, Role of tectonic denudation in warping and uplift of low angle normal faults: Geology, v. 12, p. 95–98.

Steckler, M. S., 1985, Uplift and extension at the Gulf of Suez—Indications of induced mantle convection: Nature, v. 317, p. 135–139.

Tiercelin, J. J., and others, 1987, Le demi-graben de Baringo-Bogoria, Rift Gregory, Kenya: Bulletin des Centres de Recherche, Exploration-Production Elf-Aquitaine, v. 11, p. 249–540.

Unger, J. D., 1988, A simple technique for analysis and migration of seismic reflection profiles from the Mesozoic basins of eastern North America, *in* A. J. Froelich, and G. R. Robinson, Jr., eds., Studies of the Early Mesozoic basins of the eastern United States: USGS Bulletin 1776, p. 229–235.

Van Houten, F. B., 1969, Late Triassic Newark Group, north central New Jersey, and adjacent Pennsylvania and New York, *in* S. S. Subitzki, ed., Geology of selected areas in New Jersey and eastern Pennsylvania: Rutgers University Press, p. 314–347.

Walters, R., and R. E. Linton, 1973, The sedimentary basin of coastal Kenya, *in* G. Blant, ed., Sedimentary basins of the African coast, part 2—South and east coast: Paris, Association of African Geological Surveys, p. 133–158.

Weissel, J. K., and G. D. Karner, 1989, Flexural uplift of rift flanks due to mechanical unloading of the lithosphere during extension: Journal of Geophysical Research, v. 94, p. 13919–13950.

Williams, H. H., P. A. Kelley, J. S. Janks, and R. M. Christensen, 1985, The Paleogene rift basin source rocks of central Sumatra, *in* The past, the present, the future: Indonesian Petroleum Association 14th Annual Convention, Proceedings, v. 2, p. 57–90.

Carboniferous Lacustrine Shale in East Greenland— Additional Source Rock in Northern North Atlantic?

Lars Stemmerik
Flemming G. Christiansen
Stefan Piasecki
Geological Survey of Greenland
Copenhagen, Denmark

Lacustrine organic-rich shales have been recorded at three stratigraphic levels within the uppermost Devonian-Lower Permian continental sequence of central east Greenland. The Westphalian lacustrine shales, considered here, are divided into an epilimnic association of silty shale and sandstone dominated by terrestrial organic material, and a hypolimnic association of clay shale dominated by amorphous kerogen and algae. The hypolimnic shales display hydrogen index values between 300 and 900 mg HC/g TOC and TOC between 2 and 10%, and can be characterized as potential oil-prone source rocks.

The lacustrine basins are expected to be 10–15 km wide and tens of kilometers along strike with net source rock thickness in excess of 50 m and generative potential in excess of 3 million m³/km². The shales form a hitherto overlooked potential source rock in the east Greenland basin with implications for areas offshore Norway.

INTRODUCTION

Most hydrocarbons in the North Atlantic region are believed to have been sourced from Jurassic marine shale. In the northern North Sea hydrocarbons are derived exclusively from the Kimmeridge Clay and equivalent units (Thomas et al., 1985). To the north, on the Mid-Norwegian shelf and in the Barents Sea (Figure 1), Upper Triassic-Lower Jurassic deltaic sediments have been suggested as an additional source (Elvsborg et al., 1985; Hagevang and Rønnevik, 1986; Sund et al., 1986). Studies of the onshore sequence in central east Greenland have confirmed the source potential of the Jurassic strata also on the western North Atlantic margin; in addition, Upper Permian marine shale with good source quality has been found (Surlyk et al., 1984, 1986; Surlyk, Hurst, et al., 1986).

The thick Upper Devonian-Carboniferous sequence underlying the region until recently was believed to have only negligible source potential and thus had generally been discounted except for several reports of thin, organic-rich shale intervals in Spitsbergen (Svalbard) and Greenland (Bjorøy et al., 1981; Surlyk, Hurst, et al., 1986). Renewed field investigation of Devonian and Carboniferous strata in central east Greenland showed, however, the existence of lacustrine sequences containing more abundant organic-rich black shale than previously known (Marcussen et al., 1987, 1988; Piasecki et al., 1990).

Three episodes of extensive lacustrine deposition are apparent in the east Greenland basin and are provisionally dated to the latest Devonian, Early Carboniferous, and Late Carboniferous (Westphalian). This paper focuses on Westphalian (upper Bashkirian) lacustrine sediments, their regional occurrence, tectonic and climatic setting, and implications for areas outside east Greenland.

GEOLOGICAL SETTING

Carboniferous strata of central east Greenland (Koch and Haller, 1971) historically had been poorly dated because of the absence of marine fossils and poor preservation of terrestrial fossils. New palynological studies suggest that the sequence ranges from latest Devonian (Famennian) to earliest Permian (Piasecki et al., 1990).

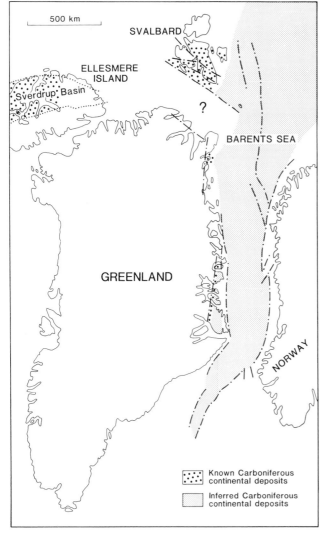

Figure 1. Late Paleozoic reconstruction of the North Atlantic with distribution of Upper Devonian to Lower Permian continental strata. Based on Ziegler (1988), Stemmerik and Worsley (1989), and Piasecki et al. (1990).

These sedimentary rocks represent the final stage of deposition of a 9-km-thick synrift sequence following the Caledonian orogeny. Rifting presumably initiated during Early Devonian time (Larsen et al., 1989) in the Jameson Land area within a regionally transtensional tectonic regime (Friend et al., 1983). Uppermost Devonian-Carboniferous strata, which unconformably overlie Devonian and older rocks, were deposited in a system of westward-tilted half-grabens (Kempter, 1961; Haller, 1970; Collinson, 1972; Surlyk et al., 1984; Surlyk, Hurst, et al., 1986) during a tectonic phase dominated by east-west extension.

In central east Greenland the Upper Devonian-Carboniferous sequence is 1500–3000 m thick. The rocks are exposed in a belt 350 km long and up to

30 km wide along the western basin margin (Figure 2). Time-equivalent strata in eastern north Greenland are believed to be the northern extension of this synrift sequence, thus linking the sequence to Upper Devonian-Carboniferous strata of Spitsbergen and the Barents Shelf (Håkansson and Stemmerik, 1984, 1989; Steel and Worsley, 1984; Surlyk, Hurst, et al., 1986; Stemmerik and Worsley, 1989).

The Upper Devonian-Carboniferous sequence in east Greenland consists entirely of continental rocks of fluvial and lacustrine origin. The provenance lay to the west, where deposition took place in a system of coalescing alluvial fans and braid plains (Collinson, 1972; Surlyk, Hurst, et al., 1986; Marcussen et al., 1987, 1988). To the east, a major flood plain with northward-draining rivers developed laterally to the fans and braid plains.

SEDIMENTOLOGY

Four types of lacustrine shale have been tentatively identified within the uppermost Devonian-Lower Permian succession—(1) thick (>10 m) sequences of gray to black, well-laminated shale with minor limestone and sandstone; (2) thin (0.5–3 m) sequences of gray, well-laminated shale associated with thin coal beds, limestone, and fine-grained sandstone (which is often the dominant lithology); (3) thin (<0.5 m) gray, green, and reddish shales; and (4) thick (>20 m) sequences of red siltstone with thin sandstone. The type 1 and 2 lacustrine sequences represent the more permanent lacustrine systems developed under humid conditions. Type 3 shale is believed to be overbank and backswamp deposits of short-lived lakes, and type 4 shale represents lake development during arid conditions that induced repeated exposure and oxidation.

From geochemical analyses it is evident that the source potential within the uppermost Devonian–Lower Permian succession is confined mainly to type 1 lacustrine deposits (Figure 3). Because the organic material of the shallower, short-lived type 2 lakes is mainly of terrestrial origin, these shales are therefore gas prone. Deep permanent lakes with type 1 depositional sequences are found in three stratigraphic intervals, of which only the Westphalian will be considered here.

Westphalian lacustrine deposits are divided into hypolimnic and epilimnic associations on basis of shale sedimentology, organic content, and geochemical signature. Sandstones are restricted to the epilimnic association where they are important. Both associations have a low-diversity fauna of ostracods; remains of freshwater fish are restricted to the hypolimnic association.

Hypolimnic Association

The hypolimnic association is dominated by black, laminated shale with moderately high content of silt-

Stemmerik et al.

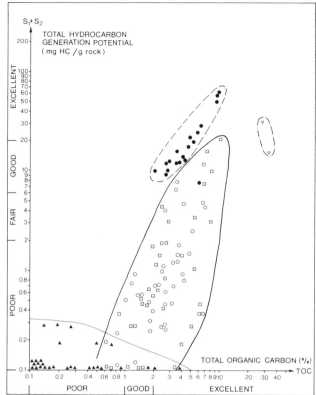

Figure 3. Source rock characterization of east Greenland lacustrine shales by TOC vs. total hydrocarbon-generation potential. Filled circles, hypolimnic shale; open circles, epilimnic shale; open squares, type 2 shallow-lacustrine shale; filled triangles, type 3 and type 4 shale; open, inverted triangles, coal-rich type 3 shale.

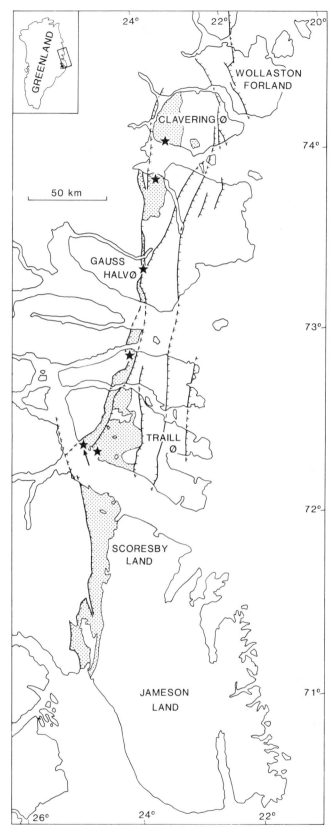

Figure 2. Generalized map of central east Greenland showing outcrop of uppermost Devonian to Lower Permian strata (stippling). Known Westphalian lacustrine sequences marked by stars.

size mica and fine-grained, windblown quartz sand. The lamination is defined by couplets of mica-rich and organic-rich laminae. Occasionally carbonate-rich laminae also occur, and triplets are developed. Associated with this facies are nonlaminated mudstone with abundant algae and shaly limestone with abundant ostracod shells and millimeter-size subspherical bodies of micritic calcite (Figure 4b).

The shales have total organic carbon (TOC) content in the range of 2-10%. Organic matter consists of 80-90% amorphous kerogen that is believed to be degraded algal material, including remains of *Botryococcus* (Figure 5a). The remaining 10-20% is composed mainly of trilete spores (dominant) and monosaccate pollen (Piasecki et al., 1990).

Geochemically, the shales show high hydrogen index (HI) values (300-800) and low oxygen index (OI) values (10-35) typical of type I and type II kerogens (Figure 6, left graph). Samples whose maturity corresponds to early generation (defined as high residual potential and high hydrocarbon content) have S_1 values of 0.5-1 mg HC/g rock and bitumen yields between 1100 and 2500 ppm (Table 1). All the shales display very low sulfur contents with

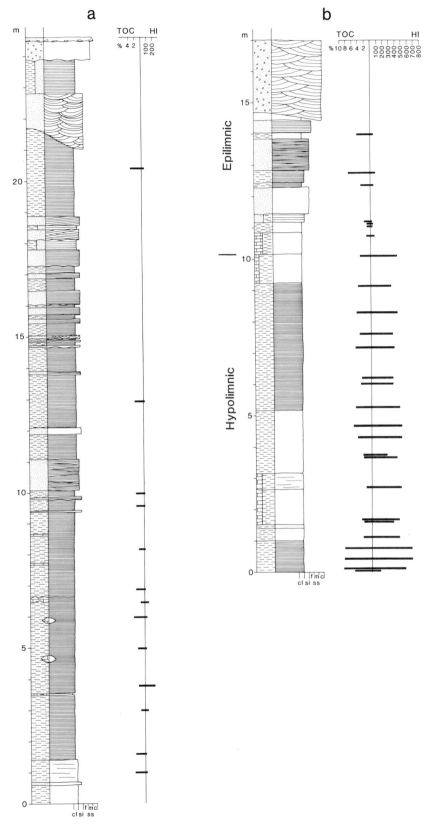

Figure 4. Representative sedimentological logs of the (a) epilimnic and (b) hypolimnic associations. Note the upward-coarsening trend in both sections and upward change from hypolimnic to epilimnic sedimentation in section b, clearly expressed in the HI index.

Stemmerik et al.

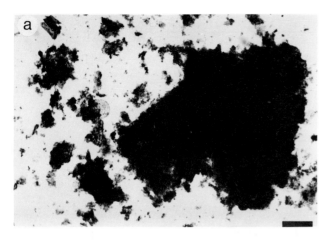 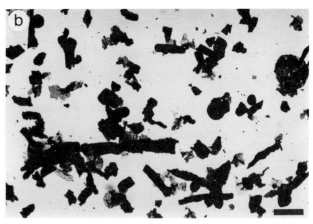

Figure 5. Photomicrographs of (a) amorphous kerogen from hypolimnic shale; (b) woody material, coaly fragments, and sporomorphs from epilimnic shale. Scale bars equal 0.1 mm.

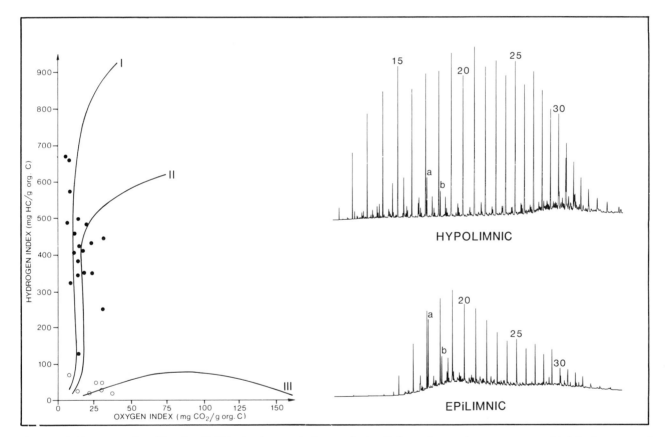

Figure 6. Kerogen classification (left) by hydrogen index vs. oxygen index (modified Van Krevelen diagram). Filled circles, hypolimnic shale; open circles, epilimnic shale. Selected gas chromatograms (right) of saturated hydrocarbons. Numbers are n-alkane carbon number; a, pristane; b, phytane.

a tendency toward increasing total sulfur with increasing organic-matter content (Figure 7). The highest recorded sulfur value, 0.58%, was determined in one of the most organic-rich shales. The generally low values clearly distinguish these materials from marine shale and support a freshwater depositional environment (e.g., Berner and Raiswell, 1984;

Duncan and Hamilton, 1988).

The hypolimnic sediments were deposited in the deep anoxic areas of the lakes. Their facies represent various parts of this geochemically and biologically rather uniform environment, with laminated shale probably representing proximity to inflowing rivers. Silt-rich and organic-rich couplets are believed to

Table 1. Selected characteristics of hypolimnic and epilimnic shales

Property	Hypolimnic Association	Epilimnic Association
Dominant lithology	Clay shale (laminated)	Laminated silty shale
Sand/shale ratio	0	0–50%
TOC	2–10%	0.5–7%
HI	300–900	<80 (170)
OI	10–35	10–40
Extractability	High	Negligible
Waxiness	High	Low
Amorphous kerogen/TOC	>80%	–
Spores/pollen	0–30%	25–45%
Macroplants	–	+
Freshwater fish	+	–
Ostracodes	+	+

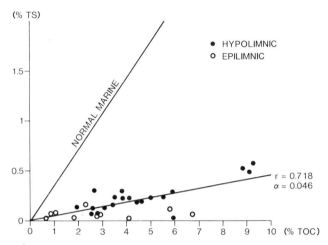

Figure 7. Total sulfur vs. TOC. Filled circles, hypolimnic shale; open circles, epilimnic shale. Normal marine trend from Berner and Raiswell (1984).

represent seasonal variations in river discharge (cf. Donovan, 1980; Halfman and Johnson, 1988). In contrast, the nonlaminated mudstone facies may represent the most distal areas of the lake lacking a supply of silt-sized material. Very limited supply of fluvial material into the hypolimnic areas is confirmed by composition of the organic material. Amorphous kerogen, *Botryococcus*, and sheet-shaped organic particles that dominate this association are regarded as autochthonous, having been deposited from suspension approximately at the habitat of living algae.

The remaining organic matter is dominated by trilete spores from lycopods, articulates, ferns, and seed ferns that lived along the lake margins. Less than 5% of the organic matter is river-transported monosaccate pollen from gymnosperms and Corda-ites from presumed drier, more distant environments (Piasecki et al., 1990).

Epilimnic Association

The epilimnic association is composed of gray silty shale and fine-grained sandstone (Figures 4, 8). In contrast to hypolimnic strata, they include abundant macroscopic plant fragments up to 10 cm long. Shales are actually gray, laminated or microcross-bedded, micaceous siltstone, whose individual laminae are graded and commonly erosive. Associated with these are occasional thin sandy siltstones with wave ripples and hummocky cross-stratification (Figure 9). Fine- to medium-grained sandstone, which usually forms erosively based, laminated or low-angle planar cross-bedded units up to 50 cm thick, can be traced laterally over the entire length of the outcrops (100–200 m). In the upper part of the section the sandstones fill channels and mark a transition into coarse-grained and pebbly fluvial deposits.

According to geochemical analyses, epilimnic shales have TOC values of 0.5–7% and are dominated by coaly fragments and woody material, which comprise 80–90% of the organic material (Figure 5b). *Botryococcus* colonies occur infrequently, and monosaccate pollen are more common than in hypolimnic shale. The organic matter is type III kerogen with HI values less than 75 and OI values between 10 and 40 (Figure 6; Table 1). Hydrocarbons are negligible, whether expressed as S_1 values or extractable material. Sulfur content is even lower than in hypolimnic shale, with the highest recorded value of 0.17% (Figure 7).

Epilimnic strata were deposited in the shallow oxic areas of the lakes. Their organic material and silt content indicate a significant contribution of river-transported material during deposition; consequently, the shales are believed to represent mainly prodeltaic deposition. This is confirmed by their association with sandstone in the upper part of the lacustrine sequences. Deposition in these small deltas apparently was controlled by fluvial processes because wave-induced structures are rare and may represent extreme conditions.

Depositional Evolution

The Westphalian lacustrine sequences, although diverse in detail, all show a common depositional pattern—lacustrine shale directly overlies coarse-grained fluvial deposits with no interbedding. All the sequences are asymmetrical with apparently deepest

Stemmerik et al.

Figure 8. Interbedded epilimnic shale and sandstone erosively overlain by fluvial coarse-grained sandstone. Vertical exposure approximately 15 m high.

water or most distal deposits at or near the base of the sequence followed by gradual upward coarsening and shallowing (Figure 4). The upper boundary is sharp and erosive, and the overlying fluvial sandstone is petrographically distinct from epilimnic sandstone.

This evidence suggests that initiation of lacustrine systems was an abrupt process that not only changed the water level locally in the basin but apparently also changed the sediment source area. Most likely lakes became established in structural depressions along the downfaulted margins of half-graben systems during times of block tilting, as lakes respond more rapidly to changes in hydrological regime than do alluvial fans and braid plains (cf. Blair and Bilodeau, 1988). Alternatively, lakes may have formed as the combined result of increased precipitation in the area and eustatic rise in sea level, which dammed the water in the former flood plain.

When established, a lake apparently underwent simple infilling. High TOC values in the hypolimnic shale indicate slow sedimentation rates during early lake history. Later, the lake became gradually shallower as deltas began to prograde. Apparently

its termination was associated with general lowering of base level, as overlying braid-plain deposits have eroded several meters into the lake deposits (Figure 8). This agrees with the hypothesis of Piasecki et al. (1990) that the east Greenland lakes represent the final phase of 200- to 300-m-thick fining-upward megasequences deposited under humid conditions. The overlying braid-plain deposits thus are associated with times of more arid climate.

SOURCE ROCK POTENTIAL

As evidenced from the geochemical analyses, different parts of the lacustrine system exhibit highly variable source rock potential. Only the hypolimnic shale, with its high TOC and high HI, is a potential oil-prone source rock. Based on a case study of a 12-m-thick hypolimnic shale sequence in the southern part of the area (Piasecki et al., 1990), we estimate a generation potential of about 0.8 million m^3 HC/km^2.

Figure 9. Wave-ripple lamination in silty, fine-grained sandstone of the epilimnic association. Pencil is 14 cm long.

No individual lacustrine shale sequence has been found to exceed 20 m in thickness. However, in most sections two or three such sequences occur with a net source rock thickness probably in excess of 50 m, which increases the generative potential to more than 3 million m³ HC/km². The areal extent of the lakes is more difficult to estimate from surface data. Individual shale sequences have been traced for more than 10 km east to west and more than 15 km north to south parallel to the basin axis. However, these are minimum distances because of limited exposures.

The proposed sizes of the lakes are highly dependent on the model used for their initiation. If one assumes nontectonic origin, their areal extent may approach the area of the entire basin, or about 50 km by 350 km (17500 km²). More likely, the lakes formed within intrabasinal structural depressions 10–15 km wide and several tens of kilometers long parallel to the basin axis.

In central east Greenland the shales are immature to early mature in exposed sections along the western basin margin (Figure 2). However, the Carboniferous sediments become more deeply buried eastward and therefore may have reached thermal maturity in large areas underlying Jameson Land, Traill Ø, and Hold-with Hope (Figure 2). Westphalian lacustrine shale

then may be regarded as a source rock in addition to the Upper Permian Ravnefjeld Formation in the Jameson Land basin (e.g., Surlyk et al., 1986; Surlyk, Hurst, et al., 1986), which is now being explored.

Implications for the North Atlantic

The Westphalian lacustrine shale is a hitherto overlooked potential source rock in central east Greenland and has implications for areas offshore central Norway and the Barents Sea. Our knowledge of the pre-Upper Permian sequence offshore Norway presently is very limited (Ziegler, 1988). The inference of Carboniferous continental deposits there is based mainly on the assumption that the stratigraphy is similar to that known from east Greenland as proven for the post-Permian succession. The occurrence of oil-prone lacustrine source rocks in the east Greenland Carboniferous succession may therefore be directly applicable to offshore central Norway during this stage of exploration.

Stratigraphy of the Barents Sea and onshore basins of Arctic Canada, Spitsbergen, and eastern north Greenland (Figure 1) farther north is well known and different from that of east Greenland (Steel and

Stemmerik et al.

Worsley, 1984; Häkansson and Stemmerik, 1984; Beauchamp et al., 1989; Stemmerik and Worsley, 1989). Because the Arctic basins were transgressed during middle Carboniferous time, the Westphalian (Bashkirian) thus is dominated by marine deposits. However, a general tectonic-climatic setting similar to that of central east Greenland was found in the Arctic basins for latest Devonian and Early Carboniferous time (e.g., Gjelberg and Steel, 1981; Steel and Worsley, 1984; Davies and Nassichuk, 1988; Häkansson and Stemmerik, 1989). Lower Carboniferous lacustrine oil-prone shales are described from the Sverdrup basin (Goodarzi et al., 1987; Davies and Nassichuk, 1988). Observations from both the Sverdrup basin and east Greenland thus imply that the organic-poor lacustrine sequences described from the Upper Devonian-Lower Carboniferous in Spitsbergen (e.g., Gjelberg, 1978; Gjelberg and Steel, 1981) may pass into lacustrine shale of better source rock quality. The few organic-rich shales from the Vesalstranda Member of the Røedvika Formation, as analyzed by Bjorøy et al. (1981), may represent examples of such shales. Furthermore, black lacustrine shale of the Lower Carboniferous Adriabukta Formation in southern Spitsbergen possibly may represent good source rocks that are now in a postmature stage (D. Worsley, 1989, pers. comm.).

SUMMARY AND CONCLUSIONS

Potential hydrocarbon source rocks within the uppermost Devonian-Lower Permian succession in east Greenland are confined mainly to thick lacustrine shale sequences. Based on sedimentological and palynological criteria, these shales can be divided into an epilimnic association with no source potential and a hypolimnic association with a good to excellent source potential. Hypolimnic shales have TOC content in the range of 2-10%, HI values between 300 and 900, S_1 values between 0.5 and 1 mg HC/g rock, and bitumen yields of 1100-2500 ppm. Organic matter consists of 80-90% amorphous kerogen, which is believed to be degraded algal material.

The lacustrine basins perhaps were 10-15 km wide and extended several tens of kilometers along strike; net source thickness may exceed 50 m, for a possible generative potential of more than 3 million m^3 HC/km^2.

We conclude that Carboniferous lacustrine shale may be regarded as a hydrocarbon source rock in the onshore basin of east Greenland and offshore central Norway during this stage of exploration. In addition, we suggest that comparable oil-prone shale may have formed in the slightly older continental sequence underlying the Barents Sea, which then adds to the overall source-rock potential of this region.

ACKNOWLEDGMENT

This contribution is published with permission of the Geological Survey of Greenland.

REFERENCES

Beauchamp, B., J. C. Harrison, and C. M. Henderson, 1989, Upper Paleozoic stratigraphy and basin analysis of the Sverdrup Basin, Canadian Arctic Archipelago, part 1—Time frame and tectonic evolution: Geological Survey of Canada Current Research, v. 89-1G, p. 105-113.

Berner, R. A., and R. Raiswell, 1984, C/S method for distinguishing freshwater from marine sedimentary rocks: Geology, v. 12, p. 365-368.

Bjorøy, M., A. Mørk, and J. O. Vigran, 1981, Organic geochemical studies of the Devonian to Triassic succession on Bjørnøya and the implications for the Barent shelf, in M. Bjorøy, and others, eds., Advances in organic geochemistry: New York, John Wiley, p. 49-59.

Blair, T. C. and W. L. Bilodeau, 1988, Development of tectonic cyclothems in rift, pull-apart, and foreland basins—Sedimentary response to episodic tectonism: Geology, v. 16, p. 517-520.

Collinson, J. C., 1972, The Røde Ø conglomerate of inner Scoresby Sund and the Carboniferous(?) and Permian rocks west of the Schuchert Flod: Bulletin Grønlands Geologiske Undersøgelse, v. 102, 48 p.

Davies, G. R., and W. W. Nassichuk, 1988, An Early Carboniferous (Viséan) lacustrine oil shale in Canadian Arctic Archipelago: AAPG Bulletin, v. 72, p. 8-20.

Donovan, R. N., 1980, Lacustrine cycles, fish ecology, and stratigraphic zonation in the Middle Devonian of Caithness: Scottish Journal of Geology, v. 16, p. 35-50.

Duncan, A. D., and R. F. M. Hamilton, 1988, Palaeolimnology and organic geochemistry of the Middle Devonian in the Orcadian Basin, in A. J. Fleet, K. Kelts, and M. R. Talbot, eds., Lacustrine petroleum source rocks: Geological Society Special Publication 40, p. 173-201.

Elvsborg, A., T. Hagevang, and T. Throndsen, 1985, Origin of the gas-condensate of the Midgard Field at Haltenbanken, in B. M. Thomas, and others, eds., Petroleum geochemistry in exploration of the Norwegian Shelf: Norwegian Petroleum Society, p. 213-219.

Friend, P. F., P. D. Alexander-Merrick, K. C. Allen, J. Nickolson, and A. K. Yeats, 1983, Devonian sediments of East Greenland, part VI—Review of results: Meddelelser om Grønland, v. 206, 96 p.

Gjelberg, J. G., 1978, Facies analysis of the coal-bearing Vesalstranda Member (Upper Devonian) of Bjørnøya: Norsk Polarinstitutt Årbok, 1977, p. 71-97.

Gjelberg, J. G., and R. J. Steel, 1981, An outline of Lower-Middle Carboniferous sedimentation on Svalbard—Effects of tectonic, climatic and sea-level changes in rift basin sequences, in J. W. Kerr, and A. J. Fergusson, eds., Geology of the North Atlantic borderlands: Canadian Society of Petroleum Geologists Memoir 7, p. 543-561.

Goodarzi, F., W. W. Nassichuk, L. R. Snowdon, and G. R. Davies, 1987, Organic petrology and Rval analysis of the Lower Carboniferous Emma Fiord Formation in Sverdrup Basin, Canadian Arctic Archipelago: Marine and Petroleum Geology, v. 4, p. 132-145.

Hagevang, T., and H. Rønnevik, 1986, Basin development and hydrocarbon occurrence offshore mid-Norway, in M. T. Halbouty, ed., Future petroleum provinces of the world: AAPG Memoir 40, p. 599-614.

Häkansson, E., and L. Stemmerik, 1984, Wandel Sea Basin—The North Greenland equivalent to Svalbard and the Barents Shelf, in A. M. Spencer, and others, eds., Petroleum geology of the north European margin: Norwegian Petroleum Society, p. 97-107.

Häkansson, E., and L. Stemmerik, 1989, Wandel Sea Basin—A new synthesis of the Late Paleozoic to Tertiary accumulation in North Greenland: Geology, v. 17, p. 683-686.

Halfman, J. D., and T. C. Johnson, 1988, High-resolution record of cyclic climatic change during the last 4 Ka from Lake Turkana, Kenya: Geology, v. 16, p. 496-500.

Haller, J., 1970, Tectonic map of East Greenland: Meddelelser om Grønland, v. 171, no. 5, 286 p., 1:500,000.

Kempter, E., 1961, Die Jungpaläozoischen Sedimente von Süd Scoresby Land (Ostgrönland, 71½°N): Meddelelser om Grønland, v. 164, no. 1, p. 1-123.

Koch, L., and J. Haller, 1971, Geological map of East Greenland 72°-76°N lat: Meddelelser om Grønland, v. 183, 26 p.

Larsen, H. C., G. Armstrong, C. Marcussen, S. Moore, and L. Stemmerik, 1989, Deep seismic data from the Jameson Land basin, East Greenland: Terra Abstracts.

Marcussen, C., F. G. Christiansen, P.-H. Larsen, H. Olsen, S. Piasecki, L. Stemmerik, J. Bojesen-Koefoed, H. F. Jepsen, and H. Nøhr-Hansen, 1987, Studies of the onshore hydrocarbon potential in East Greenland 1986-87—Field work from 72° to 74°N: Rapport Grønlands Geologiske Undersøgelse, v. 135, p. 72-81.

Marcussen, C., P.-H. Larsen, H. Nøhr-Hansen, H. Olsen, S. Piasecki, and L. Stemmerik, 1988, Studies of the onshore hydrocarbon potential in East Greenland 1986-1987—Field work from 73° to 76°N: Rapport Grønlands Geologiske Undersøgelse, v. 140, p. 89-95.

Piasecki, S., F. G. Christiansen, and L. Stemmerik, 1990, Depositional history of a Late Carboniferous organic-rich lacustrine shale from East Greenland: Bulletin of Canadian Petroleum Geology, v. 38, no. 3.

Steel, R. J. and D. Worsley, 1984, Svalbard's post-Caledonian strata—An atlas of sedimentational patterns and palaeogeographic evolution, in A. M. Spencer, and others, eds., Petroleum geology of the north European margin: Norwegian Petroleum Society, p. 3-26.

Stemmerik, L., and D. Worsley, 1989, Late Palaeozoic sequence correlations, North Greenland, Svalbard and the Barents Shelf, in J. C. Collinson, ed., Correlation in hydrocarbon exploration: Norwegian Petroleum Society, p. 99-111.

Sund, T., O. Skarpnes, L. N. Jensen, and R. M. Larsen, 1986, Tectonic development and hydrocarbon potential offshore Troms, northern Norway, in M. T. Halbouty, ed., Future petroleum provinces of the world: AAPG Memoir, v. 40, p. 615-628.

Surlyk, F., J. M. Hurst, S. Piasecki, F. Rolle, P. A. Scholle, L. Stemmerik, and E. Thomsen, 1986, The Permian of the western margin of the Greenland Sea—A future exploration target, in M. T. Halbouty, ed., Future petroleum provinces of the world: AAPG Memoir, no. 40, p. 629-659.

Surlyk, F., S. Piasecki, and F. Rolle, 1986, Initiation of petroleum exploration in Jameson Land, East Greenland: Rapport Grønlands Geologiske Undersøgelse, v. 128, p. 103-121.

Surlyk, F., S. Piasecki, F. Rolle, L. Stemmerik, E. Thomsen, and P. Wrang, 1984, The Permian basin of East Greenland, in A. M. Spencer, and others, eds., Petroleum geology of the north European margin: Norwegian Petroleum Society, p. 303-315.

Thomas, B. M., P. Møller-Pedersen, M. F. Whitaker, and N. D. Shaw, 1985, Organic facies and hydrocarbon distributions in the Norwegian North Sea, in B. M. Thomas, and others, eds., Petroleum geochemistry in exploration of the Norwegian Shelf: Norwegian Petroleum Society, p. 3-25.

Ziegler, P. A., 1988, Evolution of the Arctic-North Atlantic and the western Tethys: AAPG Memoir 43, 198 p.

Stemmerik et al.

Lacustrine and Associated Deposits in a Rifted Continental Margin—Lower Cretaceous Lagoa Feia Formation, Campos Basin, Offshore Brazil

Dirceu Abrahão
Petrobrás America Inc.
Houston, Texas, U.S.A.

John Edward Warme
Colorado School of Mines
Golden, Colorado, U.S.A.

The geologic evolution of the Campos basin, offshore southeastern Brazil, is linked to Mesozoic rifting that separated Africa and South America. Lagoa Feia is the basal Lower Cretaceous synrift formation in the stratigraphic sequence of the basin. It formed in a complexly evolving system of rift lakes of variable size and chemistry, overlying and closely related to rift volcanics. Lacustrine limestone and shale and alluvial fan volcaniclastic conglomerates dominate. The lake sequence reaches thickness of 3500 m and is capped by marine evaporites.

Organic-rich lacustrine shale of the Lagoa Feia Formation is the main source rock for the oil discovered to date in the Campos basin. Pelecypod coquinas constitute the intraformational reservoirs within a limited region. Lacustrine shale also sources reservoirs in overlying Cretaceous platform carbonates and Cretaceous and Tertiary turbidites, including those of the giant Albacora and Marlim fields. Lagoa Feia strata compare well with younger lake deposits of the east African rift system, which provide useful analogs for development of exploration models in the Lagoa Feia.

INTRODUCTION

The Campos basin is situated on the southeastern continental shelf of Brazil and covers an area of about 32500 km², based on a seaward limit at the 1000-m isobath (Figure 1). Its geological development, as well as that of other coastal Brazilian basins, stems from Mesozoic rifting that separated the continents of Africa and South America beginning in Late Jurassic-Early Cretaceous time (Emery et al., 1975). Within the basin all the synrift sedimentary sequences, in addition to a final evaporite cover, constitute the Lagoa Feia Formation. Among the several oil fields that were discovered, four were found within coquina reservoirs of the Lagoa Feia Formation; the remainder are sourced in the Lagoa Feia but produce from reservoirs in overlying postrift sequences. Oil exploration in the Campos basin began in the early 1970s. Today it is the country's most prolific basin, producing about 65% of Brazil's 650000 bbl of oil/day.

The stratigraphy of the Campos basin was initially defined by Schaller (1973), followed by several regional studies on the alluviolacustrine sequence of the Lagoa Feia Formation (Castro and Azambuja Filho, 1980; Bertani and Carozzi, 1985a, 1985b; Abrahão, 1987; Dias et al., 1987). Meister (1984) and Figueiredo et al. (1985) conducted studies to define source rocks and the timing of oil migration and entrapment in the basin. Interest in the Lagoa Feia Formation became particularly significant when its organic-rich lacustrine shale was identified as the

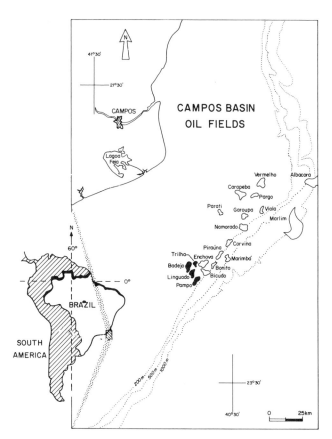

Figure 1. Location of major offshore oil fields in Campos basin, Brazil. Solid black fields produce from reservoirs within the Lagoa Feia Formation; all fields are sourced by lacustrine deposits in the formation.

GEOPHYSICAL DATABASE

For this study we had access to 1900 km of seismic lines, 137 wells with modern electric logs, and 900 m of core from 54 wells throughout the basin. In addition, lacustrine strata of similar age crop out in other coastal Brazilian basins, such as the Morro do Chaves Formation in the Sergipe-Alagoas basin, and were studied as potential analogs to the Lagoa Feia.

The Lagoa Feia is difficult to interpret seismically. Results of seismic surveys are more useful for post-Lagoa Feia strata in the Campos basin, where high-quality seismic data generate clear reflections and, through their continuity, sequence boundaries can be identified and sea level variations depicted. For the Lagoa Feia Formation, reflectors usually are not regionally significant because of high-velocity evaporitic cover, reduced thickness in large areas of the basin, and deposition in small, discontinuous subbasins with different sedimentary regimes. The deepest parts of the Campos basin are difficult to interpret because deeper reflectors have not been penetrated by drilling.

Age dating of the Lagoa Feia used in this study depends on nonmarine biostratigraphic zonations based on ostracodes, which will be discussed below.

TECTONIC FRAMEWORK AND STRATIGRAPHIC DEVELOPMENT

The geologic history of the Campos basin has evolved from the rift and spreading phases of the separation of South America and Africa. The lithostratigraphic sequence in the basin (Figure 2) reflects this tectonic evolution.

The structural contour of the framework below the Lagoa Feia Formation corresponds to the top of a sequence of basic extrusive rocks that accumulated mainly after the beginning of the rifting period. Normal faults, some with throws of more than 5000 m, trend north-northeast. Although most of their displacement occurred during rifting, some faults were reactivated later. Minor transform wrenching early in the basin's history created east-west-trending faults in limited areas of the basin. Tear faults are indicated by variations in displacement along the strike of fault planes. On the other hand, the top of the Lagoa Feia Formation is relatively flat, indicating that the horst-and-graben rift system had nearly completely filled by the time alluviolacustrine deposition had ended.

Differential subsidence was most important during Lagoa Feia deposition, with some areas accumulating thicker alluviolacustrine sediments than others. Variations in subsidence rates induced shifts in depocenters in the basin. Depocenters during rifting were contiguous between Africa and Brazil. When

basin's most important source rock (Meister, 1984; Figueiredo et al., 1985; Pereira et al., 1985). Log responses in the coquina reservoirs initially were studied by Pereira (1980). Subsequently, Abrahão (1988, 1989) undertook state-of-the-art analyses of log responses in the coquinas and the effects of organic matter on logs through the organic-rich shale.

Other lacustrine basins around the world, especially those in rift settings, provide comparisons for better understanding the tectono-sedimentary evolution of the Lagoa Feia Formation. In particular, lake basins associated with the east African rift system provide important keys to understand the geologic setting and paleolimnology of the Lagoa Feia Formation. The lacustrine Eocene Green River Formation in Wyoming, Colorado, and Utah (Cole and Picard, 1978; Surdam and Stanley, 1979; Surdam et al., 1980) is analogous to the Lagoa Feia in terms of sedimentary processes, although tectonic setting and history are different. The Green River was deposited in cratonal sags between Laramide uplifts, whereas the Lagoa Feia Formation was preserved in a subsiding continental rift that became a passive continental margin with continued rifting and seafloor spreading.

Abrahão and Warme

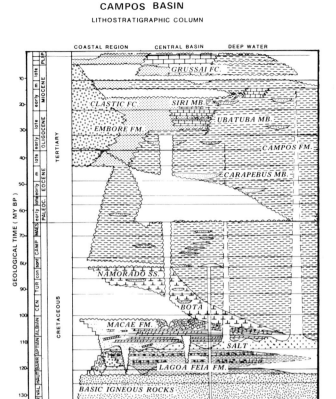

Figure 2. Upper Cretaceous-Tertiary stratigraphy of the Campos basin (from Petrobrás SECASU, 1982). Note lacustrine sequence of the Lagoa Feia Formation above basement volcanics and below evaporite cover.

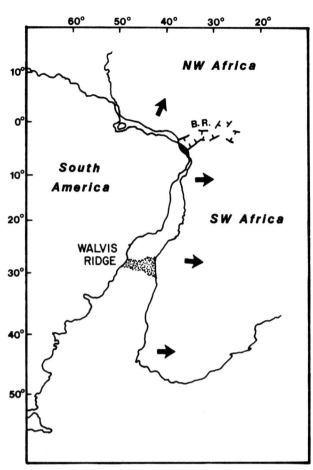

Figure 3. Reconstruction of the African-South American continental rift, showing location of the Walvis-Rio Grande ridge, during latest Aptian time (~108 Ma). Benue rift (B. R.) is an aborted rift associated with this tectonic episode (modified from Castro, 1987).

the continents drifted apart, however, some segments remained with South America and some with Africa. Along the Brazilian coast, continental separation began in the Late Jurassic in some regions but in the Early Cretaceous in others (Figure 3). The separation of South America from Africa began on the south and proceeded northward from 165 Ma to about 115 Ma. Trends of fracture zones on the Atlantic floor show that, relative to Africa, South America moved southwestward at first, then westward (Burke, 1975). This movement caused the initial South Atlantic Ocean to be narrow and fan shaped, opening to the south later. Until Late Cretaceous time, the South Atlantic was separated from the North Atlantic essentially by the sliding contact between the continental indentation of the Gulf of Guinea and the bulge of northeastern Brazil.

During Aptian time the first marine incursions occurred in the graben system formed by rifting. Relative tectonic quiescence provided little clastic influx, and a topographic barrier to the south—the Walvis-Rio Grande ridge (Figure 3)—created the restriction necessary for extensive evaporite deposition at the top of the Lagoa Feia Formation. Within

this evaporite cover, anhydrites, although thin, are common throughout the Lagoa Feia. The irregular distribution and thickness of halite can be explained by the association of two phenomena. First, when the sea initially entered the basin, a series of partly denuded ridges and small troughs created a system of marine evaporite basins. Halite deposition was thicker in the troughs. Anhydrite was deposited over the halite beds at the end of this stage, leaving evaporites formed in a Red Sea-like environment.

Second, salt tectonics created the variable thicknesses observed. Consistent with continuous evolution of the continental breakup, thermal subsidence caused gentle postdepositional eastward tilting of the entire Lagoa Feia sequence. In response to post-Early Cretaceous sedimentation, compaction, and tilting, the halite layers deformed plastically and migrated eastward, generating the salt diapirs observed in the easternmost part of the basin (Figure 2). The rate of postrift subsidence probably was controlled by thermal history of the lithosphere and this new passive margin. Subsidence was high during early drifting, when rates of lithospheric cooling were high, but exponentially declined with time as the continental margin moved away from the spreading axis and heat source (Sclater and Francheteau, 1970; LePichon and Hayes, 1971). Salt tectonics occurred at different times and was important in the distribution of facies for all sequences that overlie the Lagoa Feia Formation (Figueiredo et al., 1985).

With continued subsidence, the evaporite basins of the proto-Atlantic gulf (Figure 3) evolved into a normal marine basin during the Albian. As continuous spreading between Africa and South America generated pure oceanic conditions in the Albian, a carbonate platform developed along the basin. These marine platform carbonates comprise the Macaé Formation, which averages 900 m thick (Figure 2) and includes fluviodeltaic sandstone and shale deposited contemporaneously near the basin border. The lower part of this formation was completely dolomitized in the southern and central parts of the basin. Presumably, this area behaved as a high, where seepage reflux occurred for long periods during early deposition of the Macaé Formation. Some of these dolomites, occurring within the uppermost Lagoa Feia Formation, are laterally correlative with anhydrite levels in the evaporite cover.

Continued tilting and major subsidence controlled open-marine clastic sedimentation from the Late Cretaceous to the Holocene and led to deposition of the Campos and Emborê Formations, which form a clastic wedge that reaches a thickness of 3000 m (Figure 2). They consist of a thick, classic, passive-margin sequence of deep-water shale with interbedded siliciclastic turbidites (important reservoirs), prodeltaic shale, and deltaic sandstone. The Siri Member and the Grussaí facies consists of shelf carbonates deposited during interrupted or decreased clastic influx.

COMPARISON WITH LACUSTRINE DEPOSITS OF EAST AFRICA

The east African rift system provides a modern setting for numerous, diverse lakes within an early rift setting similar to that envisioned for the Lagoa Feia Formation. Rift faulting and volcanism, as well as important climate fluctuations during the Pleistocene, caused radical changes in area, volume, and ecology of these lakes (Beadle, 1981). Some were temporarily reduced to swamps or even desiccated completely. The Quaternary patterns of interconnection and isolation among the lakes has changed continuously, and isolated lakes developed their own limnologic and sedimentologic histories. Evidence for previous higher water levels is provided by raised beaches and strata, implying times of lowered salinities. Understanding these recent lakes can help to realistically appreciate the diversity of conditions that probably existed in the Lagoa Feia.

Similarities between the east African and the Campos basin rifts include the general setting, which is similar in area, magnitude of faulting, and sedimentary processes. Spatial distribution of tectonic elements, such as horsts with intervening lake-basin grabens, is similar to the eroding highs and lake-basin lows identified as controlling deposition in the Lagoa Feia Formation. That the sedimentary facies, described below, are generally comparable between the two areas implies that many sedimentary processes also are similar. Climatic variations, as interpreted in the sedimentary record, are recognized in both places. For example, cyclical oscillations in Lagoa Feia source rock potential are related to climatically induced deepening and shallowing events (Abrahão, 1988). Drainage systems that are alternately connected and isolated, such as in east Africa, are suspected in the Lagoa Feia, based on fossil evidence (salt-tolerant vs. nontolerant faunas) as well as sedimentary cycles, implying oscillations in source-rock quality.

Dissimilarities also arise in this comparison. The Lagoa Feia is as thick as 3500 m, which is generally greater than east African lacustrine sequences. However, the Lagoa Feia accumulated over about 20 m.y., perhaps a longer time than the east African system. Sedimentologically the Lagoa Feia is dominated by siliciclastics with volumetrically minor carbonate and evaporite compared to modern examples. This observation may be biased, first, by our subsurface sample base in the Lagoa Feia, because the wells had been drilled away from the shorelines where carbonate accumulates and, second, by the fact that carbonates and evaporites now dominate many African lakes as a result of arid postglacial climate. Older sedimentary deposits in many of the lakes are noncarbonate or nonevaporite, a condition that may be more representative overall in east Africa.

In summary, the Lagoa Feia appears to reflect long periods of relatively fresh water that supported abundant populations of bivalves; minor carbonate and evaporite accumulation; and few playa-lake characteristics. The stratigraphic sequence is dominated by coarse- and fine-grained siliciclastics that reflect little evidence of prolonged lowstands or desiccation. Future exploration may provide a more complete view of the historical variability of this important paleolacustrine system.

PALEOCHEMISTRY OF LAGOA FEIA LAKES

Water chemistry of Lagoa Feia paleolakes in the area of the oil fields had been previously characterized as saline and alkaline (Castro and Azambuja Filho, 1980; Carvalho et al., 1984). However, Bertani and Carozzi (1985a, 1985b) proposed oscillations in lake water salinity, with more saline conditions indicated by dolomite/evaporite flats and ostracode-dominated

Abrahão and Warme

microfacies, and less saline conditions by the presence of pelecypods. In their opinion, changes in biota and rock type corresponded to changes in water chemistry. Today in east Africa, variations in salinity, water chemistry, and physicochemical characteristics control sedimentation from one lake to another and within individual lakes through time (see above); the same is true for the Eocene Green River Formation (Surdam and Stanley, 1979). Similar fluctuations were likely in the lakes that constituted the Lagoa Feia system during the Early Cretaceous; however, more study is needed on vertical and lateral paleosalinity variations throughout Lagoa Feia history.

LAGOA FEIA FORMATION

The Lagoa Feia Formation initially was interpreted as nonmarine, fine- to coarse-grained siliciclastics and carbonates overlying basalt flows and covered by Lower Cretaceous evaporites (Schaller, 1973). Small basaltic dikes and basic volcaniclastics in the basal part of the section indicate that, during early stages of deposition, volcanic activity was contemporaneous with sedimentation (Bertani, 1984).

Lithofacies

Conglomerate

In the Lagoa Feia Formation, conglomerates are basaltic, commonly matrix-supported, with red and green pebbles up to 5 cm in diameter. Inverse and normal graded bedding is common. Framework grains are mostly basalt fragments (Figure 4). North of the basin more gneissic and granitic clasts appear. Pelecypod shells, which show different degrees of reworking, are common. Cementation by zeolites is frequent. Although dolomite and calcite cements are less common, dolomite replaces zeolitic cement and may replace framework grains (Bertani, 1984; Carvalho et al., 1984). Basalt fragments have been altered to chlorite and zeolites. In the matrix sand- to silt-size grains of volcanic glass have been altered to clay minerals. Other components include plagioclase, microcline, and biotite, and volumetrically minor muscovite, zircon, and opaque minerals. Volumetrically, conglomerate appears to be more important in the sequence than the sandstone and siltstone described below.

Sandstone and Siltstone

Siliciclastic rocks range from siltstone to coarse-grained sandstone, usually oxidized to pink to red but locally gray to green. Mineralogy is mainly microcline, orthoclase, plagioclase, quartz, biotite, and hornblende (Bertani, 1984). Although usually

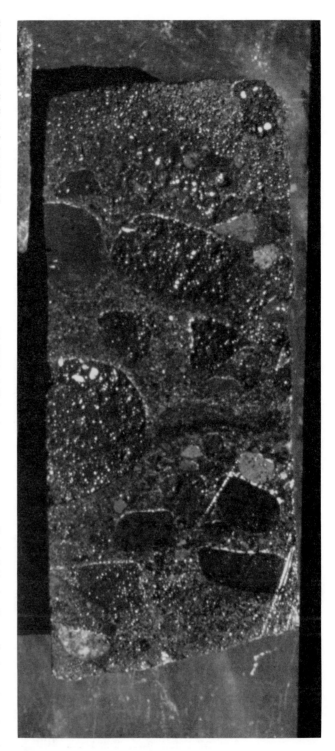

Figure 4. Core sample of volcaniclastic conglomerate containing mostly basalt fragments. Calcite rims on pebbles suggest arid-condition precipitation above the water table, perhaps in an alluvial fan environment. Width of core section is 10 cm.

poorly sorted, these clastics, when well sorted, are cemented by calcite or zeolites. The sandstone generally contains abundant lithic (primarily basalt with secondary carbonate) and feldspathic fragments but little quartz. Ripple cross-lamination and small- to large-scale cross-stratification are common, with rare scouring and graded bedding. These rocks commonly grade into highly oxidized mudstone.

Mudstone

Two groups of fine-grained siliciclastic rocks are recognized in the sequence. One is highly oxidized red mudstone composed of basic volcanic glass debris, feldspar, quartz, and minor mica (Bertani, 1984). The other is green to dark-gray shale, organic-rich at certain levels and containing chert and phosphatic nodules, which usually deform the laminae of the fine-grained beds (Figure 5).

Carbonate

Carbonates in the Lagoa Feia Formation commonly consist of fine- to coarse-grained fossiliferous grainstone to wackestone commonly referred to as coquinas. Colors are controlled by the amount of matrix and can be white (matrix absent), red, green, or gray. Cements are carbonate, silica, or zeolites. Identifiable pelecypods, ostracodes, gastropods, and associated shell hash are the principal components, ranging from 0.3 to 5 cm long (Figure 6). Although pelecypods occur in several parts of the Lagoa Feia Formation and in nearly all lithologies, their remains vary from one interval to another—from shell fragments, to disarticulated but unbroken shells (not strongly reworked), to articulated shells (Figure 7). Petrographic examination shows minor amounts of fine- to medium-grained basalt and basic glass fragments associated with the carbonates.

Although less frequent, thinly bedded algal-laminated carbonate and rare oolite beds also occurs in the stratigraphic sequence. At some horizons, it exhibits chickenwire anhydrite texture, which suggests evaporite replacement. More rarely, when interbedded with clay layers, the laminites replace crystalline intergrowths of evaporite minerals. Tepee structures are associated with algal-laminated limestones (Figure 8).

Other Features

Stevensite

A peculiar occurrence in the lower third of the Lagoa Feia stratigraphic sequence is that of stevensite oolites (Figure 9), which occur in massive beds up to 400 m thick and averaging 100 m. Stevensite is a magnesian smectite ($Mg_3Si_4O_{10}(OH)_2$), also found in the Green River Formation in Wyoming (Bradley and Fahey, 1962). Similar magnesian smectites (ghassoulite) occur in African lakes today (Baumann et al., 1975). Its high crystallinity suggests formation by chemical precipitation (Carvalho et al., 1984). Furthermore, its presence indicates magnesium-rich lake waters possibly attributable to contemporaneous volcanism. Such conditions are observed in modern African lakes and also inferred in the Green River Formation.

Figure 5. Vertical core slab of green laminated shale with phosphatic nodules. Laminae are deformed around the nodules, indicating nodule formation prior to final compaction. Width of core slab is 10 cm.

Abrahão and Warme

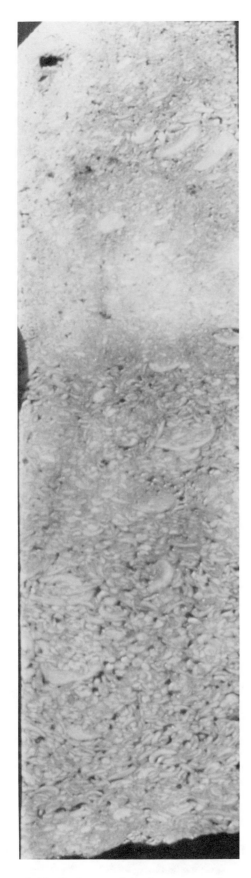

Figure 6. Vertical core slab of porous and permeable, oil-stained coquina reservoir rock from Linguado field in the Badejo high area. Width of core slab is 10 cm.

Zeolites

Zeolites are common cements in all lithofacies recognized in the Lagoa Feia Formation. They occur predominantly as cements or pseudomatrix in conglomerate and sandstone, and less commonly as cement in carbonate facies. Analcime most often occurs, but natrolite and clinoptilolite also have been recognized (Carvalho et al., 1984). Their occurrence may be related to volcanic activity or to products of this activity (Hay, 1966, 1977; Hay and Sheppard, 1977). However, additional work is necessary to identify the zeolite minerals present and to establish their associations with different lithofacies.

Diagenesis

Rocks of the Lagoa Feia Formation have been strongly altered diagenetically. Paleosols and vadose pisolites at some levels, for example, suggest periods of subaerial exposure. Carvalho et al. (1984) used CL (cathodoluminescence) and SEM to define eleven stages of diagenesis in Lagoa Feia reservoir coquinas and related beds in the Badejo high oil fields. In chronological order, they are micritization, early cementation, neomorphism, solution, late cementation, dolomitization, chlorite coating, compaction, pressure solution, silicification, and kaolinitization. Bivalve shells, the main constituent of the carbonates, when not cemented early, were affected later by pressure solution, manifested as solution seams and stylolites. Stylolites more commonly are observed where subsidence was greater, such as in the central basin. Carbonate rocks affected by pressure solution usually do not constitute reservoirs.

Reservoir Rocks

Coquinas, where porous, represent the only productive oil reservoirs within the Lagoa Feia Formation. Diagenesis has transformed the coquinas into extremely heterogeneous zones (Figure 10), wherein profound variations in reservoir quality can occur over relatively short vertical intervals. On the other hand, continuous reservoir-quality rocks do occur, reaching net pays of up to 20 m of porous and permeable rock (Figure 6). In these, porosities seldom are very high, about 12%, with maxima of about 20%. However, hydrocarbons are produced from porosities as low as 4–6% because of usually good permeabilities due to shelter (primary) porosity and vuggy and moldic (secondary) porosity (Figure 11). Silica cementation in areas where these rocks were slowly deposited seems to be important in the generation of high-quality reservoir facies. Silica cement apparently is important in preserving primary porosity because these rocks were not significantly affected by pressure solution in later diagenesis.

Interpretation of Depositional Environments

The Lagoa Feia Formation has been interpreted as forming in an alluviolacustrine complex (Schaller,

Figure 7. Vertical core slices showing variation in articulation of preserved pelecypod shells, all from one well in Linguado field in the Badejo high. Width of core slabs is 10 cm.

Abrahão and Warme

Figure 8. Algal laminites interbedded with dark-gray mudstone, stromatolites (upper left), coquinas, pseudomorphs of calcite after evaporite minerals. In third core from left, laminations have been disrupted by desiccation cracks and tepee structure. Width of core slabs is 10 cm.

1973; Bertani, 1984; Carvalho et al., 1984) based mainly on:

1. restricted fauna, primarily pelecypods and ostracodes, and secondarily gastropods, and absence of typical stenohaline marine organisms, such as brachiopods, echinoderms, cephalopods, and bryozoans;
2. characteristic lacustrine lithofacies assemblages;
3. facies geometry, distribution, and lateral relationships;
4. occurrence of unusual minerals, such as stevensite, or mineral trace elements that can only form from waters chemically different from normal seawater;
5. fluctuations in salinity that, based on ostracode and mollusk occurrences, are rapid compared to marine sequences;
6. lack of evidence for typical marine processes such as tidal currents or strong wave action;
7. subaerial emergence and consequent frequent oscillations of shorelines; and
8. comparisons with lithostratigraphic units of similar age, or with recent analogs, that have similar tectono-sedimentary settings and evolution.

From the study of open-hole logs, cuttings, and cores, three broad depositional environments, each represented by a distinct lithofacies, can be defined for the Lagoa Feia Formation—alluvial fans, exposed lake-margin mud flats, and sublacustrine deposits. The latter includes deposits from deep water to lake margins. Within these depositional environments, at the level of detailed core analysis, various subenvironments can be recognized.

Figure 9. Stevensite oolites, in calcareous matrix (upper core slab); preserved molds only (lower slab). Width of core slabs is 10 cm.

middle part of the sequence. This distribution indicates that alluvial fans were important at the beginning of the rifting phase and again at the end of Lagoa Feia deposition, times of probable stronger tectonic activity. Facies probably representing sand flats or sand aprons in the distal parts of the alluvial fans were observed in some cores. Northward in the basin, volcaniclastic content of the conglomerates, although still significant volumetrically, appears to decrease. Volcanic activity here probably was less significant, and more siliceous basement sources were exposed during the rifting period. This change in source for siliciclastics may have generated better quality reservoir rocks, although commercial production is not yet known from these areas.

Mud Flats

Mud-flat deposits include all the subordinate fine-grained facies deposited between the lake margins and alluvial fans. They are called mud flats because of the predominance of fine-grained siliciclastic deposits, although some carbonate facies are included in this broad characterization. Common rock types include sandstone and mudstone, some with desiccation cracks and evaporite mineral casts, oolites, algal laminites, and stromatolites. Terrigenous influx was temporarily low where algal laminites developed.

Periods of aridity may have induced the deposition of evaporite minerals on these flats. The chickenwire anhydrites, later replaced by calcite, probably were deposited during these arid cycles.

In some horizons oolites are found, although they are volumetrically insignificant. These rocks also formed in shallow nearshore environments, usually adjacent to the mud flats and interbedded with them.

Sublacustrine Deposits

Three groups of sublacustrine deposits have been recognized—(a) fine-grained siliciclastics, including dark-green to gray shale of anoxic origin; (b) shell-rich beds (coquinas) representing banks, bars, beaches, or storm deposits; and (c) graded sandstone and siltstone probably deposited by turbidity flows.

The green to gray shales are believed to have originated in deeper areas of the lakes. The dark, organic-rich shales are considered to be source rocks for the oil discovered to date in the Campos basin (Figueiredo et al., 1985; Meister, 1984; Pereira et al., 1985). Near the base of the unit, immediately overlying the alluvial fan deposits, the shale is predominantly green but becomes darker, organic-rich, and laminated (Figure 13) toward the top, especially in the central area of the basin.

Water depths for deposition of the lacustrine dark-green to gray shale is conjectural and has been discussed elsewhere (Surdam et al., 1980). In the Green River Formation 1- to 10-m-thick shallowing-upward cycles, with oil shale near the base and

Alluvial Fans

Lagoa Feia conglomerate and sandstone are interpreted to have formed in proximal to distal areas of alluvial fan complexes along basaltic escarpments. Pelecypod shells in the conglomerate may not be strongly reworked, indicating that fans prograded into the lakes as fan deltas. In many cases the shells are still articulated, indicating burial close to their living positions (Figure 12).

Conglomerate is most abundant in the basal Lagoa Feia Formation and close to the top of the unit below the overlying evaporites but is absent or thin in the

Abrahão and Warme

Figure 10. Heterogeneities in reservoir rocks caused by diagenetic effects. Patches of silicified rock and intervals affected by pressure solution are interbedded with favorable reservoir intervals. Width of core slabs is 10 cm.

shoreline or mud flat deposits at the top, may lead one to infer that the lakes were relatively shallow during deposition of the cycles. In the Lagoa Feia Formation, however, significant thickness of dark-green to gray shale suggests that relatively deep anoxic conditions persisted in some areas of the basin, without drying to the playa-lake stage, although cyclicity between the occurrence of source rocks and lean rocks also has been observed in the section (Abrahão, 1988). In contrast, in Lake Tanganyika, anoxic conditions do not require great depths. Although it is one of the world's deepest lakes, anoxia occurs relatively high in the water column, creating a large volume of water that is permanently anoxic. Water-column stratification is very strong, and anoxic conditions occur below depths of 100 to 150 m. We envision that similar conditions existed in Lagoa Feia lakes, and although it is difficult to judge water depths, stratified water columns existed for extended periods during deposition of the Lagoa Feia Formation.

Coquinas, the second major group of sublacustrine deposits, may in some cases represent lakeshore deposits, probably beaches and nearshore bars, or they may have accumulated in deeper water as a result of reworking by storm waves. Because of lake geometries—elongated and bordered by escarpments—these storm deposits are likely to occur as linear bodies parallel to the faulted borders of the lake basins.

Shell frequencies vary in a regular way throughout the sequence over much of the basin. Shell layers are more common in the middle than in the upper and lower parts of the section. Toward the top of the formation shells are practically absent. These large-scale variations represent shifts in environmental conditions such as changes in climate, terrigenous influx, or limnologic conditions. Variations in preserved shell articulation may reflect different species or responses to different lake paleoenvironments.

The third group of lake deposits is sandstone and siltstone deposited by turbidity flows during highstands. They have been recognized in the middle part of the section but are relatively rare. These siliciclastics, even when coarse grained, do not constitute favorable reservoir rocks because of their extremely high volcaniclastic content and diagenetic

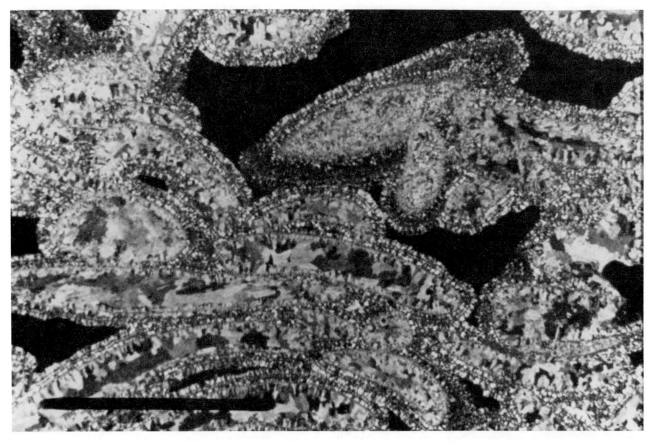

Figure 11. Photomicrograph of pore structure in coquina reservoirs. Rims of gray siliceous cement coat most pelecypod shells. Commonly pores are larger than shell clasts. Bar at lower left is 1 mm long.

alteration, usually the zeolitization of basic lithic fragments, which commonly obliterates porosity. Permeability in all cases is extremely low.

Summary of Environmental Interpretation

Figure 14 schematically summarizes the interpretation of depositional environments in the Lagoa Feia Formation, which is based on data from cuttings and log responses correlated to features observed in cores. In decreasing order of volumetric importance, the rocks are interpreted as follows:

1. Conglomerate and sandstone were deposited in proximal to distal alluvial fan complexes along basaltic or granitic/gneissic escarpments.
2. Bioclastic carbonate occurs in the lakes as layers related to periods of abundant bivalve productivity.
3. Green and dark-gray shales were deposited under anoxic conditions (when organic-rich) in sublacustrine environments.
4. Oxidized shale and siltstone were deposited on exposed mud flats close to lake shorelines and probably were associated with distal areas of alluvial fan complexes.

5. Algal-laminated limestone was deposited in flats along the lakes. Periodic arid conditions generated evaporites.
6. Turbidite sandstone and siltstone were resedimented from shallow to deep-water lacustrine environments.

Lateral Facies Relationships

Stratigraphic cross sections (Figure 15) indicate that in earlier stages of deposition, alluvial fan facies were areally common along uplifted fault blocks. They were associated laterally with lacustrine deposits and locally with mud flats. Probably because of the small size of initial catchment basins, lacustrine deposition was limited. Later, as rifting deepened and widened the basins, lacustrine deposition became much more widespread, as evidenced by higher proportions of lake deposits in the middle of the sequence.

Figure 16 is a schematic distribution of the Lagoa Feia facies that generated most of the geologic record observed in this formation. Facies relationships are strongly controlled by the distribution of normal faults along the basin and consequently by the geographic distribution of grabens and horsts.

Abrahão and Warme

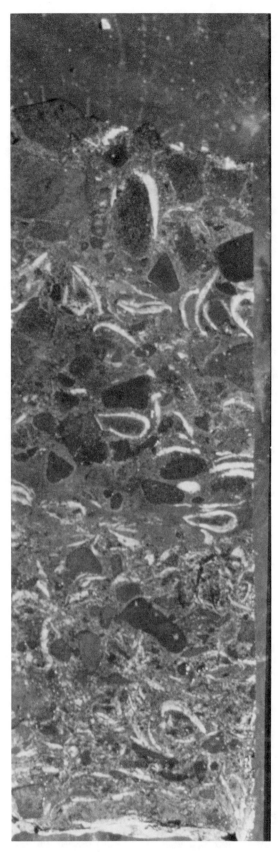

Figure 12. Volcaniclastic (basalt) conglomerate containing articulated pelecypod shells that suggest preservation in a shoreline fan-delta environment. Width of core section is 10 cm.

Figure 13. Laminated, black, organic-rich shale near base of core sample represents sublacustrine deposits under anoxic conditions. Width of core section is 10 cm.

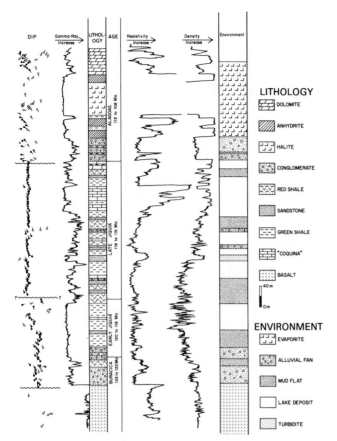

Figure 14. Composite lithologic column of Lagoa Feia Formation, showing typical responses for dipmeter, gamma-ray, resistivity, and density logs. Average thickness of Lagoa Feia, between basal basalts and capping evaporites, is about 1500 m, reaching 3500 m in places. Ages correspond to stages used in Brazil (see Figure 17). Right-hand column indicates depositional environments interpreted for lithologies and log responses shown.

Spreading of the lake areas occurred more along the structural troughs than in the confined space between the abrupt fault-bounded escarpments that parallel the basin axes. Because of this geometry, alluvial fan facies rapidly grade laterally into deep-water lacustrine facies within short distances of the fault walls—a phenomenon observed in cores and logs from wells in the Lagoa Feia Formation. Fan deltas, which should occur in these situations, are recognized in the stratigraphic sequence, to the exclusion of flats.

TIME-STRATIGRAPHIC EVOLUTION

Because of the diversity of Brazilian ostracode fauna, it was possible to develop a reasonably accurate biostratigraphic zonation for nonmarine rocks that accumulated in the Brazilian coastal basins during rifting. Based on their endemism within continental deposits, six local stages have been established—Dom João (Late Jurassic), Rio da Serra, Aratu, Buracica, Jiquiá, and Alagoas (Neocomian to Aptian) (Vianna et al., 1971).

In the Lower Cretaceous Lagoa Feia Formation, only the Alagoas, Jiquiá, Buracica, and Aratu exist (Moura, 1985), in contrast to the Recôncavo-Tucano and Sergipe-Alagoas basins (northeastern Brazilian coast) where the older stages are present (Vianna et al., 1971). Although some local stages defined in these basins have problematic correlation with international stages, a tentative correlation is presented in Figure 17.

Based on existing ostracode zonation, the geologic evolution of the Lagoa Feia stratigraphic sequence can be analyzed through four time slices of deposition—Aratu/Buracica, early Jiquiá, late Jiquiá, and Alagoas.

Aratu/Buracica Time (~130–120 Ma)

Only wells that penetrate deep into the Lagoa Feia, some to basement, cross this lowest section. Wells drilled in areas where the Lagoa Feia is thicker, in the northern or central grabens, did not penetrate this stage. The lake basins formed during Aratu/Buracica time appear to be relatively small. Sediments that accumulated during this period are thin and dominated by volcaniclastic conglomerate and derived lacustrine siliciclastics. Fossils are uncommon, and coquinas are minor. Because the lakes were smaller and perhaps intermittent, dark organic-rich shale also is absent. Rocks in this sequence commonly show high dips, indicating either higher tectonic instability during deposition or significant structural modification during or after deposition.

Early Jiquiá Time (~120–118 Ma)

During this period lake basins became better defined. Lacustrine shale is more common, indicating probably deeper and longer lived lakes. Coquinas also are more significant, and abundant stevensite suggests continued volcanic activity nearby. Dips in this section, although not as high as in the Aratu/Buracica sequence, are higher than in overlying strata, which indicates significant synsedimentary structural movement within the troughs during deposition.

Late Jiquiá Time (~118–115 Ma)

This period represents the climax of lacustrine deposition within the basin. Climatic changes affected the areal extent of the lakes. Overall, sedimentation rates were higher than in other periods. Sediment volume was greater because of

Abrahão and Warme

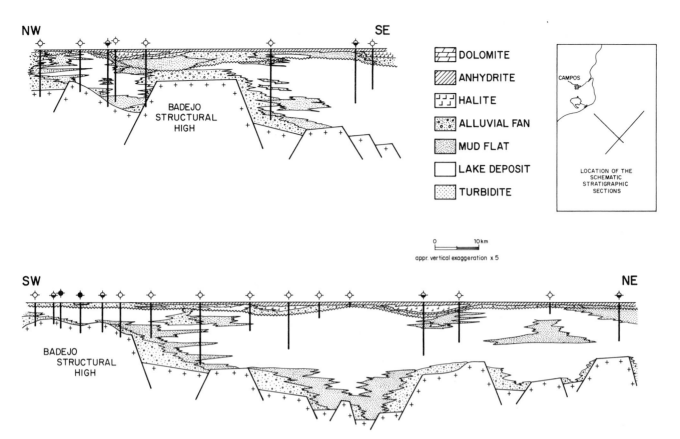

Figure 15. Schematic stratigraphic cross sections along two perpendicular directions through Campos basin, showing distribution of lithofacies across structural blocks, and producing wells associated with Badejo high.

increased subsidence and areal extent of the lake basins. During long wet periods, with more fresh water in the system, the lakes expanded and became interconnected. Life also flourished, as evidenced by thicker coquina beds. The positions of shell layers probably correlate with climatic changes or alterations in water chemistry, which provided favorable overall conditions for mollusks. This conclusion is reinforced by the presence of large molluscan accumulations around such east African lakes as Shala, Mobutu, and Turkana (Baumann et al., 1975; Beadle, 1981), indicating that these faunas developed during periods of higher lake level.

That the dips within this section are low compared with underlying time slices shows that at this time tectonic activity had significantly decreased.

Alagoas Time (~115–108 Ma)

At the beginning of Alagoas time a regionally significant tectonic pulse generated a regional unconformity that is recognized throughout the basin. Probable arid conditions persisted in the basin during this time; sedimentation rates decreased; and lake sediments are practically absent. Volcaniclastic conglomerate and sandstone that dominated the sedimentary pattern represent a long interval of

denudation after a tectonic pulse followed by remarkable subsidence, or lowered base level, or both. Consequently, conditions were created for marine invasion into the basin. Dip patterns usually are chaotic, characteristic of conglomeratic sections, and probably associated with steeper gradients.

Summary and Exploration Strategies

The geologic evolution of the Lagoa Feia sequence is shown diagrammatically in Figure 18. The major tectonic pulse initiating rifting that eventually separated Africa from South America probably occurred at the end of the Jurassic. This pulse resulted in basement uplift and block faulting. Volcanic activity, especially at the beginning, was significant (stage 1). Continued rift faulting created a system of grabens that subsided form block-faulted subbasins (stage 2). With subsequent erosion of the higher blocks these subbasins filled with alluvial fan deposits and lake sediments to form sequences of nonmarine carbonate, shale, sandstone, and conglomerate (stage 3). At times, intrarift highs became submerged, and when the lakes were deep enough, anoxic conditions developed. Continued basement faulting created discontinuities within the lake sequences (stage 4). The lacustrine cycle ended with

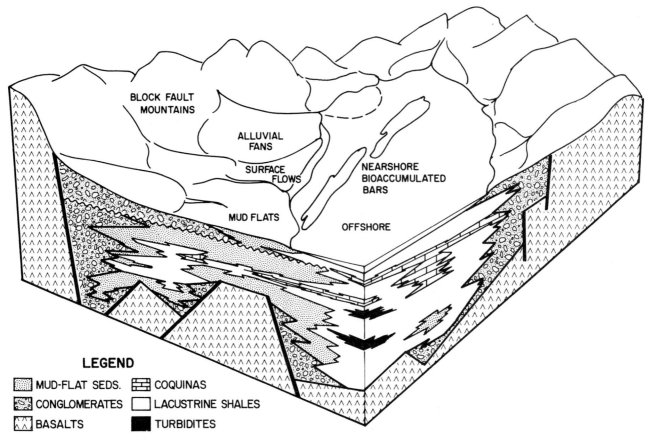

LEGEND

▦ MUD-FLAT SEDS.	⊞ COQUINAS
⬚ CONGLOMERATES	☐ LACUSTRINE SHALES
⋀ BASALTS	■ TURBIDITES

Figure 16. Schematic distribution of facies showing relationships to tectonic blocks during Lagoa Feia time.

CHRONOSTRATIGRAPHY

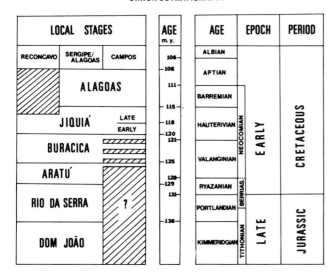

Figure 17. Local stages used in Upper Jurassic and Lower Cretaceous nonmarine strata of three Brazilian basins (Recôncavo, Sergipe-Alagoas, and Campos), showing absolute ages and correlation with standard global stages (according to Moura, 1985).

a pulse of basement faulting, uplift, and a regional erosional unconformity. Arid conditions inhibited deposition of extensive lacustrine sequences after this period (stage 5). With continued subsidence through the Aptian, marine saline waters invaded from the south-southeast, culminating in evaporite deposition under tectonically quiescent conditions (stage 6).

The knowledge of facies relationships and distribution, and comparison with east African lake basins, led to the development of exploration models for the stratigraphic sequence in the Lagoa Feia Formation. Figure 19 shows where lacustrine deposits were present during most of Lagoa Feia time. Although the positions of individual lakes almost certainly shifted within the general contours, the areas outlined contain the thickest lacustrine sequences.

The areas between the solid and dotted lines contain thinner lacustrine deposits and underwent less subsidence during deposition. During times of higher water level, the lake system probably was shallowly interconnected across these areas. Coquina reservoirs developed here and were cemented early apparently because they were undiluted by siliciclastics. The highs had better chances for develop-

Abrahão and Warme

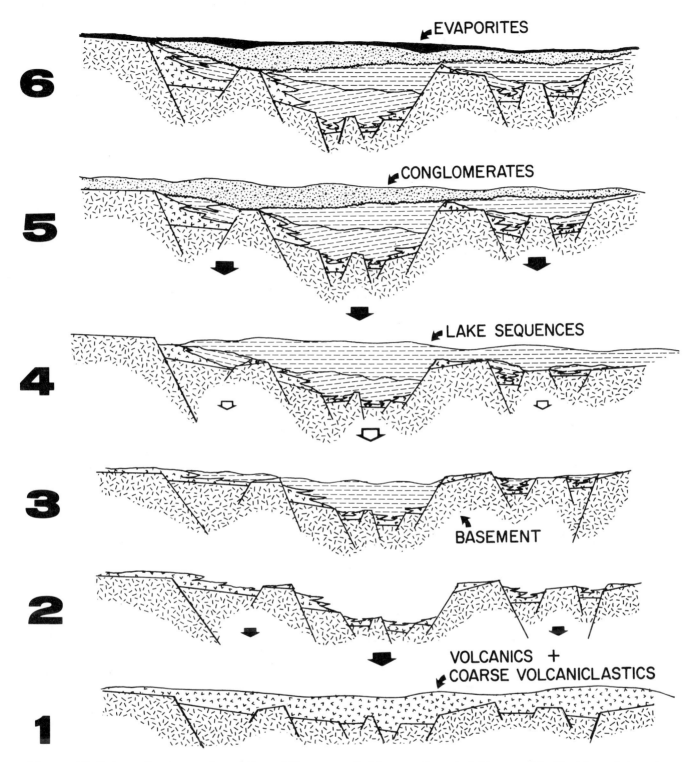

Figure 18. Schematic stages of evolution of Campos basin and Lagoa Feia Formation. 1, Rio da Serra(?); 2, Aratu/early Buracica; 3, early Jiquiá; 4, late Jiquiá; 5, Alagoas; 6, end of Alagoas. Modified from Brice et al. (1982).

ment of carbonate reservoirs similar to those that produce oil in the Badejo high fields; therefore, petroleum prospects should focus on these areas.

Siliciclastic reservoirs drilled to date show poor reservoir quality. They usually have high volcaniclastic content and low permeability due to zeolite diagenesis. Nevertheless, better quality reservoirs may exist in the northern part of the basin, where the provenance is gneissic and granitic.

In the northern part of the basin most wells were drilled along structural highs. However, paleostructural lows in that area may hold better promise for

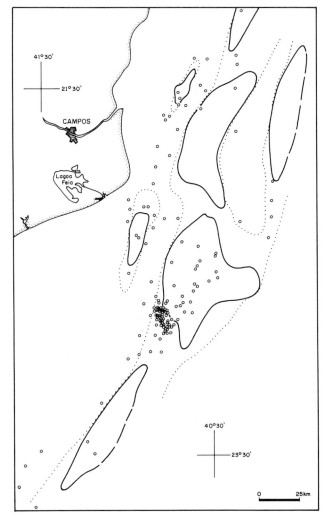

Figure 19. Distribution of Lower Cretaceous lacustrine deposits within Lagoa Feia Formation, based on subsurface and seismic information. Solid lines delimit areas wherein lacustrine deposits dominate. Areas between solid lines and dotted lines represent interbedding of lacustrine, mud flat, and alluvial fan deposits. Other areas are predominantly alluvial fan deposits. Circles represent well locations where Lagoa Feia was penetrated.

oil accumulations because of proximity to source rocks that probably accumulated along the foundered rift-trough axes.

The development of stratigraphic-type reservoirs bordering structural lows throughout the basin also may be successful; in these locations potential reservoir rocks are juxtaposed with thick lacustrine sequences with source rocks that can generate large volumes of oil.

ACKNOWLEDGMENTS

This study summarizes a Master of Science thesis defended at the Colorado School of Mines in 1987. Information in the database was released by Petróleo Brasileiro S/A-Petrobrás. We thank the management of the Exploration Department at Petrobrás for its support and for authorization of publication. We also thank Jefferson L. Dias for reviewing the manuscript.

REFERENCES CITED*

Abrahão, D., 1987, Lacustrine and associated deposits in a rifted continental margin—The Lagoa Feia Formation, Lower Cretaceous, Campos basin, offshore Brazil: M. S. Thesis 3318, Colorado School of Mines, Golden, Colo., 193 p.

Abrahão, D., 1988, Well-log signatures of alluvio-lacustrine reservoirs and source rocks of the Lagoa Feia Formation, Lower Cretaceous, Campos basin, offshore Brazil: Ph. D. Thesis 3589, Colorado School of Mines, Golden, Colo., 279 p.

Abrahão, D., 1989, Well-log evaluation of lacustrine source rocks of the Lagoa Feia Formation, Lower Cretaceous, Campos basin, offshore Brazil: Society of Professional Well Log Analysts 30th Annual Logging Symposium, Transactions, v. 1, Paper I, 22 p.

Baumann, A., U. Forstner, and R. Rohde, 1975, Lake Shala—Water chemistry, mineralogy, and geochemistry of sediments in an Ethiopian rift lake: Geologische Rundschau, v. 64, p. 593-609.

Beadle, L. C., 1981, The inland waters of tropical Africa, 2d ed.: New York, Longman, 475 p.

Bertani, R. T., 1984, Microfacies, depositional models and diagenesis of Lagoa Feia Formation (Lower Cretaceous), Campos Basin, Offshore Brazil: Ph.D. thesis, University of Illinois at Urbana-Champaign, 199 p.

Bertani, R. T., and A. V. Carozzi, 1985a, Lagoa Feia Formation (Lower Cretaceous), Campos Basin, Offshore Brazil—Rift valley stage lacustrine carbonate reservoirs, I: Journal of Petroleum Geology, v. 8, no. 1, p. 37-58.

Bertani, R. T., and A. V. Carozzi, 1985b, Lagoa Feia Formation (Lower Cretaceous), Campos Basin, Offshore Brazil—Rift valley stage lacustrine carbonate reservoirs, II: Journal of Petroleum Geology, v. 8, no. 2, p. 199-220.

Bradley, W. H., and J. J. Fahey, 1962, Occurrence of stevensite in the Green River Formation of Wyoming: American Mineralogist, v. 47, p. 996-998.

Brice, S. E., M. D. Cochran, G. Pardo, and A. D. Edwards, 1982, Tectonics and sedimentation of the South Atlantic rift sequence—Cabinda, Angola, in J. S. Watkins, and C. L. Drake, eds., Studies in continental margin geology: AAPG Memoir 34, p. 5-18.

Burke, K., 1975, Atlantic evaporites formed by evaporation of water spilled from Pacific, Tethyan, and Southern oceans: Geology, v. 3, p. 613-616.

Carvalho, M. D., M. Monteiro, A. M. Pimentel, H. A. A. A. Rehim, and A. J. Dultra, 1984, Microfácies, diagénese e petrofisica das coquinas da Formacão Lagoa Feia em Badejo, Linguado e Pampo-Bacia de Campos—Projeto 03.01.02, Evolucão diagenética dos reservatórios carbonáticos da Formacão Lagoa Feia, Bacia de Campos: Rio de Janeiro, Petrobrás unpublished internal report, 130 p.

Castro A. C. M., Jr., 1987, The Northeastern Brazil and Gabon basins—A double rifting system associated with multiple crustal detachment surfaces: Tectonics, v. 6, p. 727-738.

Castro, J. C., and N. C. Azambuja Filho, 1980, Fácies, análise estratigráfica e reservatórios da Formacão Lagoa Feia, Cretáceo Inferior da Bacia de Campos: Rio de Janeiro, Petrobrás unpublished internal report, 110 p.

Cole, R. D., and Picard, M. D., 1978, Comparative mineralogy of nearshore and offshore lacustrine lithofacies, Parachute Creek Member of the Green River Formation, Piceance Creek basin, Colorado, and western Uinta basin, Utah: GSA Bulletin, v. 89, p. 1441-1454.

---

* Petrobrás unpublished internal reports referred to here are available from Petrobrás/CENPES, Ilha do Fundão s/n, Rio de Janeiro-RJ, Brazil 20000.

Dias, J. L., J. C. Vieira, A. J. Catto, A. Q. Oliveira, V. Guazelli, L. A. F. Trindade, R. O. Kowsmann, C. H. Kiang, U. T. Mello, A. M. P. Mizusaki, and J. A. Moura, 1987, Estudo regional da Formacão Lagoa Feia: Rio de Janeiro, Petrobrás unpublished internal report, 143 p.

Emery, K. O., E. Uchupi, J. Phillips, C. Bowin, and J. Mascle, 1975, Continental margin of western Africa—Angola to Sierra Leone: AAPG Bulletin, v. 59, p. 2209–2265.

Figueiredo, A. M. F., M. J. Pereira, W. U. Moriak, P. C. Gaglianone, and L. A. F. Trindade, 1985, Salt tectonics and oil accumulation in Campos Basin, offshore Brazil: Rio de Janeiro, Petrobrás unpublished internal report, 70 p.

Hay, R. L., 1966, Zeolites and zeolitic reactions in sedimentary rocks: GSA Special Paper 85, 122 p.

Hay, R. L., 1977, Geology of zeolites in sedimentary rocks, in F. A. Mumpton, ed., Mineralogy and geology of natural zeolites: Mineralogical Society of America Short Course Notes, v. 4, p. 53–64.

Hay, R. L., and R. A. Sheppard, 1977, Zeolites in open hydrologic systems, in F. A. Mumpton, ed., Mineralogy and geology of natural zeolites: Mineralogical Society of America Short Course Notes, v. 4, p. 93–102.

LePichon, X., and D. E. Hayes, 1971, Marginal offsets, fracture zones, and early opening of the South Atlantic: Journal of Geophysical Research, v. 76, p. 6283–6293.

Meister, E. M., 1984, A geologia histórica do petróleo na Bacia de Campos: Rio de Janeiro, Petrobrás unpublished internal report, 116 p.

Moura, J. A., 1985, Ostracodes from non-marine Early Cretaceous sediments of Campos Basin, Brazil: International Symposium on Ostracoda, Preprint.

Pereira, M. J., 1980, Avaliacão das coquinas da Formacão Lagoa Feia, Bacia de Campos: Rio de Janeiro, Petrobrás unpublished internal report, 36 p.

Pereira, M. J., L. A. F. Trindade, and P. C. Gaglianone, 1985, Origem e evolucão das acumulacões de petróleo na Bacia de Campos: Anais do Terceiro Congresso Brasileiro de Petróleo.

Petrobrás SECASU, 1982, Coluna estratigráfica da Bacia de Campos: Rio de Janeiro, Petrobrás unpublished internal report, 1 p.

Schaller, H., 1973, Estratigráfia da Bacia de Campos, in Anais do XXVII Congresso Brasileiro de Geologia, Aracaju-SE: Sociedade Brasileira de Geologia, v. 3, p. 247–258.

Sclater, J. G., and J. Francheteau, 1970, The implications of terrestrial heat flow observations and current tectonic and geochemical models of the crust and upper mantle of the Earth: Royal Astronomical Society Geophysical Journal, v. 20, p. 509–542.

Surdam, R. C., and K. O. Stanley, 1979, Lacustrine sedimentation during the culminating phase of Eocene Lake Gosiute, Wyoming (Green River Formation): GSA Bulletin, v. 90, p. 93–110.

Surdam, R. C., K. O. Stanley, and H. P. Buccheim, 1980, Depositional environment of the Laney Member of the Green River Formation, southwestern Wyoming: AAPG/SEPM/EMD Annual Convention, SEPM Trip 5 Field Guide 72 p.

Vianna, C. F., E. G. Gama, Jr., I. A. Simões, J. A. Moura, J. R. Fonseca, and J. R. Alves, 1971, Revisão estratigráfica da Bacia de Recôncavo-Tucano: Boletim Técnico da Petrobrás, v. 14, nos. 3/4, p. 157–192.

Stratigraphic Development of Proto-South Atlantic Rifting in Cabinda, Angola—A Petroliferous Lake Basin

Tim R. McHargue
Chevron Overseas Petroleum Inc.
San Ramon, California, U.S.A.

During the Early Cretaceous, a thick section of nonmarine sediment accumulated in the developing rift between South America and Africa. Extensive lake systems occupied much of the rift throughout its history. The rift sequence of west Africa is economically important because the basinal lake shales include exceptional source rocks for petroleum. A large majority of the oil reserves of Gabon, Congo, Zaire, and Angola are sourced from this lacustrine section. Cabinda is a small, detached enclave of Angola that produces most of that country's oil. An understanding of this important oil province requires an understanding of its stratigraphy. Abrupt facies changes among fluvial and alluvial sandstone, lake-margin sandstone, and carbonate and basinal lacustrine shale, diamictites, and turbidites are crucial to successful exploration in the basin. Synrift stratigraphy developed in response to active faulting, geographically variable subsidence rates, and variable topographic relief during deposition. Rift strata can be subdivided into tectono-stratigraphic packages that correspond to the rift's structural history.

Early rifting (fault phase, Neocomian) was characterized by rapid subsidence adjacent to major faults such that the rift became structurally subdivided into subbasins. Each subbasin developed a stratified lake and had a similar subsidence history, although the stratigraphy of each subbasin varied depending on influx of detrital sediment. Rapid transitions from alluvial, lake-margin sand to lacustrine shale are typical of this phase. In lacustrine units, diamictites, turbidites, and contorted bedding are common.

During the early sag phase (early Barremian), as fault-related subsidence gradually ended, regional rift-basin subsidence ensued. Lakes expanded and submerged former alluvial deposits, coalescing to form a single lake. Synchronous lake-level fluctuations can be recognized in all subbasins in Cabinda. During lowstands, carbonate deposition extended into basinal settings, and organic-carbon content decreased. During highstands organic carbon increased, and deposition of basinal organic shale expanded across former sites of carbonate deposition. Influx of coarse-grained extrabasinal clastics into the rift virtually stopped.

During the late sag phase (late Barremian), rates of regional subsidence gradually diminished, and faulting became rare. The lake became shallower as it filled with sediment, and the water column became fully oxygenated. Rhythmic fluctuations of lake level and chemistry are preserved as laterally persistent cyclical alternations (typically 10 m thick) of carbonate-rich and carbonate-poor mudstone. Thick carbonates accumulated in the shallowest parts of the lake.

During the drift phase, near the beginning of the Aptian, the entire region was uplifted and subjected to erosion. Uplift is attributed to rebound of the rift shoulders after crustal rupture. Following uplift, the pattern of renewed subsidence was that of a passive margin rather than a rift.

Throughout the history of the rift, sedimentation was controlled by the rate of subsidence relative to sediment influx rate. Superimposed on the tectonic evolution of the basin were climatic variations that affected lake level and chemistry to induce depositional events essential for correlation.

INTRODUCTION

As South America and Africa rifted apart during the Neocomian and Barremian (Lower Cretaceous), a thick nonmarine section accumulated in the developing graben (Reyre, 1966; Franks and Nairn, 1973). Extensive lacustrine systems were established early in rift history (Brice et al., 1982; Reyment and Dingle, 1987) and persisted through the Barremian.

Synrift sedimentation ended during the late Barremian as the entire rift was subjected to uplift and erosion. Subsequently, during the Aptian, a thin, regionally extensive blanket sand was deposited on the erosional unconformity. Drift-related subsidence initiated the first marine transgression, which resulted in accumulation of a thick evaporitic sequence (e.g., Belmonte et al., 1966; Franks and Nairn, 1973; Evans, 1978; Brice et al., 1982).

The presalt sequence of west Africa is economically important because the basinal lake shales include exceptional source rocks for petroleum (Brice et al., 1982). Most of the oil reserves of Gabon, Congo, Zaire and Angola are sourced from this lacustrine section. In addition, the same lacustrine section on the South American side of the rift has been proven to contain source rocks for nearly all of Brazil's petroleum reserves.

Cabinda is a small, detached enclave of Angola (Figure 1) that contains most of that country's oil reserves. To date, approximately 25% of the oil in Cabinda has been found within presalt reservoirs of marginal lacustrine carbonate and sandstone. As in other areas of the Lower Congo basin, basinal lacustrine shale is the source of this oil.

In Cabinda, lithofacies within the presalt section include fluvial and alluvial sandstone, lake-margin sandstone and carbonate, and basinal lacustrine shale, diamictites, and turbidites. Facies changes within the rift section developed in response to active faulting, geographically variable subsidence rates, and variable topographic relief during deposition. Facies boundaries are abrupt, both laterally and vertically, and are crucial to successful exploration in the basin. Wells that penetrate over 1000 ft (305 m) of continuous oil-saturated reservoir can be offset 3000 ft (914 m) by a well that encounters no reservoir at all. Traps are formed by both structure and stratigraphy.

In their overview of the stratigraphy of Cabinda, Brice et al. (1982) used a tectono-stratigraphic approach similar to that advocated here, although subsequent drilling and extensive additional palynology have necessitated revisions of their conclusions. The most important revision is that the prerift sedimentary section, initially believed to be Jurassic in age, now is considered part of the Neocomian fault phase. Brice and others' synrift I unit corresponds to part of the fault phase and all of the early and late drift phases of this report. The early synrift II unit is approximately equivalent to the early drift phase of this report. Other revisions of their correlations are important only on a local scale.

This study reevaluates the stratigraphic relationships and nomenclature of presalt sedimentary rocks of Cabinda based on approximately 200 wells (Figure 2). Palynomorphs from critical intervals of key wells provide a time framework for extensive, detailed reevaluation of log correlations.

PRESALT STRUCTURAL FEATURES

Figure 3 shows major presalt structural features that are most likely to influence regional sedimentation patterns. Their location, size, and orientation are based on well control, regional gravity, magnetics, and seismic data. Presalt structure is dominated by prominent north-south- or northwest-southeast-trending highs that terminate against northeast-southwest-trending transfer fault zones. The effect of these structures was the creation of subbasins that acted as separate depocenters or sediment traps, especially during early rifting. An isopach map of total presalt strata clearly reveals the presence of these subbasins (Figure 4).

CHRONOSTRATIGRAPHIC FRAMEWORK

Time correlations in this study are based on two lines of evidence—palynology and key log markers. Palynological zonation has been developed by N. J. Norton (Figure 5) based on a study of wells in Gabon and numerous wells in Cabinda and Zaire. Zones K13, K15, K17, and K18 are believed to be temporally significant, whereas the organic palynofacies is a facies of other zones and cannot be used for time correlations. Zone K16 has been recognized in only two Cabinda wells, and in both cases, K16 flora was recovered from sandy intervals beneath shales that contain organic palynofacies. These occurrences suggest that zone K16 is time significant but rare because of widespread anoxia. Samples that contain abundant black, angular, light-reflective amorphous residue (Balrar facies) have not been demonstrated to be time significant, although they occur in sands at the base of sedimentary sequences in Zaire, Gabon,

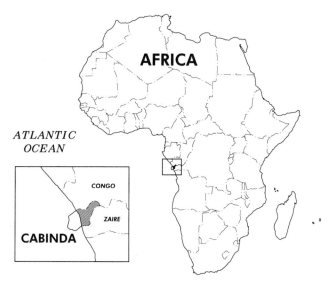

Figure 1. Location map of Cabinda, west Africa, and outline of offshore study area.

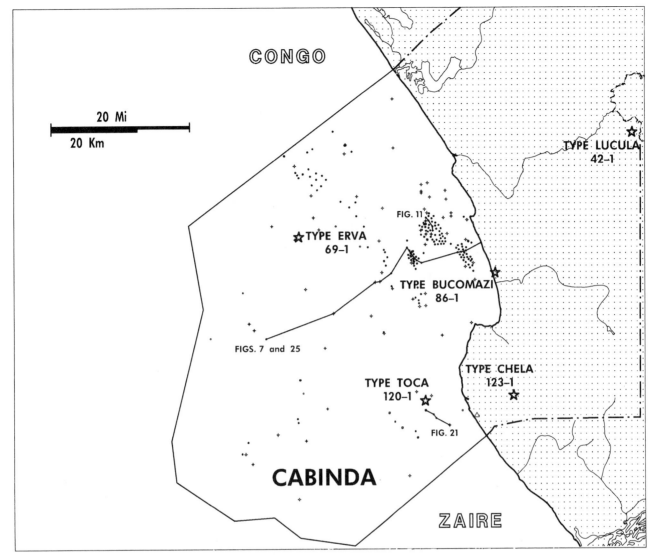

Figure 2. Index map of offshore study area of Cabinda, Angola, with wells, type sections (stars), and cross section locations.

and the Recôncavo basin of Brazil. J. P. Popek (pers. comm.) suggests that the Balrar facies is, at least in part, an artifact of incomplete processing of samples that are barren of organic residue.

Key log markers within the Bucomazi Formation have long been used for correlation. These markers are the result of fluctuations in the ratio of organic material to carbonate within the middle Bucomazi and fluctuations in carbonate content within the upper Bucomazi. Those markers that are believed to be reliable and useful for regional correlation have been used extensively in this study. Consistency of log character and relative stratigraphic position suggest that they represent time-stratigraphic events across an extensive lake system.

STRATIGRAPHIC SUCCESSION

In this section, regional depositional patterns and boundary relationships for each presalt formation are illustrated with regional isopach maps and paleogeographic reconstructions. Type sections and supplemental information necessary for formal definition of these lithostratigraphic units are given in the appendix.

The presalt stratigraphic section in Cabinda and Zaire is dominated by synrift lacustrine, alluvial, and fluvial strata. Age control is sparse, and stratigraphic relationships are well documented in only a few areas. The lessons learned in these areas are extrapolated elsewhere so that the resulting interpretation is the

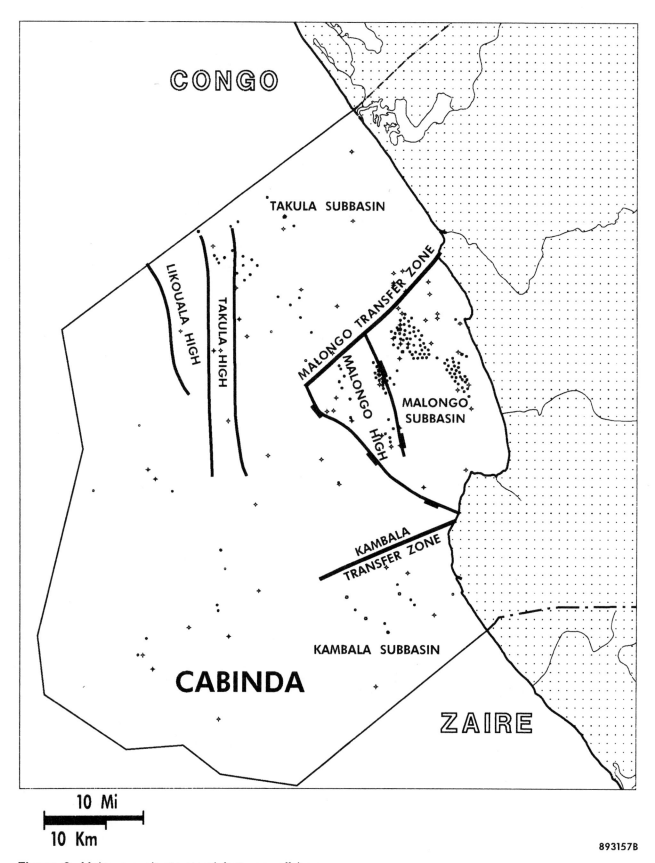

Figure 3. Major presalt structural features, offshore Cabinda.

McHargue

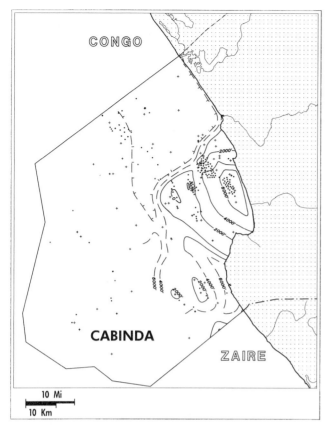

Figure 4. Isopach of total presalt strata thickness, offshore Cabinda. Contour interval 2000 ft (610 m).

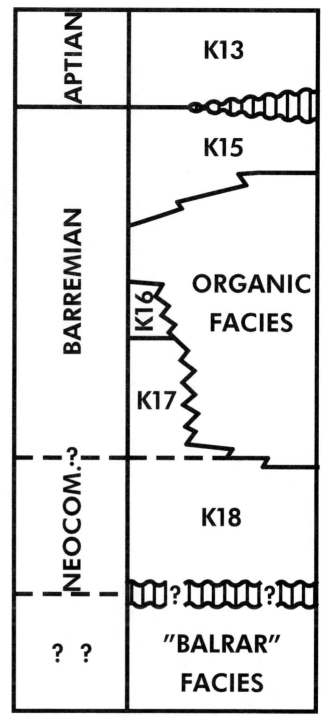

Figure 5. Palynological zonal scheme for Lower Cretaceous strata of Cabinda.

simplest required to account for the observations. These stratigraphic relationships are summarized in Figure 6. Regional relationships of presalt units are evident in a cross section that traverses the center of Cabinda (Figure 7).

Lucula Sandstone

Typically the Lucula Sandstone is white to light-gray, perhaps red near its base, moderately to poorly sorted, very fine-grained to pebbly, subrounded to subangular (Figures 8, 9). Its composition ranges from arkose to subarkose, occasionally quartz arenite, with variable amounts of potassium feldspar and mica; plagioclase is rare. Brown to black fissile shale and thin-bedded siltstone may be common near the top of the formation (Figure 8). Texture and composition mature upsection.

The age of the Lucula Sandstone is Neocomian at its type locality and Barremian to Neocomian elsewhere. Palynomorphs of zone K18 have been recovered throughout the Lucula at the type locality and from several locations around the study area. K17 palynomorphs have been recovered from upper Lucula Sandstone at the margin of the Malongo subbasin, and a K16 flora has been recovered on the flank of the Kambala subbasin. All known examples of the Lucula are interpreted to be older than log marker X, the oldest regionally persistent log marker.

The Lucula Sandstone lies directly on granitic gneiss, which defines its lower boundary (Figure 6). The upper boundary is defined by the lowest occurrence of either massive shale, marl, or siltstone of the Bucomazi Formation or the Toca Formation.

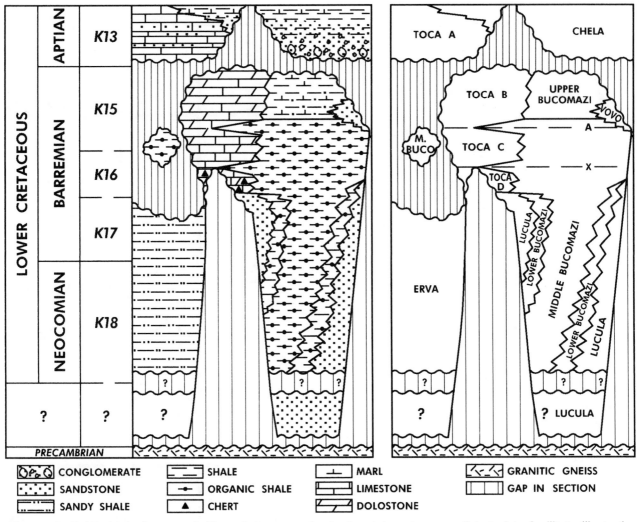

Figure 6. Palynological zones (left), rock types and facies relationships (center), and stratigraphic terminology (right) of presalt units of Cabinda. Vertical and horizontal scales are distorted to facilitate illustration of unit distributions and lateral relationships.

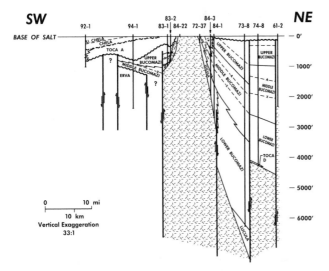

Figure 7. Regional stratigraphic cross section (flattened on base of Loeme Salt) of presalt units, offshore Cabinda. Pattern indicates granitic basement. See Figure 2 for location of section.

The Bucomazi Formation in several wells contains lenses or tongues of sandstone of typical Lucula composition.

The Lucula Sandstone is thickest along the eastern rift margin and near the Zaire border, areas apparently where sand was introduced into the rift. An implication is that isolated thicks within the rift probably are derived locally from exposed intrarift highs. This explains the thick sand accumulations along the eastern margin of the Malongo high (Figures 7, 10) and in the Malongo North field (Figure 11). The Lucula Sandstone interfingers with lower Bucomazi and is believed to shale out completely in farthest basinward locations.

The upper Lucula commonly gives a lobate gamma-ray log response, especially around the Malongo subbasin. Core samples, when correlated to logs, suggest that clean, well-sorted, mature quartzose sand, although commonly micaceous, is present within the lobate intervals, referred to as L-Lucula (Figure 9A). On the other hand, the Lucula is

McHargue

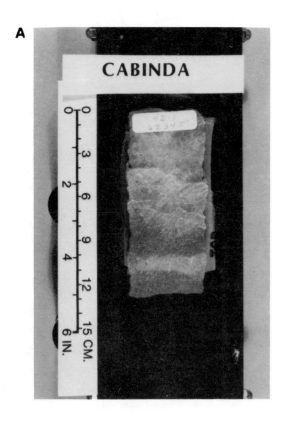

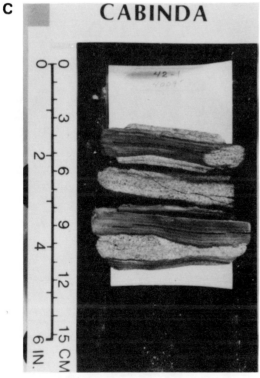

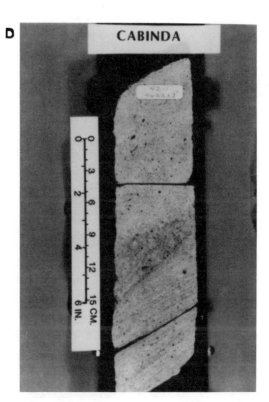

Figure 8. Representative core samples from Lucula Sandstone type section, well 42-1, onshore Cabinda. A, thinly interbedded sandy carbonate and red, argillaceous, arkosic sandstone, 6234.5 ft (1900 m). B, thinly interbedded subarkosic, medium- to well-sorted sandstone with gray, silty shale and siltstone, 4004– 4008 ft (1220–1221.6 m). C, thinly interbedded subarkosic sandstone with gray, silty shale, 4009 ft (1221.9 m). D, medium- to very coarse-grained, moderately to poorly sorted, cross-bedded arkose, 4622–4623 ft (1408.7–1409 m).

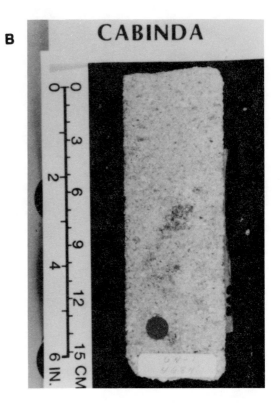

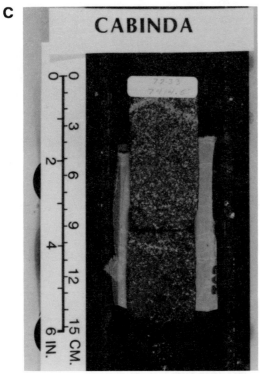

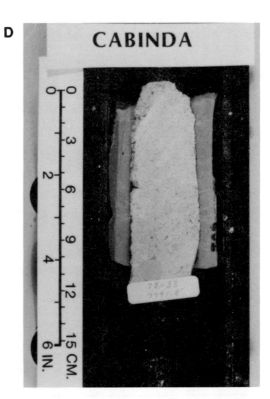

Figure 9. Additional core samples of Lucula Sandstone from various wells. A, fine- to medium-grained, moderately well-sorted, subarkose (photographed wet) from an interval with lobate gamma-ray log pattern, 8331 ft (2539.2 m). B, poorly sorted, coarse- to very coarse-grained arkose from interval with irregular gamma-ray log pattern, 4684 ft (1427.6 m). C, poorly sorted, coarse-grained argillaceous arkose with residual oil from interval with straight to serrate gamma-ray log pattern, 7414.5 ft (2259.8 m). D, poorly sorted, coarse-grained argillaceous arkose from interval with straight to serrate gamma-ray log pattern, 7791.8 ft (2374.8 m).

McHargue

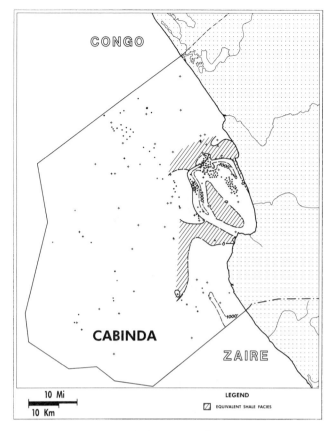

Figure 10. Isopach of Lucula Sandstone, offshore Cabinda. In patterned areas the Lucula is interpreted to be absent because of facies change to shale. Over the Malongo high, Lucula is absent due to erosion and nondeposition. Contour interval 1000 ft (305 m).

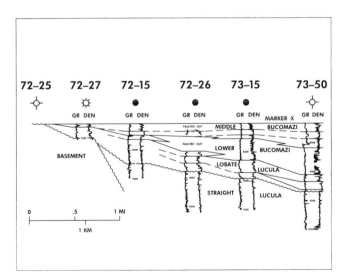

Figure 11. Representative stratigraphic cross section through Malongo North field, offshore Cabinda. Interfingering relationship between upper Lucula and Bucomazi is evident in well 73-50. Bucomazi shale that is laterally equivalent to lobate Lucula Sandstone is greenish and organic lean. Section is hung on the X marker in the middle Bucomazi. See Figure 2 for location of section.

texturally and mineralogically immature within intervals with straight or serrate (S-Lucula) gamma-ray responses (Figures 8D, 9B-D).

Log character of the Lucula can be explained through detailed correlations in areas of abundant well control. In the Malongo North field, L-Lucula overlies S-Lucula. In wells 73-15 and 72-14, tongues of L-Lucula interfinger with a green shale facies of the lower Bucomazi (Figure 11). This relationship strongly suggests that the L-Lucula was deposited as a shoreline facies of the Bucomazi. This also accounts for the increased maturity of the L-Lucula because shoreline sands would be subjected to considerable sorting and abrasion at their site of deposition. Most likely, S-Lucula was derived directly from adjacent basement exposures and was deposited as an alluvial apron. L-Lucula was derived both from basement and from underlying S-Lucula and accumulated in a high-energy marginal-lacustrine setting (e.g., deltas, beaches, and coastal bars).

The integrated interpretation of regional Neocomian paleogeography summarized in Figure 12 is applicable to most of Lucula deposition except for the last stages. Small, probably deep, lakes were present during most of Lucula deposition in the Malongo and Kambala subbasins. Subsequently, in the early Barremian (K16 and K17) lake levels rose and drowned most of the Lucula depositional areas. Land areas between basins were submerged to produce a single huge, stratified lake. As lake levels rose and alluvial S-Lucula was transgressed by lake-margin environments, sands were winnowed and cleaned by high-energy currents and deposited as lobate tongues that interfinger with lake shale. These L-Lucula accumulations are the most productive Lucula reservoirs in Cabinda.

Erva Formation

The Erva Formation, Early Cretaceous (Neocomian to Barremian) in age (zones K17 and K18), consists of granular mudstone and argillaceous granular conglomerate. In two cores from the Erva, the dominant lithology is matrix-supported coarse-grained argillaceous sandstone. These rocks include light- to dark-gray, thick- to thin-bedded, poorly sorted, subangular to angular, medium-grained to granular, micaceous and feldspathic (including plagioclase) lithic wacke or arenite, often suspended in a mud matrix (Figure 13). This lithology almost certainly represents accumulations of debris flows. Their occurrence in both the Erva Formation and the lower Bucomazi probably indicate that these units accumulated in deep lacustrine basins with high topographic relief. Distinction of the two units is based on significantly more sand throughout the Erva. Subordinately interbedded are black, poorly to well-laminated, organic-rich shale and uncommon laminated, moderately well-sorted, fine- to medium-grained sandstone.

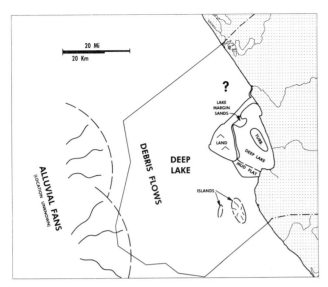

Figure 12. Neocomian paleogeographic reconstruction of offshore Cabinda in relation to modern shoreline and concession outline. Turb., area of possible turbidite deposition; ?, area with no well control. Alluvial fans are postulated to the west because of thick, extensive subaqueous debris flows in Erva Formation.

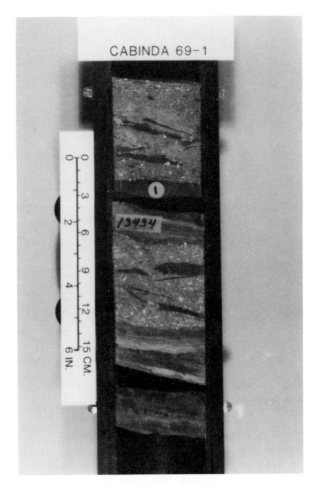

Figure 13. Representative core sample of Erva Formation from 13434 ft (4094.5 m), consisting of interbedded coarse-grained sandy diamictite, laminated black shale, and thinly laminated sandstone. Diamictite includes clasts of laminated black shale.

Because the base of the Erva has not been penetrated in any well, underlying units are unknown. The top of the Erva is defined by the highest occurrence of sandy mudstone and argillaceous, micaceous, feldspathic wacke, sandstone, and conglomerate (Figure 6). Any of various stratigraphic units may unconformably overly the Erva, including the Bucomazi, Toca, Chela, Loeme, and Pinda Formations. The Erva is overlain by the middle member of the Bucomazi Formation at some locations, but it is unclear whether or not the two units are conformable. The Erva Formation is temporally equivalent, at least in part, to the Lucula and Bucomazi Formations.

The maximum thickness of the Erva Formation is unknown but exceeds 3860 ft (1176 m) in well 45-1. Although its lateral extent also is poorly known, it does occur across large areas of offshore Cabinda west of the Malongo high (Figure 14). No definitive Erva has been penetrated east of the Malongo or Kambala highs, where it probably was never deposited. This conclusion is reflected in the Neocomian paleogeographic reconstruction (Figure 12). A western source is proposed for Erva sediment, perhaps an intrarift high or the western rift margin, because the Erva is so thick in the west but appears to be absent in eastern areas of the rift; however, this proposal is conjectural. Rift-axial sediment transport for the Erva is an alternative hypothesis.

The presence of significant porosity in the Erva is unusual. The only production from the formation comes from well 95-3 in the Kali field. Porosity may be a result of winnowing on the Kali high block during deposition of the Erva or, more likely, from fractures associated with nearby faults.

Bucomazi Formation

The Lower Cretaceous (Neocomian to Barremian) Bucomazi Formation consists of shale, calcareous shale, and marl, often micaceous, that is brown, gray, or black, poorly to well-laminated, occasionally contorted, and poorly to extremely organic rich (Figure 15). Thin carbonate and micaceous sandstone are common in the upper member. The lower member contains some brown to gray, poorly bedded, matrix-supported granular mudstone and occasional thick sandstone.

The upper boundary of the Bucomazi is defined as the top of shale beneath either coarse-grained sandstone and conglomerate of the Chela Formation or carbonate of the Toca Formation. The lower

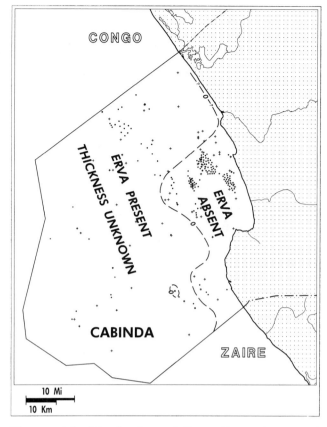

Figure 14. Distribution of Erva Formation; total thickness is unknown.

boundary is defined by the lowest occurrence of mudstone lithology. The Bucomazi's subdivision into upper, middle, and lower members is informal because lithologic gradation and variability make consistent, objective recognition difficult. The upper member consists of green to brown, micaceous mudstone with occasional gray shale (Figure 15A). The middle Bucomazi consists of organic-rich, brown to gray to black, laminated shale, carbonate-rich shale, and marl (Figures 15B–C) interbedded with geographically variable amounts of organically lean brown and gray shale and carbonate mudstone or marl. The lower Bucomazi is absent at the type section but regionally consists of brown to gray, laminated shale, granular mudstone, and calcareous shale and marl with rare to abundant sandstone (Figure 15D). The three members are most easily identified by log response. The upper and lower members have relatively low resistivity, high sonic velocity, and high density, whereas the middle member mostly has high resistivity, low velocity, and low density. The upper and lower members are distinguished by stratigraphic position relative to the middle member.

The Bucomazi Formation either unconformably overlies basement or conformably overlies the Lucula Sandstone (Figure 6). It is separated by an unconformity from the overlying Chela Sandstone. Toca

carbonates overlie the Bucomazi conformably in some places and unconformably in others.

Thickness of the Bucomazi Formation varies considerably, from 0 to more than 7000 ft (2133 m) (Figure 16). The upper, middle, and lower members may be absent because of erosion or nondeposition. The amount of sandstone in the lower member is variable, ranging from insignificant to several hundred feet of virtually continuous sandstone. Carbonate content also is variable, especially in the middle and upper members, and lateral gradation with the Toca Formation is probable in places (Figure 6).

The lower Bucomazi facies is laterally equivalent to the middle Bucomazi. Except near the extreme eastern margin of the rift basin (onshore Cabinda), the lower Bucomazi member occurs only beneath the X marker. It exceeds 1000 ft (305 m) in thickness in the Malongo subbasin and is present in the Kambala subbasin. The lower Bucomazi forms a higher proportion of total Bucomazi thickness near basin margins, although deep areas of the Malongo and Kambala subbasins may contain significantly thick lower Bucomazi. On a large scale, black, laminated shale in the middle Bucomazi interfingers with brown shale of the lower Bucomazi.

Compared to the middle member, the lower Bucomazi is less anoxic because it is located near the point of inflow of oxygenated water into the stratified lake. Usually, this inflow was accompanied by an influx of sand; consequently, the lower Bucomazi tends to fringe the upper Lucula Sandstone as well as sands within the Bucomazi (e.g., Figures 11, 12).

Thickness of the middle Bucomazi changes rapidly from basinal to positive areas because of the combined effects of structural relief and facies changes from middle Bucomazi in subbasin centers to lower Bucomazi and Lucula Sandstone toward subbasin margins. However, thickness variation of the youngest middle Bucomazi, the interval between the X and A markers (Figure 6), is more subdued. During its deposition, tectonic activity apparently diminished, and subsidence became more regional and less controlled by growth faults. Seismic data support this conclusion because few faults offset the A marker. Before deposition of the A marker, regional subsidence and burial of interbasinal highs allowed expansion of the stratified lake over most of the study area (Figure 17).

During the late Barremian, the Bucomazi lake developed a fully oxygenated water column and ceased to accumulate organic-rich shale. This change represents the boundary between the middle and upper Bucomazi. This change from a permanently stratified lake to an unstratified lake appears to have been synchronous across the study area. Bathymetric relief within the basin probably was small and gentle resulting in only subtle thickness variations within the upper Bucomazi. However, this is difficult to confirm because subsequent to Bucomazi deposition and prior to deposition of the Chela Formation, a

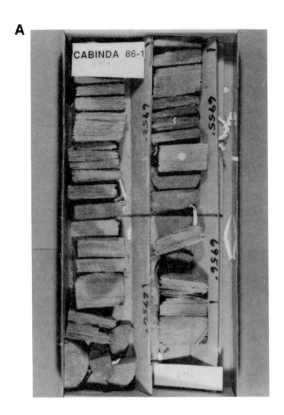

A

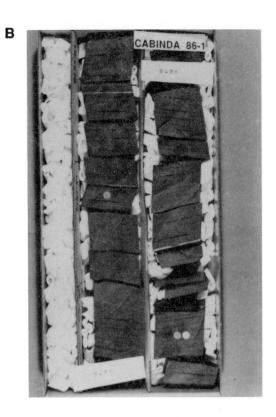

B

C

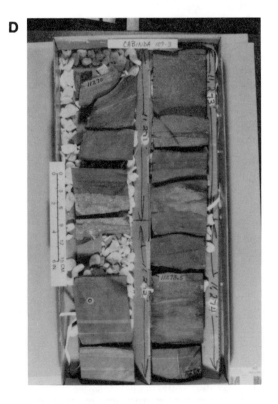

D

Figure 15. Representative intervals of Bucomazi Shale. A, upper member, gray-brown, laminated marl from 6953–6956 ft (2119–2120 m), well 86–1, type section. B, middle member, dark-gray, laminated calcareous shale from 8680–8685 ft (2645.5–2647 m), well 86–1, type section. C, middle member, dark-gray, laminated calcareous shale from 8684 ft (2646.7 m), well 86–1, type section. D, lower member, interbedded diamictite, laminated dark shale, and contorted, thin, fine-grained sandstone from 11270–11274 ft (3434.9–3436.1 m), well 109–3.

McHargue

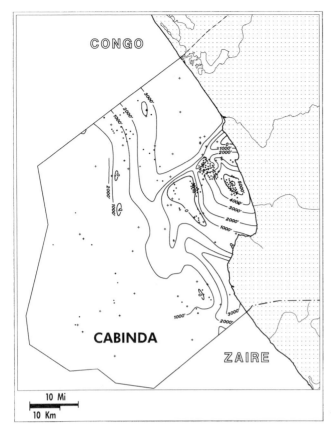

Figure 16. Isopach of total Bucomazi Shale, offshore Cabinda. Contour interval 1000 ft (305 m).

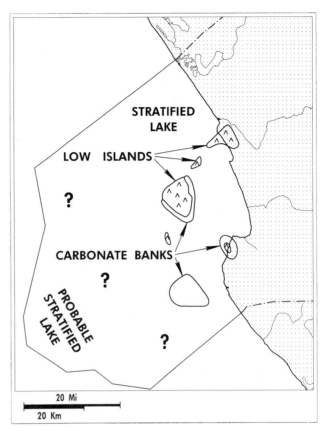

Figure 17. Barremian paleogeographic reconstruction of offshore Cabinda in relation to modern shoreline and concession outline.

substantial but variable amount of the Bucomazi Formation were eroded.

Toca Formation

Limestone, dolomite, argillaceous carbonate, and subordinate shale and sandy carbonate comprise the Toca Formation of Barremian to Aptian age (Figure 18). Its limits are defined by the highest and lowest occurrences of massive carbonate beneath the Loeme Salt. Four informal members of the Toca (A, B, C, and D) are recognized based on stratigraphic position relative to regional log markers. At this time, poor understanding of the lateral variability of the Toca precludes formal subdivision based on lithology.

The Toca usually is overlain unconformably by the Chela Sandstone but may be overlain conformably by the Bucomazi Formation or unconformably by the Loeme Salt. Immediately beneath the Toca, Bucomazi shale is most common, but the Lucula Sandstone, Erva Formation, or basement rocks also are found in places.

Continuity of the Toca Formation between its several known locations is unlikely. With a thickness of 0 to about 1000 ft (305 m) (Figure 19), the formation, as defined here, includes at least two unrelated packages with different distribution patterns—Barremian carbonates that apparently fringe high

blocks around the Malongo, Kambala, and Takula subbasins (Toca B, C, D; Figure 20); and an Aptian carbonate that is present only west of the Kambala, Malongo, and Takula highs (Toca A). Laterally, members B, C, and D interfinger abruptly with the Bucomazi Formation (Figure 6), a facies change well illustrated near the Kambala high. Distinction among members B, C, and D is based partly on lithology but more on stratigraphic position. For example, member D is cherty and older than the X marker (this may include several carbonates of slightly different ages); member C, a dolomite in the Malongo West and Limba areas (lithologically variable elsewhere), lies between the A and X markers; and member B is a limestone in the Malongo West and Limba areas (lithologically variable elsewhere) and lies between the A marker and the pre-Chela unconformity.

The A and X horizons are more than convenient markers to subdivide a carbonate section. Both probably represent lake highstands that allowed abrupt expansion of anoxic conditions across formerly shallow, oxygenated, carbonate-rich lake-margin settings that developed during lowstands. A lowstand followed by a highstand (probably climatically controlled) provided the opportunity first for subaerial exposure and diagenesis of lake-margin

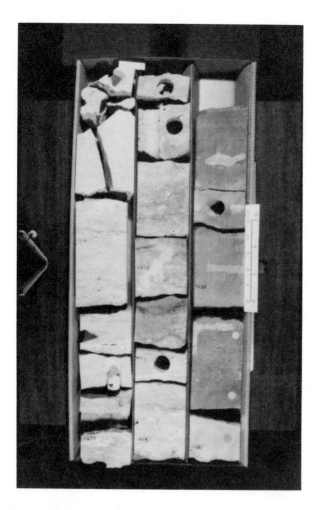

Figure 18. Representative interval of Toca Formation—white, fossiliferous dolopackstone with occasional vuggy porosity over dark wackestone, 11672.5–11681 ft (3557.6–3560.2 m).

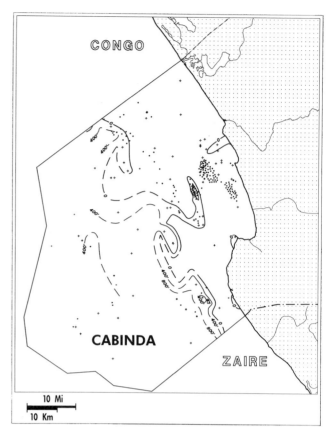

Figure 19. Isopach of total Toca Formation, offshore Cabinda. Contour interval 400 ft (122 m).

carbonates followed by their burial beneath dark, often anoxic, shale. This sequence of events is best expressed for the A marker in the Kambala field (Figure 21). Farther north in the Malongo West and Takula fields, the A marker shale is not preserved across the entire area of carbonate accumulation, but disruption of carbonate deposition and patterns of diagenesis suggest the location of an equivalent horizon.

The Toca A member is approximately the same age as the Chela Sandstone, although the exact relationship between the two units is unclear. In some wells the Chela appears to conformably overlie Toca A carbonate, but in other areas, such as the Takula field, the Chela may be absent. In Figure 6, the Toca A is treated as a western facies of the Chela Sandstone. This implies an onshore-offshore relationship between the Chela and Toca A, respectively, that presages the postsalt basinal configuration and signals initiation of drift or spreading rather than rift subsidence. Support for this hypothesis is

provided by an isopach map of presalt Aptian units (Toca A and Chela combined; Figure 22). This map reveals the development of a sedimentary wedge in the west that corresponds to distribution of the Toca A. The paleogeographic interpretation of this time interval is shown in Figure 23. The possible occurrence of a single dinoflagellate from the Toca A in well 94-1 (N. J. Norton, pers. comm.) suggests possible marine influence on this unit.

Chela Sandstone

White to gray-green, very fine- to very coarse-grained quartzose sandstone and conglomerate are overlain by dark-brown, often carbonate-rich, micaceous shale in the Chela Sandstone.

The top of the Chela (Aptian) is recognized as the first clastic rocks below either the Loeme Salt or younger units, when the Loeme is absent. The base of the Chela unconformably overlies basement rocks or the Bucomazi, Toca, Lucula or Erva Formations. The Chela may occur conformably above the Toca A member.

Informally, the Chela has been divided into an upper shale/carbonate member and a lower sandstone/conglomerate member. In southern Cabinda and Zaire the upper member is widespread and distinctive—gray shale with types III and IV gas-

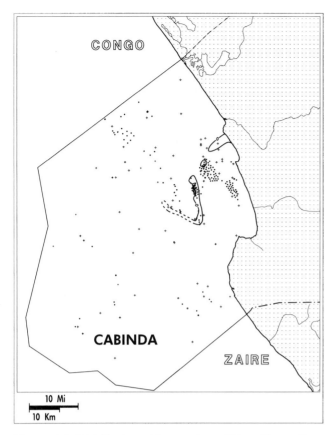

Figure 20. Distribution of D member of Toca Formation, offshore Cabinda.

prone organic material (B. J. Huizinga, pers. comm.). In northern Cabinda the upper member either is gradational with the lower member, absent, or too thin to resolve with logs.

The Chela appears to grade westward laterally into the Toca A member (Figures 22, 23), but this relationship is not well understood because of sparse well control in key areas. Westward thickening of the Aptian section presages shelf/slope relationships in younger postsalt strata and is interpreted to indicate initiation of subsidence associated with spreading. The upper boundary of the Chela is gradational with the Loeme Salt, but the lower boundary is unconformable above the Bucomazi, Toca (B, C, and D), Erva, or Lucula. In some areas the Chela lies directly on basement rocks. Thickness of the Chela ranges from 0 to about 500 ft (0–52 m) (Figure 24).

TECTONO-STRATIGRAPHIC MODEL

The distribution, thickness, and lithology of synrift sedimentary rocks, together with adequate chrono-stratigraphic control, serve as the basis for recon-structing the history of rifting in the lower Congo basin of Cabinda. Subsidence rate appears to have decreased gradually from a peak early in the rift's history, a trend that is compatible with some models of rifting (e.g., Keen, 1987). Furthermore, based on the geographic pattern of subsidence, rift history can be subdivided into three phases—(1) fault phase during early rifting, (2) sag phase during late rifting, and (3) drift or spreading phase after rifting ended. The paleogeographic reconstructions previously introduced for episodes of each phase of rifting (Figures 12, 17, 23) will be summarized more fully here. In addition, Figure 7 has been redrawn in Figure 25 to illustrate development of the stratigraphic section phase by phase.

Fault Phase

The fault phase, the initial stage of rifting, occurred during the Neocomian (Figures 12, 25). Subsidence at this time resulted primarily from lithospheric extension, and subsidence rates were highest adjacent to major active faults. As a result, several subbasins within the rift were established that accumulated distinct stratigraphic sections depend-ing on access to external sediment sources. For example, the sections in subbasins adjacent to the eastern rift boundary are composed almost entirely of coarse-grained clastics derived from outside the rift. In contrast, during the same time interval the Malongo subbasin accumulated a thick succession of shale and marl but comparatively less sand adjacent to exposed, high blocks to the north and west. The Malongo basin received less coarse-grained clastics because surrounding subbasins and adjacent basin-margin highs served as barriers to sediment influx.

In western offshore Cabinda, several thousand feet of debris-flow deposits and interbedded organic-rich lacustrine shale of the Erva Formation accumulated. The Erva Formation indicates a major source of diamictic sediment, presumably large fan deltas, somewhere to the west, northwest, or southwest. Although its exact location is unknown, it may have been a large intrarift high or the western rift margin, now in Brazil.

An isolated lake formed within each subbasin during the fault phase. The lakes became deep and stratified and accumulated organic-rich sediment (middle Bucomazi and Erva) in those subbasins where subsidence rate exceeded sedimentation rate—e.g., the Malongo and Kambala subbasins (middle Bucomazi) and areas west of the Malongo high (Erva). Lakes were shallow or absent in subbasins where sedimentation nearly equaled or exceeded subsi-dence. These basins filled with fluvial or alluvial sandstone.

Few faults offset the X marker which indicates that the fault phase ended by the time the X marker was deposited.

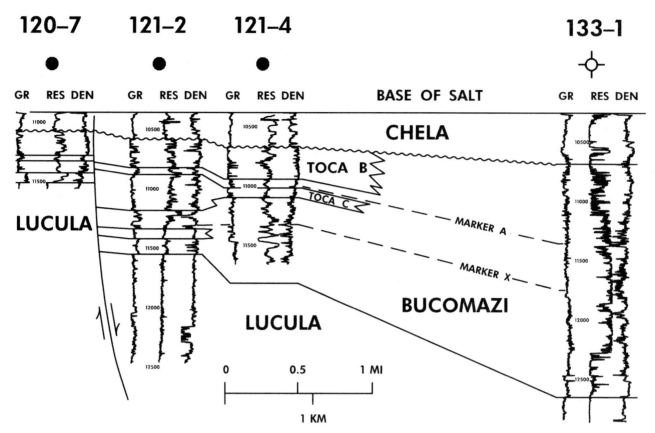

120–7   **121–2**   **121–4**   **133–1**

GR RES DEN   GR RES DEN   GR RES DEN   **BASE OF SALT**   GR RES DEN

CHELA

TOCA B

TOCA C

MARKER A

LUCULA

MARKER X

BUCOMAZI

LUCULA

0   0.5   1 MI

1 KM

Figure 21. Representative cross section through Kambala field, offshore Cabinda. Interfingering of the Toca and Bucomazi is evident. Episodes of organic accumulation in the Bucomazi (e.g., marker A) are represented as shale breaks in Toca carbonates. Intervals of maximum expansion of the Toca are represented by carbonate-rich, organic-lean intervals in the Bucomazi. Section is hung on base of salt. See Figure 2 for location of section.

Sag Phase

As fault-bounded subsidence became less pronounced, it was replaced by moderate to slow, broad, regional subsidence (sag), presumably due to lithospheric cooling. During this phase of rifting, once isolated lakes coalesced into a single, deep, stratified lake that covered most of the study area (Figures 17, 25). Organic shale of the middle Bucomazi was deposited, even across most of the highs and perhaps beyond the rift shoulders. In fact, during this phase, the extent of organic shale probably was even greater than it is today because subsequent pre-Chela erosion removed the middle and upper Bucomazi from higher areas.

Coarse-grained clastic influx was nil across most of the rift during the sag phase because the expanded lake covered former highlands. Although topography was low, differential compaction, perhaps augmented by differential subsidence, allowed thicker columns of sediment to accumulate in basinal areas.

Carbonates (Toca C and D) formed where the water was clear, and the lake bottom was elevated into the oxygenated part of the water column. These conditions were present around the flanks of intrarift highs, such as the Malongo and Kambala highs.

During the sag phase, fluctuations in lake level and chemistry increasingly influenced sediment composition. Based on regional well correlations, it appears that during lake lowstands, the carbonate content of Bucomazi shale increased at the expense of organic content. These lowstand intervals are represented on logs as high-density, low-resistivity units, probably marls, that are widespread across Cabinda. A prominent example occurs immediately below the X marker in most wells. In contrast, during highstands, organic content was high and carbonate content was low, as suggested by low-density, high-resistivity log patterns. The two most prominent examples of this situation, the A and X markers, have been used as convenient mapping events, but numerous similar events have been recognized and correlated regionally. Lowstand marl probably formed during atypically arid conditions when salinity increased in response to higher evaporation. Highstand conditions probably formed during unusually humid periods when deepening of the lake caused anoxic water to expand over formerly oxic locations, and when salt concentrations were diluted by the increased volume of the lake.

McHargue

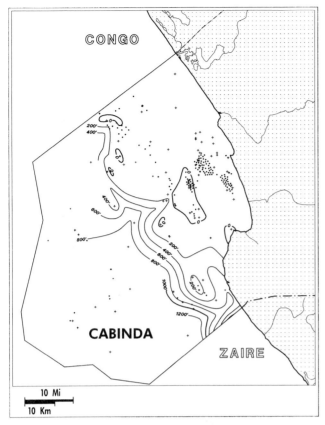

Figure 22. Isopach of presalt Aptian rocks, offshore Cabinda. Interval includes Chela Sandstone and A member of Toca Formation. Contour interval 200 ft (61 m).

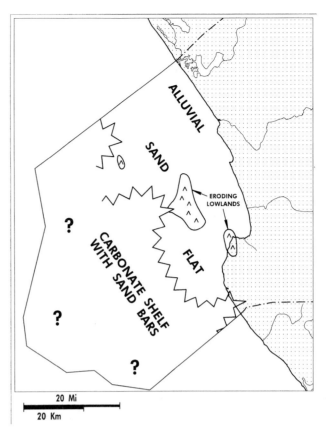

Figure 23. Early Aptian (presalt) paleogeographic reconstruction of onshore-offshore relationships between Chela Sandstone and A member of Toca Formation, offshore Cabinda.

As the lake gradually filled with sediment, the entire lake bottom became oxygenated. The last widespread anoxic event is synchronous across much of the study area and corresponds to the TO marker (top of the middle Bucomazi). Toca carbonate accumulation expanded geographically but still in association with bathymetric highs.

As the lake became shallower, its sediment became increasingly sensitive to subtle fluctuations in lake level and chemistry. Even though the entire water column eventually became oxygenated, carbonate content continued to fluctuate in response to climatic variations. A result was the episodic accumulation in basinal areas of numerous, thin, laterally persistent carbonate marker beds in the upper Bucomazi. These markers allow for extremely precise correlation within the upper Bucomazi.

In wells that preserve the youngest interval of upper Bucomazi (73–8, 36–1, and 58–2), the top of the Bucomazi section is silty, and no carbonate marker beds can be recognized. Presumably this interval represents transition to lake-margin, deltaic, or subaerial deposition as the lake finally filled with sediment.

Drift Phase

When the attenuated continental crust ruptured and oceanic crust began to form within the rift, drift or spreading began. The change from rifting to spreading is interpreted to correspond with development of the pre-Chela unconformity. When sedimentation resumed following development of this unconformity, subsidence no longer was focused within subbasins (Figure 16) but occurred west of a north-south-trending structural hinge. An isopach of strata immediately above the unconformity (Figures 22, 25), as well as facies distributions within this section (Figure 23), suggest development of a shelf and continental slope as in the Upper Cretaceous rather than rift-related subsidence of older units. Some models predict that rupture of the continental crust triggers crustal rebound (uplift) near the rift margin (Steckler, 1985; Buck, 1986; Keen, 1987). This uplift would have induced erosion, even of formerly basinal areas, until passive-margin subsidence overcame the rebound effect.

Two major lithofacies occur within the presalt part of the spreading phase—coarse-grained clastics

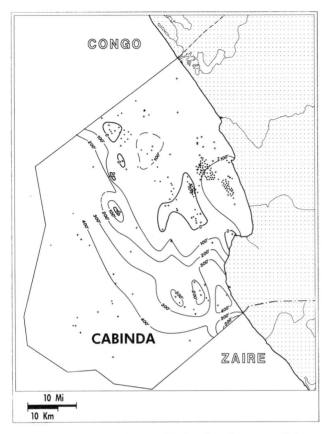

Figure 24. Isopach of Chela Sandstone, offshore Cabinda. Contour interval 100 ft (30.5 m).

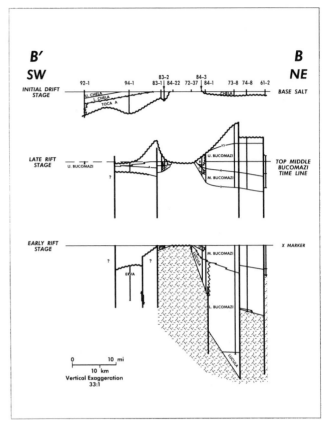

Figure 25. Regional cross section of presalt strata, offshore Cabinda (same section as Figure 7) with sediments of each rift stage isolated and flattened on separate datums. Early rift stage is flattened on the X marker, late rift stage is flattened on top of middle Bucomazi Formation, and initial drift stage is flattened on base of salt. See Figure 2 for location of section.

(Chela Sandstone) and carbonate and mixed carbonate and clastics (Toca A member). These two units tend to occupy opposite sides of the structural hinge line that follows the western margins of the Kambala, Kali, and Malongo highs. East of this hinge line, subsidence rate was low, and deposition was dominated by siliciclastics (Chela Sandstone) that had been reworked from the eroded section. West of the hinge line, carbonate and interbedded sandstone (Toca A), accumulated as a westward-thickening wedge. It is unknown if these carbonates are lacustrine or restricted marine.

CONCLUSIONS

1. Prerift sedimentary rocks have not been found in Cabinda.
2. Rifting began in the Neocomian and evolved through three phases—fault phase during early rifting, sag phase during late rifting, and spreading after rifting ended.
3. During the fault phase, subsidence was focused along major subbasin-bounding normal faults. Isolated subbasins with closed drainage developed

deep stratified lakes. Sediments in each subbasin are characterized by rapid facies changes from alluvial or fluvial sandstone (Lucula), to lake-margin sandstone (Lucula) and oxic mudstone (lower Bucomazi), to basinal, laminated, organic-rich lacustrine marl (middle Bucomazi). Lucula Sandstone was derived either locally from erosion of emergent horsts within the rift or regionally from outside the rift. To the west, a thick sequence of diamictites accumulated at this time.

4. During the sag phase, subsidence was dispersed across the rift rather than concentrated along faults. Isolated lakes coalesced into one large, well stratified lake that expanded to the rift margin and submerged all but the highest intrarift horsts. Organic-rich laminated marl accumulated on the lake bottom in basinal settings at this time (middle Bucomazi). Alluvial, fluvial, and lake-margin sands were rare during the sag phase because source areas were submerged. However, lake-margin carbonates (Toca members C and D) episodically were well developed.

5. As the sag phase progressed, subsidence rate decreased, the water column became fully oxygen-

McHargue

ated, and the lake gradually filled with sediment. Massive lake-margin carbonates (Toca B member) expanded toward the centers of subbasins. Greenish shales (upper Bucomazi) that accumulated in basinal settings preserved rapid cyclical fluctuations in carbonate abundance that are correlatable across the rift. The sag phase ended with uplift and erosion across the entire rift.

6. The spreading phase began in the Aptian with renewed subsidence and sedimentation. The pattern of subsidence during Chela deposition suggests development of a slowly subsiding shelf on which a thin, blanket sand (Chela Sandstone) accumulated, and a rapidly subsiding slope on which a mixture of sand, shale, and carbonate accumulated (Toca A member).

ACKNOWLEDGMENTS

Without the work of Chevron and Gulf staff past and present, this study would not have been possible. I especially would like to thank N. J. Norton, M. R. Cassa, A. R. Pytte, G. Pardo, S. E. Brice, P. E. C. Reed, L. Caflisch, E. L. Couch, R. R. Terres, M. J. Hanou, L. D. Littlefield, and J. A. Lopes.

REFERENCES

Belmonte, Y., P. Hirtz, and R. Wenger, 1966, The salt basins of the Gabon and the Congo (Brazzaville)—A tentative palaeogeographic interpretation, *in* Salt basins around Africa: London, Institute of Petroleum, p. 55–74.

Brice, S. E., M. D. Cochran, G. Pardo, and A. D. Edwards, 1982, Tectonics and sedimentation of the South Atlantic rift sequence, Cabinda, Angola, *in* J. S. Watkins, and C. L. Drake, eds., Studies in continental margin geology: AAPG Memoir 34, p. 5–18.

Buck, W. R., 1986, Small-scale convection induced by passive rifting—The cause for uplift of rift shoulders: Earth and Planetary Science Letters, v. 77, p. 362–372.

Evans, R., 1978, Origin and significance of evaporites in basins around Atlantic margin: AAPG Bull., v. 62, p. 223–234.

Franks, S., and A. E. M. Nairn, 1973, The equatorial marginal basins of west Africa, *in* A. E. M. Nairn, and F. G. Stehli, eds., The ocean basins and margins—The South Atlantic: New York, Plenum Press, v. 1, p. 301–350.

Keen, C. E., 1987, Some important consequences of lithospheric extension, *in* M. P. Coward, J. F. Dewey, and P. L. Hancock, eds., Continental extensional tectonics: Geological Society Special Publication 28, p. 67–73.

Reyment, R. A., and R. V. Dingle, 1987, Paleogeography of Africa during the Cretaceous Period: Palaeogeography, Palaeoclimatology, Palaeoecology, v. 59, p. 93–116.

Reyre, D., 1966, Particularités géologiques des bassins côtiers de l'ouest africain (essai de récapitulation), *in* D. Reyre, ed., Sedimentary basins of the African coast, part 1—Atlantic Coast: Paris, Association of African Geological Surveys, p. 253–304.

Steckler, M. S., 1985, Uplift and extension at the Gulf of Suez—Indications of induced mantle convection: Nature, v. 317, p. 135–139.

APPENDIX 1. TYPE SECTIONS (LOCATIONS ON FIGURE 2)

Lucula Sandstone

Well: 42-1 (formerly Lucula-1). This well is not an ideal type section but is acceptable. Because it is located at the extreme margin of the basin, the texture of the Lucula at its type section is coarser grained and less well-sorted than at most other sections. However, historical precedent and abundant core dictate selection of this well as the type locality.

Completion date: June 11, 1960

Location: 5°9'10"S, 12°30'43.5"E
Kelly bushing elevation: 139 ft (42.4 m)

TD: 6245 ft [-6106 ft true vertical depth (TVD)] [1903.4 m (-1861 m TVD)]

Interval: 4000–6204 ft (-3861 to -6065 ft TVD) [1219–1891 m (-1176.8 to -1848.5 m TVD)]

Lithologic samples: Cuttings plus conventional core for the intervals (log depths) 4004–4024, 4191–4211, 4617–4628, and 5501–5521 ft (1220.4–1226.5, 1277.4–1283.4, 1407.2–1410.5, and 1676.6–1682.7 m).

Derivation and history of name: Named for the Lucula-1 well (now 42-1).

Erva Formation

Well: 69-1

Location: 5°22'26.96"S, 11°48'43.56"E

Completion date: July 18, 1982

Kelly bushing elevation: 78 ft (23.8 m)

TD: 13443 ft (-13365 ft TVD) [4097.2 m (-4073.5 m TVD)]

Interval: 10067–13443 ft (-9989 to -13365 ft TVD) [3068.3–4097.2 m (-3044.5 to -4073.5 m TVD)]

Lithologic samples: Cuttings and 30 ft (9.1 m) of conventional core from 13413–13443 ft (4088.1–4097.2 m).

Derivation and history of name: Selection of name is arbitrary.

Bucomazi Formation

Well: 86-1 (formerly Bucomazi-1)

Location: 5°26'35"S, 12°14'30"E

Completion date: June 25, 1959

Kelly bushing elevation: 398 ft (121.3 m)

TD: 9435 ft (-9037 ft TVD) [2875.6 m (-2754.3 m TVD)]

Interval: 6677-9137 ft (-6279 to -8739 ft TVD) [2035-2784.8 m (-1913.7 to -2663.5 m TVD)]

Lithologic samples: Cuttings, plus conventional cores from the intervals 6677-6703, 6833-6839, 6953-6964, 6983-6987, 7031-7038, 7197-7205, 7249-7270, 7417-7425, 7590-7617, 7641-7652, 7797-7820, 7940-7962, 8152-8155, 8345-8359, 8441-8461, 8677-8692, 8886-8893, 8901-8904, 8918-8933, and 9052-9065 ft (2035-2043, 2082.6-2084.4, 2119.2-2122.5, 2128.3-2129.5, 2142.9-2145.1, 2193.5-2196, 2209.4-2215.8, 2260.6-2263, 2313.3-2321.5, 2328.9-2332.2, 2376.4-2383.4, 2420-2426.7, 2484.6-2485.5, 2543.4-2547.7, 2572.7-2578.8, 2644.6-2649.2, 2708.3-2710.5, 2712.9-2713.8, 2718.1-2722.6, and 2758.9-2762.8 m).

Derivation and history of name: Named for the Bucomazi-1 well (now 86-1). The type section contains only the upper and middle members. A good reference section for the lower Bucomazi is in the 73-8 well from 10006-12880 ft (3049.7-3925.6 m).

Toca Formation

Well: 120-1 and 120-1 (sidetrack)

Location: 5°42′28.5″S, 12°4′10.6″E

Completion date: March 21, 1972

Kelly bushing elevation: 109 ft (33.2 m)

TD: 11342 ft (-11262 ft TVD) [3456.9 m (-3432.5 m TVD)] for first borehole; 10952 ft (-10827 ft TVD) [3338 m (-3299.9 m TVD)] for sidetrack.

Interval: 10377-10930 ft (-10297 to -10852 ft TVD) [3162.8-3331.3 m (-3138.4 to -3307.5 m TVD)] in first borehole; 10403-10952 ft (-10286 to -10827 ft TVD) [3170.7-3338 m (-3135 to -3299.9 m TVD)] for sidetrack.

Reference sections for informal members of the Toca Formation are as follows (log depths): Toca A, well 94-1, 11295-11949 ft (3442.5-3641.9 m); Toca B, well 120-1, 10240-10377 ft (3121-3162.8 m); Toca C, well 120-1, 10377-10930 ft (3162.8-3331.3 m); Toca D, well 72-33, 7083-7272 ft (2158.8-2216.4 m).

Lithologic samples: Cuttings and 88 ft (26.8 m) of conventional core from well 120-1 sidetrack from 10550-10593, 10783-10812, and 10854-10870 ft (3215.5-3228.6, 3286.5-3295.6, and 3308.1-3313 m).

Derivation and history of name: Selection of the name is arbitrary. One could argue that because the type section is located within Kambala field, the formation should be called the Kambala Carbonate rather than an arbitrary name like Toca. However, widespread use of the name Toca, not only by Chevron but by other companies in several countries, argues for retention of the name Toca.

Chela Sandstone

Well: 123-1 (formerly LeLe-1)

Location: 5°41′41″S, 12°15′35″E

Completion date: July 18, 1964

Kelly bushing elevation: 397 ft (121 m)

TD: 8071 ft (-7674 ft TVD) [2460 m (-2339 m TVD)]

Interval: 7792-7954 ft (-7395 to -7557 ft TVD) [2374.9-2424.3 m (-2253.9 to -2303.3 m TVD)]

Lithologic samples: Cuttings and 95 ft (29 m) of conventional core from 7792-7890 ft (2374.9-2404.7 m).

Derivation and history of name: Chela is used for two different units in Angola—an Aptian sandstone in the lower Congo basin, and an outcropping, Precambrian unmetamorphosed sandstone (established in 1935 by field mapping). The origin of the name Chela as used for Cretaceous sandstone is unknown. Perhaps early in exploration efforts, the Cretaceous unit was confused with the Precambrian Chela. It certainly is inconsistent with the International Stratigraphic Guide for two units in the same country with the same gross lithology but dramatically different age to have identical names. The Angolan Stratigraphic Commission has decided not to change the name of either unit.

Hydrocarbon Accumulation in Meso-Cenozoic Lacustrine Remnant Petroliferous Depressions and Basins, Southeastern China

Li Desheng
Research Institute of Petroleum Exploration and Development
Beijing, People's Republic of China

Luo Ming
Department of Geological Sciences
University of Texas at El Paso
El Paso, Texas, U.S.A.

More than 140 small lacustrine Meso-Cenozoic remnant petroliferous depressions and basins are distributed along deep fault zones in southeastern China. Some contain abundant oil and gas reserves in high-yield traps, although areal extent of this type of petroliferous basin is only 800-1000 km². Among the structural and lithologic characteristics of hydrocarbon accumulations in these depressions and basins are (1) distribution along deep fault zones, (2) widely distributed, thick source rocks, (3) well developed sand bodies, (4) high geothermal gradient, and (5) multiple-trap styles.

Meso-Cenozoic remnant depressions and basins in southeastern China contain thick, deep-water lacustrine shale, the primary source rocks for the region's hydrocarbon accumulations. These dark, organic-rich, fine-grained strata were deposited in the deeper parts of lake depressions and basins. Consequently, the principal source rocks have been well preserved over much of their original extent, although uplift and erosion following shale deposition have resulted in the destruction of depression and basin margins.

Most of these small remnant structures occur as half-grabens resulting from successive development of faults along one side of the basin; however, some also may appear as full grabens. Well developed remnant depressions and basins consist of three characteristic structural elements—(1) step-faulted belt on the steep flank, (2) central, low buried-hill belt, and (3) monoclinal slope belt on the gentle flank. Examples of these types of depressions and basins include the Damintun, Lanpu, and Sulu depressions, and the Baishe and Sanshui basins. Poorly developed remnant basins consist of (1) a step-faulted belt on the steep flank, (2) central trough, and (3) monoclinal slope belt on the gentle flank. Examples of these types include the Biyang, Changwei, Qianjiang, and Yuanjiang depressions, and the Maoming basin.

INTRODUCTION

Southeastern China covers about one-third of the entire country and involves 14 provinces. Within this huge area many continental Meso-Cenozoic basins have been proved to be rich in oil and gas, such as the Bohai, Jianghan, and Subei basins (Li Desheng, 1982). In addition, many small continental, lacustrine Meso-Cenozoic depressions and basins have good petroleum potential. In recent years many of these features, such as the Damintun, Biyang, and Chaluhe depressions and the Baishe basin, have been successfully drilled (Cai Shuntiau and Guo Yiqiu, 1984). However, exploration of the lacustrine depressions and basins is still in an early stage. This paper attempts to explain the significant petroliferous characteristics of these small continental lacustrine depressions and basins in southeastern China.

TECTONIC EVOLUTION OF DEPRESSIONS AND BASINS

Southeastern China is a mosaic of four interlocking plates—North China plate in the northeast, Yangtze plate in the center, East China plate in the southeast, and Qingzhang plate in the west. The North China and South China plates form the southeastern frontal margin of the Eurasian plate (Li Cuengyu, 1975). During the Yanshan and Himalayan orogenies, this frontal edge was transformed from an Andes-type continental margin into a west Pacific island-arc-type marginal sea under the influence of northward compression of the East Pacific and West Indian plates (Guo Lingzhi, 1980). A series of Meso-Cenozoic half-grabens or graben-type rift depressions and basins formed as a result of upwelling of the upper mantle and subsequent taphrogenesis. In addition, the Tethys Ocean and Himalayan Trench closed in response to strong northward compression of the West Indian and Eurasian plates along the suture zone between the Yaly Tsangpo River and Ailao Mountains in the west Yiulan. Another series of Meso-Cenozoic half-graben or graben-type depressions and basins and small piedmont depressions were created by left-lateral shear stress and/or compressive stress in western China. These two genetically distinct types of small, oil-bearing depressions and basins have been affected by uplifting late in their development. Despite significant erosion around their margins, most good source and reservoir sequences are well preserved. These depressions and basins are referred to as *remnant petroliferous depressions and basins*.

HYDROCARBON OCCURRENCE

Important characteristics of hydrocarbon accumulations in Meso-Cenozoic remnant petroliferous depressions and basins in southeastern China include (1) distribution along deep fault zones, (2) thick, widely distributed source rocks, (3) well developed sand bodies, (4) high geothermal gradient, and (5) multiple-trap styles.

Deep Fault Zones

The formation and development of remnant petroliferous depressions and basins was controlled by the presence of long, deep fault zones, most trending northeast-southwest. Long-term and successive Meso-Cenozoic movement along these fault zones has resulted in formation of faulted troughs 3000-10000 m deep in which thick source, reservoir, and seal rocks were developed. For example, three remnant petroliferous depressions lie along the Tanlu deep fault zone—Chaluhe (Jielin

Province), Damintun (Liaoning Province), and Changwei (Shandong Province). Other remnant structures and their respective deep fault zones are shown in Figure 1.

Basement rocks in these depressions and basins vary greatly in age and lithology and include Archean to Proterozoic granite gneiss, Proterozoic metamorphic rocks, and Mesozoic basalts and carbonates. Tertiary sedimentary rocks in these depressions and basins vary in thickness from 3200 to 7000 m (Table 1).

Thick Source Rocks

The Meso-Cenozoic remnant depressions and basins contain thick sequences of deep-water, lacustrine shale, which is the principal source for the region's hydrocarbon accumulations. These widely distributed dark shales range from 1000 to 2200 m thick and contain abundant organic matter. Organic geochemical analysis indicates a total organic carbon content of 0.375-2.260%, with chloroform "A" extracts ranging from 0.012-0.203%, and total hydrocarbon content of 52-1718 ppm (Table 1). Kerogen in these source rocks is primarily a mixed sapropelic-humic type.

Commonly, the lake depressions and basins have large primary areas and underwent successive subsidence during their development. Because the organic-rich, fine-grained sediments were deposited in the deeper parts of the lake basins, many source rocks have been well preserved over a wide area, although the basin margins were eroded as a result of uplift.

Sand Bodies

Generally continental rift lake depressions and basins are characterized by rapid subsidence and filling. The remnant depressions and basins contain abundant clastic materials eroded from surrounding mountains and uplifts. These clastics include Archean granite gneiss from northern China and Yanshanian granite from southern China. In the basin centers, sedimentation rates for Tertiary sequences are estimated to have ranged from 0.1-0.25 mm/yr (Li Desheng, 1982), an order of magnitude higher than rates associated with marine depressions and basins.

Because of the complex of lacustrine environments, sedimentary models of these depressions and basins are of multisource type. Characteristic sand bodies occur in three principal structural and depositional regimes within the depressions and basins. First, on the steeper flanks of the wedge-shape depressions and basins, piedmont and alluvial fan deposits exist in gravity-flow deposits. Second, on the gentler flanks of the depressions and basins, deltaic systems or fluvial and channel bars are developed. Third, the deeper parts of the depressions and basins are filled with lacustrine turbidites. These types of sand bodies

Li and Luo

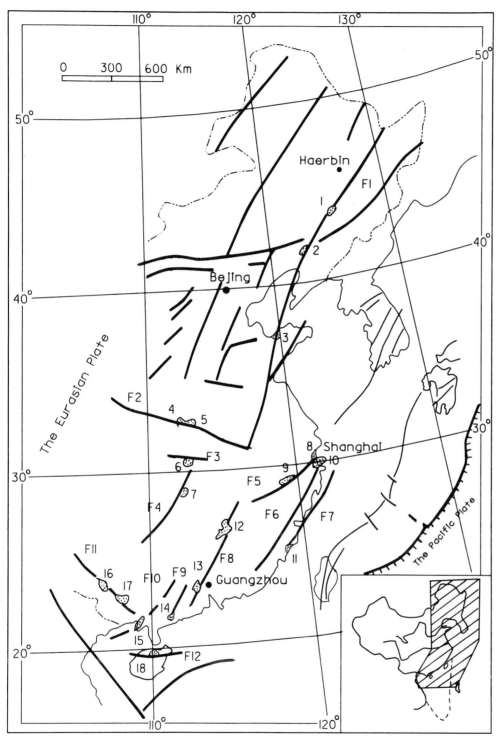

Figure 1. General map of deep fault zones and Meso-Cenozoic remnant petroliferous depressions and basins in southeastern China. Faults—F1, Tanlu; F2, Tonbai; F3, Qianbei; F4, Xiupu; F5, Jiangshan; F6, Ninghai; F7, Changle; F8, Enping; F9, Wuchuan; F10, Wuzhou; F11, Youjiang; F12, Dingan. Depressions— 1, Chaluhe; 2, Damintun; 3, Changwei; 4, Nanyang; 5, Biyang; 6, Qianjiang; 7, Yianjiang; 8, Chenghe. Basins— 9, Jingqu; 10, Ningbo; 11, Fotan; 12, Jitai; 13, Sanshui; 14, Maoming; 15, Hepu; 16, Baishe; 17, Nanning; 18, Fushan.

Lacustrine Basin Exploration: Case Studies and Modern Analogs

Table 1. The characteristics of the lacustrine Tertiary remnant petroliferous depressions and basins in southeast China

Basin Name	Damintun	Biyang	Baishe	Changwei
Area (km)	800	800	830	780
Basement (Age)	Granite Gneiss (Ar)	Metamorphic (Pt)	Carbonate (T)	Basalt (Mz)
Tertiary thickness (m)	7000	3500	3200	5000
Source bed (m)	1000	2000	1000	2200
Kerogen type	Mixed	Mixed & Sapropel	Mixed	Mixed & Humic
C%	1.5–2.26	1.66	0.37–1.87	0.5–1.0
Chloroform "A"%	0.06–0.149	0.2	0.029–0.1	0.012–0.203
HC ppm	168–1010	870–1718	210–1300	52–1029
Geothermal gradient (°C/100m)	4.0	4.1	3.8	3.5
Threshhold depth (m)	2200	1700	1500	1700

Note: Ar—Archean, Pt—Proterozoic, T—Triassic, Mz—Mesozoic, HC—Total Hydrocarbon Content

commonly form good reservoir rocks, with porosities of 18–25.9% and permeabilities of 220–1863 md.

High Geothermal Gradient

In general, intense magmatism has accompanied movement along the deep fault zones where the remnant depressions and basins are developed. In northern China several eruptions of andesitic and basaltic magma occurred during the Mesozoic and Cenozoic. These magmas were differential-melting products of the subducted Pacific plate. In Yunan, Guizhou, and Guangxi provinces, southern China, extensive sea-floor basaltic magmatism occurred during the Permian. Triassic sequences in this area contain intrusive ultramafic rocks that originated from upwelling of the upper mantle as well as subsequent taphrogenesis. Over large areas of Zhejiang, Fujian, Guangdong, and Hunan provinces in eastern China, granitic magmatism occurred along most of the deep fault zones during the Yanshanian orogenic event (Tertiary). A gradual transition between granite and eruptive rhyolite indicates that these melts are recycled (remelted) continental crust.

These intense magmatic actions resulted in increased heat and chemical energies along the deep fault zones; therefore, the geothermal gradients in these remnant depressions and basins are higher than in depressions and basins elsewhere in China. Geothermal gradients in these depressions and basins range from 3.50 to 4.10°C/100 m (Table 1) and are favorable for the maturation of organic matter.

Multiple-Trap Style

Most remnant depressions and basins occur as half-grabens that result from successive development of faults along one side of the basin, but full grabens also develop (Li Desheng, 1986). Both types of basins show a kind of zonation. Well developed remnant depressions and basins can be divided into three structural units—(1) step-faulted belt on the steep flank; (2) central, low buried-hill belt (Li Desheng, 1985); and (3) monoclinal slope belt on the gentle flanks. Examples of these structures include the

Damintun (Figure 2A), Lanpu, and Sulu depressions, and the Baishe (Figure 2B) and Sanshui basins. Poorly developed remnant depressions and basins also are typified by step-faulted and monoclinal slope belts but instead have a central trough. Examples of these are the Biyang (Figure 2C), Changwei (Figure 2D), Qianjiang, and Yianjiang depressions, and the Maoming basin.

In the remnant depressions and basins, oil and gas traps are controlled by local geological factors such as faults, unconformities, and depositional environment distribution. Known oil and gas traps include (1) structural traps (rollover, draping, and compressional anticlines, step-faulted traps, and structural noses); (2) stratigraphic-lithologic traps (unconformity, updip pinch-out sand bodies, channel sand bars, deep lacustrine turbidites, delta-front sand bodies, and subaqueous alluvial fans); and (3) buried-hill traps (reservoirs include weathered and fractured Mesozoic carbonate and igneous rocks, Paleozoic-Proterozoic carbonate and clastic rocks, and Precambrian granite gneiss and metamorphics). These types of traps in the various structural belts can become stacked vertically to form multiple-pay horizons and multiple-trap types (Figure 2).

Depression and Basin Cases

To illustrate the complexity of multiple traps, we summarize four examples—Damintun (well developed depression, Liaoning province), Biyang (poorly developed depression, Henan province), Baishe (well developed basin, Guangxi province), and Changwei (poorly developed depression, Shandong province).

Six pay zones in the Damintun depression yield high production over an area of 800 km² (Cheng Yixian, 1980). The buried-hill reservoirs of Archean granite and middle to upper Proterozoic carbonates are of critical importance in hydrocarbon accumulation within the depression (Figure 2A). The Dongshengpu buried-hill reservoir, for example, has an oil column of up to 400 m. The Archean granite reservoir is characterized by fractures and secondary dissolution pores. Wells yielding 1000 tons oil/day have been drilled on the fracture zones. Tertiary

Li and Luo

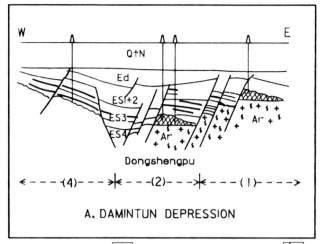

A. DAMINTUN DEPRESSION

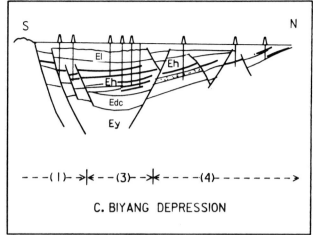

C. BIYANG DEPRESSION

`+s` `s+` Granite Gneiss   Carbonates   `v v v` Basalt

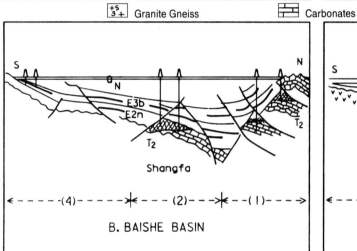

B. BAISHE BASIN

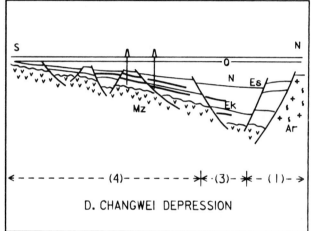

D. CHANGWEI DEPRESSION

Figure 2. Cross sections of small Meso-Cenozoic remnant petroliferous depressions and basins in southeastern China. A, Damintun depression, Liaoning province; B, Biyang depression, Henan province; C, Baishe basin, Guangxi province; D, Changwei depression, Shandong province. Extent of principal structural zones shown below sections—1, step-faulted belt; 2, central low buried-hill belt; 3, central trough; 4, monoclinal slope belt. Q+N, Quaternary and Neogene; N, Neogene; Ed, Eogene Dongying Formation; El, Eogene Liaozhuang Formation; Eh, Eogene Hetaoyuan Formation; Edc, Eogene Dachangfang Formation; Ek, Eogene Kongdian Formation; Es, Eogene Shahejie Formation; Esl+2, Eogene Shahejie Formation (Members I and II); Es3, Eogene Shahejie Formation (Member III); Es4, Eogene Shahejie Formation (Member IV); E3b, Eogene Baigang Formation; E2n, Eogene Nadu Formation; T2, Middle Triassic.

strata overlying the Archean-Proterozoic buried hills form anticlines, updip pinch-out sandstone traps, lenticular lithologic traps, and rollover anticlines, which comprise the Dongshengpu-Jinganpu composite oil and gas trap belt. Annual oil production in the Damintun depression reaches up to 2.5 million tons.

The four pay zones in the Baishe basin (830 km²) include two carbonate and two clastic reservoirs (Cai Shuntiau and Guo Yiqiu, 1984). In the Shangfa area of the central basin, wells with production of 250–418 tons oil/day have been drilled in buried hills of Middle Triassic limestone (Figure 2B). A 12- to 23-m-thick fossiliferous limestone at the base of the Nadu Formation (Paleogene) shows oil potential in the Nasheng uplift area. Reservoirs of the Nadu and

Baigang Formations, distributed along the northern step-faulted belt, on a low, central buried hill, and on the southern monoclinal slope belt, tend to develop as multiple oil-bearing horizons.

Stratigraphic-lithologic traps are the dominant hydrocarbon traps in the Biyang depression, which has an area of 800 km² in Henan province (Zhu Shuian et al., 1981). In terms of sand-body distribution, alluvial fans have developed on the southern steep flank of the depression, and deltaic deposits have formed on the northern gentle flank (Figures 2C, 3). Some sand bodies cover more than 120 km². Trap types include updip pinch-out (Shuanghe field), rollover anticline (Xiaermen field), fault-block (Wangji field), faulted nose (Gucheng field), and lenticular sand body (Anpeng field). Recently some

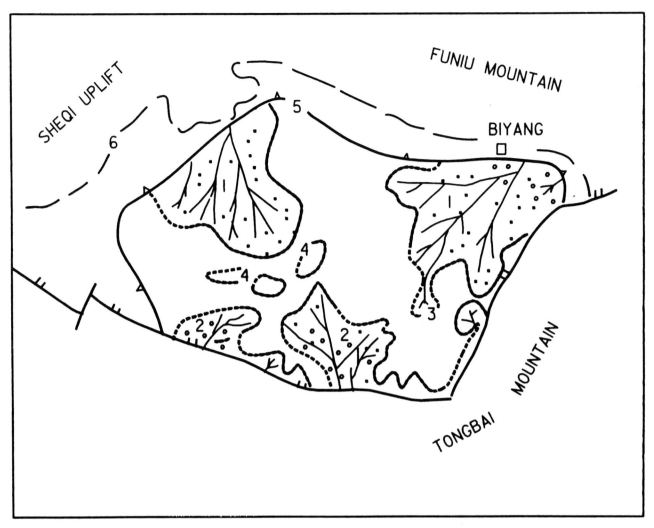

Figure 3. Schematic diagram of sand-body distribution in fault-bounded Biyang depression, Henan province. 1, delta; 2, alluvial fan; 3, turbidite sand-body; 4, sheet or lenticular sand-body; 5, line of hiatus; 6, primary or ancient basin margin.

unconformity and faulted anticlinal traps were found in the Jinglou area, from which shallow heavy oil is produced. A fractured lacustrine dolomite reservoir with dissolution pores has been developed in the central part of the depression.

The Changwei depression, covering an area of 780 km² in Shandong province, is dominated by fault-block traps (Figure 2D). The lacustrine Paleogene Kongdian Formation (members 1 and 2) together with underlying Mesozoic basalts are the pay zones in this depression. In general, exploration results in the Kongdian Formation have not been very good. Two major problems are that some of the single pays are too thin (1-2 m), and their lateral extent is limited. However, petroleum potential in the underlying andesitic basalt appears promising. These purple-red andesitic basalts are widely distributed and have a thickness of about 700 m. They have been fractured, and the zeolitic fillings of amygdules have been partly or totally leached, leaving dissolution pores 5-15 mm in diameter. Both phenomena have enhanced the

basalt's permeability and porosity. The Cheng 39 well in this reservoir type yielded about 9 tons oil/day during testing at the 1536.5-1565.98 m interval.

CONCLUSIONS

More than 140 lacustrine Meso-Cenozoic remnant petroliferous depressions and basins are scattered throughout southeastern China. Despite their relatively small size, they exhibit a variety of trap styles in different structural settings, which typically can become stacked vertically to form multiple-pay horizons and multiple-trap types. Cost-effective development of digital seismic techniques and improvement of seismic stratigraphic interpretation will improve the ability to forecast the types and distribution of these oil and gas traps and consequently increase exploration efficiency.

Li and Luo

The Biyang, Nayang, Qianjiang, and Gaoyiou depressions still maintain good petroleum production and potential after many years of exploration and development. High oil flows recently have been obtained in exploratory wells into buried hills of Triassic limestone in the Baishe basin. The prospect for increasing oil and gas reserves and production is very promising in these remnant petroliferous depressions and basins.

ACKNOWLEDGMENTS

We would like to thank David V. LeMone, Barry Katz, and Louis Liro for their helpful comments and critical reviews. William C. Cornell and Arthur Harris also gave good suggestions on the manuscript.

REFERENCES CITED

Cai Shuntiau, and Guo Yiqiu, 1984, Petroleum geology of the Baishe basin, Guangxi, China: Shiyóu Yu Tiārānqi Dizhi [Oil & Gas Geology], v. 5, no. 4, p. 362–371.

Cheng Guoda, 1965, A new type of mobile area at post-platform stage, in Tectonic problems in China: Beijing, Sciences Press.

Cheng Yixian, 1980, On the formation mechanism and geological problems of the Paleogene Liaohe rift-valley: Petroleum Exploration and Development, no. 2, p. 20–31.

Guo Lingzhi, 1980, Tectonic framework and crust evolution in south China: International Geological Congress, p. 109–116.

Li Cuengyu, 1975, The primary analyses of tectonic development of some areas in China: Geophysics, v. 8, p. 52–76.

Li Desheng, 1982, Tectonic types of oil and gas basins in China: Acta Petrolei Sinica, v. 3, no. 3, p. 1–12.

Li Desheng, 1985, Tilted fault block-buried hill traps—A new type of oil/gas trap in rift-related tensional basins: Shiyóu Yu Tiārānqi Dizhi [Oil & Gas Geology], v. 6, no. 4, p. 386–394.

Li Desheng, 1986, The development prospects on the composite megastructural oil and gas fields of Bohai Gulf basin of China: Acta Petrolei Sinica, v. 7, no. 1, p. 1–12.

Zhu Shuian, Xu Shirong, Zhu Shaobi, and Wang Yixian, 1981, The petroleum geological characteristics of Biyang Depression, Henan: Acta Petrolei Sinica, v. 2, no. 2, p. 21–28.

Index

Index